Modern Earth Science

William L. Ramsey
Clifford R. Phillips
Frank M. Watenpaugh
Robert Nanney
Carolyn Sumners
Warren E. Yasso

HOLT, RINEHART AND WINSTON, INC.
Austin, New York, San Diego, Chicago, Toronto, Montreal

William L. Ramsey
Former Head of the Science Department
Helix High School
La Mesa, California

Clifford R. Phillips
Former Head of the Science Department
Monte Vista High School
Spring Valley, California

Frank M. Watenpaugh
Former Chairman of the Science Department
Helix High School
La Mesa, California

Robert Nanney
Former Curriculum Associate in Mathematics
and Science
College Park High School
Pleasant Hill, California

Carolyn Sumners
Director of Astronomy and Physics
Houston Museum of Natural Science
Houston, Texas

Warren E. Yasso
Department of Mathematics and Science Education
Teacher's College, Columbia University
New York, New York
Assistant Professor of Geology
Virginia Polytechnic and State University

Developmental assistance by Ligature, Inc.

Printed in the United States of America

ISBN 0-03-004449-9

4 5 6 7   041   14 13 12 11

Acknowledgments

Consultants

David Hammond
San Juan Unified School District
Carmichael, California

George Ladd
School of Education
Boston College
Chestnut Hill, Massachusetts

Robert Ridky
Department of Geology
University of Maryland
Adelph, Maryland

Fact Checkers

Saul Ash
Department of Geology
Hunter College
New York, New York

Hany Doss
Department of Geology
Hunter College
New York, New York

Teacher Reviewers

Roger Anselmino
Elizabeth Forward Senior High School
Elizabeth, Pennsylvania

Bobby Boykin
Northwood High School
Shreveport, Louisiana

Robert Hallihan
Doherty High School
Worcester, Massachusetts

Nancy S. Harms
Mount Diablo Unified School District
Concord, California

Ronald Louchren
Bennett High School
Buffalo, New York

Dennis R. Snyder
Upper St. Claire High School
Pittsburgh, Pennsylvania

Thomas Urchuk
Weehawken High School
Weehawken, New Jersey

Neil Wintringham
Scotch Plains-Fanwood High School
Scotch Plains, New Jersey

Contents

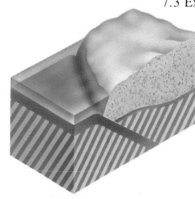

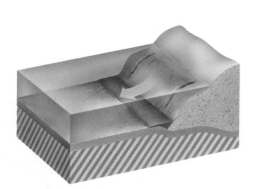

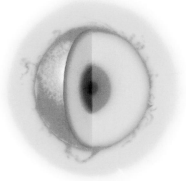

Investigation

Impact

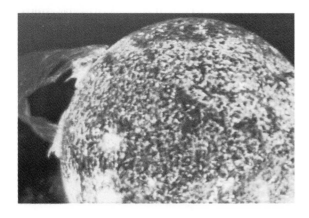

Career Focus

Horizons

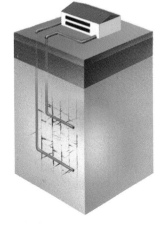

Introducing Earth Science

Our planet is very special. Its distance from the sun is just far enough away to allow sufficient heat to sustain life; its orbit around the sun and its tilt toward the sun distributes the heat over the earth's surface so that each hemisphere receives neither too much nor too little heat. The planet earth also has sufficient gravity to hold an atmosphere. Its atmosphere contains the correct blend of gases and shields us enough to support life. Then, there is an envelope of water to distribute heat, to moderate climate, and to make life possible.

I can remember saying to my students—"earth science— the only science." That statement usually sparked some discussion! Another statement would follow—"If it weren't for the earth, we wouldn't be here." While this statement may sound overly dramatic, it is true.

Whether you choose to be a technician, a plumber, a banker, a teacher, or a scientist—one thing remains clear— you need the earth and its resources to survive. Did you know, however, that the earth contains a limited amount of resources—its minerals, its fuels, its fresh water, and its soils? The natural resources of the earth are important to everyone. Moreover, as we use resources, we produce many tons of wastes each year and as countries become more industrialized, even more wastes will be produced. There is also a limit to the number of inhabitable terrains on the earth. Many areas of the earth are uninhabitable due to the geology or climate of the

area. The planet earth needs people who understand how to protect the planet, use its resources wisely, make wise decisions about its future, and seek solutions to its problems.

The study of the earth puts into perspective and gives meaning to the concepts of ecology, life science, and physical science. Earth science also provides an alternate approach to scientific study—one that is different from the experimentation used in physics and chemistry. For example, the evolution of life as evidenced by the fossils found in a highway cut or in the Grand Canyon cannot be repeated in a laboratory. Yet, the historical method of study is an important key to the study of the earth.

Obviously, the earth is a planet worthy of study.

Andrew J. Verdon Jr.

Andrew J. Verdon, Jr.
Director of Education
American Geological Institute
Alexandria, Virginia

Using Modern Earth Science

Welcome to the *Modern Earth Science* Expedition! An expedition is a journey with a specific purpose, and that is just what you are about to begin. The purpose of your journey is to learn about the matter and forces that make up the earth and its neighbors in space, and the forces that shape them. Your *Modern Earth Science* textbook is your expedition guide.

Units

Each of the eight units of *Modern Earth Science* is introduced by an outstanding earth scientist. His or her message to you appears alongside an illustrated list of the chapters in the unit. Before you begin each unit, take a few minutes to study the illustrations and to read the message on the unit introduction pages. You will receive a broad preview of the unit.

Chapters

The beginning of each of the 30 chapters in *Modern Earth Science* is distinguished by a color photograph that illustrates a major topic discussed in the chapter. To get an overview of the chapter, study the photograph and read both the brief message and chapter outline that appear on the page. Having an idea of where you are headed helps make a journey easier and more enjoyable.

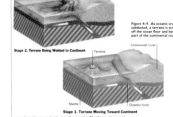

Section Objectives

- List two types of stream deposition and explain the differences between them.
- Describe the change in a stream that causes flooding.
- Identify direct and indirect methods of flood control.

13.3 Stream Deposition

The total load a stream can carry is greatest when a large volume of water is flowing swiftly. When the velocity of the water decreases, the ability of the stream to carry its load generally also decreases. As a result, part of the stream load is deposited as sediment. When the velocity increases again, some or all of the deposited sediment is carried away. In some cases, stream deposits remain in place for only a brief time. In other cases, these deposits can become permanent features of the land.

Deltas and Alluvial Fans

Most of the load carried by a stream is deposited w[hen it] reaches a large body of water. As a stream empties [into a] stream such as an ocean, a gulf, or a lake, the vel[ocity of the] stream decreases greatly. The load then is deposited [by] the stream in a fan shape with its tip facing upstrea[m, as shown] in Figure 13–8. This fan-shaped deposit at the mou[th is] called a **delta**. The exact shape and size of a delta [depends on] the local waves and tides.

When a stream descending a steep slope reache[s a plain, the] speed of the stream is suddenly reduced. As the stre[am slows] rapidly, it deposits some of its load on the level pla[in at the base of] the slope. This deposit forms a fan-shaped heap [of material] upstream. This deposit is called an **alluvial fan**. In [arid or semiarid] desert regions, temporary streams often form alluvi[al fans. An allu-]vial fan differs from a delta in the following three [ways. First, the] sediment that forms an alluvial fan is deposited on [dry land. Sec-]ond, an alluvial fan is made up of coarse, angular [material.] Third, the surface of an alluvial fan is sloping, whi[le the surface of] a delta is relatively flat.

Figure 13-8. When a stream meets an ocean or a gulf, the sediment carried by the stream is deposited in a delta (below). Runoff from slopes in desert regions deposits sediment in an alluvial fan (inset).

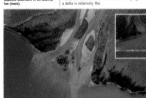

come from the ocean depths near the equator. The theory of suspect terranes explains how coral atolls and equatorial ocean fossils became part of the geology of northern California.

According to the suspect terrane theory, blocks of terranes are carried along on the ocean floor by the action of seafloor spreading to a lithospheric plate boundary where subduction is occurring. As the plate with oceanic crust moves under the plate with continental crust, the terranes are scraped off the descending ocean floor, as shown in Figure 4–9. Some terranes may form mountains, while others simply add to the surface area of a continent.

When Alfred Wegener first proposed his hypothesis of continental drift, he could not have imagined the explosion of scientific inquiry it would inspire. Like many hypotheses, continental drift asked more questions than it answered. The theories of plate tectonics and suspect terranes are attempts to answer some of those questions.

Figure 4–9. As oceanic crust is subducted, a terrane is scraped off the ocean floor and becomes part of the continental crust.

Stage 2. Terrane Being Welded to Continent

Stage 1. Terrane Moving Toward Continent

Section 4.2 Review

1. Summarize the theory of plate tectonics.
2. Name and describe the three types of plate boundaries.
3. Describe the three types of plate collisions that occur along convergent boundaries.
4. How might convection currents cause plate movement?
5. Explain how mountains on land can be composed of rocks that contain fossils of animals that lived in the ocean.

Chapter 4 Plate Tectonics **67**

Chapter Organization

Each chapter in *Modern Earth Science* is divided into two to four main sections. The basic outline structure of the Chapter helps you to understand how the various pieces of chapter information relate to one another. In addition, the headings summarize the chapter and can be used as a study guide.

Vocabulary

Terms that are basic to your understanding of the material presented in each chapter are printed in boldface and are followed by a definition. Key terms with difficult pronunciations are phonetically respelled.

Illustrations

In *Modern Earth Science* the text and illustrations work together to give you a clear understanding of the information presented. To receive the greatest benefit from the illustrations, take time to stop and study them as you read the corresponding text. In many cases both a photograph and a labeled drawing are provided. The photograph lets you see the object as it actually appears. The drawing simplifies structure and highlights details.

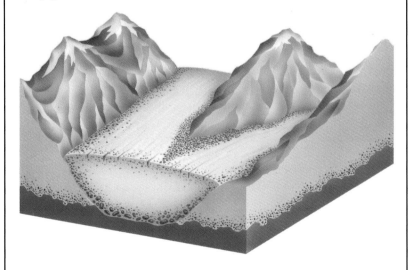

xiii

Features

In addition to your main earth science journey, *Modern Earth Science* has provided interesting side trips in the form of the special features described on this and the following page. When you come to one of these features, finish the subsection you are reading and then study the feature before going on with the main text.

Investigation

Lithospheric Plate Boundaries

The movement of lithospheric plates has created many of the earth's topographical features. You can demonstrate the results of plate movement by using clay models of lithospheric plates.

Science Process Skills Focus: constructing models, demonstrating, describing, explaining

Materials
ruler, paper, scissors, rolling pin or rod, modeling clay, plastic knife

Procedure
1. Draw two rectangles 10 × 20 cm on your paper and cut them out.
2. Use a rolling pin to flatten out two pieces of clay until they are about 1 cm thick. Cut each piece into a rectangle 10 × 20 cm. Place one square of paper on each piece of clay.
3. Place the two clay models side by side on a flat surface, paper side down. Push the models together until the edges begin to buckle and rise off the surface of the table.
4. Turn the clay models around so that the unbuckled edges face one another.
5. Place one hand on each clay model. Apply only slight pressure toward the seam. Slide

one clay model forward and the other model backward about 7 cm.
6. Repeat Step 5 three more times, alternating the direction in which you push each model.

Analysis and Conclusions
1. What type of plate boundary are you demonstrating when you push the models head-on?
2. What type of plate boundary are you demonstrating when you slide the models against each other?
3. How does the appearance of the facing edges of the models in the two processes compare? How do you think these processes might affect the appearance of the earth's surface?

Investigation

Sometimes the best way to learn is by doing. Each *Investigation* helps you explore an important chapter topic. Investigations require only common household items and most can be completed in about 30 minutes. The *Investigation* may be completed at home or in class.

Career Focus: Earth Structure

"As a geologist, you know that no matter what problem you're trying to solve, the answer is out there just waiting to be discovered."

Field Geologist

Talking to Denver-based field geologist Ken Pierce is one thing. Finding him is quite another.

"During the field season, I can be out in the field area for as long as four months at a time," he said. "Though I mostly drive a jeep and hike, I have made some month long pack trips on horseback as well."

Like other field geologists, Pierce, pictured left, studies the composition, structure, and history of the earth's crust. Pierce specializes in the study of the youngest part of geologic time.

After collecting samples including sand, gravel, and glacial deposits, Pierce, who works for the U.S. Geological Survey, determines which samples merit further study. He brings them to the lab, where they are X-rayed, studied under high-powered microscopes, and subjected to chemical analysis. Pierce also performs experiments to test his geologic theories.

"Although I have always liked being outdoors, it wasn't until I was in college that I really became interested in learning why the surface of the earth looks the way it does and knew I wanted a career in geology."

Pierce holds undergraduate and graduate degrees in geology with an emphasis on geomorph...

Horizons

Technology: Landsat Maps the World

In July 1982, *Landsat 4,* a remarkable mapping satellite, was launched. The satellite, illustrated at left, circles the earth at an altitude of 700 km. *Landsat 4* is the fourth in a series of earth-scanning satellites, the first of which was launched in 1972.

With a scanning sensor system called a *thematic mapper* (TM), *Landsat* records highly detailed images of the earth's surface. Each *Landsat* image covers nearly 33,800 km² of the earth's surface. Since 1972 these satellites have recorded more than a million images.

Landsat images are not ordinary photographs, however. For example, the *Landsat* photo at the far right shows

the Baltimore-Washington, D.C., area with bright-red vegetation, dark-blue business districts, and pink suburbs.

The *Landsat* photos are multispectral images. The TM sensors see not only visible

48 Unit 1 Studying the Earth

Career Focus

People are a major part of earth science. Each *Career Focus* lets you meet a person active in a specific branch of earth science and also provides information about two related careers.

Horizons

Earth science is an active and ongoing field of study. Each *Horizons* describes a current hypothesis, theory, or technological development in the field of earth science.

Science Notebook

Life Cycle of a Lake

All precipitation does not either evaporate or immediately flow from the land to the ocean. Sometimes this water collects in a depression in the land and creates some of the most beautiful formations of water on the earth—lakes.

Lakes occur most frequently in high latitudes and in mountainous areas. However, they also are common in low areas near the ocean, such as Florida, and near rivers with low gradients. Most of the water in lakes comes from precipitation and melting ice and snow. Springs, rivers, and runoff directly from the land also are sources of lake water.

Like rivers, lakes have a life cycle. Unlike rivers, most lakes are relatively short-lived. Many lakes eventually disappear because too much of their water drains away or evaporates. The most common

cause of excess drainage is an outflowing stream that cuts its bed below the lake basin. A lake also may lose its water if the climate becomes drier and evaporation exceeds precipitation.

Lake basins may be destroyed if they are filled with sediments. Streams that feed a lake carry and deposit sediments in the lake.

IMPACT

Living on the Mid-Atlantic Ridge

At 2 A.M. on January 23, 1973, a fiery fissure opened in the land less than a mile from the town of Vestmannaeyjar on the island of Heimaey off the south coast of Iceland. The island, part of the Mid-Atlantic ridge, had been split open by sea-floor spreading. Ash, molten rock, and poisonous gases spewed forth from the fissure. Within six hours, more than 5,000 panic-stricken islanders had been evacuated to the mainland.

During the next two months, 300 buildings were lost to fire, and another 65 were buried beneath black volcanic ash. Lava pouring out of the fissure and flowing to the ocean threatened to block the entrance to the harbor and deprive the islanders of their livelihood—fishing the

waters of the north Atlantic. When the volcanic activity ceased four months later and the people of Vestmannaeyjar returned to reclaim their town, they found a dramatically altered landscape. Nearly one third of the town had been buried by lava, and the rest lay under a thick blanket of ash. The lava had added 1 km² of new land to the island as well as an excellent breakwater in the harbor. The cleanup resulted in more than 1.3 million m³ of volcanic ash

being cleared from roofs, gardens and streets.

At the present time, sea-floor spreading adds an average of 2.5 cm of new material each year to Iceland. At this rate, Iceland will grow 400 k in width during the next million years. Iceland and its neighboring islands are actually a continent in the making.

If geologists want to locate the youngest rocks on Iceland, where should they look? Where should they look to find the oldest rocks? Why?

Science Notebook

Sometimes you want to know a little more about something. Each *Science Notebook* gives you details, background, or related information about points raised or procedures discussed in the main text.

Impact

Earth science is relevant to daily life. Each *Impact* describes the impact that a particular earth science event or process had or is having on people.

Chapter Review

Checking your understanding is an important step toward successful learning. Each *Chapter Review* helps you check to see that you have successfully completed this part of your earth science expedition. The chapter review lists the key ideas and key terms presented in the chapter.

Reviewing Chapter 16

Key Concepts

- The wind erodes the land by weathering and by removing sediments. See page 255.
- The two types of wind deposits are made up of particles of different sizes. See page 257.
- Beaches result from the deposition of sediments by waves. See page 262.
- Longshore currents move sediments along a shoreline. See page 264.
- Coastlines are exposed or submerged as sea level changes. See page 265.
- Barrier islands are unstable coastline features. See page 268.
- Coral reefs that surround volcanic islands undergo

- Waves weather and erode the shoreline, producing characteristic features. See page 263.

changes as the islands sink. See page 269.
- Human activities affect land along the coasts. See page 269.

Key Terms

atoll (269)	desert pavement (256)	loess (259)	submergent coastline (266)
barchan (258)	dune (257)	sand bar (263)	tidal flat (268)
barrier island (268)	emergent coastline (268)	sea arch (261)	tombolo (264)
barrier reef (269)	estuary (266)	sea cave (261)	ventifact (256)
beach (262)	fiord (268)	sea cliff (260)	wave-built terrace (261)
berm (263)	fringing reef (269)	sea stack (261)	wave-cut terrace (261)
coral reef (269)	headland (264)	shoreline (260)	
deflation (256)	lagoon (268)	spit (264)	
deflation hollow (256)			

Review

On your paper, write the letter of the term that best completes each of the following statements.
1. Loose fragments of rock and minerals measuring between 0.07 mm and 5 mm are referred to as (a) pollen (b) dust (c) sand (d) desert pavement.
2. Sand moves by (a) saltation (b) abrasion (c) emergence (d) depression.
3. The most common form of wind erosion is (a) migration (b) abrasion (c) saltation (d) deflation.
4. Dunes move primarily by (a) abrasion

(b) deflation (c) migration (d) submergence.
5. Unlayered, yellowish, fine-grained deposits are called (a) beaches (b) loess (c) dunes (d) desert pavement.
6. The most common type of dune is a (a) barchan (b) transverse dune (c) deflation hollow (d) parabolic dune.
7. All of the following shoreline features are produced by the erosion of sea cliffs except (a) dunes (b) sea stacks (c) wave-cut terraces (d) sea arches.
8. The composition of beach deposits depends

on (a) the climate (b) the source rock (c) the time of year (d) wave action.
9. Deposition of sand at the end of a headland produces a (a) sand bar (b) dune (c) spit (d) sea cliff.
10. Sea level is now (a) stationary (b) falling about 1 mm/yr (c) rising about 1 cm/yr (d) rising about 1 mm/yr.
11. A coastline resulting from a rise in sea level or subsidence of the land is (a) emergent (b) submergent (c) glaciated (d) volcanic.
12. Barrier islands tend to migrate (a) seaward

(b) along the shore (c) in the summer (d) toward the shore.
13. Coral reefs are produced by living organisms that (a) swim in circles (b) extract minerals from seawater to form hard skeletons (c) attach themselves to sand bars (d) build nests of sand.
14. If shoreline resources continue to be used as in the past, (a) coastal land will be greatly improved (b) the damage can be repaired (c) sea level will fall (d) much of the existing coastal land will be unlivable.

Application

On your paper, write answers to the following questions.
1. The deserts of the southwestern United States contain many tall, sculpted rock formations that are the result of weathering and erosion. Was wind or water the more likely agent responsible for these formations? Explain your answer.
2. Beautifully colored sunsets and sunrises are the result of dust in the atmosphere. Such sunsets and sunrises were visible around the world for two years after Krakatoa, a volcanic island in the Indonesian chain, erupted in 1883. Explain how this is possible.

wave-built terrace have on erosion of the shoreline? What phenomenon might counteract this effect?
5. Suppose that scientists have observed that the size of the sand bars in a particular area has been decreasing steadily over the past five years. What does this observation suggest about the climate of the area? Explain why.
6. As you approach a large landmass by ship, you notice an island connected to the shore by tombolos. What type of shoreline do you predict the mainland will have? Explain your answer.
7. Describe two ways that the melting of continents

270 Unit 4 Reshaping the Crust

Chapter Review (continued text)

ative level of the
ands form paral-

chusetts coast
ers.
ist. You have
nment of a
for preserving
should consider
ill as the unrec-
es. Present your
a discussion,
t.

d Waves **271**

Reference

Sometimes you need additional specific information. The *Reference* section at the end of your textbook contains useful facts and information, including a complete glossary and index.

Reference

Contents

550 Reference

Studying the Earth

Astronaut Spring on space station E.A.S.E. (background); Coral reef, Red Sea (inset)

Introducing Unit One

Communication is the key to progress in science. Without a way to build on past knowledge and pass new knowledge on to others, each new generation would have to rediscover facts about the planet on which we live.

When I was a student majoring in urban and economic geography, I discovered one of our most valuable aids to scientific communication—maps. Maps are a delightful blend of art and science. They give us an efficient, graphic way to present volumes of knowledge in a single picture. By looking at a map, you can view the world or part of it at a glance. By making a map, you can preserve your knowledge for future use.

In this unit, you will discover how other earth scientists use both words and pictures to communicate important information about our amazing planet.

Elizabeth Ann Olson

Elizabeth Ann Olson
Cartographer
U.S. Geological Survey

1 Introduction to Earth Science

2 The Earth in Space

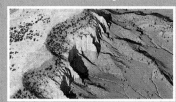

3 Models of the Earth

Chapter 1

Introduction to Earth Science

The natural world challenges our understanding with questions such as why seasons change, rivers flood, and planets spin through space. Unlocking the secrets of the natural world takes earth scientists from the deserts of equatorial Africa to ice caves such as this one in Canada.

This chapter describes the origins of earth science and some of the methods earth scientists use to learn about the world around us.

Chapter Outline

These climbers are poised at the mouth of an ice cave in Canada.

1.1 What Is Earth Science?

Section Objectives

- **Name the four main branches of earth science.**
- **Discuss the relationship between earth science and ecology.**

Since the beginning of history, people have observed the world around them and wondered about the forces that shaped that world. As early humans watched a volcano erupt, felt the earth tremble beneath them, or saw the moon darken during an eclipse, they asked why. To explain these natural phenomena, ancient people forged myths and legends, attributing such events to powerful supernatural forces. Angry goddesses hurled fire from volcanoes; giants wrestled underground, causing the earth to shake.

Not until people began to make careful observations and to search for natural causes to natural phenomena did the scientific study of the earth begin. The ancient Chinese, for example, began keeping written records of earthquakes as early as 780 B.C. The ancient Greeks compiled a catalogue of rocks and minerals in the third century B.C. Other ancient people, such as the Mayas in Central America, kept track of the movements of the sun, the moon, and the planets. They used these observations to create accurate calendars.

At first, scientific discoveries were limited to observations made with the unaided eye. Then, in the seventeenth century, the invention of instruments such as the microscope and the telescope extended human observation to previously hidden worlds.

Eventually, people accumulated an organized body of knowledge about the earth, and the field of **earth science** was born. Earth science is the study of the earth and of the universe around it. Earth science, like other modern sciences, is based on the assumption that the causes of natural phenomena can be discovered through careful observation and experimentation.

Branches of Earth Science

As the technology for studying the earth improved, the range of human observation increased dramatically. With the help of special equipment, scientists began to explore the ocean depths, the earth's

Figure 1-1. El Caracol, an observatory built by the ancient Mayas of Mexico, is the oldest known observatory in the Americas.

Science Notebook

The Earth's Circumference

More than two thousand years ago, a Greek mathematician named Eratosthenes used careful observations and simple geometry to determine the circumference of, or distance around, the earth.

Eratosthenes read of a well in Syene—modern-day Aswan—Egypt, where the sun's rays reached the bottom once each year, at noon on June 21. Furthermore, the sun cast no shadows in Syene at that time. Eratosthenes thought these facts meant that the sun was directly overhead at Syene on June 21. Eratosthenes knew that the sun *did* cast shadows in Alexandria, a city to the north of Syene, on June 21, indicating that the sun's rays were striking that city at an angle.

Eratosthenes determined that the angle of the sun's rays in Alexandria at noon on June 21 was 7.2°. That meant, he thought, that the distance between the two cities was also 7.2°, or 1/50th of the 360° circumference of the earth. Eratosthenes knew that the distance between Syene and Alexandria was about 5,000 stadia (925 km). By multiplying the known distance between Alexandria and Syene by 50, Eratosthenes calculated the circumference of the earth to be 250,000 stadia (46,250 km).

Currently scientists with the aid of modern technology and sophisticated instruments calculate the earth's circumference to be about 40,000 km. This figure, which is considered accurate, differs from Eratosthenes' calculation of 46,250 km by only 6,250 km.

Suppose that city A is 3,335 km north of city B and that on June 21 the angle of the sun's rays in city A is 0° and in city B, 30°. Use Eratosthenes' method to find the earth's circumference.

interior, and the realms of space. Their discoveries have created an immense body of knowledge about the earth.

Because one person cannot keep up with the developments in all areas of earth science, most earth scientists today specialize. Currently, earth scientists specialize in one of the following four major areas of study: the solid earth, the oceans, the atmosphere, and the universe beyond the earth. *Career Focus*, a special feature that appears in each unit of this book, offers detailed information about careers in earth science.

Geology

The study of the origin, history, and structure of the solid earth and the processes that shape it is called **geology.** Geology is a broad field and includes many areas of specialization. Some geologists explore the earth's crust in search of new deposits of coal, oil, gas, and other valuable resources. Some geologists study the forces within the earth in order to predict earthquakes and volcanic eruptions. Other geologists study fossils to learn more about the earth's past. Units 1, 2, 3, 4, and 5 of this book deal with topics of primary concern to geologists.

Oceanography

Vast oceans cover nearly three fourths of the earth's surface. The study of the earth's oceans is called **oceanography.** Many oceanographers work on research ships equipped with special instruments for studying the sea. Some oceanographers study waves, tides, and other movements of the sea. Some oceanographers explore the ocean floor for clues to the earth's history and to locate mineral deposits. Other oceanographers study marine plant and animal life. A discussion of the earth's oceans is presented in Unit 6.

Meteorology

The study of the earth's atmosphere is called **meteorology.** Using satellites and other modern equipment, meteorologists study the variations in atmospheric conditions that produce weather. Many meteorologists work as weather observers, measuring such factors as wind

Figure 1–2. A geologist collects ore samples (left). Oceanographers prepare to enter a submersible to study the ocean floor (above).

Figure 1-3. Two astronomers use a telescope to observe distant galaxies (left). A meteorologist uses a computerized system to study the weather (right).

speed, temperature, and rainfall. This weather information is then used to prepare detailed weather maps. Other meteorologists use the maps and computer data to make weather forecasts. You will learn more about meteorology in Unit 7 of this book.

Astronomy

The study of the universe beyond the earth is called **astronomy.** It is one of the oldest branches of earth science. In fact, the ancient Babylonians charted the positions of planets and stars nearly 2,000 years ago. Modern astronomers use earth-based and space-based telescopes and other instruments to study the universe. Space probes, such as *Mariner, Pioneer, Viking,* and *Voyager,* have also provided much useful data. Unit 8 of this book presents information about the moon, the planets, the sun, the stars, and the universe.

The Importance of Earth Science

Powerful forces are at work on the earth. Volcanoes erupt and earthquakes shake the ground. These events not only shape the earth, but they also affect the plant and animal life on the earth. A volcanic eruption may bury a town in ash. An earthquake may cause a huge wave of water that destroys a shoreline. By discovering how natural forces shape our environment, earth scientists can predict potential disasters and help save lives and property.

Observations made by earth scientists have contributed greatly to our knowledge of the world around us. For example, information gathered by astronomers studying distant galaxies has led to theories about the origins of this solar system. Geologists studying rock layers have found clues to the earth's past and to the origins of life on this planet.

The earth also provides many valuable resources that enrich the quality of people's lives. For example, the fuel that powers a jet, the metal used to make surgical instruments, and the paper and ink in this book all come from the earth. The study of earth science can help people gain access to the earth's resources and teaches them to use those resources wisely.

Ecology

Earth scientists primarily study the **geosphere**—the solid earth; all of its water, or hydrosphere; and its atmosphere—the gases surrounding the earth. Other scientists, called biologists, study the living world. An area of science in which biology and earth science are closely linked is called **ecology.** Ecology is the study of the complex relationships between living things and their environment. Most ecologists have backgrounds in either earth science or biology.

Organisms on the earth inhabit many different environments. A community of organisms and the environment they inhabit is called an **ecosystem.** The terms *ecology* and *ecosystem* come from the Greek word *oikos,* meaning "house." Each ecosystem is a physically distinct, self-supporting system. An ecosystem may be as large as a desert or a city or as small as a tide pool or a rotting log.

The largest ecosystem is called the **biosphere.** The biosphere encompasses the entire world of life on earth and the physical environment that supports it. The biosphere extends from the ocean depths to the atmosphere a few kilometers above the earth's surface.

A tropical rain forest is one example of a large ecosystem within the biosphere. Plants in the rain forest use sunlight to produce food through a process known as *photosynthesis*. The plants are then eaten by animals, which are in turn eaten by other animals. When

Figure 1-4. The small tidal pool on the shore of Acadia National Park and the vast ocean are both ecosystems.

Figure 1-5. People help protect the environment by cleaning up the shore after an oil spill (top). Pollution also harms wildlife. Volunteers remove oil from a bird's feathers (bottom).

rain forest plants and animals die, their bodies are decomposed by microorganisms. The resulting chemicals enter the soil to nourish other plants and animals. Thus, the system is totally self-supporting. What other examples of ecosystems can you name?

Environmental Pollution

Each ecosystem is delicately balanced. When that fragile ecological balance is upset, the survival of the ecosystem, and in some cases the entire biosphere, is threatened. One serious threat to ecosystems today is **pollution,** the contamination of the environment with waste products or impurities.

Some waste products are **biodegradable.** As such, they can be broken down by microorganisms into harmless substances that can then be used by other organisms. Biodegradable waste products pose little threat to the environment, and in many cases they contribute to the well-being of the environment. For example, the chemicals found in such biodegradable wastes as orange peels and eggshells make excellent plant fertilizer.

Many modern waste products, such as most plastics, however, are not biodegradable. They cannot be broken down into harmless substances. Many ecosystems are threatened by the large quantities of nonbiodegradable wastes.

For example, plastic wastes dumped in oceans or lakes can harm the fish life there. When particles of plastic are ingested, they clog the digestive tracts of the fish. Ducks and other animals can starve to death when they become tangled in plastic litter and can not move.

Protecting the Environment

Pollution poses serious problems for all living organisms. To help protect the environment from pollution, ecologists often work together with earth scientists in other fields such as meteorology.

For example, in the early 1970's, meteorologists found that the level of ozone, a form of oxygen, in the upper atmosphere was decreasing. This discovery was alarming to ecologists and earth scientists. They knew that ozone helps protect the earth's plant and animal life from the harmful ultraviolet rays of the sun. Further research revealed that the ozone layer was being destroyed by fluorocarbons, chemical compounds commonly used as propellants in aerosol sprays. To reduce this threat to the environment, the U.S. government passed regulations in 1978 limiting the use of fluorocarbon aerosols.

Section 1.1 Review

1. What are the four major branches of earth science?
2. Describe the work of meteorologists.
3. What is ecology?
4. Give an example of an ecosystem and explain how it is self-supporting.
5. How might the study of earth science contribute to the survival of the biosphere?

1.2 Paths to Discovery: Scientific Methods

Section Objectives

- **Identify the steps that make up scientific methods.**
- **Explain how the meteorite-impact hypothesis developed.**

Through research, scientists seek to explain natural phenomena and solve mysteries of the earth. Over the years, the scientific community has developed organized, logical approaches to scientific research, called **scientific methods.** Scientific methods are not a set of sequential steps that scientists invariably follow. Rather, they are guides to scientific problem solving.

State the Problem

Scientific inquiry often begins as a result of **observation.** Simply put, observation is using the senses of sight, touch, taste, hearing, and smell to gather information about the world. When you notice thunderclouds forming in the summer sky, that is an observation. So, too, is feeling the cool, smooth surface of polished marble, or hearing the roar of white water around the bend of a river.

Observations often lead to questioning. Why do thunderclouds form? What is the chemical composition of marble? What causes a river to change its course? Asking questions like these is one way of stating the problem to be investigated through scientific methods.

One problem that has long puzzled scientists is the sudden extinction of the dinosaurs. For more than 135 million years these huge reptiles dominated the earth. Then suddenly, about 65 million years ago, the dinosaurs and three fourths of all the other species on the earth died out. Scientists wondered what could have caused such a mass extinction.

Gather Information

To investigate a problem, such as the extinction of the dinosaurs, scientists gather information. An important means of gathering information is **measurement.** Measurement involves the comparison of

Figure 1-6. One means of gathering information is through careful measurement. In this photo, geologists are measuring a crack in the earth's surface caused by an earthquake.

Figure 1–7. Geologists Bruce Bohor, Pete Modreski, and Eugene Foord (left to right) found a layer of high-iridium content clay at Brownie Butte, Montana.

some aspect of an object or phenomenon with a standard unit, such as a meter, a Celsius degree, or a kilogram. For example, when you measure a rock and find that it is 20 cm long, you are comparing the length of the rock with a unit of measure—one centimeter. What other units of measure can you name?

Accuracy is important in scientific measurements. Imprecise measurements can lead to an incorrect conclusion. Scientists often use special tools such as micrometers and calipers to help them make precise measurements.

In the case of the dinosaurs, scientists examined the fossil record for clues to what happened 65 million years ago. They studied rock layers throughout the world that date from the time when the dinosaurs disappeared. The scientists discovered that in certain locations these layers contain iridium, a substance that is uncommon in earth rocks, but common in meteorites. Scientists then measured the amount of iridium in the rock layers. They found that the rock layers in these particular locations contained nearly 160 times the amount of iridium normally found in earth rocks. The scientists searched for an explanation that would relate the iridium measurements to the disappearance of the dinosaurs.

Horizons

Theory: The Gaia Hypothesis

As satellites such as Voyager 2 send back information about other planets, one overriding discovery looms large. Earth appears to be the only planet that has been able to maintain an environment supportive of life over time. Earth has supported life, without pause, for almost 4 billion years.

Some scientists think that the living organisms on earth may somehow interact with earth's nonliving matter to keep the planet alive. James Lovelock, a British scientist, has suggested that the earth's rock, water, and air work with its plant and animal life to form "a complex system that can be seen as a single organ-ism and that has the capacity to keep our planet a fit place for life." Lovelock, below, named this organism after the Greek goddess Gaia, Mother Earth. Lovelock's ideas are known as the *Gaia Hypothesis*.

According to Lovelock, Gaia acts as a single organism, unconsciously making adjust-

Form a Hypothesis

Once a problem has been stated and information gathered, a scientist may propose a **hypothesis,** (HIE-POTH-uh-sus, pl. hypotheses), a possible explanation or solution to the problem. A hypothesis is based on facts, which are often established through observation.

For example, the scientists who discovered the iridium-laden rock layers proposed the meteorite-impact hypothesis to explain the extinction of the dinosaurs. This hypothesis states that about 65 million years ago a giant meteorite crashed into the earth. The impact of the collision raised enough dust to block the sun's rays for many years. The earth became colder, plant life began to die, and many animal species, including the dinosaurs, became extinct. As the dust settled over the earth, it formed a layer of iridium-laden rock.

Test the Hypothesis

Once a hypothesis has been proposed, it should be tested. A hypothesis will not be accepted by the scientific community unless there is evidence to support it.

ments that maintain the conditions necessary for life.

In humans, the ability to regulate body processes and adjust to changes is called *homeostatic ability.* The state of balance that the body maintains is *homeostasis.*

When a person's body temperature heats up, the body perspires and cools itself. When a person runs, that person's heart and respiration rates increase to supply the increased need for oxygen. These events are examples of homeostasis.

According to Lovelock, the earth, too, is homeostatic. Although scientists have not been able to identify and explain the earth's mechanisms

for homeostasis, they do offer evidence that the mechanisms are in effect. For example, the earth's temperature has remained stable, even though the sun is 30 percent hotter today than it was 4 billion years ago. Somehow the earth

has reacted to the increase in the sun's temperatures and maintained the temperatures necessary to sustain life.

Which two branches of science are likely to devote the most time to testing the Gaia Hypothesis?

A hypothesis is tested by **experimentation.** An experiment is a scientific procedure carried out according to certain guidelines. An experiment enables scientists to test each **variable** that might prove or disprove the hypothesis. A variable is a factor in an experiment that can be changed. An experiment set up to test a variable is called a *controlled experiment.*

To ensure that only one variable is tested in an experiment, scientists will also run a control. The control will have the same conditions as the experiment except for the variable being tested. For example, to test the effects of sunlight on a green plant, a scientist would grow two identical plants. To control the experiment, the scientist would vary the amount of sunlight reaching one plant, while keeping the amount of sunlight constant on the other plant. The scientist would then observe both plants and record the observations. What is the variable in this experiment?

In the study of earth science, setting up controlled experiments to test a hypothesis is often difficult, and sometimes impossible. The scientists studying the disappearance of the dinosaurs, for instance, cannot bombard the earth with a giant meteorite to see if it produces life-threatening conditions.

Recently, however, scientists have developed computer models that enable them to test hypotheses by simulating certain conditions. For example, scientists have entered information into a computer about the possible climatic conditions during the period when the dinosaurs became extinct. They found that a dust cloud resulting

Figure 1-8. According to the meteorite-impact hypothesis, a huge meteorite crashed into the earth 65 million years ago. Dust from the impact blocked out the sun and led to the extinction of the dinosaurs.

Figure 1-9. The Barringer Meteorite Crater in Arizona is 1,300 m in diameter and nearly 200 m deep. The crater is visible proof of the explosive power of a meteorite hitting the earth.

from the collision of a meteorite 10 km in diameter would have been sufficient to lower the earth's temperature considerably.

When experimentation is impossible, scientists often use observation to gather evidence that will either support or discredit the hypothesis. The hypothesis is then tested by examining how well it fits or explains all the known observations.

To test the meteorite-impact hypothesis further, scientists had to find additional evidence that the iridium in the rock layers on earth had once come from meteorites. Scientists again examined the rock layers and this time they found strangely deformed particles of the mineral quartz. Previously this type of quartz had been found only near meteorite craters, at nuclear-testing sites, and in moon rocks. Scientists concluded that such quartz particles could only have been produced by an extremely powerful explosion. The collision of the earth with a huge meteorite, they reasoned, could have produced such an explosion.

State a Conclusion

After many experiments and observations, scientists generally reach conclusions regarding the correctness of the hypothesis being considered. Depending on how well the hypothesis fits the known facts, it may be accepted as stated, altered slightly, or discarded altogether.

The fossil evidence for the meteorite-impact hypothesis does not prove that a meteorite was responsible for the extinction of the dinosaurs. The evidence does show, however, that an abnormally high amount of meteorite dust reached the earth at that time. Thus, until new evidence is found or a better hypothesis is proposed, the meteorite-impact hypothesis serves as one possible explanation of why the dinosaurs disappeared.

Section 1.2 Review

1. What are scientific methods?
2. Define *hypothesis*.
3. How do scientists test hypotheses?
4. Summarize the evidence scientists found to support the meteorite-impact hypothesis.
5. How have scientific methods contributed to the development of modern science?

Section Objectives

- **Distinguish between a hypothesis, a theory, and a scientific law.**
- **Describe the Doppler effect.**
- **Summarize the big bang theory of the origin of the universe.**
- **List evidence for the big bang theory.**

1.3 Birth of a Theory: The Big Bang

Scientific methods are useful tools for the study of earth science. However, the development and testing of a hypothesis is just one step along the way to scientific understanding. Once a hypothesis has been tested and generally accepted, it may lead to the development of a **theory.** A theory is a hypothesis or a set of hypotheses that is supported by the results of experimentation and observation. A theory provides a general explanation for scientific observations that is consistent with known facts.

Once a theory is well established through research and experimentation, it may become a **scientific law.** A scientific law is a rule that correctly describes a natural phenomenon. To become a law, a theory must be proven correct every time it is tested. For example, the law of conservation of energy, which states that energy can neither be created nor destroyed, has been tested again and again. It has never been found to fail during these experiments.

Light and the Doppler Effect

One of the most exciting theories of modern science—how the universe began—has its roots in observations made more than 300 years ago. In 1665, the British scientist Isaac Newton observed that sunlight passing through a glass prism produced a rainbow of colors: red, orange, yellow, green, blue, and violet. Newton named this display of colors the **spectrum** (pl. spectra).

Light travels in waves. The distance from the crest of one wave to the crest of the next is a **wavelength.** Each color in the spectrum has a different wavelength. Red light has the longest wavelength, and violet light has the shortest. As light passes through a prism, each wavelength is bent to a different degree, and the band of colors results.

Figure 1-10. A prism bends the different wavelengths of light, separating white light into a band of colors.

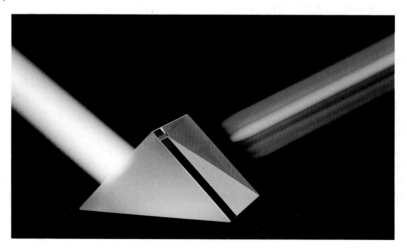

BARIUM

HELIUM

HYDROGEN

| | | | | | | | | | | | | | | | | | | | | | | | | | | | | | | | | | | | | | | | |
7500   7000   6500   6000   5500   5000   4500   4000

Adapted from the SPECTRUM CHART, Welch Scientific Company

Figure 1–11. Each element produces a bright-line spectrum.

In the late nineteenth century, research revealed that when chemical **elements** are heated, they too produce spectra. An element is a substance, such as hydrogen or iron, that cannot be broken down into a simpler form by ordinary chemical means. Instead of a full spectrum of continuous colors, like that produced by sunlight, a heated element produces only a series of thin colored lines spaced at uneven intervals. This series of colored lines, called a *bright-line spectrum,* indicates that the light source is sending out only certain wavelengths of light. An example of a bright-line spectrum is shown in Figure 1–11. Each element produces its own bright-line spectrum, as unique as a set of fingerprints.

Scientists also discovered that when a light source is moving toward an observer, the wavelengths of the light produced appear shorter to the observer. As a result, the spectral lines of the light source appear to shift slightly toward the shorter wavelengths, or the blue end of the spectrum. When a light source is moving away from an observer, the light waves appear longer to the observer. The spectral lines of the light source appear to shift toward the longer wavelengths, or the red end of the spectrum. The faster a light source is moving, the greater the shift of its spectrum. The apparent shift in the wavelengths of light emitted by a light source moving away from or toward an observer is called the **Doppler effect.**

Evidence: Red Shift

Using an instrument called a **spectroscope,** scientists studied starlight to determine what elements were present in the stars. A spectroscope contains a prism, which splits starlight into a spectrum of different colors and wavelengths. By comparing the spectrum produced by each star with the spectra of known elements, scientists were able to determine the chemical makeup of various stars. The sun, for example, was found to be about 94 percent hydrogen and almost 6 percent helium, with traces of nearly 100 other elements.

The study of starlight spectra revealed surprising information about our universe. Scientists found that the spectrum of every

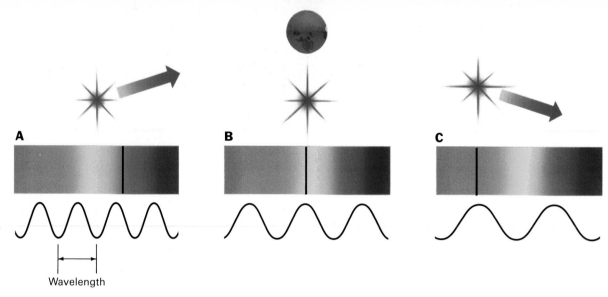

Figure 1-12. The wavelengths of light produced by star A, moving toward the earth, appear shorter. Therefore, the spectral lines of star A are shifted toward the blue end of the spectrum. The wavelengths of light produced by star C, moving away from the earth, appear longer. Therefore, the spectral lines of star C are shifted toward the red end of the spectrum. Star B is stationary.

galaxy, or large system of stars, tested was shifted toward the red end of the spectrum. No galaxy, however, showed a shift toward the blue end of the spectrum. The red shift indicates that all the galaxies in the universe are moving away from the earth.

By examining the degree of red shift, scientists were also able to determine the speed at which the galaxies were traveling. Scientists found that the most distant galaxies showed the greatest red shift and thus were moving away the fastest. From these observations of starlight spectra, scientists concluded that the universe is expanding.

A Theory Emerges

Observations of red shift led scientists to propose a hypothesis to explain the expanding universe. The hypothesis states that billions of years ago all the matter and energy in the universe was compressed into one extremely hot, dense sphere. About 17 billion years ago, this sphere exploded in what is called the *big bang*, sending all the matter and energy hurtling outward in a giant cloud. As the cloud expanded, some of the matter gathered into clumps that evolved into galaxies. Today the universe is still expanding, and the galaxies continue to move apart from one another. This movement causes the apparent red shift in the spectra of galaxies.

Despite the evidence of red shift in the spectra of the galaxies, a number of scientists did not accept the big bang hypothesis. They argued that if the big bang had taken place as the hypothesis proposed, the energy left from the explosion would be found evenly distributed throughout the expanding universe. If this energy could not be found, they insisted, then there was little reason to accept the big bang hypothesis.

An important discovery in the 1960's finally convinced most scientists who had doubted the evidence of red shift. Using radio telescopes, researchers detected low levels of energy, called **background radiation,** evenly distributed throughout the universe. The presence of this energy convinced most scientists that the big bang hypothesis was correct. Because of the abundant evidence and

Investigation

The Big Bang Theory

According to the big bang theory, all stars and galaxies are moving outward from all other stars and galaxies. You can demonstrate the principles of this expansion with a simple model.

Science Process Skills Focus: constructing models, interpreting data, hypothesizing

Materials

large, uninflated round balloon; water-based felt-tip pen; string, 15 cm long; ruler

Procedure

1. Mark a pair of dots 0.5 cm apart across the middle of the balloon. Label them *A* and *B*. Mark a third dot 5 cm away from *B*. Label this dot *C*.
2. Blow into the balloon for 2–3 seconds. Record your time. Keep the balloon inflated but do not tie the neck.
3. Use the string and ruler to measure the distance between *A* and *B* and between *C* and *B*.
4. Calculate the rate of change in the distances between *A* and *B* and between *C* and *B*. To calculate the rate, subtract the original distance between the dots from the latest distance. Divide this number by the number of seconds

you blew into the balloon.

5. Inflate the balloon further for 2–3 seconds.
6. Measure and calculate the rate of change in the distances between *A* and *B* and between *C* and *B*. To calculate the rate, use the distance measured in Step 3 as the ''original'' distance.

Analysis and Conclusions

1. Did the distance between *A* and *B* or between *C* and *B* show the greatest rate of change?
2. Did the rate of change for either set of dots differ in Steps 4 and 6?
3. Suppose dots *C* and *A* represent galaxies and dot *B* represents the earth. How does the distance between the galaxies and the earth relate to the rate at which they are moving apart?

widespread acceptance of the big bang hypothesis, this explanation of the origin of the universe became known as the **big bang theory.**

Like any theory, the big bang theory must continue to be tested against each new discovery about the universe. As new information about our universe emerges, the big bang theory may be revised, or a new theory may take its place.

Section 1.3 Review

1. What is a scientific theory?
2. Describe the Doppler effect.
3. Explain the big bang theory.
4. What evidence supports the big bang theory?
5. If scientists discovered blue shift in the spectra of some galaxies, how might this information affect the big bang theory?

Reviewing Chapter 1

Key Concepts

- The four main branches of earth science are geology, oceanography, meteorology, and astronomy. See page 3.
- Earth scientists and ecologists can work together to protect the environment. See page 8.
- Scientific research attempts to solve problems logically through scientific methods. See page 9.
- The meteorite-impact hypothesis provides a possible explanation for the mass extinction of the dinosaurs. See page 12.

- Through observation and experimentation, scientists develop hypotheses, theories, and laws to describe natural phenomena. See page 14.
- The Doppler effect describes the shift in the spectrum of a moving light source. See page 14.
- The big bang theory provides a possible explanation of the origin of the universe. See page 16.
- Evidence for the big bang theory comes from the study of starlight spectra and background radiation. See page 16.

Key Terms

astronomy (6)	earth science (3)	hypothesis (11)	scientific methods (9)
background radiation (16)	ecology (7)	measurement (9)	spectroscope (15)
	ecosystem (7)	meteorology (5)	spectrum (14)
big bang theory (17)	elements (15)	observation (9)	theory (14)
biodegradable (8)	experimentation (12)	oceanography (5)	variable (12)
biosphere (7)	geology (4)	pollution (8)	wavelength (14)
Doppler effect (15)	geosphere (7)	scientific law (14)	

Review

On your paper, write the letter of the term that best completes each of the following statements.

1. The study of the solid earth is called (a) geology (b) oceanography (c) meteorology (d) astronomy.
2. The earth scientist most likely to study a tornado is (a) a geologist (b) an oceanographer (c) a meteorologist (d) an astronomer.
3. The study of the complex relationships between living things and their environment is called (a) geology (b) meteorology (c) ecology (d) astronomy.
4. The largest ecosystem is called the (a) biosphere (b) geosphere (c) atmosphere (d) hydrosphere.
5. An example of a nonbiodegradable waste product is (a) an orange peel (b) a plastic

milk jug (c) a pile of rotting leaves (d) an egg shell.
6. Usually the first step in scientific problem solving is to (a) form a hypothesis (b) state the problem (c) gather information (d) state a conclusion.
7. A possible explanation for a scientific problem is called (a) an experiment (b) an observation (c) a theory (d) a hypothesis.
8. The development of the meteorite-impact hypothesis began with the observation of (a) blue shift in the spectra of stars (b) red shift in the spectra of stars (c) background radiation (d) iridium in earth rocks.
9. A statement that consistently and correctly describes some natural phenomenon is a scientific (a) hypothesis (b) observation

(c) law (d) control.
10. The display of colors that results when sunlight passes through a prism is called (a) a spectrum (b) a wavelength (c) the Doppler effect (d) a spectroscope.
11. The apparent change in the wavelengths of a moving light source is called (a) the big bang (b) the Doppler effect (c) the spectrum (d) background radiation.
12. Scientists have found that as a light source moves toward a stationary observer, the wavelengths of the light source appear (a) longer (b) shorter (c) higher (d) lower.
13. The big bang theory states that the galaxies in the universe are (a) moving away from one another (b) moving towards one another (c) remaining stationary (d) being bombarded by meteorites.
14. To study the chemical composition of stars, scientists use a (a) geosphere (b) microscope (c) spectroscope (d) biosphere.
15. Evidence for the big bang theory includes (a) iridium in earth rocks (b) deformed quartz particles in earth rocks (c) blue shift in the spectra of galaxies (d) red shift in the spectra of galaxies.

Application

On your paper, write answers to the following questions.
1. A meteorite lands in your backyard. Which earth science specialists would you call to study the meteorite? Why?
2. A stream that feeds a small pond gradually dries up. How might this change affect the ecosystem of the pond?
3. You find a yellow rock and wonder if it is gold. How could you apply scientific methods to this problem?
4. Some scientists have hypothesized that meteorites have periodically bombarded the earth, causing mass extinctions every 26 million years. How might this hypothesis be tested?
5. A scientist observes that each eruption of a volcano is preceded by a series of small earthquakes. The scientist then makes the following statement: Earthquakes cause volcanic eruptions. Is the scientist's statement a hypothesis or a theory? Why?
6. If you detected a blue shift in the sun's spectrum, what might you infer about the distance between the earth and the sun? Why?
7. Imagine you are on another planet in a galaxy far from the earth. If you used a spectroscope to examine the spectrum of the sun, would you expect to find red shift, blue shift, or no shift at all? Why?
8. According to the big bang theory, the original big bang took place about 17 billion years ago. How might scientists have been able to determine this?

Extension

1. Find information about three recent technological developments in different branches of earth science. Write a short essay explaining the impact each development will have on the study of earth science.
2. Choose an ecosystem and draw a diagram showing some of the interrelationships between the living things and their environment. Identify the living things according to their roles as producers, consumers, or decomposers.
3. Find out about the organizations in your community that are working to protect the environment. Write a short profile describing the work of each organization. Put the information together in a directory.
4. Research related to the meteorite-impact hypothesis also led to the development of the nuclear-winter hypothesis. Find information on the nuclear-winter hypothesis and report your findings to the class.
5. Find more information about computer models and write a report explaining how they can be used to test hypotheses.

Chapter 2

The Earth in Space

During the first Apollo mission to the moon in 1969, an astronaut snapped a picture of the earth suspended in the blackness of space. For the first time in history, people could view the earth as a whole. Since that time, space exploration has added much to the knowledge of the earth. This chapter describes the earth, its movements through space, and the use of spacecraft to explore the earth.

Chapter Outline

The earth, viewed from space, is a unique planet.

2.1 Earth: A Unique Planet

Photos of the earth from space reveal a blue-green sphere covered with white clouds. Surrounded by the blackness of space, the earth appears fragile and alone. In fact, the earth is unique in the solar system. It is the only known planet with liquid water on its surface and an atmosphere that contains a large amount of oxygen. Most importantly, the earth is the only planet known to support life.

Statistics

Viewed from space, the earth appears to be a perfect sphere. On a perfect sphere, the circumference, or distance around, is the same no matter where it is measured. However, careful measurements reveal that the earth's circumference varies, depending on where it is measured. The circumference measured around the poles is 40,007 km. The circumference measured around the *equator* is 40,074 km. The equator is an imaginary line that divides the earth into the Northern and Southern hemispheres.

Because of these differences in the earth's circumference, the earth is more accurately described as an *oblate spheroid,* a slightly flattened sphere. The spinning of the earth on its **axis** causes the polar regions to flatten and the equatorial zone to bulge. The earth's axis is an imaginary straight line running through the earth from the North Pole to the South Pole.

Seen from space, the earth's surface also appears to be smooth. Given the size of the planet, its surface is relatively smooth. The difference between the height of the tallest mountains and the depth of the deepest ocean trenches is only about 20 km. This distance is very small compared to the earth's average diameter of 12,735 km.

The Hydrosphere and the Atmosphere

About 97 percent of the earth's water is salt water in the oceans. The remaining 3 percent of earth's water is fresh water found in lakes, rivers, and streams, and frozen in glaciers and the polar ice caps. All the earth's water makes up the **hydrosphere.**

The earth is surrounded by a thick blanket of gases called the **atmosphere.** Hundreds of kilometers thick, the atmosphere provides the air you breathe and shields the earth from the sun's heat and harmful radiation. The atmosphere is made up of 78 percent nitrogen and 21 percent oxygen. The remaining 1 percent includes other gases such as argon, carbon dioxide, and helium.

The Earth's Interior

Because direct observation of the earth's interior is impossible at this time, scientists must rely on indirect methods of study. For example, scientists have made important discoveries about the earth's interior through studies of **seismic** (SIZE-mik) **waves.** Seismic waves are

Section Objectives

- List the characteristics of the earth's three major layers.
- Explain how studies of seismic waves have provided information about the earth's interior.
- Define *magnetosphere* and identify the possible source of the earth's magnetism.
- Summarize Newton's law of gravitation.

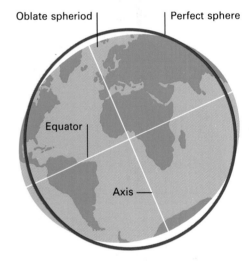

Oblate spheriod   Perfect sphere

Equator

Axis

Figure 2–1. The earth's shape is really an oblate spheroid, not a perfect sphere. The red line shows the shape of a perfect sphere.

vibrations that travel through the earth. Earthquakes and explosions on or near the earth's surface produce seismic waves. By studying seismic waves as they travel through the earth, scientists have determined that the earth is made up of three major layers.

Layers of the Earth

The thin, solid, outermost layer covering the earth is the **crust.** The crust makes up only 1 percent of the earth's volume. Beneath the oceans, the crust is called *oceanic crust*. Oceanic crust is 5 km to 10 km thick. On the continents, the crust is called *continental crust*. Continental crust varies in thickness from 32 km to 70 km. Continental crust is the thickest beneath mountains.

Below the crust lies the **mantle,** a layer of rock nearly 2,870 km thick. The mantle makes up about 80 percent of the earth's volume and almost two thirds of the earth's **mass.** Mass is the amount of matter in an object.

The uppermost part of the mantle is solid. This solid portion of the mantle and the crust above it make up the *lithosphere,* a rigid layer 65 km to 100 km thick. Just below the lithosphere is a region of the mantle called the *asthenosphere*. The asthenosphere is approximately 200 km thick. Because of enormous heat and pressure, the solid rock of the asthenosphere has the ability to flow. The ability of a solid to flow is called *plasticity.*

Below the mantle is the **core,** the center of the earth. Studies of seismic waves indicate that the core is made mostly of iron. Scientists think the outer portion of the core is a dense liquid layer about 2,190 km thick. The inner portion is a dense solid about 2,680 km in diameter. The inner and outer core together make up 19 percent of the earth's volume and nearly one third of its mass.

Seismic Wave Studies

Two types of seismic waves, primary waves, or *P waves* and secondary waves, or *S waves,* are useful to scientists exploring the earth's interior. P waves and S waves behave differently. P waves travel through liquids, solids, and gases. S waves travel only through solids. P waves also travel faster than S waves. The speed and the direction of both types of waves are affected by the composition of the material through which they travel. Both P waves and S waves travel faster through more-rigid materials. The speed and direction of seismic waves reveal much about the earth's interior.

The Moho   In 1909 Andrija Mohorovičić (MOE-huh-ROE-vuh-CHICH), a Yugoslavian scientist, discovered that the speed of seismic waves increases abruptly 32 km to 64 km beneath the earth's surface. As Figure 2–2 shows, this change in the speed of the waves marks the boundary between the crust and the mantle. This boundary is called the *Mohorovičić discontinuity,* or the **Moho.** The increase in speed at the Moho indicates that the earth's mantle is more rigid than its crust.

Below the Moho, at a depth of about 100 km, a decrease in seismic-wave speed marks the boundary between the lithosphere and

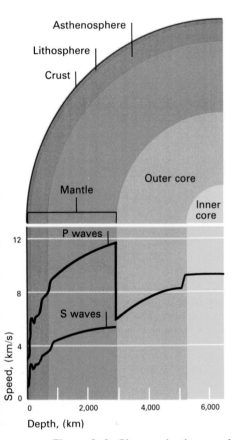

Figure 2–2. Changes in the speed of seismic waves were used to determine the location of the earth's layers.

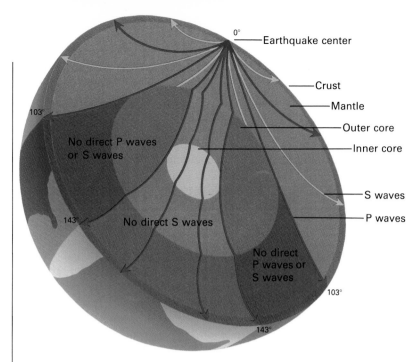

0° — Earthquake center

Crust

Mantle

Outer core

Inner core

103°

No direct P waves
or S waves

143°

No direct S waves

No direct
P waves or
S waves

S waves

P waves

103°

143°

Figure 2–3. No direct S waves can be detected in locations more than 103° from the earthquake epicenter. No direct P waves can be detected in locations between 103° and 143° from the earthquake center.

the less rigid asthenosphere. Seismic waves then increase in speed until, at a depth of about 2,900 km, P waves slow down again, while S waves disappear entirely. These changes in the seismic waves mark the boundary between the mantle and the outer core. Since S waves cannot travel through liquids and P waves slow down in less-rigid materials, scientists think the outer core may be a dense liquid. At a depth of 5,150 km, P waves speed up again, marking the boundary between the outer core and the inner core. This increase in speed suggests that the inner core is a dense, rigid solid.

Shadow Zones   Recordings of seismic waves around the world reveal **shadow zones** on the earth's surface. Shadow zones are locations on the earth's surface where neither S waves nor P waves are detected or where only P waves are detected.

Shadow zones occur because the materials that make up the earth's interior are not uniform in rigidity. When seismic waves travel through materials of differing rigidities, their speed changes, causing the waves to bend and change direction.

As Figure 2–3 shows, a large S-wave shadow zone covers the side of the earth that is opposite an earthquake. S waves do not reach the S-wave shadow zone because they are blocked by the liquid outer core. Although P waves can travel through all the layers, the speed and direction of the waves change as the waves pass through each layer. The waves bend in such a way that a P-wave shadow zone forms.

The Earth as a Magnet

If you have ever used a magnetic compass to find direction, you know that the earth acts as a giant magnet. Like a bar magnet, the earth has two magnetic poles. As Figure 2–4 shows, the lines of force of the earth's magnetic field extend between the north magnetic pole and the south magnetic pole. The earth's magnetic field

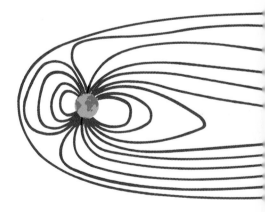

Figure 2–4. The flow of charged particles from the sun compresses and shapes the magnetosphere so that it flares out on the side of the earth away from the sun.

also affects an area that extends beyond the atmosphere. The part of the atmosphere that is affected by the earth's magnetic field is called the **magnetosphere** (mag-NEET-uh-SFIR).

The source of the earth's magnetic field may be the liquid iron in the earth's outer core. Because iron is a good conductor, scientists hypothesize that motions within the core produce electrical currents that in turn create the earth's magnetic field.

Recent research, however, indicates that there may be another source for the magnetic field. Scientists have found that both the sun and the moon have magnetic fields. Yet the sun contains

Career Focus: *Earth Structure*

"As a geologist, you know that no matter what problem you're trying to solve, the answer is out there just waiting to be discovered."

Field Geologist

Talking to Denver-based field geologist Ken Pierce is one thing. Finding him is quite another.

"During the field season, I can be out in the field area for as long as four months at a time," he said. "Though I mostly drive a jeep and hike, I have made some month long pack trips on horseback as well."

Like other field geologists, Pierce, pictured left, studies the composition, structure, and history of the earth's crust. Pierce specializes in the study of the youngest part of geologic time.

After collecting samples including sand, gravel, and glacial deposits, Pierce, who works for the U.S. Geological Survey, determines which samples merit further study. He brings them to the lab, where they are X-rayed, studied under high-powered microscopes, and subjected to chemical analysis Pierce also performs experiments to test his geologic theories.

"Although I have always liked being outdoors, it wasn't until I was in college that I really became interested in learning why the surface of the earth looks the way it does and knew I wanted a career in geology."

Pierce holds undergraduate and graduate degrees in geology with an emphasis on geomorphology, the study of the origin of the earth's surface features. He says he enjoys the never-ending challenge of his work.

"You know if you do enough research, make the right observations, and ask the right questions, you'll find your answer. It is mostly a matter of perseverance."

no iron, and the moon does not have a liquid core. Discovering the sources of the magnetic fields of the sun and moon may help scientists identify the source of the earth's magnetic field.

The Earth's Gravity

The seventeenth-century British scientist Isaac Newton made many contributions to the fields of mathematics, physics, and astronomy. Among the most important were his studies of **gravity,** the force of attraction that exists between all objects in the universe.

Surveyor

Surveyors measure the height and shape of the land, marking waterways, boundary lines, and distances. Surveyors usually work in small groups called *field parties.* One surveyor may operate a transit, which is a small telescope that measures angles and distances. Meanwhile, another member of the field party holds the rod that marks the height of land being surveyed.

The results of the survey are recorded, and the data verified. Surveyors then prepare the sketches, maps, and reports needed to establish official land and water boundaries.

Although some states require surveyors to have a four-year college degree, most people enter the field through technical training programs.

Instrumentation Technician

Instrumentation technicians, such as the one pictured on the left, help develop the instruments that earth scientists use to study the earth. Many of these devices aid in the search for fossil and geothermal sources of power. Others are used in the fields of soil conservation, pollution control, and oceanography.

In addition to helping design new machinery, instrumentation technicians may also test, install, operate, and service earth-science equipment.

Instrumentation technicians generally have completed a technical degree program or received on-the-job training.

For Further Information

For more information on careers that deal with the earth's structure, write the American Geological Institute, 4220 King St., Alexandria, VA 22302.

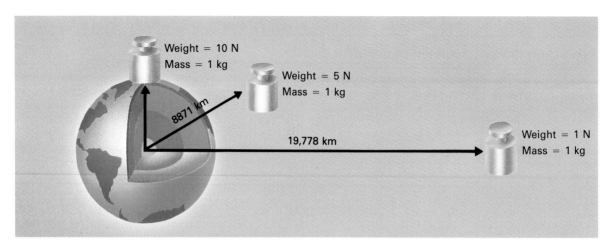

Figure 2–5. As the distance from the earth's center increases, the weight of an object decreases. The mass of an object stays the same.

Newton described the effects of gravity in his **law of gravitation.** This law states that the force of attraction between any two objects depends upon their masses and the distance between them. The larger the masses of two objects and the closer together they are, the greater will be the force of gravity between them.

Because the mass of the earth is so great, its force of gravity pulls objects toward the center of the earth. **Weight** is a measure of the strength of the pull of gravity on an object. Weight is measured in *newtons* (N). On the earth's surface, a kilogram of mass weighs about 10 N. What is the weight of a 2-kg mass on the earth's surface? A 10-kg mass?

Weight and mass are not the same. Mass is the amount of matter in an object. Weight, however, is a force. The mass of an object does not change with location, but its weight does change. The weight of an object on earth depends on its mass and its distance from the earth's center. As the law of gravitation states, the force of gravity decreases as the distance from the earth's center increases. For example, at 8871 km from the center of the earth (2500 km above the surface), a 1-kg mass weighs about 5 N, half of its weight on the earth's surface. At 19,778 km from the center of the earth (13,400 km above the surface), a 1-kg mass weighs only about 1 N.

Weight also varies according to location on the earth's surface. As you have read, the earth spins on its axis, and this motion causes the earth to bulge out slightly near the equator. Therefore, the distance between the earth's surface and its center is greater at the equator than it is at the poles. This difference in distance means that your weight at the equator would be about 0.3 percent less than your weight at the North Pole.

Section 2.1 Review

1. How thick is the mantle?
2. Describe the earth's core.
3. Compare the behavior of P waves and S waves.
4. Define magnetosphere.
5. Explain why astronauts are weightless when they are in space.

2.2 Movements of the Earth

Section Objectives

- **Describe the earth's revolution and rotation.**
- **Tell why the seasons change.**
- **Explain how the sun is used as a basis for measuring time.**

The earth is constantly in motion. You are traveling with the earth around the sun at an average speed of 106,000 km/hr. The movement of the earth around the sun is called **revolution.** Each complete revolution takes 365.24 days, or about one year.

As it revolves around the sun, the earth also spins on its axis. This spinning motion is called **rotation.** Each complete rotation takes 23 hours and 56 minutes, or about one day.

The Rotating Earth

For everyone on earth, the most observable effects of the earth's rotation on its axis are day and night. As the earth rotates from west to east, the sun appears to rise in the east in the morning. It travels across the sky and sets in the west in the evening. At any given moment, it is daytime on the side of earth facing the sun. It is nighttime on the side of the earth facing away from the sun.

The Revolving Earth

The earth's orbit, or path around the sun, is slightly elliptical, or oval-shaped. Therefore, the earth is not always the same distance from the sun. At its closest point to the sun, the earth is said to be at **perihelion** (PER-uh-HEEL-yuhn). At its farthest point, the earth is at **aphelion** (a-FEEL-yuhn). As shown in Figure 2–6, the earth reaches perihelion on about January 3 and aphelion on about July 4.

The earth's aphelion distance is 152 million km. Its perihelion distance is 147 million km. The average distance between the earth and the sun is 150 million km. When the orbit of the earth is drawn to correct scale, as in Figure 2–6, it appears to be circular. However, it is actually slightly elliptical.

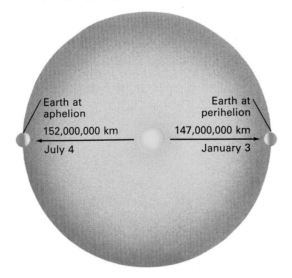

Figure 2–6. Each year, the earth is farthest from the sun about July 4 and closest to the sun about January 3.

Earth at
aphelion
152,000,000 km
July 4

Earth at
perihelion
147,000,000 km
January 3

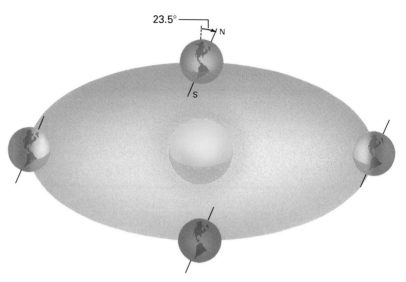

23.5°

N

S

Figure 2–7. The earth's orbit forms an elliptical plane. The axis is tilted 23.5° from the perpendicular to the orbital plane. The direction of tilt of the earth's axis remains the same throughout the earth's orbit.

As shown in Figure 2–7, the earth's orbit lies in a plane. The earth's axis is tilted 23.5° from the perpendicular (90°) to the plane of its orbit. As the earth revolves around the sun, the direction of the tilt of its axis does not change. Consequently, during each revolution, the North Pole points at times toward the sun, and at other times away from it.

When the North Pole is pointed toward the sun, the Northern Hemisphere has longer periods of daylight. When the North Pole is pointed away from the sun, the Southern Hemisphere has longer periods of daylight.

The sun's rays are nearly parallel to one another as they reach the earth. Because the earth's surface is curved, however, the sun's rays strike different parts of the earth at different angles. The amount of solar heat an area receives depends on the angle at which the sun's rays strike that area of the earth's surface.

When the sun is directly overhead, its rays strike the earth at a 90° angle. The closer to 90° the rays are, the more concentrated they are and the greater is the heat they produce on the earth's surface. As the angle of the rays decreases, the rays spread out. When this happens, the solar heat reaching the surface of the earth becomes less intense.

The angle at which the sun's rays strike each part of the earth's surface changes as the earth moves through its orbit. When the North Pole is pointed toward the sun, the sun's rays strike the Northern Hemisphere at angles close to 90°. When the North Pole is pointed away from the sun, the sun's rays strike the Southern Hemisphere at angles close to 90°.

The Seasons

Changes in the angle at which the sun's rays strike the earth's surface and changes in the amount of daylight cause the seasons. For example, when the North Pole is tilted away from the sun, the angle of the sun's rays falling on the Northern Hemisphere is lower. As a result, there are fewer hours of daylight. The weak rays of the sun and the short hours of daylight produce the cool winter season in the Northern Hemisphere. At the same time, the angle of the sun's rays striking the Southern Hemisphere is higher, and there are more hours of daylight. The sun's strong rays and the longer hours of daylight produce the warm summer season there.

Summer and Winter Solstices

Because of the earth's orbit, each year on June 21 or 22, the North Pole points toward the sun. On this day the sun's rays strike the earth at a 90° angle along the Tropic of Cancer. This day is called the **summer solstice.** It marks the beginning of summer in the Northern Hemisphere. *Solstice* means ''sun stop'' and refers to the fact that in the Northern Hemisphere the sun appears to follow its highest path across the sky. Figure 2–8 shows this path as it appears in the Northern Hemisphere at the summer solstice.

The Northern Hemisphere has most hours of daylight at the summer solstice. The farther north of the equator, the longer is the period of daylight. North of the Arctic Circle, there are 24 hours of weak daylight at the summer solstice. At the other extreme, south of the Antarctic Circle, there are 24 hours of darkness.

By December the earth is halfway through its orbit, and the North Pole points away from the sun. On December 21 or 22, the sun's rays strike the earth at a 90° angle along the Tropic of Capricorn. This day is called the **winter solstice.** It marks the beginning of winter in the Northern Hemisphere.

At winter solstice, the Northern Hemisphere has fewest daylight hours. The sun follows its lowest path across the sky, as Figure 2–8 shows. Regions within the Arctic Circle have 24 hours of darkness; those inside the Antarctic Circle, 24 hours of daylight.

Autumnal and Vernal Equinoxes

On September 22 or 23 of each year, the sun's rays strike the earth at a 90° angle along the equator. This day is called the **autumnal equinox** and marks the beginning of the fall season in the Northern Hemisphere. *Equinox* means ''equal night'' and refers to the fact

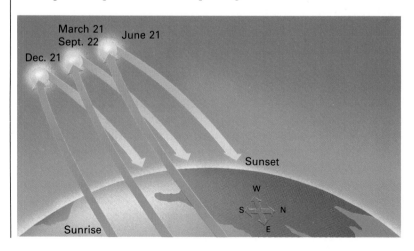

Figure 2–8. In the Northern Hemisphere, the sun appears to follow its highest path across the sky on the summer solstice.

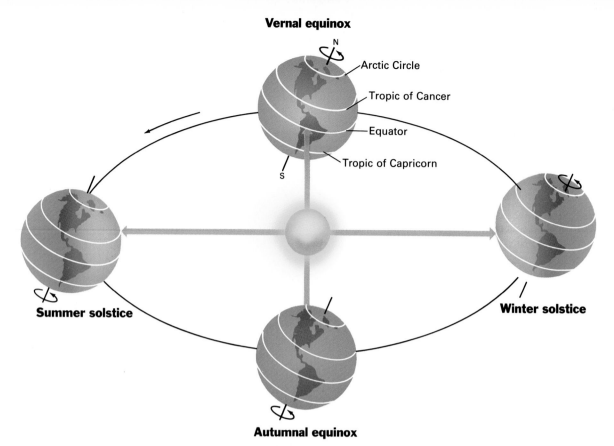

Vernal equinox

Arctic Circle

Tropic of Cancer

Equator

Tropic of Capricorn

Summer solstice

Winter solstice

Autumnal equinox

Figure 2–9. At the solstices, the sun's rays hit the Tropic of Cancer or the Tropic of Capricorn at a 90° angle. At the equinoxes, they hit the equator at a 90° angle.

that the hours of daylight and darkness are equal everywhere on the earth on that day. The hours of daylight and darkness are equal because at the equinox the North Pole points neither toward nor away from the sun. This position of the North Pole in relation to the sun is illustrated in Figure 2–9.

On March 21 or 22, the sun's rays again strike the earth at a 90° angle along the equator. This day is called the **vernal equinox** and marks the beginning of spring in the Northern Hemisphere. As during the autumnal equinox, the hours of daylight and darkness are equal everywhere on the earth. On this day the North Pole points neither toward nor away from the sun.

Precession

A year is defined as the length of time it takes the earth to complete one revolution around the sun. The most common way to measure a year is to determine the exact length of time between two consecutive vernal equinoxes. This period has proved to be 365 days, 5 hours, 48 minutes, and 46 seconds. However, the actual time it takes the earth to make one revolution is about 20 minutes longer than the time between two consecutive vernal equinoxes. The difference in time results from the slow wobbling of the earth as it turns on its axis. This wobbling motion, called **precession,** is caused by the gravitational pull exerted on the earth by the moon, the sun, and the other planets. Precession causes the earth's axis to move slowly in a circle. The earth's axis completes one turn in this circle every 26 thousand years.

Time Zones

Using the sun as the basis for measuring time, 12:00 noon is defined as the time when the sun is highest in the sky. Because of the sun's apparent movement from east to west, the sun appears highest over different locations at different times. For example, assume the sun is highest over New York City at 12:00 noon. In Philadelphia, a short distance west of New York City, the sun would appear highest a few minutes later. In Baltimore, just west of Philadelphia, the sun would appear highest a few minutes after that. If these communities were to set their clocks precisely by the sun, the clocks in each community would mark slightly different times. To avoid problems created by different local times, the earth's surface is divided into 24 **standard time zones.** In each zone, noon is set as the time when the sun is highest over the center of that zone.

IMPACT

Biological Clocks

Our earliest ancestors lived in harmony with the cycles of the sun. They were awakened by the sun's first rays, and they ended their workday at sunset. With the development of artificial light sources, however, humans have become less dependent on the availability of sunlight to set their daily routines. Yet research reveals surprising evidence that human behavior is still closely tied to solar rhythms.

Scientists have discovered that many body processes occur in 24-hour cycles called *circadian rhythms. Circa* is Latin for "about"; *dies* is Latin for "day." No one understands exactly what controls circadian rhythms, but the human body seems to have a number of internal clocks. They regulate patterns of sleeping and waking, daily changes in body temperature, hormone secretions, heart rate, and blood pressure. Even moods, coordination, and memory have their own circadian rhythms.

Studies indicate that the cycle of darkness and light caused by the earth's rotation sets and resets the clocks of the body. It more or less keeps vital processes on a 24-hour schedule.

When the internal clocks get out of synch with the sun's cycle, problems arise. One graphic example is jet lag, the combination of exhaustion, irritability, and insomnia that travelers often suffer after a long flight across several time zones.

Based on your knowledge of time zones and circadian rhythms, explain the cause of jet lag.

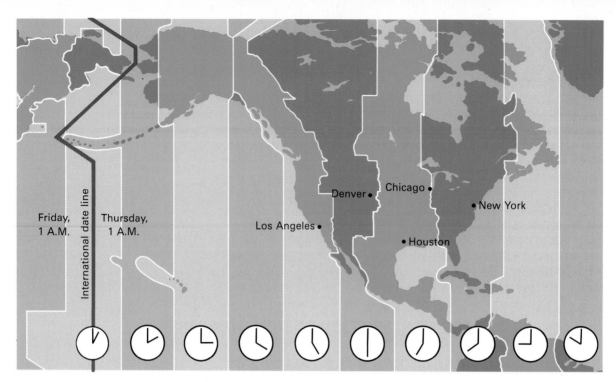

International date line

Friday,
1 A.M.

Thursday,
1 A.M.

Denver•   Chicago•

Los Angeles•   •New York

•Houston

Figure 2–10. The earth has been divided into 24 standard time zones. Going east, travelers must set their clocks ahead one hour for each time zone crossed. Going west, travelers must set their clocks back one hour.

During the summer, most of the United States uses **daylight saving time.** Under this system, in April clocks are set one hour ahead of standard time, which provides an additional hour of daylight during the evening. For example, if the sun sets at 7 P.M. standard time, it would set at 8 P.M. daylight saving time. In October clocks are set back one hour, returning to standard time.

Because the earth is spherical, its circumference equals 360°. Dividing 360° by the 24 hours needed for one rotation, you find that the earth rotates at a rate of 15° each hour. Therefore, each of the earth's 24 standard time zones covers about 15°. The time in each zone is one hour earlier than the time in the zone to its east. Figure 2–10 shows the standard time zones in the United States. At six o'clock in New York City, what time is it in Los Angeles?

There are 24 standard time zones and 24 hours in a day. But there must be some point on the earth's surface where the date changes from one day to the next. To prevent confusion, the **international date line** has been established. The international date line is an imaginary line running from north to south through the Pacific Ocean. When it is 8:00 A.M. Friday west of the international date line, it is 8:00 A.M. Thursday east of the line. The international date line is drawn so that it does not cut through islands and continents in the Pacific. Thus, the people living within one country have the same date.

Section 2.2 Review

1. Describe the position of the earth during the summer solstice. Where do the sun's rays strike the earth at a 90° angle?
2. What causes winter in the Northern Hemisphere?
3. What are the advantages in using daylight saving time?

2.3 Artificial Satellites

If you drop a ball, it will fall straight down because the force of gravity pulls objects toward the center of the earth. If you throw a ball horizontally, it will follow a curving path. The force of gravity still pulls the ball toward the earth, but the ball also moves horizontally as it falls. The greater the speed at which the ball is thrown, the farther it will travel before gravity pulls it to the earth's surface. For example, a ball thrown at about 8 km/s and not slowed down by air resistance would follow the curve of the earth. As shown in Figure 2–11, the ball would fall toward the earth but never reach the earth's curved surface. Instead the ball would become an artificial earth **satellite.** A satellite is any object in orbit around another body with a larger mass.

Satellites and Orbits

Today many artificial satellites orbit the earth. They range in size from about 15 cm to nearly 30 m in diameter. Meteorological satellites gather and transmit weather information. Communications satellites relay radio, telephone, and television signals to and from various locations on earth. Navigation satellites send out radio signals that help pilots of ships and aircraft determine their locations. Scientific satellites, such as orbiting telescopes, are outfitted with special instruments that allow scientists to study the distant reaches of space.

Satellites are put into orbit by powerful computer-guided rockets. A rocket carries the satellite to an appropriate altitude above the earth. The rocket then aims the satellite at the angle necessary for the desired orbit. Once the satellite is in place, the rocket automatically detaches from the satellite. A smaller rocket on the satellite may provide the extra speed necessary to send the satellite orbiting around the earth. The earth's gravity holds the satellite in orbit.

The altitude of a satellite determines the speed necessary to keep it in orbit. As the distance from the earth increases, the force of earth's gravity decreases. Air resistance is also reduced in the earth's thin upper atmosphere. Therefore, the higher the orbit of a satellite, the lower will be the speed needed for it to stay in orbit.

At an altitude of 36,100 km, a satellite completes one revolution in 24 hours. At this altitude, a satellite in orbit directly above the earth's equator and moving in the direction of the earth's rotation is in **geosynchronous orbit.** A satellite in geosynchronous orbit always remains at the same point above the equator and appears to be stationary in the sky. Satellites used for communications are usually put into geosynchronous orbits. High above the earth, these satellites can act as antennas, relaying messages over great distances.

A satellite can also be placed in a **polar orbit.** A polar orbit carries the satellite over the earth's North and South poles. As the earth rotates beneath it, a satellite in polar orbit passes over a differ-

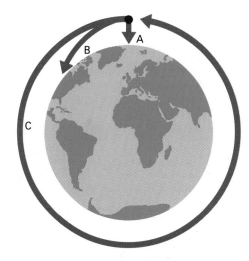

Figure 2–11. *A* is the path of a ball that is dropped. *B* is the path of a ball thrown horizontally. *C* is the path of a ball moving fast enough to go into orbit.

ent portion of the earth's surface during each revolution. After a certain number of revolutions, the satellite will have surveyed the entire surface of the earth. A polar orbit is especially useful for weather satellites. Why might a polar orbit be better than a geosynchronous orbit for a weather satellite?

Most satellites follow elliptical orbits around the earth. The point closest to the earth in the orbit is called *perigee*. The point farthest from the earth in the orbit is called *apogee*. A satellite travels fastest at perigee and slowest at apogee.

To stay in orbit, a satellite must maintain a speed adequate for its altitude. If the orbit comes too close to the earth at perigee, friction with the earth's atmosphere will slow the satellite down. This decrease in speed means that each successive orbit will bring the satellite closer to the earth at perigee. Eventually, the increasing friction of the earth's atmosphere will generate too much heat, and the satellite will burn up.

Exploring the Earth by Satellite

Satellites have given scientists new ways of studying the earth. Some of the most fascinating information has come from a series of scientific satellites called *Landsat*. Each *Landsat* orbits the earth in a polar orbit and collects information, using both television cameras and electronic sensors. The resulting images are used to identify features on the earth's surface, such as population centers and areas of vegetation.

Figure 2–12. This *Landsat* image shows a portion of the Eel River, Humboldt Bay, and Coast Ranges south of Eureka, California.

Investigation

Gravity and Orbits

A moving body tends to move in a straight line at constant speed unless some outside force acts on it. Gravity is the outside force that acts on satellites and keeps them in orbit around the earth. You can investigate the effect of gravity on a moving body with a simple model.

Science Process Skills Focus: constructing models, analyzing, inferring

Materials
strip of flexible cardboard, 3 × 30 cm; transparent tape; sheet of white paper; a marble

Procedure
1. Form a hoop with the cardboard strip and fasten the two ends together with tape.
2. Place the hoop in the center of your paper and trace a circle around it. Mark four points at equal distances around the circle. Number the points as shown in the Figure.
3. Place a marble inside the cardboard hoop. Slowly swirl the hoop clockwise until the marble rolls around the edge of the hoop. Stop swirling the hoop as the marble approaches point *1*, then quickly lift the hoop, allowing the marble to escape. You may have to prac-

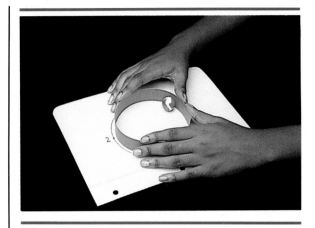

tice this step several times to get the marble released at the right time.
4. Observe and record the path of the marble.
5. Repeat Steps 3 and 4, stopping the hoop and releasing the marble at points *2*, *3*, and *4*.

Analysis and Conclusions
1. What path does the marble take when the hoop is removed?
2. In what direction does the hoop push the marble? What force does the hoop represent?
3. Compare the motion of the marble with the motion of a satellite around the earth.

Another satellite, called *Seasat,* launched in 1978, orbited the earth for three months. It revealed previously unknown features of the ocean floor, such as underwater mountain ranges and valleys.

Sarsat is a network of navigational satellites in polar orbit. It picks up signals from ships that are lost or in trouble.

Most satellites are designed to send information back to earth. The space shuttle is an exception. It is designed to carry cargo, orbit the earth, and then return to the earth's surface. While in orbit, the shuttle can release or pick up other satellites.

Section 2.3 Review
1. Compare a geosynchronous orbit with a polar orbit.
2. How are satellites used to study the earth?
3. Why is a polar orbit useful for surveying purposes?

Reviewing Chapter 2

Key Concepts

- The solid earth consists of three major layers: the crust, the mantle, and the core. See page 22.
- Studies of seismic waves provide information about the thickness and composition of the layers of the earth's interior. See page 22.
- The earth has magnetic properties that may originate in the earth's core. See page 23.
- Newton's law of gravitation states that the strength of the gravitational force between two objects depends upon their mass and the distance between them. See page 26.
- The earth revolves around the sun and rotates on its axis. See page 27.
- The apparent motion of the sun across the sky is the basis for measuring time. See page 31.
- Many satellites orbit the earth in geosynchronous orbits or in polar orbits. See page 33.

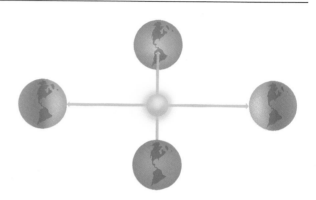

- **The seasons of the year are caused by two factors. See page 28.**
- Satellites collect information on features of the earth's surface, using equipment such as television cameras and other electronic sensors. See page 34.

Key Terms

aphelion (27)	geosynchronous orbit	mantle (22)	satellite (33)
atmosphere (21)	(33)	mass (22)	seismic wave (21)
autumnal equinox (29)	gravity (25)	Moho (22)	shadow zone (23)
axis (21)	hydrosphere (21)	perihelion (27)	standard time zone (31)
core (22)	international date line	polar orbit (33)	summer solstice (29)
crust (22)	(32)	precession (30)	vernal equinox (30)
daylight saving time	law of gravitation (26)	revolution (27)	weight (26)
(32)	magnetosphere (24)	rotation (27)	winter solstice (29)

Review

On your paper, write the letter of the term that best completes each of the statements.

1. The layer that makes up nearly two thirds of the earth's mass is the (a) crust (b) mantle (c) core (d) lithosphere.
2. The boundary between the earth's crust and mantle is called the (a) shadow zone (b) asthenosphere (c) Moho (d) magnetosphere.
3. Both P waves and S waves can travel through (a) liquids and solids (b) solids (c) liquids d) gases.
4. The possible source of the earth's magnetism is the earth's (a) crust (b) mantle (c) core (d) lithosphere.

5. The amount of matter in an object is the object's (a) mass (b) weight (c) gravity (d) plasticity.
6. As the distance from the center of the earth increases, the force of gravity (a) decreases (b) increases (c) stays the same (d) doubles.
7. The measured weight of an object is slightly less at the equator than it is at the poles because of the earth's (a) orbit (b) axis (c) shape (d) tilt.
8. The point closest to the sun in the earth's orbit is called (a) apogee (b) perigee (c) aphelion (d) perihelion.
9. At noon on the winter solstice, the sun's vertical rays strike the earth along the (a) Tropic of Cancer (b) Tropic of Capricorn (c) equator (d) North Pole.
10. At noon on the vernal equinox, the sun's vertical rays strike the earth along the (a) Tropic of Cancer (b) Tropic of Capricorn (c) equator (d) North Pole.
11. When the sun's rays strike the Northern Hemisphere at nearly 90°, the season is (a) spring (b) summer (c) fall (d) winter.
12. The wobbling motion made by the earth as it turns on its axis is called (a) precession b) gravity (c) plasticity (d) equinox.
13. A person crossing the international date line gains or loses (a) 2 hours (b) 8 hours (c) 12 hours (d) 24 hours.
14. A satellite in geosynchronous orbit is always directly above the (a) equator (b) North Pole (c) South Pole (d) international date line.
15. *Landsat* provides information about the (a) earth's surface (b) moon's surface (c) sun's surface (d) stars.

Application

On your paper, write answers to the following questions.
1. Is a hard-boiled egg a good model of the earth's layers? Why or why not?
2. A scientist proposes the hypothesis that the moon has a liquid core. How might this hypothesis be tested?
3. If scientists discovered that the earth's magnetic field had weakened substantially, what might they suspect to be the cause of this change?
4. Explain why the weight of an object might increase when the object moves from point *A* to point *B* on the earth's surface.
5. Although the earth's orbit brings it closest to the sun on about January 3, the Northern Hemisphere is having winter at that time of the year. Explain.
6. If the earth ceased rotating as it revolved around the sun, how would periods of daylight and surface temperatures on the earth be affected?
7. If the North Pole were always pointed toward the sun, how would the seasons be affected?
8. Suppose the earth's rotation slowed to one rotation every 48 hours. How might that change affect timekeeping systems?
9. A group of scientists is planning to launch a satellite into orbit around the earth. The satellite will be used to survey the entire earth's surface in search of oil deposits. Which type of orbit would you recommend for such a satellite? Why?
10. How does the space shuttle differ from other satellites? What advantages might the space shuttle offer?

Extension

1. Scientists made recent discoveries about the earth's core using seismic tomography. Research seismic tomography and write a report on its uses.
2. Find out how the Lapps and the Eskimo have adapted to periods of continuous daylight and continuous darkness.
3. Draw a simple map of the world. Label Washington, D.C. with the time and date *1:00 P.M., January 1*. Label the correct time and date for five other cities.
4. Gather information about the *Meteosat* weather satellite and write an article describing the satellite and its uses.

Chapter 3
Models of the Earth

Aerial photographs, such as this one, reveal much about the earth's surface. For example, an aerial photograph can show the course of a river, the curve of a shoreline, or the general shape of other landforms. Sometimes, however, photographs do not provide the detailed information that earth scientists need. As a result, earth scientists often rely on maps as models of the earth's surface. This chapter explains how various types of maps are made and used.

Chapter Outline

3.1 Finding Locations on the Earth
Latitude
Longitude
Great Circles
Finding Direction

3.2 Mapping the Earth's Surface
Map Projections
Reading a Map

3.3 Topographic Maps
Making a Topographic Map
Interpreting a Topographic Map

This photograph shows the northern ridge of the Grand Canyon.

3.1 Finding Locations on the Earth

Section Objectives

- **Distinguish between latitude and longitude.**
- **Explain how latitude and longitude can be used to locate places on the earth.**
- **Explain how a magnetic compass can be used to find directions on the earth.**

The earth is very nearly a sphere. A sphere has no top, bottom, or sides to use as reference points for finding locations on its surface. The earth does rotate, however. The points on the surface of the earth that are intersected by the earth's axis of rotation are used as reference points for establishing direction. These reference points are called the *North* and *South geographic poles*. Halfway between the poles, an imaginary line called the *equator* divides the earth into the Northern and Southern hemispheres. Based on these reference points, an entire system of imaginary lines has been established to locate places on the earth's surface.

Latitude

One set of lines describes positions north and south of the equator. These imaginary lines are called **parallels.** They are circles that run east and west around the world parallel to the equator.

The angular distance north or south of the equator is called **latitude.** Latitude is measured in degrees, beginning at the equator with 0°. A full circle has 360°. Since the distance from the equator to either of the poles is one quarter of a circle, the latitude of both the North and South poles is ¼ of 360°, or 90°. See Figure 3–1. In actual distance, one degree of latitude equals 1/360 the earth's circumference (over 40,000 km), or about 111 km.

Parallels of latitude north of the equator are labeled *N;* those south of the equator are labeled *S.* For example, in the Northern Hemisphere, Washington, D.C., is located near a parallel of latitude that is 39° north of the equator. The latitude of Washington, D.C., is thus about 39° N. In the Southern Hemisphere, Melbourne, Australia, has a latitude close to 39° S.

Each degree of latitude consists of 60 equal parts, called **minutes.** One minute (symbol: ′) of latitude equals 1.85 km. A more precise latitude for Washington, D.C., is 38°53′ N. For even greater precision, each minute is divided into 60 equal parts, called **seconds** (symbol: ″). Using degrees, minutes, and seconds, the latitude of Washington, D.C., is 38°53′51″ N. How much distance on the earth's surface does one second of latitude equal?

Longitude

The latitude of a particular place indicates only its position north or south of the equator. To determine the specific location of a place, you also need to know how far east or west the location is along its line of latitude. Imaginary lines, called **meridians,** are used to establish east-west location. As shown in Figure 3–2, each meridian is a semicircle running from pole to pole.

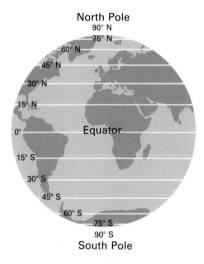

Figure 3–1. Parallels are imaginary lines describing positions north and south of the equator. Each parallel forms a complete circle around the globe.

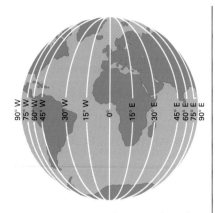

Figure 3–2. Meridians are imaginary lines reaching around the earth from pole to pole.

One meridian has been selected to be 0°. This meridian, called the **prime meridian,** passes through Greenwich, England. **Longitude** is the angular distance, measured in degrees, east or west of the prime meridian. Since a full circle is 360°, the meridian opposite the prime meridian, halfway around the world, is labeled *180°*.

All locations east of the prime meridian have longitudes between 0° and 180° and are labeled *E*. All locations west of the prime meridian have longitudes between 0° and 180° and are labeled *W*. Washington, D.C., lies west of the prime meridian; its longitude is about 77° W. As with latitude, longitude can be expressed in degrees, minutes, and seconds. Therefore, a more precise location for Washington, D.C., is 38°53′51″ N, 77°0′33″ W.

The distance covered by a degree of longitude depends upon where the degree is measured. At the equator, 0° latitude, a degree of longitude equals approximately 111 km. However, meridians meet at the poles. The distance measured by a degree of longitude decreases as you move from the equator to the poles. For example, at a latitude of 60° N, one degree of longitude equals about 55 km. At 80° N, one degree of longitude equals only about 20 km.

Great Circles

An imaginary line called a **great circle** is often used in navigation. A great circle is any line that circles the earth and divides the globe into halves. Any circle formed by two meridians of longitude directly across from each other on opposite sides of the globe is a great circle. The equator, however, is the only parallel of latitude that is a great circle. Great circles can also run diagonally around the globe. Just as a straight line is the shortest distance between two points on a plane, a great-circle route is the shortest distance between two points on a sphere. As a result, air and sea routes are often plotted along great circles.

Figure 3–3. As the illustration shows, a great-circle route from point A to point B is much shorter than a route following a parallel. Great-circle routes, such as this one between points C and D, can run diagonally around the earth.

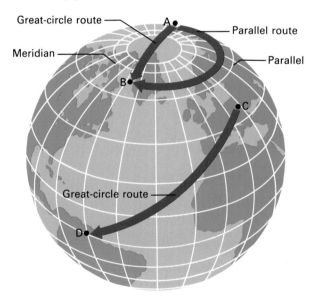

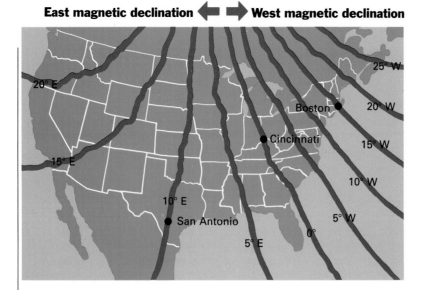

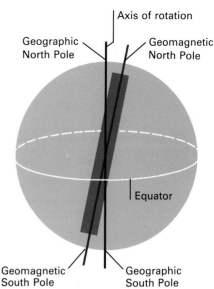

East magnetic declination ⬅ ➡ West magnetic declination

Figure 3–4. The red lines on the map (left) connect points with the same magnetic declination. Note that the earth's magnetic poles are at an angle of about 11° to the earth's axis of rotation (right).

Finding Direction

One way to find direction on the earth is to use a magnetic compass. A magnetic compass can indicate direction because the earth has magnetic properties, as if a powerful bar-shaped magnet were buried inside. As you can see in Figure 3–4, the earth's imaginary magnet is at an angle to the earth's axis of rotation.

The points on the earth's surface just above the poles of the imaginary magnet are called the **geomagnetic poles.** Because of the tilt of the imaginary magnet inside the earth, the geomagnetic poles and the geographic poles are located in different places. A compass needle points to the geomagnetic North Pole.

The angle between the direction of the geographic pole and the direction in which the compass needle points is called **magnetic declination.** In the Northern Hemisphere, magnetic declination is measured in degrees east or west of the geographic North Pole. As Figure 3–4 shows, a compass needle at Boston, Massachusetts, points 15° west of **true north,** the direction of the geographic North Pole. At Cincinnati, Ohio, the compass needle lines up with the geographic and geomagnetic North poles, so the declination is zero. At San Antonio, Texas, the declination is 10° east of true north.

Magnetic declination has been determined for points all over the earth. By adjusting a measurement of magnetic north, a person can determine geographic north for any place on earth. Locating geographic north is important in navigation and mapmaking. The magnetic declination for most of the United States is shown in Figure 3–4. What is the magnetic declination for western Illinois?

Section 3.1 Review

1. Define parallels of latitude.
2. What reference points are used for establishing lines of latitude and longitude?
3. Name one way of determining direction on the earth.
4. Why is a great-circle route often used in navigation?

3.2 Mapping the Earth's Surface

A globe is a familiar model of the earth. Because a globe is spherical like the earth, the locations of surface features and their relative sizes can be represented accurately. A globe is especially useful in studying the larger surface features, such as continents and oceans. However, most globes are too small to show many details of the earth's surface, such as streams and highways. For that reason, a variety of maps have been developed for studying the earth.

A map is a flat representation of the earth's curved surface. Transferring a curved surface to a flat map, however, causes distortion. For example, if you remove the peel from an orange and attempt to flatten the peel, it will stretch and tear. The larger the piece of peel, the more its shape is distorted as it is flattened. Also distorted are distances between points on the orange peel. Similarly, an area shown on a map may be distorted in size, shape, distance, and direction. The larger the surface area being shown, the greater is the distortion. A map of the entire earth, like the peel of an entire orange, would show the greatest distortion. A map of a small area, such as a city, would show only slight distortion.

Map Projections

Over the years, mapmakers have developed several ways of transferring the curved surface of the earth onto flat maps. A flat map that represents the three-dimensional curved surface of a globe is called a **map projection.** To understand how map projections are made, imagine the earth as a transparent globe lighted from within. If you hold a piece of paper against the globe, shadows appear on the paper. The shadows reflect markings on the globe, such as continents, oceans, meridians, and parallels. The way the paper is held against the lighted globe determines the kind of projection made.

The three most common types of map projections are **Mercator, gnomonic** (noe-MON-ick), and **conic projections.** None of these

Figure 3–5. A light at the center of a transparent globe would project lines on a cylinder of paper (left), producing a Mercator projection (right).

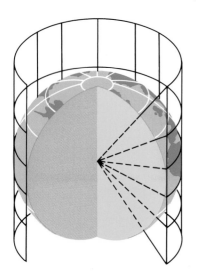

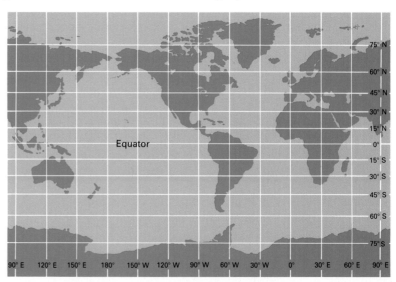

map projections is an entirely accurate representation of the earth's surface. However, each kind of projection has certain advantages, often related to its use.

Mercator Projection

If you were to wrap a cylinder of paper around a lighted globe, a Mercator projection like the one shown in Figure 3–5 would result. The meridians on a Mercator projection appear as straight parallel lines with an equal amount of space between lines. On a globe, however, the meridians come together at the poles. A Mercator projection distorts distances between regions of land and distorts the sizes of areas near the poles. Only areas near the equator are shown with relative accuracy.

Although some areas on the map are distorted, a Mercator projection has several advantages. As Figure 3–5 shows, all compass directions appear as straight lines. The four main points of the compass are located along the four sides of the map. All meridians and parallels are shown clearly. Also, the shapes of the coastlines are shown correctly. These qualities make the Mercator projection a valuable tool for navigation.

Gnomonic Projection

A sheet of paper touching a lighted globe at only one point produces a gnomonic projection, as shown in Figure 3–6. On a gnomonic projection, little distortion occurs at the point of contact, which is usually one of the poles. However, the unequal spacing between parallels causes great distortion in both direction and distance. As you can see in Figure 3–6, the distortion increases as the distance from the point of contact increases.

Despite distortion, a gnomonic projection is a great help to navigators in plotting routes used in air travel. As you know, the great-circle route is the shortest distance between two points on the globe. A great circle, when projected onto a gnomonic projection, appears as a straight line. Therefore, by drawing a straight line between any two points on a gnomonic projection, navigators can readily find the great-circle route between those points. They can then determine the latitude and longitude of various points along the great-circle route.

Figure 3–6. This gnomic projection (left) is produced as points on a globe are projected onto a sheet of paper in contact with the North Pole of the globe.

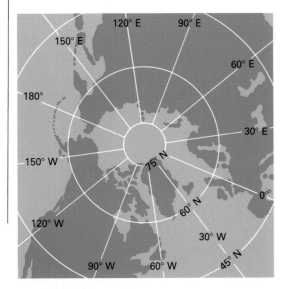

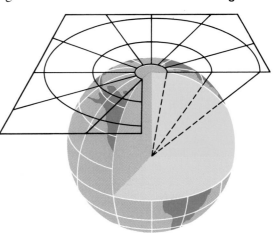

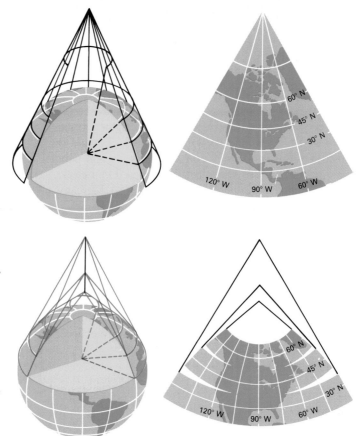

Figure 3–7. In the top left illustration, the cone comes in contact with the globe along the parallel of latitude at 30° N, producing the conic projection shown top right. In the bottom left illustration, a conic projection is made using a series of cones, each touching the globe at a different parallel. The conic projections are then assembled to form the polyconic projection shown bottom right.

Conic Projection

A paper cone placed over a lighted globe so that the axis of the cone aligns with the axis of the globe produces a conic projection. The cone touches the globe along one parallel of latitude. As shown in Figure 3–7, areas along the parallel of latitude where the cone and globe are in contact are distorted least.

A series of conic projections may be used to map a number of neighboring areas. Each cone touches the globe at a slightly different latitude, as Figure 3–7 shows. Fitting the adjoining areas together then produces a continuous map. Maps made in this way are called **polyconic projections.** The relative size and shape of small areas on the map are nearly the same as those on the globe.

Reading a Map

Maps are models of the earth's surface. They provide information through the use of symbols. To read a map, you must understand the symbols and be able to find directions and calculate distances.

Symbols

Maps often have symbols for features such as cities and rivers. The symbols are explained in the map **legend,** a list of the symbols and their meanings. Some symbols resemble the features they represent. Others are more abstract, such as those for towns and urban areas. In Figure 3–8, which symbols resemble the features they represent?

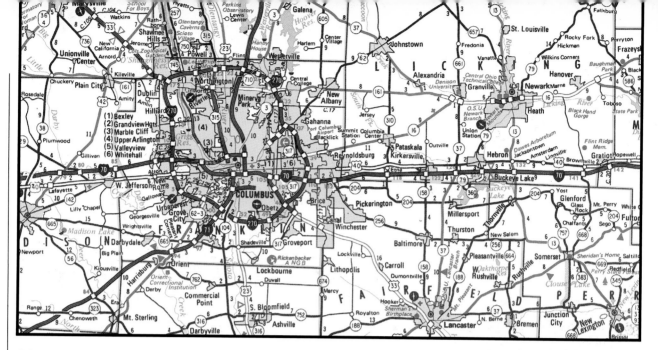

Direction on a Map

Once you understand the legend, you must determine the compass directions. Most maps are drawn with north at the top, east at the right, west at the left, and south at the bottom. Meridians run from top to bottom, and parallels, from side to side. Directions should always be determined in relation to the meridians and parallels.

Map Scale

To be accurate, a map must be drawn to **scale.** The scale of a map indicates the relationship between distance as shown on the map and actual distance. As Figure 3–8 shows, a map scale can be expressed as a graphic scale, a fractional scale, or a verbal scale.

A graphic scale is a printed line divided into equal parts and labeled. The line represents a unit of measure, such as kilometers or miles. Each part of the scale represents a specific distance on the earth. To find the actual distance between two points on the earth, you first measure the distance between the points as shown on the map. Then you compare that measurement with the map scale.

A second way of expressing scale is a ratio, or a fractional scale. For example, a fractional scale such as 1:25,000 indicates that 1 unit of distance on the map represents 25,000 of the same unit on the earth. A fractional scale remains the same with any system of measurement. In other words, the scale 1:100 could be read as 1 in. equals 100 in. or as 1 cm equals 100 cm.

A verbal scale may be the statement, "One centimeter equals one kilometer." This means that 1 cm on the map represents 1 km on the earth. What would the fractional scale be on such a map?

Legend

Roads and related symbols

- ▬▬ Freeway
- ═══ Multilane divided
- ▬ Two or more lane
- ▬▬ Multilane toll road
- ═══ Future location
- Interchange
- Interstate route
- U.S. route
- State route
- Scenic highway

Other symbols

- ┼┼┼ Railroad
- Airport
- ▲ State forest
- ● Park
- University
- ■ Hospital
- ◉○ Towns

Scale

0   10   20

Kilometers

1: 76,923
1.3 centimeters is equal to one kilometer

Figure 3–8. When using the type of map shown above, a map scale is needed. Three types of map scale are shown on this legend. The map legend also explains the other symbols used on this map.

Section 3.2 Review

1. Name three common map projections.
2. What is a map legend?
3. Explain why all maps are inaccurate representations in some way.

3.3 Topographic Maps

A type of map that is especially useful in earth science is called a **topographic map.** Topographic maps show the surface features, or **topography,** of the earth. Most topographic maps show both natural features, such as rivers and hills, and constructed features, such as buildings and roads.

The top illustration in Figure 3–9 shows a drawing of an island in the ocean. The drawing shows a hill on the island, but it does not indicate the size of the island or the height of the hill. The middle illustration shows the same island on a political map. The political map shows the shape of the island at sea level and the relative length and width of the island. However, the political map gives no information about the height of the island, the steepness of its slopes, or the shape of the land above sea level. A topographic map provides more-detailed information about the surface of the island than either the drawing or the political map does.

Making a Topographic Map

A topographic map of the island shows the **elevation,** or height above sea level, of various island locations. Elevation is measured from **mean sea level,** the point midway between the highest and lowest tide levels of the ocean. The elevation at mean sea level is zero. Other elevations are measured as distances above or below mean sea level.

Contour Lines

On topographic maps **contour lines** are used to show changes in elevation. Each contour line connects all points on the map that have the same elevation. For example, one contour line would connect all points on the map that have an elevation of 100 m. Another line would connect all points with an elevation of 200 m. Because all points at any given elevation are connected, the shape of the contour lines reflects the shape of the land.

Figure 3–9. A drawing gives little information about the surface of the island (top). A political map shows only the shape of the island (middle). To start making a topographic map of the island, a topographer connects points at an elevation of 20 m above sea level to form a contour line (bottom). In the completed map (right), additional contour lines have been drawn. A bench mark (symbol: X) marks the highest point.

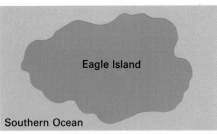

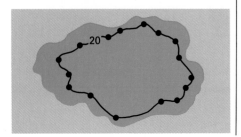

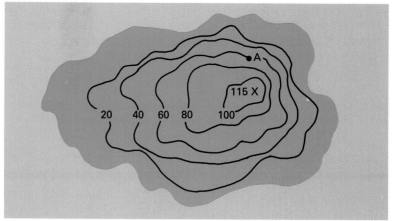

Contour Intervals

The difference in elevation between one contour line and the next is called the **contour interval.** A mapmaker chooses a contour interval suited to the size of the map and the **relief** of the land. Relief is the difference in elevation between the highest and lowest points of the area being mapped. On maps of very mountainous areas where the relief is high, the contour interval may be as large as 50 m or 100 m. Where the relief is low, the interval may be only 1 m or 2 m.

In mapping the island shown in Figure 3–9, the mapmaker chooses a contour interval of 20 m. The mapmaker marks a series of points surveyed at 20 m above sea level and connects the points with a contour line. Next the mapmaker marks points and contour lines for elevations of 40, 60, 80, and 100 m. The shoreline serves as the contour line for points at sea level. The completed topographic map in Figure 3–9 shows the elevation of the island and the general shape of the land above sea level. Find point *A* on the map. What is its elevation?

Interpreting a Topographic Map

Just as printed words on a page transmit ideas, contour lines and other symbols on topographic maps give a picture of the earth's surface. Because of the specialized nature of topographic maps, some training and practice are needed in order to read and interpret these maps accurately.

The United States Geological Survey (U.S.G.S.), a branch of the federal government, has made topographic maps of most of the United States. These detailed maps are called *topographic sheets* or *quadrangles*. One series of U.S.G.S. maps represents quadrangles

Figure 3–10. This portion of a topographic sheet produced by the U.S.G.S. shows the area around Aspen, Colorado.

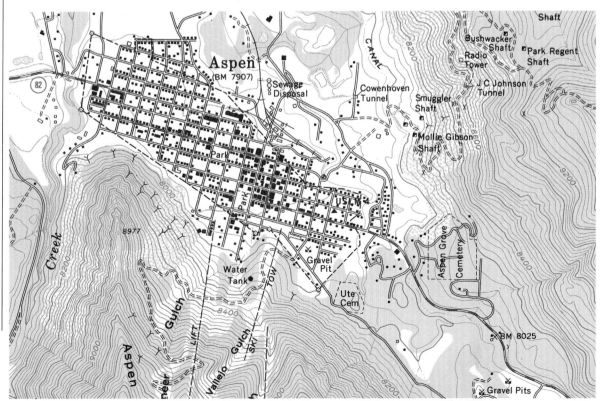

Legend

■ ▭ ■	Buildings
▟ ✚	School, church
═══════	Road or highway
— — — — —	Trail
┼─┼─┼─┼─┼	Railroad
✕	Bridge
⊗	Bench mark
∿	Stream
⬭	Lake or pond
⬭	Depression
⸱⸱⸱	Swamp

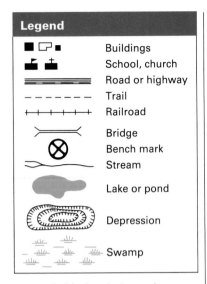

Figure 3–11. Symbols used on topographic maps represent both natural and constructed features.

that cover 15′ of latitude and 15′ of longitude. Another series of maps represents quadrangles that cover 7.5′ of latitude and 7.5′ of longitude. The 7.5′ topographic sheets show an area in greater detail than the 15′ sheets do.

As shown in Figure 3–11, symbols are used to show certain features on topographic maps. Different colors are used for different types of symbols. In general, constructed features, such as buildings and railroads, are shown in black. Major highways are shown in red. Bodies of water are usually shown in blue; and woodland areas are printed in green. Contour lines are shown in brown or red.

On most topographic maps, direction is found by following lines of latitude and longitude. On maps published by the U.S.G.S., north is located at the top of the map and marked by a parallel of latitude. The southern boundary, at the bottom of a map, is also marked by a parallel. At least two additional parallels are usually drawn in or indicated by cross hairs (+) at 2.5′ intervals.

Meridians of longitude indicate the eastern and western boundaries of U.S.G.S. maps. Additional meridians may also be shown. All meridians and parallels shown on these maps are labeled in degrees and minutes.

Horizons

Technology: Landsat Maps the World

In July 1982, *Landsat 4,* a remarkable mapping satellite, was launched. The satellite, illustrated at left, circles the earth at an altitude of 700 km. *Landsat 4* is the fourth in a series of earth-scanning satellites, the first of which was launched in 1972.

With a scanning sensor system called a *thematic mapper* (TM), *Landsat* records highly detailed images of the earth's surface. Each *Landsat* image covers nearly 33,800 km² of the earth's surface. Since 1972 these satellites have recorded more than a million images.

Landsat images are not ordinary photographs, however. For example, the *Landsat* photo at the far right shows

the Baltimore-Washington, D.C., area with bright-red vegetation, light-blue business districts, and pink suburbs.

The *Landsat* photos are multispectral images. The TM sensors see not only visible

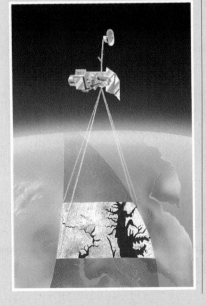

48   Unit 1   Studying the Earth

Distance on Topographic Maps

As on other maps, distance on a topographic map is determined by referring to the map scale. A common scale used on U.S.G.S. maps is 1:25,000. Based on this scale, 1 cm on the map is equal to 250 m, or ¼ km, on the earth's surface. You can use a ruler to measure distances on the map and then convert the centimeters to meters or kilometers. If a graphic scale is used, you can mark off distance on a piece of paper and then compare it with the scale.

Elevation on Topographic Maps

On a topographic map, the contour interval determines the elevation at which each contour line will be drawn. If the contour interval is 10 m, contour lines will be shown for elevations of 10, 20, 30, 40, 50, 60 m and so on. To make reading the map easier, every fifth contour line is printed darker than the others. These lines are called **index contours** and are labeled by elevation. A point located between two contour lines has an elevation somewhere between the elevations of the two lines. For example, if a point is located halfway between the 50-m and 100-m contour lines, its elevation is approximately 75 m.

light, the light recorded by an ordinary camera, but also infrared light, light waves that the human eye cannot see. This capability gives *Landsat* images much more detail than a conventional photograph would have.

The images recorded by TM sensors are radioed back to earth. There they are processed by computers, which assign bright artificial colors to the various light waves. Because of these vivid colors, the images can be more easily interpreted. Vegetation appears pink or red; land that is rich in iron oxides appears yellow or orange; and urban areas appear grayish blue.

The images from *Landsat 4* enable cartographers, such as the one shown at left, to prepare thematic maps, maps intended to illustrate a particular subject or theme. A thematic map might highlight geologic faults, the extent of snow cover, or different types of vegetation. Thematic maps allow agronomists—people involved in crop production and soil management—to inventory crops and forecast yields. They also enable geologists to locate mineral deposits.

How might Landsat *images help cartographers prepare accurate topographic maps?*

Landforms on Topographic Maps

By studying the spacing and the direction of contour lines, you can also determine the shapes of landforms shown on a topographic map. Contour lines spaced widely apart indicate that the change in elevation is gradual and that the land is relatively level. Closely spaced contour lines indicate that the change in elevation is rapid and that the slope is steep. Contour lines that are almost touching indicate a very steep slope or cliff. Evenly spaced contour lines indicate that the slope increases at about the same angle over a great distance. Find point *B* on the map in Figure 3–12. How would you describe the change in elevation in the area around point *B?*

Contour lines that bend to form a V-shape indicate a valley. The V points toward the higher end of the valley. If a stream or river flows through the valley, the V in the contour lines will point upstream, or in the direction from which the water flows. A river al-

Investigation

Topographic Maps

Contour lines show elevation and landforms on topographic maps. You can use contour lines to make a topographic map of a model mountain.

Science Process Skills Focus: constructing models, observing, analyzing, comparing

Materials
modeling clay; paper clip; large waterproof container, at least 8 cm deep; ruler; adhesive tape

Procedure
1. Make a mountain 6–8 cm high out of modeling clay. Make the mountain slightly steeper on one side.
2. Run a paper clip down one side of the model to gouge out a valley.
3. Place the model in the center of the container. Tape the ruler upright in the container, as shown in the figure. Make sure the container is resting on a level surface.
4. Add water to the container to a depth of 1 cm. With a sharp pencil, trace around the model along the waterline as shown in the figure.
5. Raise the water level 1 cm at a time until you reach the top. Each time you add water to the container, mark another contour line on the

model.
6. Remove the model from the container.

Analysis and Conclusion
1. What is the contour interval on your model?
2. Observe your model from above. Try to duplicate the size and spacing of the contour lines on a sheet of paper.
3. Compare the contour lines on a steep slope with those on a gentle slope.
4. How is a valley represented on your topographic map?

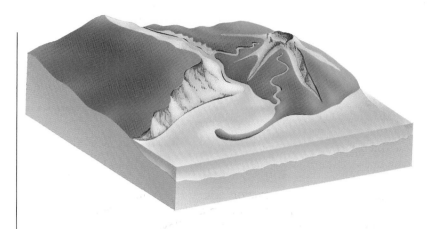

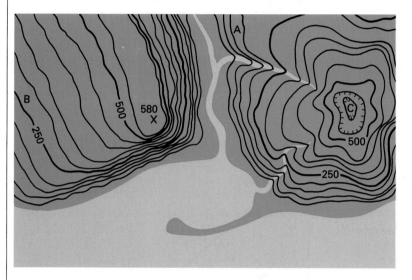

Figure 3–12. The features of the coastal valley shown in profile (top) can be represented by contour lines on a topographic map (bottom). What is the elevation at the bench mark?

ways flows from higher to lower elevation. The steeper the course of the river or stream, the closer together are the contour lines that cross it. The width of the valley is shown by the width of the V formed by the contour lines.

Contour lines that form closed loops indicate a hilltop or a depression. To avoid confusion, a depression is usually marked with **hachure lines,** which are short, straight lines. These lines are drawn along the inside of the loop and point toward its center, indicating the direction of depression. Find point *C* on the map in Figure 3–12. Is it located on a hilltop or in a depression?

Section 3.3 Review

1. How is elevation shown on a topographic map?
2. Define *contour interval*.
3. How would you determine the elevation of a point that falls between two contour lines?
4. How is a depression shown on a contour map?
5. Why are topographic maps useful to someone who wishes to study earth science?

Reviewing Chapter 3

Key Concepts

- Longitude and latitude are a system of imaginary lines used to locate places on the earth's surface. See page 39.

- Parallels of latitude run east-west around the earth. Meridians of longitude run north-south from pole to pole. See page 39.

- Because of the earth's magnetic properties, a magnetic compass can be used to find directions on the earth. See page 41.

- Three common map projections are the Mercator, gnomonic, and conic projections. See page 42.

- Map scale is used to find distances on a map. See page 45.

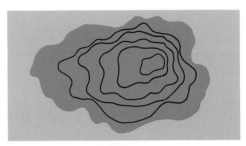

- **Contour lines can be used to show topography on a map. See page 46.**

- The spacing and direction of contour lines on a topographic map indicate the shapes of landforms. See page 46.

Key Terms

conic projection (42)	hachure line (51)	mean sea level (46)	prime meridian (40)
contour interval (47)	index contour (49)	Mercator projection (42)	relief (47)
contour line (46)	latitude (39)		scale (45)
elevation (46)	legend (44)	meridian (39)	second (39)
geomagnetic pole (41)	longitude (40)	minute (39)	topographic map (46)
gnomonic projection (42)	magnetic declination (40)	parallel (39)	topography (46)
great circle (40)	map projection (42)	polyconic projection (44)	true north (40)

Review

On your paper, write the letter of the term that best completes each of the statements.

1. A point whose latitude is 0° is located on the (a) North Pole (b) South Pole (c) equator (d) prime meridian.

2. One degree of latitude equals (a) ¹⁄₉₀ the earth's circumference (b) ¹⁄₁₈₀ the earth's circumference (c) ¹⁄₃₆₀ the earth's circumference (d) ¹⁄₇₂₀ the earth's circumference.

3. A point whose longitude is 0° is located on the (a) North Pole (b) South Pole (c) equator (d) prime meridian.

4. A point halfway between the equator and the South Pole has a latitude of (a) 45° N (b) 45° S (c) 45° E (d) 45° W.

5. The distance in degrees east or west of the prime meridian is (a) latitude (b) longitude (c) declination (d) projection.

6. The distance covered by a degree of longitude (a) is ¹⁄₁₈₀ the earth's circumference (b) is ¹⁄₃₆₀ the earth's circumference (c) increases as you approach the poles (d) decreases as you approach the poles.

7. In the Northern Hemisphere the needle of a magnetic compass points toward the (a) geomagnetic North Pole (b) geographic North

Pole (c) parallels (d) meridians.
8. In the Northern Hemisphere, a declination of 10° E indicates that the compass needle points 10° east of the (a) geomagnetic North Pole (b) geographic North Pole (c) equator (d) prime meridian.
9. On a Mercator projection, distortion is greatest near the (a) poles (b) great circles (c) meridians (d) parallels.
10. Compass directions are shown as straight lines on a (a) gnomonic projection (b) conic projection (c) Mercator projection (d) polyconic projection.
11. The shortest distance between any two points on the globe is along (a) the equator (b) a line of latitude (c) the prime meridian (d) a great circle.

12. A navigator can find the shortest distance between two points by drawing a straight line between any two points on a (a) Mercator projection (b) gnomonic projection (c) conic projection (d) polyconic projection.
13. The relationship between distance on a map and actual distance on the earth is called the (a) legend (b) scale (c) elevation (d) relief.
14. If 1 m on a map equals 1 km on the earth, the fractional scale would be written (a) 1:1 (b) 1:10 (c) 1:100 (d) 1:1,000.
15. On a topographic map, elevation is shown by means of (a) hachure lines (b) contour lines (c) verbal scale (d) fractional scale.
16. Closely spaced contour lines indicate a (a) gradual slope (b) flat area (c) steep slope (d) valley.

Application

On your paper, write answers to the following questions.
1. What is wrong with the following locations: 135° N, 185° E?
2. As you move from point A to point B in the Northern Hemisphere, the length of a degree of longitude progressively decreases. In which direction are you moving?
3. Imagine you are at a location where the magnetic declination is 0°. Describe your position in relation to magnetic north and true north.
4. One expedition is preparing to explore the South Pole; another, to explore the equator. To which expedition would you recommend the Mercator projection? Explain why.

5. A mapmaker has to draw one map for use in three different countries that do not share a common unit of measure. Which type of scale should the mapmaker use? Why?
6. You examine a topographic map on which the contour interval is 100 m. In general, what type of terrain is shown on the map?
7. You are using a topographic map to plan a hike. Along path A the contour lines are widely spaced. Along path B the contour lines are almost touching. Which path would probably be the easier and safer? Why?
8. How could you use contour lines on a topographic map to help you locate the source of a river?

Extension

1. Do some research on navigation devices such as the sextant and octant. Make a diagram explaining how these devices can be used to determine latitude.
2. Use a local street map and a magnetic compass to determine the magnetic declination at your school.
3. On a piece of unlined paper, draw an estimated topographic map of an area near your school or home. If possible, choose an area that includes various surface features, such as

hills, cliffs, or a river. If you wish, you may draw a map of an imaginary area that includes a hilltop, a depression, a stream, a railroad, and a road. Be sure to use the correct colors for the features on your map. Refer to the chart of symbols on page 48.
4. Use the library to research a report on the use of side-looking imaging radar (also called synthetic-aperture radar). This technologically advanced form of radar is used for topographic mapping.

2
The Dynamic Earth

Great Smoky Mountains National Park, North Carolina (background); Wind River Range, Wyoming (inset)

Introducing Unit Two

Earthquakes are one of the many mysteries of the planet Earth that we are beginning to solve. During one year alone, 13,400 earthquakes were located throughout the world. In each case, the only thing we knew in advance was that the potential for damage to lives, property, and our natural surroundings was catastrophic. As seismologists, we don't start earthquakes and we can't stop them. No one can. What we can do is save lives and minimize suffering by providing early alerting services.

When the alarms go off like they did during the Mexican earthquake in September 1985, we're ready—24 hours a day, every day—to respond to one of the earth's most violent forces.

The unit you are about to study discusses earthquakes and other forces that shape the earth. Understanding these forces is one key to unlocking the mysteries of our planet.

Waverly J. Person

Waverly J. Person
Seismologist
National Earthquake Information Service

4 Plate Tectonics

5 Deformation of the Crust

6 Earthquakes

7 Volcanoes

Chapter 4
Plate Tectonics

The earth is in constant motion. The planet spins on its axis and orbits the sun. Now earth scientists believe that the surface of the earth is also in motion, broken up into plates that drift slowly around the planet, a process called plate tectonics. Plate tectonics has shaped the San Francisco Bay area, which includes crust from the ocean floor as well as from now distant continents. In this chapter you will learn how plate tectonics causes changes in the earth's crust.

Chapter Outline

San Francisco is built on crust from many areas of the earth.

4.1 Continental Drift

The development of one of the most exciting theories in earth science began with observations made more than 400 years ago. As explorers such as Christopher Columbus and Ferdinand Magellan sailed the oceans of the world, they brought back information about new continents and their coastlines. Mapmakers used the information to chart the new territories and to make the first reliable world maps.

As people studied the maps, they were impressed by the similarity of the continental shorelines on either side of the Atlantic Ocean. The continents looked as though they would fit together, like parts of a giant jigsaw puzzle. The east coast of South America seemed to fit perfectly into the west coast of Africa. Greenland seemed to fit between North America and northwestern Europe.

These observations soon led to questions. Were the continents once part of the same huge landmass? If so, what caused this landmass to break apart? What caused the continents to move to their present locations? These questions eventually led to the formulation of hypotheses.

In 1912, a German scientist, Alfred Wegener, proposed a hypothesis called **continental drift,** which stated that the continents had moved. Wegener hypothesized that the continents once formed part of a single landmass, which he named **Pangaea** (pan-JEE-uh), meaning ''all lands.'' Surrounding Pangaea was a huge ocean, **Panthalassa,** meaning ''all seas.'' According to Wegener, about 200 million years ago, Pangaea began breaking up into smaller continents, which drifted to their present locations. Wegener speculated that this motion may have crumpled the crust in places, producing mountain ranges such as the Andes on the western coast of South America.

Evidence of Continental Drift

In addition to the similarities in the coastlines of the continents, Wegener soon found other evidence to support his hypothesis. If the continents had once been joined, he reasoned, research should

Section Objectives

- **Explain Wegener's hypothesis of continental drift.**
- **List evidence for Wegener's hypothesis of continental drift.**
- **Describe seafloor spreading.**

Figure 4–1. The map on the left shows the present position of the continents. The map on the right shows Pangaea as Alfred Wegener envisioned it.

Africa	Eurasia	North America
Antarctica	Greenland	South America
Australia	India	

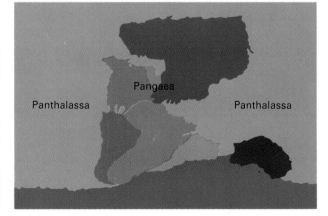

Atlantic Ocean

Pacific Ocean

Panthalassa

Pangaea

Panthalassa

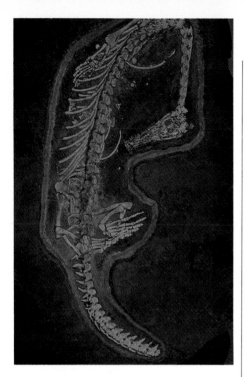

Figure 4–2. These *Mesosaurus* bones were discovered in Sao Paulo, Brazil. Identical fossil records of *Mesosaurus*, found in both eastern South America and western Africa, are evidence of continental drift.

uncover fossils of the same plants and animals in areas that had been adjoining parts of Pangaea. Wegener knew that identical fossil remains of *Mesosaurus*, a small reptile that lived 270 million years ago, had already been found in both eastern South America and western Africa. Wegener knew it would have been impossible for these reptiles to have swum across the Atlantic. There was also no evidence of any land bridges that might have connected the continents at some earlier time. Wegener thus concluded that the eastern coast of South America and the western coast of Africa must have been joined at one time.

Geologic evidence also supported Wegener's hypothesis of continental drift. The age and type of rocks in the coastal regions of widely separated areas, such as western Africa and eastern Brazil, matched closely. Mountain chains that ended at the coastline of one continent seemed to continue on landmasses across the ocean. The Appalachians, for example, extend northward along the eastern United States, while mountains of similar age and structure are found in Greenland and northern Europe. If these three landmasses are assembled in a model of Pangaea, the mountains fit together in one continuous chain.

Evidence of changes in climatic patterns added strength to Wegener's hypothesis. Geological research revealed layers of debris from glaciers in southern Africa and South America, areas that today have tropical climates. Other fossil evidence—such as the coal deposits in the eastern United States, Europe, and Siberia—indicated that tropical swamps covered much of the land area in the Northern Hemisphere. If the continents were once joined and positioned over the South Pole, Wegener suggested, these climatic differences would be easy to explain.

Despite the evidence supporting the hypothesis of continental drift, Wegener's ideas met with strong opposition. Many scientists rejected the hypothesis because it did not satisfactorily explain the force causing continental drift. In an effort to convince the scientific community that his hypothesis was valid, Wegener spent the rest of his life searching for evidence of that force. Wegener died in 1930 without having found an explanation of what caused continents to move.

Seafloor Spreading

The conclusive evidence that Wegener sought to support his hypothesis of continental drift was finally discovered nearly two decades after his death. The evidence lay on the ocean floor.

In 1947, a group of scientists set out to map the **Mid-Atlantic Ridge,** an undersea mountain range with a steep, narrow valley running down its center. The Mid-Atlantic Ridge is one part of an entire system of **mid-ocean ridges** 73,600 km long that wind their way around the earth. As the scientists examined rock samples that they brought up from the ocean floor, they made a startling discovery. Contrary to most scientists' assumptions, the ocean floor was relatively young compared with the age of continental rocks. None of the rocks found were more than 150 million years old. The oldest continental rocks are 4 billion years old. Why do you suppose scientists found this information surprising?

The Renewal of the Ocean Floor

After analyzing the data gathered about the Mid-Atlantic Ridge, Harry Hess, a geologist at Princeton University, suggested the following hypothesis. Suppose, he said, that the valley at the center of the ridge was actually a fracture, or rift, in the earth's crust and that molten rock, or magma, from deep inside the earth was welling up through the rift. This upwelling would be possible, Hess reasoned, if the ocean floor was moving away from either side of the ridge. As the ocean floor moved away from the ridge, it was replaced by magma that cooled and solidified into rock. When the spreading ocean floor reached a barrier, such as the edge of a continent, it was

Science Notebook

Evidence of Seafloor Spreading

Deep under the Atlantic Ocean lies a mountain range so vast that the Himalayas are dwarfed by comparison. This mountain range is the Mid-Atlantic Ridge, the mid-ocean ridge at the diverging boundary

between the North American and Eurasian plates and the South American and African plates.

Explorations of the Mid-Atlantic Ridge have enabled scientists to obtain firsthand evidence of seafloor spreading. In 1968, scientists aboard the *Glomar Challenger*, shown left, took a series of core samples, from the rift valley outward. Scientists examining the fossils embedded in the samples discovered that cores drilled closest to the rift valley had the youngest fossils. The fossils proved to be progressively older as the samples were drilled farther from the rift valley. This find supported the idea that new crust forms along the rift as older crust moves aside.

In 1973, the first manned dive into the rift was made in a submarine called *Archimede*. The crew on board the craft were among

the first to witness magma bubbling up from the mantle.

In 1974, cameras on board the submarine *Alvin* took photographs of lava formations that had emerged from the mantle and hardened. One of these photographs is shown above. The lava formations provided further evidence of seafloor spreading.

If you had six core samples in sequence, how many samples would you have to test to determine whether the samples on the right or the left were closer to a mid-ocean ridge? Why?

forced beneath the barrier and under the earth's crust, where it melted again. Geologist Robert Dietz named this movement **seafloor spreading.** Hess suggested that if the ocean floor was moving, the continents might also be moving. Perhaps seafloor spreading was the force that Wegener had failed to find to support his hypothesis of continental drift.

Still, Hess's ideas were just hypotheses. The proof would come years later, in the mid-1960's, and would be discovered through paleomagnetics, the study of the magnetic properties of rocks.

Paleomagnetism of the Ocean Floor

If you have ever used a magnetic compass to determine direction, you know that the earth acts as a giant magnet, with both a north and a south magnetic pole. The compass needle aligns with the field of magnetic force that extends from one pole to the other.

IMPACT

Living on the Mid-Atlantic Ridge

At 2 A.M. on January 23, 1973, a fiery fissure opened in the land less than a mile from the town of Vestmannaeyjar on the island of Heimaey off the south coast of Iceland. The island, part of the Mid-Atlantic ridge, had been split open by sea-floor spreading. Ash, molten rock, and poisonous gases spewed forth from the fissure. Within six hours, more than 5,000 panic-stricken islanders had been evacuated to the mainland.

During the next two months, 300 buildings were lost to fire, and another 65 were buried beneath black volcanic ash. Lava pouring out of the fissure and flowing to the ocean threatened to block the entrance to the harbor and deprive the islanders of their livelihood—fishing the

waters of the north Atlantic.

When the volcanic activity ceased four months later and the people of Vestmannaeyjar returned to reclaim their town, they found a dramatically altered landscape. Nearly one third of the town had been buried by lava, and the rest lay under a thick blanket of ash. The lava had added 1 km² of new land to the island as well as an excellent breakwater in the harbor. The cleanup resulted in more than 1.3 million m³ of volcanic ash

being cleared from roofs, gardens and streets.

At the present time, seafloor spreading adds an average of 2.5 cm of new material each year to Iceland. At this rate, Iceland will grow 400 km in width during the next million years. Iceland and its neighboring islands are actually a continent in the making.

If geologists want to locate the youngest rocks on Iceland, where should they look? Where should they look to find the oldest rocks? Why?

A similar phenomenon occurs when magma cools and solidifies. Certain iron-bearing minerals within the rock become magnetized. When the rock hardens, the magnetic orientation of the minerals becomes permanent and should always point to the north.

Scientists discovered, however, that this was not always the case. From the beginning of the nineteenth century, they had been finding rocks with magnetic orientations that pointed south. Some scientists concluded that the earth's magnetic field must have reversed itself at times during the earth's history. This conclusion was verified by dating rocks with different magnetic orientations. All the rocks with magnetic fields pointing north fell into the same time periods—periods of normal polarity. All the rocks with magnetic fields pointing south also fell into similar time periods—periods of reverse polarity. The scientists discovered that throughout the earth's history the magnetic field has reversed itself nine times.

At the same time these discoveries about the earth's magnetic field were being made, scientists were also finding puzzling magnetic patterns on the ocean floor. These patterns, when drawn in on maps of the ocean floor, showed alternating bands of normal and reversed magnetism. As molten rock rises from the rift in a mid-ocean ridge, it quickly cools and hardens and its magnetic orientation becomes fixed. Its magnetic orientation will reflect either the normal or reversed polarity of the earth's magnetic field at that time.

Scientists' confidence in the validity of this idea grew when they discovered that the stripes on one side of a ridge are mirror images of the stripes on the other side of the ridge. This discovery supported Hess's idea that as the molten rock from a rift cools and hardens, it splits. The fractured rock then moves away in opposite directions on both sides of the ridge. The ocean floor, it seemed, was indeed spreading.

Finally, in 1965, two groups of scientists working independently of each other discovered a previously unknown reversal in the earth's magnetic field. One group discovered the reversal in rocks on land, and the other group discovered the reversal in rocks on the ocean floor. The dates of both reversals were exactly the same. This was clear evidence that the earth's magnetic polarity does reverse itself and that the ocean floor does spread. Scientists reasoned that seafloor spreading provides a way for the continents to be moved over the surface of the earth. Here, at last, were the discoveries that Wegener had sought, the scientific evidence he needed to verify his hypothesis of continental drift.

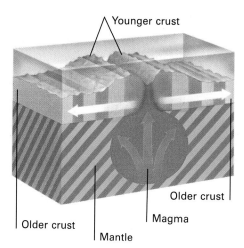

Figure 4–3. The gray stripes shown in the crust are not actually visible on the ocean floor. They are shown here to illustrate the earth's alternating magnetic field. Dark gray stripes represent the ocean floor with normal polarity.

Section 4.1 Review

1. What observation first led to Wegener's hypothesis of continental drift?
2. What types of evidence support Wegener's hypothesis?
3. Describe the process of seafloor spreading.
4. Explain how scientists know that the earth's magnetic poles have reversed themselves nine times during the earth's history.

Section Objectives

- **Summarize the theory of plate tectonics.**
- **Compare the characteristic geologic activities that occur along the three types of plate boundaries.**
- **Explain the possible role of convection currents in plate movement.**
- **Summarize the theory of suspect terranes.**

4.2 The Theory of Plate Tectonics

By the 1960's, accumulated evidence supporting the hypothesis of continental drift and seafloor spreading led to the formulation of a more far-reaching theory. This theory is called **plate tectonics.** The theory of plate tectonics not only describes continental movement but also proposes a possible explanation of why and how continents move. The term *tectonics* comes from the Greek word *tektonikos*, meaning "construction." Tectonics is the study of the formation of features in the earth's crust.

The earth's crust consists of two types—**oceanic crust** and **continental crust.** Material on the ocean floor forms oceanic crust. Continental crust makes up the continental landmasses.

The oceanic and continental crust of the earth and part of the earth's upper mantle make up the **lithosphere,** the solid, relatively rigid outer shell of the earth. Beneath the lithosphere lies the **asthenosphere,** a layer of plastic rock—that is, solid rock that flows slowly when under pressure. According to the theory of plate tectonics, the lithosphere is broken into separate plates that float on the denser asthenosphere much like blocks of wood float on water. The continents and oceans are carried along as "passengers" on the moving lithospheric plates. Most lithospheric plates are composed of both continental and oceanic crust.

To date, about 7 major and 13 minor plates have been identified. Some plates are moving toward each other, some are moving apart, and some are sliding past each other. This constant movement has created many of the earth's surface features, such as mountain chains and deep-sea trenches.

Figure 4–4. This map shows the major lithospheric plates.

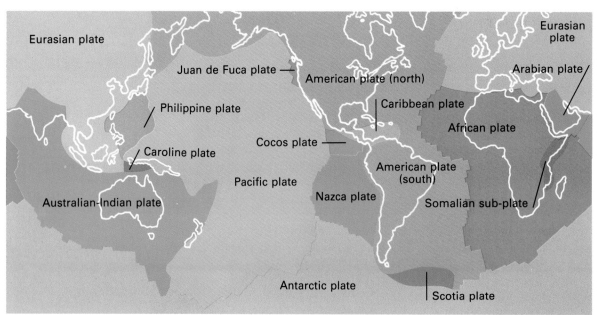

Eurasian plate

Juan de Fuca plate

Philippine plate

Caroline plate

Australian-Indian plate

American plate (north)

Cocos plate

Pacific plate

Nazca plate

Caribbean plate

American plate (south)

Antarctic plate

Scotia plate

Eurasian plate

Arabian plate

African plate

Somalian sub-plate

Lithospheric Plate Boundaries

Many changes in the earth's crust originate along lithospheric plate boundaries. The boundaries of the plates are not always easy to identify. As you can see in Figure 4–4, the familiar outlines of the continents and oceans depicted on maps do not closely resemble the outlines made by the plate boundaries. Plate boundaries are in the middle of the ocean floor, around the edges of continents, and within continents. There are three types of plate boundaries, each of which is associated with a characteristic type of geological activity.

Divergent Boundaries

The geologic activity that occurs along plate boundaries differs according to the way plates move in relation to each other. For example, two plates moving away from each other form a **divergent** (duh-VUR-junt) **boundary.** As the plates move apart, molten rock from the asthenosphere rises and fills the space between the plates. As the molten rock cools, it hardens onto the edges of the separating plates and creates new oceanic crust. Most divergent boundaries are found on the ocean floor. The locations of these spreading boundaries are marked by mid-ocean ridges.

In the center of a mid-ocean ridge is a narrow valley formed as the plates separate. This formation is called a **rift valley.** Other rift valleys may form where continents are separated by plate movement. For example, the Red Sea occupies a huge rift valley formed by the separation of the African plate and the Arabian plate.

Convergent Boundaries

As seafloor spreading pulls plates apart at one boundary, those plates push into neighboring plates at other boundaries. The direct collision of one plate with another makes another type of plate boundary—a **convergent** (kun-VUR-junt) **boundary.**

Three kinds of collisions can occur at convergent boundaries. One kind occurs when a plate with oceanic crust at its leading edge collides with a plate with continental crust at its edge. Because the plate with the oceanic crust is the denser of the two, it is subducted, or forced under the plate with the less dense continental crust, as shown in Figure 4–5. Scientists refer to the region along a plate boundary where one plate is forced under another plate as a **subduction** (sub-DUK-shun) **zone.** Scientists think that some of the plate pushed down into a subduction zone melts and becomes part of the mantle material. The rest of the melted plate rises to the surface on the continental crust and forms volcanic mountains.

A second kind of collision along convergent boundaries occurs when two plates with continental crust at their leading edges come together. During this type of collision, neither plate is subducted Instead, the colliding edges are crumpled and uplifted, producing large mountain ranges. Scientists are convinced that the Himalaya Mountains were formed by this type of collision.

The third type of collision along convergent boundaries occurs when oceanic crust collides with oceanic crust. A deep trench forms

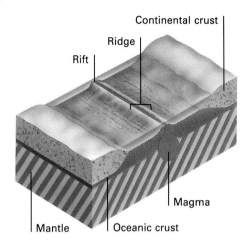

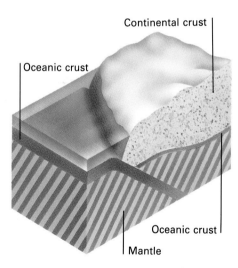

Figure 4–5. At divergent boundaries (top), plates separate. Plates collide at convergent boundaries (bottom).

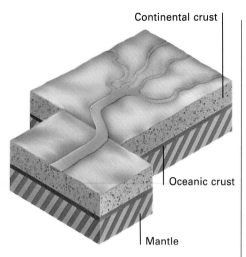

Figure 4–6. Plates scrape past each other at transform fault boundaries. Note the change in the course of the river as the plates move past each other.

Continental crust

Oceanic crust

Mantle

when one of these plates is subducted. Part of the subducted plate melts, and the resulting molten rock rises to the surface along the trench to form a chain of volcanic islands, called an **island arc.**

Transform Fault Boundaries

A plate boundary called a **transform fault boundary** forms where two plates are grinding past each other. The plate edges usually do not slide along smoothly. Instead, they scrape together and move in a series of sudden spurts of activity separated by periods of little or no motion. A major transform fault boundary is the San Andreas fault in California.

Causes of Plate Motion

Many earth scientists think that the movement of lithospheric plates is due to **convection,** the transfer of heat through the movement of heated fluid material. This same process occurs when you place a pot of water on the stove to boil. As the water at the bottom of the pot heats up, it expands and becomes less dense than the cool water above it. The cool water, which is now denser than the warm water, sinks and forces the warm water to the surface. This cycle of warm water rising and cool water sinking to replace it is called a **convection current.**

Scientists think a similar process results in convection currents within the asthenosphere. Heat from the earth's core and mantle causes some material in the lower asthenosphere to become hotter and therefore less dense than the material above it. The hot material rises. When this hot material reaches the base of the lithosphere, it cools. When the molten material cools, it becomes more dense and starts to sink. The cooling material is pushed to the side by new hot

Figure 4–7. Convection currents drive the mechanism believed by scientists to move lithospheric plates.

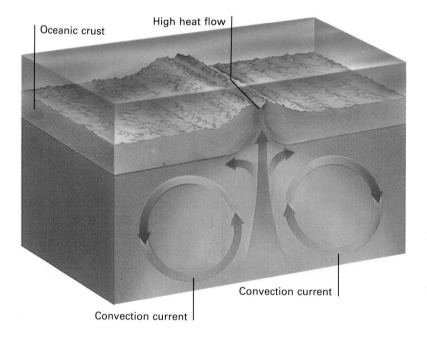

Oceanic crust

High heat flow

Convection current

Convection current

material that rises. As the process continues, the lithospheric plate is carried along with the moving material, as shown in Figure 4–7.

Evidence for the existence of convection currents in the asthenosphere comes from recent studies of the ocean floor. Scientists have measured the amount of heat leaving rocks at various points in the lithosphere. They have found this heat flow to be higher along plate boundaries where two plates are moving apart than it is elsewhere on the ocean floor. If hot convection currents are rising along these plate boundaries as the theory suggests, these temperature differences can be explained.

Though convection currents can explain some aspects of plate movement, questions remain. Scientists asked whether convection currents alone are strong enough to move the plates at the rates suggested by geological evidence. If not, they speculated, another mechanism may be responsible.

Investigation

Lithospheric Plate Boundaries

The movement of lithospheric plates has created many of the earth's topographical features. You can demonstrate the results of plate movement by using clay models of lithospheric plates.

Science Process Skills Focus: constructing models, demonstrating, describing, explaining

Materials
ruler, paper, scissors, rolling pin or rod, modeling clay, plastic knife

Procedure
1. Draw two rectangles 10 × 20 cm on your paper and cut them out.
2. Use a rolling pin to flatten out two pieces of clay until they are about 1 cm thick. Cut each piece into a rectangle 10 × 20 cm. Place one square of paper on each piece of clay.
3. Place the two clay models side by side on a flat surface, paper side down. Push the models together until the edges begin to buckle and rise off the surface of the table.
4. Turn the clay models around so that the unbuckled edges face one another.
5. Place one hand on each clay model. Apply only slight pressure toward the seam. Slide

one clay model forward and the other model backward about 7 cm.
6. Repeat Step 5 three more times, alternating the direction in which you push each model.

Analysis and Conclusions
1. What type of plate boundary are you demonstrating when you push the models head-on?
2. What type of plate boundary are you demonstrating when you slide the models against each other?
3. How does the appearance of the facing edges of the models in the two processes compare? How do you think these processes might affect the appearance of the earth's surface?

Suspect Terranes

Alfred Wegener's hypothesis of continental drift was an attempt to explain how the continents arrived at their present locations. The theory of plate tectonics refined Wegener's hypothesis, suggesting the actual mechanisms by which the continents might move. Neither continental drift nor plate tectonics, however, can explain how the continents were formed.

New discoveries are providing some possible explanations of how continents formed. These new discoveries provide the basis for the **theory of suspect terranes.** Simply put, this theory suggests that the continents are actually a patchwork of **terranes**—pieces of land, each with its own distinct geologic history. Each terrane has three identifying characteristics. First, a terrane contains rock and fossils that differ from the rock and fossils of neighboring terranes. Secondly, there are major faults at the boundaries of a terrane. Finally, the magnetic properties of a terrane do not match those of neighboring terranes.

Geologists have found evidence to support the suspect terrane theory. Northern California is a good place to observe this evidence. Geologists have found ten different terranes in the San Francisco Bay area alone. For example, in the hills of Palo Alto, there is fossil evidence of coral atolls—ocean islands made up of coral, the skeletons of sea organisms. Farther south lies a terrane called Permanette. The limestone of Permanette contains fossils that could only have

Figure 4–8. The rocks of the Cache Creek terrane, located in British Columbia, Canada, are shallow-water limestones that were deposited on oceanic crust.

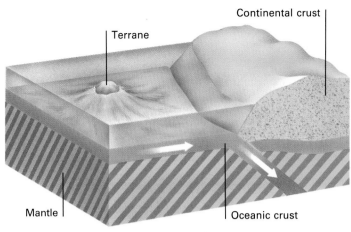

Continental crust

Terrane

Mantle

Oceanic crust

Stage 1. Terrane Moving Toward Continent

Figure 4–9. As oceanic crust is subducted, a terrane is scraped off the ocean floor and becomes part of the continental crust.

Stage 2. Terrane Being Welded to Continent

come from the ocean depths near the equator. The theory of suspect terranes explains how coral atolls and equatorial ocean fossils became part of the geology of northern California.

According to the suspect terrane theory, blocks of terranes are carried along on the ocean floor by the action of seafloor spreading to a lithospheric plate boundary where subduction is occurring. As the plate with oceanic crust moves under the plate with continental crust, the terranes are scraped off the descending ocean floor, as shown in Figure 4–9. Some terranes may form mountains, while others simply add to the surface area of a continent.

When Alfred Wegener first proposed his hypothesis of continental drift, he could not have imagined the explosion of scientific inquiry it would inspire. Like many hypotheses, continental drift asked more questions than it answered. The theories of plate tectonics and suspect terranes are attempts to answer some of those questions.

Section 4.2 Review

1. Summarize the theory of plate tectonics.
2. Name and describe the three types of plate boundaries.
3. Describe the three types of plate collisions that occur along convergent boundaries.
4. How might convection currents cause plate movement?
5. Explain how mountains on land can be composed of rocks that contain fossils of animals that lived in the ocean.

Reviewing Chapter 4

Key Concepts

- Wegener's hypothesis of continental drift states that the continents were once a single landmass. See page 57.

- Scientists have found fossil, geological, and climatic evidence to support the hypothesis of continental drift. See page 58.

- The ocean floor constantly renews itself through seafloor spreading. See page 59.

- The theory of plate tectonics proposes that changes in the earth's crust are caused by the very slow movement of large lithospheric plates. See page 62.

- The geological activity that occurs along the three types of plate boundaries differs according to the way plates move in relation to each other. See page 63.

- **Convection currents may be responsible for plate movements. See page 64.**

- The theory of suspect terranes proposes that the continents are a patchwork of terranes scraped off subducting lithospheric plates. See page 66.

Key Terms

asthenosphere (62)
continental crust (62)
continental drift (57)
convection (64)
convection current (64)
convergent boundary (63)

divergent boundary (63)
island arc (64)
lithosphere (62)
Mid-Atlantic Ridge (58)
mid-ocean ridges (58)

oceanic crust (62)
Pangaea (57)
Panthalassa (57)
plate tectonics (62)
rift valley (63)
seafloor spreading (60)
subduction zone (63)

suspect terrane (66)
terrane (66)
transform fault boundary (64)

Review

On your paper, write the letter of the term that best completes each of the following statements.

1. The German scientist Alfred Wegener proposed the existence of a huge landmass called (a) Panthalassa (b) rift valley (c) Mesosaurus (d) Pangaea.

2. Support for Wegener's hypothesis of continental drift includes evidence of changes in (a) climatic patterns (b) Panthalassa (c) terranes (d) subduction.

3. The ocean floor is constantly renewing itself through the process known as (a) subduction (b) continental drift (c) seafloor spreading (d) terranes.

4. An underwater mountain chain formed by molten rock is called a (a) divergent boundary (b) subduction zone (c) mid-ocean ridge (d) convergent boundary.

5. The term *tectonics* comes from a Greek word meaning (a) "movement" (b) "plate"

(c) "continent" (d) "construction."

6. The layer of the earth with a plastic structure that supports the moving plates is called the (a) lithosphere (b) asthenosphere (c) oceanic crust (d) terrane.
7. To date, scientists have identified approximately (a) 5 plates (b) 20 plates (c) 10 plates (d) 50 plates.
8. Two plates moving away from each other form a (a) transform fault boundary (b) convergent boundary (c) fracture (d) divergent boundary.
9. The collision of one lithospheric plate with another forms a (a) convergent boundary (b) transform fault boundary (c) rift valley (d) divergent boundary.
10. The region along lithospheric plate boundaries where one plate is pushed under another is called a (a) rift valley (b) transform fault

boundary (c) subduction zone (d) convergent boundary.
11. Two plates grinding past each other is called a (a) transform fault boundary (b) convergent boundary (c) subduction zone (d) divergent boundary.
12. Convection occurs because heated material becomes (a) less dense and rises (b) more dense and rises (c) more dense and sinks (d) less dense and sinks.
13. Scientists think that the convection currents that are responsible for the movement of lithospheric plates are found in the (a) lithosphere (b) asthenosphere (c) terranes (d) rift valleys.
14. Geologists claim that the continents are made up of formerly separate pieces of land called (a) terranes (b) plates (c) continental crust (d) oceanic crust.

Application

On your paper, write answers to the following questions.
1. In what ways might the concept of continental drift be compared to a jigsaw puzzle?
2. If Alfred Wegener had found identical fossil remains of plants and animals that had lived no more than 10 million years ago in both eastern Brazil and western Africa, what might he have concluded about the breakup of Pangaea?
3. Assume that the total surface area of the earth is not changing. If new material is being added to the earth's crust at one boundary,

what would you expect to find happening at another boundary?
4. Explain the role of technology in the progression from the hypothesis of continental drift to the theory of plate tectonics.
5. Explain the following statement: Due to the action of convection currents, the ocean floor is constantly renewing itself.
6. If you wanted to prove that an alien terrane had been scraped onto the North American plate, what kind of evidence would you search for?

Extension

1. Copy a map showing the outlines of the continents. Cut out the continents and assemble a model of Pangaea. Compare your model to Wegener's model of Pangaea that is shown on page 57.
2. In the late 1800's, Edward Suess, an Austrian scientist, developed the concept of a supercontinent called Gondwanaland. Look up information about Suess's ideas and compare them with the hypothesis of continental drift

proposed by Wegener.
3. Plate tectonics is a relatively new theory. Conduct library research to find out how scientists accounted for changes in continental landmass before the introduction of plate tectonics. Report your findings.
4. The process of convection is used to heat many homes. Do research and draw a diagram showing the convection currents in such a heating system.

Chapter 5
Deformation of the Crust

If you were to look carefully among the rocks on the peaks of some mountains, you might be surprised to find fossils of animals that lived in the sea. A closer inspection might also reveal waves of folded, twisted, or fractured rock. The forces that deform the earth's crust and make mountains out of ocean beds are mainly the result of plate tectonics, the movement and collision of lithospheric plates. In this chapter you will learn how plate movement builds and alters the topographic features of the earth.

Chapter Outline

The action of plate tectonics raised these Colorado mountains.

5.1 How the Crust Is Deformed

The Himalayas, the Grand Tetons, the Andes—these are the names of some of the earth's majestic mountain ranges. These imposing landforms are visible reminders that the shape of the earth's crust is always changing. Such changes result from **deformation,** the bending, tilting, and breaking of the earth's crust. Plate tectonics, the movement of the earth's lithospheric plates, is the major cause of crust deformation. Plate movement, however, is not the only force that shapes the earth's crust.

Isostatic Adjustment

Some changes in the earth's crust occur because sometimes there is a change in the weight of some part of the crust. The crust floats on the mantle. When parts of the crust are made heavier, they will sink more deeply into the mantle. If the crust is made lighter, it will rise higher on the mantle.

These movements can be compared to the behavior of a block of wood floating on water, as illustrated in Figure 5–1. When a weight is placed on the block, the weight forces the block to float more deeply in the water. What will happen to the block of wood when the weight is lifted?

The up-and-down movement of the crust occurs because of two opposing forces. The crust presses down on the mantle. The mantle presses up on the crust. When these two forces are balanced, the crust moves neither up nor down. However, when weight is added to the crust, the weight increases the force with which the crust presses on the mantle. The crust will sink until a balance of the forces is again reached. The balancing of these two forces is called **isostasy.** The up-and-down movements of the crust to reach isostasy are called **isostatic adjustments.** As these isostatic adjustments occur, areas of the crust are bent up and down. Pressure created by this bending causes the rocks in that area of the crust to deform.

- **Predict isostatic adjustments that will result from changes in the thickness of the earth's crust.**

- **Identify sources of stress in crustal rock.**

Figure 5–1. The thicker, heavier parts of the crust float more deeply in the asthenosphere than the thinner, lighter parts (left). The illustration to the right shows isostatic adjustment. Adding weight to the wooden block on the left forces it to float more deeply in the water.

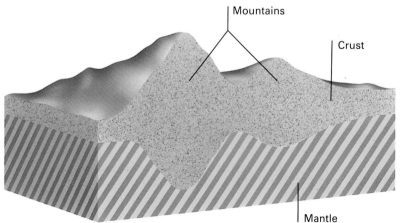

Mountains

Crust

Mantle

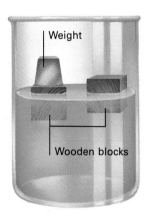

Weight

Wooden blocks

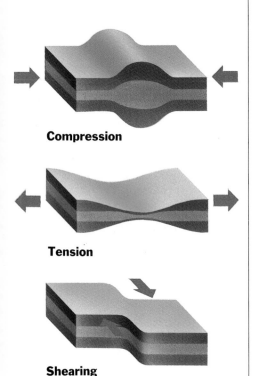

Compression

Tension

Shearing

Figure 5–2. The forces that deform rock are compression, tension, and shearing.

Isostatic adjustments are constantly occurring in areas of the crust with mountain ranges. Over a period of many thousands of years, the wearing away of rocks can significantly reduce the overall height of a mountain range and, thus, the weight of the crust. As the crust becomes lighter, it will rise.

Another example of isostatic adjustment can be found in areas where rivers flow into large bodies of water. Rivers carry large amounts of sand, gravel, and large rocks. When a river flows into a lake or an ocean, this material is deposited on the lake or ocean floor near the shore. The added weight of the material causes the floor to sink, allowing more material to be deposited. A very thick accumulation of these deposits is called a **geosyncline.**

Isostatic adjustments can also be found in areas where glaciers—large, slow-moving bodies of ice—once covered the land. The weight of the ice caused the crust underneath it to sink. Much of the glacial ice has now retreated. The land that earlier was covered with ice is slowly rising again in response to its reduced weight. Thus the land is undergoing another isostatic adjustment.

Stress

Isostatic adjustment and plate movement cause **stress** in the rocks that make up the earth's crust. Stress is a force that applies pressure to the rocks in the crust. Changes in the shape and volume, or size, of rocks that occur due to stress are called **strain.**

For example, when the weight of an icecap causes part of the crust to sink, stress is applied to the rocks of the sinking crust. A rise in the crust also would cause stress. Similarly, crustal stress occurs when lithospheric plates collide, separate, or rub together. This stress causes strain in crustal rocks, changing the shape or volume of the rocks by twisting them or pulling them apart.

There are three main types of stress, as illustrated in Figure 5–2. **Compression** occurs when crustal rocks are squeezed together. Compression reduces the volume of rocks and tends to push them higher up and deeper down in the crust.

Another type of stress is **tension.** Tension pulls rocks apart. When rocks are pulled apart by tension, they tend to become thinner in the middle than they are at the ends.

A third type of stress is called **shearing.** Shearing pushes rocks in two opposite horizontal directions. Sheared rocks bend, twist, or break apart.

Section 5.1 Review

1. Explain the principle of isostatic adjustment.
2. Define *stress* and *strain.*
3. Name and describe the three main types of stress.
4. Contrast the isostatic adjustment that might result from the melting of glacial ice with the isostatic adjustment that a geosyncline might cause.

5.2 The Results of Stress

Section Objectives

- **Compare folding and faulting as responses to stress.**
- **Describe four types of faults.**

The high pressure and temperatures caused by stress generally deform rocks. When stress is applied slowly, the deformed rock usually returns to its original shape as the force is removed. There is a limit, however, to the amount of force each type of rock can withstand and still retain its shape. If the force exceeds that limit, the shape of the rock changes permanently. In some cases, rock is deformed until it breaks.

Folding

When rock responds to stress by becoming permanently deformed without breaking, the result is **folding.** Folding is most easily observed when flat layers of rock are squeezed inward from the sides and the layers move into new positions without breaking. Cracks may appear, but the rock layers remain intact.

Folds, which appear as wavelike structures in rock layers, vary greatly in size. Some folds are small enough to be contained in a hand-held rock specimen. Others cover thousands of square kilometers and can be viewed best from the air.

The three general types of folds—**monocline** (MON-uh-kline), **anticline** (ANT-ih-kline), and **syncline** (SIN-kline)—are shown in Figure 5–3. Monoclines are gently dipping bends in horizontal rock layers. Anticlines are upcurved folds in the layers, and synclines are downcurved folds.

You might expect that wherever large folds occur, an anticline would form a ridge and a syncline would form a valley. Actually, just the opposite occurs. Because anticlines are the first folds to be exposed to wind and rain, they are worn down more rapidly than are synclines. This more rapid wearing down of the anticlines leaves the anticlines as valleys and the syncline as a mountain or ridge. Landforms of this type are commonly found among the peaks and valleys of the Appalachian Mountains.

Figure 5–3. Study the rock fold shown in the photo below on the left. Compare it with the three types of rock folds illustrated below right. Which type of rock fold is shown in the photo?

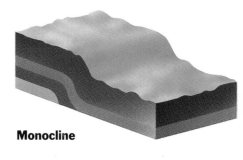

Monocline

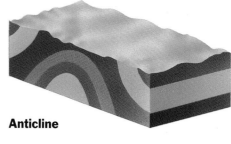

Anticline

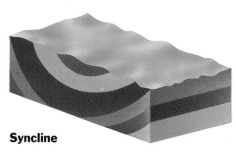

Syncline

Faulting

Rock does not always respond to stress just by folding. Cooler temperatures and lower pressure near the earth's surface often cause rock to respond to stress by breaking. The difference in the ways rock near the surface and rock deep in the interior respond to stress can be compared to the behavior of a heated glass rod. If the rod is heated until it is red hot, it can be bent easily. What would happen if you tried to bend a glass rod without heating it?

Breaks in rock structure are divided into two categories. When there is no movement in the rocks along either side of a break, it is called a **fracture.** When the rocks do move, it is called a **fault.** A **normal fault** is one where the **fault plane**—the break in the rock—is almost vertical. The rocks above the normal fault plane, called the **hanging wall,** move down relative to the rocks below the fault plane, called the **footwall.** Normal faults occur along divergent boundaries, where the crust is being pulled apart due to tensional stress. Normal faults usually occur in a series of parallel fault lines, forming steplike landforms. The great rift valleys of eastern Africa are areas of large-scale normal faulting.

A second type of nearly vertical fault is called a **reverse fault.** A reverse fault occurs when compression causes the hanging wall to move up relative to the footwall, as shown in Figure 5–4.

A **thrust fault** is a special type of reverse fault. The fault plane of a thrust fault is nearly horizontal rather than vertical. Because of

Figure 5–4. The illustration below shows four basic types of faults.

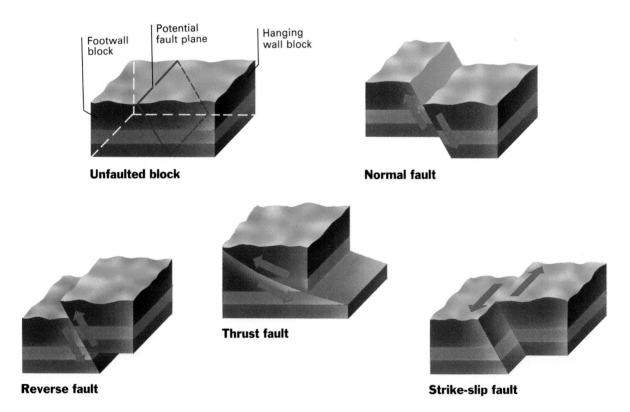

Footwall block

Potential fault plane

Hanging wall block

Unfaulted block

Normal fault

Reverse fault

Thrust fault

Strike-slip fault

Investigation

Folds and Fractures

You can use some common objects to demonstrate the factors that govern the ways rock responds to stress from plate movements.

Science Process Skills Focus: observing, identifying cause-and-effect relationships

Materials
safety glasses; balsa wood dowel, 2 mm × 15 cm; 2 books; plastic play putty

Procedures
1. **Put on the safety glasses.** Lay the dowel on a table. Place a book at each end of the dowel.
2. Place one hand on each book and gently and slowly slide the books against the ends of the dowel until the dowel bends slightly.
3. Move the books back to their original position. Observe and record what happens to the dowel as the books are moved.
4. Again move the books toward each other with the dowel between them. This time move the books quickly and forcefully. Record what happens to the dowel. Remove the safety glasses.
5. Roll the putty into a cylinder about 15 cm long and about the same diameter as the dowel.

6. Repeat Steps 1 through 3, using the putty in place of the dowel.
7. Re-form the putty into a cylinder. Grasp one end of the cylinder in each hand and pull quickly and sharply on both ends of the putty.

Analysis and Conclusions
1. Compare the responses of the dowel and the putty in Step 3. What two responses of rock to stress are represented?
2. Compare the response of the dowel in Step 4 with the response of the putty in Step 7.
3. What two factors influence the way the items respond to stress in this investigation? How do these factors influence the way rock responds to stress? Explain your answer.

---

the low angle of the fault plane, the rocks in the hanging wall are pushed up and on top of the rocks in the footwall.

Along a **strike-slip fault,** the rock on either side of the fault plane moves horizontally. These faults often occur at transform fault boundaries, where one plate is sliding past another.

Section 5.2 Review
1. What results when rock responds to stress by permanently deforming without breaking?
2. Why is faulting more likely to occur near the surface than deep within the earth?
3. Describe the four types of faults.
4. Why are strike-slip faults most often found at transform-fault boundaries?

5.3 Mountain Formation

Mount Everest is one of the tallest mountains on the earth, rising 8 km above sea level. Mount St. Helens became the most newsworthy mountain in the United States after it erupted in May of 1980.

Despite the fame of these mountains, neither stands alone. Each is part of a **mountain range,** a group of adjacent mountains with the same general shape and structure. Mount Everest is part of the Himalaya range and Mount St. Helens is part of the Cascade range.

Just as a group of individual mountains make up a range, a group of adjacent mountain ranges make up a **mountain system.** The Great Smoky, the Blue Ridge, the Cumberland, and the Green mountain ranges make up the Appalachian mountain system in eastern United States.

The largest mountain systems are part of two still larger systems called **mountain belts.** The two major mountain belts on earth, the **circum-Pacific belt** and the **Eurasian-Melanesian belt,** are shown in Figure 5–5. The circum-Pacific belt surrounds the Pacific Ocean, and the Eurasian-Melanesian belt runs through Asia, northern Africa, and southern Europe.

Plate Tectonics and Mountains

Both the circum-Pacific mountain belt and the Eurasian-Melanesian mountain belt are located along converging plate boundaries. Scientists think that the location of these two mountain belts is evidence that most mountains were formed when lithospheric plates collided. Some mountains, such as the Appalachians, do not lie along currently converging plate boundaries. However, evidence indicates that these ranges formed where plates collided in the past.

Figure 5–5. Most of the great mountain ranges of the world lie along two mountain belts.

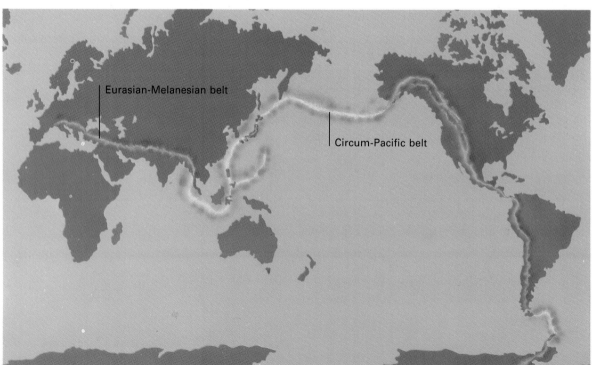

Eurasian-Melanesian belt

Circum-Pacific belt

Collisions Between Continental and Oceanic Crust

Some mountains are formed when oceanic crust and continental crust collide at converging plate boundaries. When the moving plates collide, the oceanic crust subducts, or plunges beneath, the continental crust. This type of collision produces such large-scale deformation of rock that high mountains are pushed up. In addition, the subducted oceanic crust partially melts, producing the magma that might eventually erupt and form mountains on the earth's surface. The mountains of the Cascade range in the Pacific Northwest were formed in this way.

Some mountains at the boundary between oceanic crust and continental crust may have been formed by a different process. As the oceanic crust was subducted, pieces of crust called *terranes* were scraped off. According to the suspect terrane theory, they became part of the continental crust. Some of these terranes formed mountains. To learn more about terranes, see Chapter 4.

Collisions Between Oceanic Crust and Oceanic Crust

Volcanic mountains sometimes form where two plates with oceanic crust at their edges collide. The oceanic crust of one plate subducts the oceanic crust of the other plate. As the oceanic crust plunges deeper into the mantle, the intense heat melts the crustal material, forming magma. The magma rises and breaks through the oceanic crust. These eruptions form volcanic mountains on the ocean floor. The Fiji Islands are the peaks of volcanic mountains that rose above sea level.

Collisions Between Continents

Mountains can also be formed when two continents collide. The Himalaya Mountains were formed by just such a collision. According to the theory of plate tectonics, India was at one time a separate

Figure 5–6. The Himalayas, shown below, were formed when India collided with Asia.

continent riding the Indian plate, which was moving north toward Asia. It continued to move until the continental crust of India collided with the continental crust of Asia. The oceanic crust of the Indian plate was subducted beneath the Asian plate. This collision produced intense deformation of the crust of both continents. As a result, the Himalayas were formed and India was connected to the Eurasian continent. Severe earthquakes still occur in the Himalayas. What can you infer about the movement of the Indian plate?

Types of Mountains

Mountains are more than just elevated parts of the earth's crust. They are complicated structures with rock formations that yield evidence of the forces that created them. Scientists classify mountains according to the way in which the crust was deformed and shaped by mountain-building forces.

Folded Mountains and Plateaus

The largest mountain ranges in the world are made up of **folded mountains** and are commonly found where continents have collided.

Horizons

Technology: Laser Tracking

In what appears to be a scenario right out of a science fiction movie, scientists are aiming and firing laser beams at a satellite orbiting high above the earth. These scientists—from the United States, Italy, Egypt, Peru, and several other countries—do not intend to disable the satellite. Rather, they hope to gather data on the speed of lithospheric plate movement. The information the scientists obtain will help them to better understand how plate tectonics shapes the crust of this planet.

The satellite, called *Lageos* (Laser Geophysical Satellite), resembles the spinning mirrored globe used to reflect dots of light around a room at dance parties. *Lageos,* shown left, was launched by **NASA** in 1976 into an orbit 5900 km above the earth. The surface of *Lageos* is covered with 426 light retroreflectors called *corner cubes.* A corner cube is a specially designed mirror that reflects light back on exactly

In a folded mountain, tectonic movements have squeezed rock layers together like an accordion. Parts of the Alps, the Himalayas, the Appalachians, and the Ural Mountains consist of very large and complex folds. These long mountain chains also show evidence of faulting and volcanic activity.

The same forces that uplift folded mountains also form **plateaus.** Plateaus are large areas of flat-topped rocks high above sea level. Most plateaus are formed when thick, horizontal layers of rock are slowly uplifted. The area is pushed up gently enough so that the layers remain flat rather than faulting and folding as do mountains. Most plateaus are found next to mountain ranges. For example, the Tibetan Plateau is next to the Himalayan Mountains, and the Colorado Plateau is next to the Rockies. A number of plateaus also were formed when layers of molten rock hardened and piled up on the earth's surface.

Fault-block Mountains and Grabens

Some mountains have been formed by faults where parts of the earth's crust have been broken into large blocks. These blocks were then lifted above the surrounding crust. Faulting tilted the blocks,

Figure 5–7. The Alps are an example of folded mountains.

the same path along which it came.

When scientists at a laser station such as the one shown below, fire a beam of light at *Lageos,* the corner cubes reflect the beam directly back to the station. This process is depicted in the illustration at right. The scientists carefully calculate the length of time it took for the pulse of light to reach the satellite and return. Then they determine the orbit of *Lageos* and pinpoint its exact location at the time they fired their laser. Using this information, plus the precise location on earth of their laser, the scientists can determine the speed at which the laser is drifting and the direction in

which it is traveling. The laser, of course, rests on a continent, and the continent rests on a lithospheric plate. Thus, the scientists have also determined the speed and direction of movement of the plate.

Scientists are interested in this information for several reasons. Knowing which way the earth's plates are moving will enable scientists to predict the future geography of the earth. Also, this data could help scientists predict future geologic activity along plate boundaries, including mountain formation.

What are some specific predictions scientists might make when they know the direction and speed of plate movements?

creating **fault-block mountains,** as shown in Figure 5–8. The Grand Teton Mountains of Wyoming are an example of fault-block mountains. Much of Nevada, Arizona, western Utah, southern Oregon, northern New Mexico, and southeastern California is covered by fault-block mountains. These mountains form nearly parallel ranges that average 80 km in length.

The same type of faulting that forms fault-block mountains also forms long, narrow valleys called **grabens.** Grabens develop when faulting breaks the crust into blocks and a block slips downward relative to the surrounding blocks. Grabens are usually found in groups. The grabens are separated by minor faults with major faults occurring at the edges of the entire group. The Imperial Valley in California is a graben.

Volcanic Mountains

Mountains that form when molten rock erupts onto the earth's surface are called **volcanic mountains.** They may develop either on land or on the ocean floor. The Cascade Mountains of Washington, Oregon, and northern California are volcanic mountains. Some of the largest volcanic mountains are found along mid-ocean ridges on the ocean floor because that is where most of the earth's volcanic activity occurs. These mid-ocean ridges are actually huge volcanic mountain chains that run through the centers of the Atlantic, southern Pacific, and Indian oceans. The peaks of the highest mountains sometimes rise above sea level to form islands, such as Iceland and the Azores.

Other large volcanic mountains are formed on the ocean floor over hot spots. Hot spots are pockets of magma beneath the earth's crust that erupt onto the surface. The Hawaiian Islands are the tips of high volcanic mountains that were formed over a hot spot. The main island of Hawaii is actually a mountain that reaches almost 9 km above the ocean floor with a base that is more than 160 km wide. Almost 4 km of this mountain is above sea level.

Dome Mountains

An unusual type of mountain is formed when molten rock rises through the crust and pushes up the rock layers above it. The result is a circular dome on the earth's surface. The molten rock that

Figure 5–8. In addition to folded mountains, there are fault-block mountains such as the Sierra Nevadas (top), volcanic mountains such as the North Cascades (below, left), and dome mountains such as the Black Hills (below, right).

Science Notebook

The Disappearing Mediterranean

Two of the more famous regions in the world are the Alps and the Mediterranean Sea. The Alps, considered to be among the earth's most beautiful mountains, have become a vast, natural playground for skiers, hikers, and climbers. The Mediterranean plays host to travelers from around the world who wish to sample its diverse cultures, balmy climate, and famous resorts. More importantly, the people who live nearby depend on the sea for their economic well-being.

The same natural forces that produced the Alps are slowly swallowing up the Mediterranean. The Alps were formed, and are still being shaped, by the collision of two lithospheric plates. Italy, part of which rides on the African plate, collided sometime in the past with Eurasia. The collision formed the Alps, but it did not stop the movement of the African plate. The northern oceanic crust of the African plate, which is actually the sea floor of the Mediterranean, is still subducting beneath the continental crust of Eurasia. As more oceanic crust subducts, the Mediterranean Sea will become smaller. Italy, which continues to be pushed into Eurasia, will eventually cease to exist as we know it. When the northern coast of the African continent finally collides with Eurasia, the Mediterranean Sea will become just a memory, a photograph, perhaps, in some history book.

What do you think will happen to the Alps as the African plate continues to push northward?

pushed up the rock layers eventually cools and forms hardened rock. When the pushed-up rock layers are worn away over time, the hardened rock is exposed. This rock wears away in places, leaving separate high peaks, or **dome mountains.** The Black Hills of South Dakota, and the Adirondack Mountains of New York State are dome mountains.

Section 5.3 Review

1. Describe the types of lithospheric plate collisions that build mountains.
2. Name the four types of mountains and explain how each is formed.
3. Explain how plateaus are formed.
4. How do volcanic mountains grow?

Reviewing Chapter 5

Key Concepts

- Stress can squeeze rocks together, pull them apart, and bend and twist them. See page 72.
- Rock responds to stress by bending into folds and by fracturing or faulting. See pages 73-74.
- Four major types of faults occur in rock: normal faults, reverse faults, thrust faults, and strike-slip faults. See page 74.
- Mountain building is often the result of the collision of lithospheric plates. See page 76.
- Mountains are classified according to the way in which the crust was deformed and shaped by mountain-building forces. See page 78.

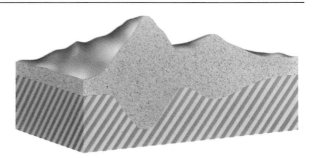

- **The movements of lithospheric plates and isostatic adjustments are sources of stress that causes deformation in crustal rock. See pages 71-72.**

Key Terms

anticline (73)	fault plane (74)	monocline (73)	syncline (73)
circum-Pacific belt (76)	folded mountain (78)	mountain belt (76)	tension (72)
compression (72)	folding (73)	mountain range (76)	thrust fault (74)
deformation (71)	footwall (74)	mountain system (76)	volcanic mountain (80)
dome mountain (81)	fracture (74)	normal fault (74)	
Eurasian-Melanesian belt (76)	geosyncline (72)	plateau (79)	
fault (74)	graben (80)	reverse fault (74)	
fault-block mountain (80)	hanging wall (74)	shearing (72)	
	isostasy (71)	strain (72)	
	isostatic adjustment (71)	stress (72)	
		strike-slip fault (75)	

Review

On your paper, write the letter of the term that best completes each of the following statements.

1. The state of balance between the thickness of the crust and the depth at which it floats on the asthenosphere is called (a) stress (b) isostasy (c) strain (d) shearing.
2. The increasing weight of mountains causes the crust to (a) sink (b) fold (c) rise (d) fracture.
3. A thick, ocean-floor deposit of sand, gravel, and rocks from a river is called a (a) footwall (b) reverse fault (c) geosyncline (d) monocline.
4. The force that changes the shape and volume of rocks is (a) footwall (b) isostasy (c) rising (d) stress.
5. The type of stress that squeezes rocks together is (a) compression (b) tension (c) shearing (d) faulting.
6. The type of stress that pulls rocks apart until they are thinner in the middle than they are at their ends is (a) folding (b) compression

(c) tension (d) isostasy.
7. Shearing (a) bends, twists, or breaks rocks
   (b) squeezes rock together (c) causes rock to
   melt (d) pulls rock apart.
8. High pressure and high temperature will cause
   rocks to (a) fracture (b) adjust (c) plateau
   (d) deform.
9. Upcurved folds in rock are called (a) anticlines (b) monoclines (c) geosynclines
   (d) synclines.
10. Downcurved folds in rock are called (a) geo-
    synclines (b) monoclines (c) anticlines
    (d) synclines.
11. Gently dipping bends in rock formations with
    horizontal layers are called (a) monoclines
    (b) geosynclines (c) synclines (d) anticlines.
12. When no movement occurs along the sides of
    a break in a rock structure, the break is called

a (a) normal fault (b) fracture (c) fold
   (d) hanging wall.
13. The rock above the fault plane makes up the
    (a) tension (b) footwall (c) hanging wall
    (d) compression.
14. A nearly vertical fault in which the rock on
    either side of the fault plane moves horizon-
    tally is called a (a) normal fault (b) reverse
    fault (c) strike-slip fault (d) thrust fault.
15. The largest mountain systems are part of still
    larger systems called (a) continental margins
    (b) ranges (c) belts (d) synclines.
16. Mount St. Helens in Washington State is an
    example of a (a) folded mountain (b) vol-
    canic mountain (c) fault-block mountain
    (d) dome mountain.

Application

On your paper, write answers to the following
questions.
1. Suppose glaciers, which are vast fields of
   slow-moving ice, were to cover much of the
   earth's surface once again. What would you
   expect to happen to those parts of the continents that were covered by ice? Explain your
   answer.
2. When the Indian plate collided with the Eurasian plate, producing the Himalayan Mountains, which type of stress most likely
   occurred? Which type of stress is most likely
   occurring along the Mid-Atlantic Ridge?
   Which type of stress would you expect to find
   along the San Andreas fault? Use your knowledge of stress and plate tectonics to explain
   your answers.

3. If the force that is causing a rock to be
   slightly deformed begins to ease, what might
   happen to the rock? What would happen if
   the force causing the deformation became
   greater?
4. Suppose that a new highway is being
   planned. This proposed road would intersect a
   transform-fault boundary. What would happen
   to the highway if a strike-slip fault occurred
   along the boundary? Why?
5. A geologist discovers that part of a mountain
   along the west coast of the United States contains the fossil remains of animals that live on
   the ocean floor. What is the most likely explanation for this phenomenon?
6. Why do you suppose dome mountains do not
   become volcanic mountains?

Extension

1. Devise a laboratory activity to demonstrate
   isostatic adjustment. Include written directions. Ask a classmate to try your activity.
2. Using modeling clay to represent crustal rock,
   create examples of rock that have been subjected to the forces of compression, tension,
   and shearing.
3. Look for examples of folding and faulting in

the area where you live. Take photographs of
the examples you find. Use your photos, or
drawings if you prefer, to create a poster explaining folding and faulting.
4. Do some research about different types of
   mountains in the United States. Create an illustrated list of folded, dome, volcanic, and
   fault-block mountains in this country.

Chapter 6

Earthquakes

Strain causes rocks along a fault to fracture and shift. The result is an earthquake. Some earthquakes are mild, but others are exceedingly violent and destructive. In 1556, for example, in Shaanxi Province in north central China, more than 800,000 people perished during a single earthquake.

In this chapter you will learn about the causes and effects of earthquakes, the way they are measured, and scientists' attempts to predict future earthquakes.

Chapter Outline

An earthquake split this Hawaiian Island road.

6.1 Earthquakes and Plate Tectonics

Vibrations of the earth's crust are called **earthquakes.** Earthquakes usually occur when rocks that have fractured under stress suddenly shift along a fault. As you read in Chapter 5, stress is a force that can change the size and shape of rocks.

Normally the rocks along both sides of a fault are pressed together tightly. For a time, friction prevents the rocks from moving past each other. In this immobile state, a fault is said to be "locked." Parts of a fault remain locked until the strain becomes so great that the rocks suddenly grind past each other. This slippage causes the trembling and vibration of an earthquake.

Elastic Rebound Theory

Geologists explain many earthquakes by the **elastic rebound theory.** According to this theory, the rocks on each side of a fault are moving slowly. If the fault is locked, strain in the rocks increases. When they are strained past a certain point, however, the rocks fracture, separate at their weakest point, and spring back to their original shape, or rebound.

In fracturing and slipping into new positions, rocks along a fault release energy in the form of vibrations called *seismic waves.* This release of energy often increases the strain in other rocks along the fault, causing them to fracture and spring back. This chain reaction is the reason that major earthquakes are usually followed by a series of smaller tremors called **aftershocks.**

As you can see in Figure 6–1, the area along a fault where slippage first occurs is called the **focus,** or hypocenter, of an earthquake. The point on the earth's surface directly above the focus is called the **epicenter** (EP-ih-SENT-ur). When an earthquake occurs,

Section Objectives
- **Discuss the elastic rebound theory.**
- **Explain why earthquakes generally occur at plate boundaries.**

Figure 6–1. The epicenter of an earthquake is the point on the surface directly above the focus.

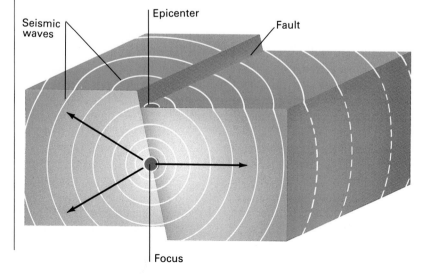

Seismic waves

Epicenter

Fault

Focus

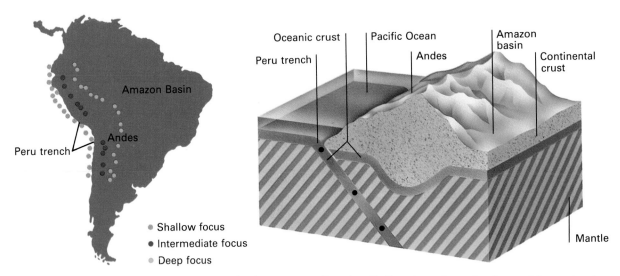

Figure 6–2. Earthquakes at various depths have occured along the west coast of South America in nearly parallel lines (left). The illustration (right) shows a subduction zone along the west coast of South America. The pattern of the earthquakes is characteristic of an area where one plate has been subducted.

seismic waves radiate in all directions from the focus. The action is similar to what occurs when you drop a stone into a pool of still water and circular waves ripple out from the center.

Although the depths of earthquake focuses vary, about 90 percent of continental earthquakes have a shallow focus. **Shallow-focus earthquakes** occur within 70 km of the earth's surface. **Intermediate-focus earthquakes** occur at a depth of between 70 km and 300 km. **Deep-focus earthquakes** occur in subduction zones at a depth of between 300 km and 650 km. As you can see in Figure 6–2, deep-focus earthquakes usually occur farther inland than do shallow-focus or intermediate-focus earthquakes. Why do earthquakes not occur deeper than 650 km within the earth?

By the time the vibrations from an intermediate-focus or deep-focus earthquake reach the surface, much of their energy has been used up. For this reason, the earthquakes that cause the most damage have a shallow focus.

Major Earthquake Zones

When you study a map that shows the locations of earthquakes worldwide, you can see the link between earthquakes and plate tectonics. Most earthquakes occur along or near the edges of the earth's lithospheric plates. It is at these moving plate boundaries that stress is greatest and rock is subjected to the most strain.

The earth has three major earthquake zones, as shown in Figure 6–3. The first is a large area known as the **Ring of Fire.** The Ring of Fire includes the western coasts of North America and South America and the eastern coast of Asia. This area forms a ring around the Pacific Ocean. Along this ring, some plates are being subducted, while other plates are scraping past each other. Many earthquakes occur along the Ring of Fire because the plate movements cause strain to build up in rocks. Eventually the rocks fracture and shift, and an earthquake occurs.

The second major earthquake zone is the Mid-Atlantic Ridge. Earthquakes occur along the Mid-Atlantic Ridge because oceanic crust is pulling away from the sides of the ridge. This spreading motion creates strain in the rocks along the ridge.

The third major earthquake zone is the Eurasian-Melanesian mountain belt. You read in Chapter 5 that the mountains along this belt were formed by the collision of the Eurasian plate with the African and the Indian plates. These plates are still colliding, and the same forces that are pushing up the mountains also produce numerous earthquakes.

At some plate boundaries there exists a group of interconnected faults called a **fault zone.** Fault zones form at plate boundaries because of the intense strain that results when the plates collide, move apart, subduct, or slide past each other. One such fault zone includes the San Andreas Fault, which extends almost the length of California. The San Andreas Fault zone has formed where the edge of the Pacific plate slips northward against the North American plate. Movement could occur along one or more of the individual faults in the San Andreas Fault zone. Any such movement could cause a major earthquake within the zone.

Not all earthquakes, however, result from movement along plate boundaries. For example, the most widely felt series of earthquakes in the history of the United States did not occur near any plate boundary. Rather, these earthquakes occurred in the middle of the continent, near New Madrid, Missouri, in 1812. The vibrations from the earthquakes that rocked New Madrid were so strong that church bells rang as far away as Boston, Massachusetts. Scientists were puzzled by earthquake activity in the center of the United States. After all, they reasoned, New Madrid is far away from any active plate boundary.

Figure 6–3. The map below illustrates the earth's three major earthquake zones: the Ring of Fire, the Mid-Atlantic Ridge, and the Eurasian-Melanesian belt.

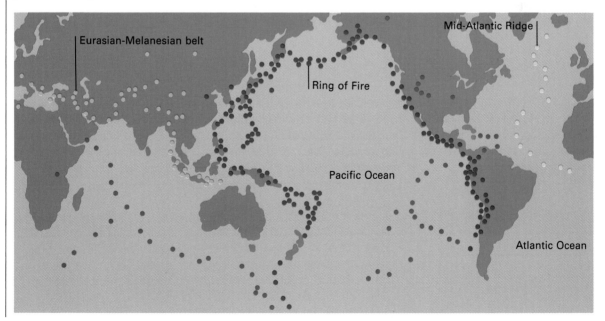

Eurasian-Melanesian belt

Mid-Atlantic Ridge

Ring of Fire

Pacific Ocean

Atlantic Ocean

The Big Squeeze

Scientists think that about 600 million years ago the North American continent began to break apart into two or three pieces. For reasons yet unknown, this tectonic activity stopped and the continent remained intact.

The only clue to this activity is a fault zone buried deep underground in the central United States. This fault zone, discovered in the late 1970's, consists of three interlocking faults that stretch about 240 km from northern Arkansas to southern Illinois. Scientists think this fault zone was responsible for the 1812 earthquake that shook New Madrid, Missouri.

When these faults were first discovered, one question puzzled scientists. How, they wondered, could a fault zone far from any plate boundary accumulate enough strain to become active—to fracture and rebound, causing an earthquake. The answer seems to lie in plate tectonics.

Scientists have found that the North American plate is being compressed—squeezed together—by the force of the spreading Atlantic ocean floor. This stress may eventually be powerful enough to activate the centrally located fault zone.

China has also had earthquakes that do not originate at plate boundaries, and a similar hypothesis has been suggested to explain those quakes. China is being squeezed together on three sides, by the North American plate, the Indian plate, and the Pacific plate.

If strain continues to increase along the fault zone in the central United States, what might result?

Then, in the late 1970's, studies of the Mississippi River region revealed an ancient fault zone deep within the earth. This zone appears to be part of a major fracture in the North American plate. Scientists have calculated that the fracture formed at least 600 million years ago and was later buried under many layers of sediment and rock.

Section 6.1 Review

1. Explain the elastic rebound theory.
2. In earthquakes that cause the greatest damage, at what depth would slippage most likely occur?
3. What four types of plate movements can cause earthquakes?
4. If an earthquake occurs in the center of Brazil, what can you infer about the geology of that area?

6.2 Recording Earthquakes

Seismic waves can be detected and recorded by using an instrument called a **seismograph** (SIZE-muh-GRAF). A seismograph consists of three separate sensing devices. One device records the vertical motion of the ground. The other two record horizontal motion in the east-west and north-south directions. A seismograph records motion by tracing wave-shaped lines on paper or by translating the motion into electronic signals. The electronic signals can be recorded on magnetic tape. This tape can then be loaded into a computer that analyzes the seismic waves.

Types of Seismic Waves

Scientists have determined that every earthquake produces three major types of seismic waves. Each type of wave travels at a different speed and causes different movements in the earth's crust.

Primary waves, or **P waves,** move the fastest and are therefore the first to be recorded by a seismograph. P waves moving through the earth can travel through solids, liquids, and gases. The more rigid the material is, the faster the P waves travel through it. P waves cause rock particles to move together and apart along the direction of the waves.

Secondary waves, or **S waves,** are the second waves to be recorded on a seismograph. Unlike P waves, S waves can travel only through solid material. You may recall that S waves cannot be detected on the side of the earth opposite that on which the earthquake occurs. The S waves cannot be detected because they do not penetrate the liquid part of the earth's outer core. S waves cause rock particles to move at right angles to the direction in which the waves are traveling.

When P waves and S waves reach the earth's surface, their energy is converted into a third type of seismic wave. *Surface waves,* also called *long waves* or **L waves,** are the slowest-moving waves

Section Objectives

- **Compare the three types of seismic waves.**
- **Discuss the method scientists use to pinpoint an earthquake.**
- **Discuss the method most commonly used to measure the magnitude of earthquakes.**

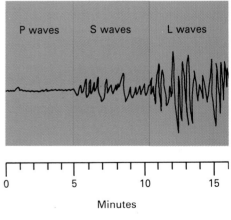

Figure 6–4. Most seismograph stations, like the one shown (left), have banks of seismographs to record earth tremors. Each type of seismic wave (right) leaves a unique "signature" on a seismograph.

and therefore the last to be recorded on a seismograph. L waves travel slowly over the earth's surface in a movement similar to that of ocean waves. L waves, which cause the surface to rise and fall, are particularly destructive when traveling through loose earth. Why do L waves cause the greatest damage during an earthquake?

Locating an Earthquake

To determine the focus of an earthquake, scientists analyze the difference between the arrival times of the P waves and the S waves. P waves travel about 1.7 times faster than S waves. Thus, if the S waves arrive soon after the P waves, the earthquake must have originated fairly close to the seismograph station. If the S waves arrive a long time after the P waves, the earthquake must have occurred farther away. To determine how far an earthquake is from a given seismograph station, scientists plot the difference between the arrival times of the two waves. Then they consult a standard graph that translates the difference in arrival times into distance from the focus.

Once the distance from the focus has been determined, scientists can locate the epicenter of the earthquake. To do so, they need information about the focus from at least three seismograph stations. Using the example shown in Figure 6–5, suppose that the focus of an earthquake is 560 km away from one station, 2,000 km away from another, and 900 km away from a third. Three circles are drawn on a map that shows all three stations. Each station is at the center of a circle. The distance from a particular station to the focus is the radius of that circle. The epicenter of the earthquake is very near the point at which the three circles intersect.

Earthquake Measurement

Scientists use the **Richter scale** to express the magnitude of an earthquake. Magnitude is a measure of the energy released by an earthquake. Each increase of one whole number of magnitude represents the release of 31.7 times more energy than that of an earthquake measuring one whole number lower. Thus, an earthquake with a magnitude of 8 releases 31.7×31.7—or about 1,000 times—as much energy as an earthquake with a magnitude of 6.

The largest earthquake so far recorded registered a magnitude of 8.9. A major earthquake, one that causes widespread damage, has a magnitude of 7 or above. A moderate earthquake has a magnitude between 6 and 7, and a minor earthquake, between 2.5 and 6. Earthquakes with magnitudes of less than 2.5 are called **microquakes** and usually are not felt by people.

The **Mercalli scale** expresses the **intensity** of an earthquake, or the amount of damage it causes, by a Roman numeral and a description. The Mercalli scale describes an earthquake with a rating of II, which is an earthquake with a low intensity, as follows: "Felt only by a few persons at rest, especially on upper floors of buildings. Delicately suspended objects may swing." An earthquake with the

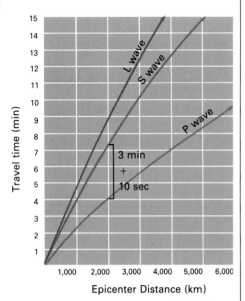

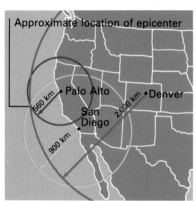

Figure 6–5. The time-distance graph (top) is a standard graph used to find the distances of an earthquake from seismograph stations. The distances can be plotted on a map (bottom) to find the location of the earthquake.

Investigation

Seismographic Record

A seismograph records the energy release of an earthquake. You can observe energy release by making a model seismograph.

Science Process Skills Focus: constructing models, observing, comparing, describing

Material
shoe box, sand, felt-tip pen, rubber band, pad of paper, objects with various masses, newspaper

Procedure
1. Fill the box with sand. Put on the lid.
2. Fasten the felt-tip pen to the lid of the box with the rubber band, as shown in the figure.
3. Mark an X near the center of the lid.
4. Have a partner hold the pad of paper so that it just touches the pen.
5. Hold your first object directly over the X at a height of about 30 cm. As your partner slowly moves the paper horizontally past the pen, drop the mass on the X. It may take one or two trials to get your timing right.
6. Label the resulting line on the paper with the mass of the object dropped and the material in the box.
7. Repeat Steps 4 through 6 with three more ob-

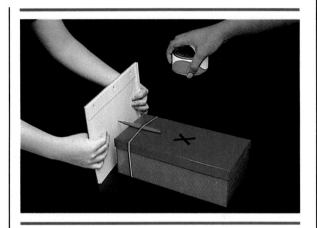

jects of different masses.
8. Empty the box and refill with crumpled newspaper. Repeat Steps 1 through 7.

Analysis and Conclusions
1. What do the lines on the paper represent?
2. What do the sand and newspaper represent?
3. Compare the lines made with sand in the box and the lines made with newspaper. What could explain any differences?
4. What can you infer from your seismographic records about the way seismographs are used to measure energy release?

---

highest intensity, one designated by Roman numeral X, is described as follows: "Some well-built wooden structures destroyed; most masonry and frame structures destroyed with foundations; ground badly cracked. Rails bent. Landslides considerable from riverbanks and steep slopes. Shifted sand and mud. Water splashed over banks."

Section 6.2 Review

1. What instrument is used to record seismic waves?
2. Explain the three types of seismic waves.
3. How are the focus and the epicenter of an earthquake located?
4. How do scientists measure the intensity of an earthquake?
5. Why do P waves travel faster through the lithosphere than through the asthenosphere?

- **Describe possible effects of a major earthquake on buildings.**
- **Discuss the relationship of tsunamis to earthquakes.**
- **List safety rules to follow when an earthquake strikes.**
- **Identify changes in the earth's crust that may signal earthquakes.**

6.3 Earthquake Damage

During a severe earthquake, you would be much safer in an open, level field than in a large city. Movement of the ground itself seldom causes many deaths or injuries. Instead, most injuries and deaths result from the collapse of buildings and other structures. Falling objects and flying glass may also cause injuries. Other dangers include landslides, fires, and explosions caused by broken electric and gas lines, and floodwaters released from collapsing dams. Duration can affect the intensity of an earthquake. A moderate earthquake that continues for a long time often causes more damage than a major one that lasts only a short time.

Destruction to Buildings and Property

Most buildings are not designed to withstand the swaying motion caused by earthquakes. Buildings with weak walls may completely collapse. Very tall buildings may sway so violently that they tip over and fall onto lower neighboring structures.

The type of ground beneath a building can affect the way the building will respond to seismic waves. For example, a building constructed on loose soil is much more likely to be damaged during an earthquake than one built on more solid ground. During an earthquake, the loose soil can vibrate like jelly. Buildings constructed on top of this kind of soil absorb this exaggerated motion and sway violently. On what kind of soil should a tall building in a fault zone be constructed?

Tsunamis

A giant ocean wave that often occurs after a major earthquake with an epicenter located on the ocean floor is called a **tsunami** (tsoo-NAHM-mee). Scientists think that most tsunamis are caused by two events related to undersea earthquakes: faulting and underwater landslides. Faulting may bring about a sudden drop or rise in the level of a part of the ocean floor. A large mass of water drops or rises with the ocean floor. The mass of water churns up and down as it adjusts to the change in sea level. This violent water movement sets into motion a series of long, low waves that develop into tsunamis.

An earthquake may trigger a severe underwater landslide. The water above the landslide is thrown into an up-and-down motion, thereby creating a series of tsunamis.

The tsunami that accompanied the 1964 Alaskan earthquake caused heavy damage to towns near the Gulf of Alaska. Most of the fishing fleet off Kodiak, Alaska, was destroyed. Many fishing vessels were swept into the business district of the town by the tsunami. The tsunami caused 107 deaths, whereas only 9 persons died as a direct result of the earthquake vibrations.

Disastrous earthquakes and tsunamis have encouraged the expansion and improvement of the Seismic Sea Wave Warning System (SSWWS). This network of seismograph stations around and in the Pacific Ocean alerts scientists to the location and magnitude of earthquakes. If a tsunami seems possible, scientists estimate its arrival times at different locations. They can then issue warnings immediately to these areas. However, there may not be enough time to issue warnings to areas very near the site of a predicted earthquake.

Earthquake Safety

A destructive earthquake may occur in any region of the United States. However, scientists feel this type of earthquake is more likely to occur in certain areas. People living or visiting near active faults should be ready to follow a few simple earthquake-safety rules.

IMPACT

Mexico: 1985

At 7:18 A.M. on Thursday, September 26, 1985, the early morning calm of Mexico City was shattered by an earthquake that produced a thunderous roar throughout the city.

In the downtown area, hotels, office towers, schools, and hospitals began vibrating wildly. Some buildings swung back and forth so violently that they struck nearby buildings or lifted off their foundations and toppled over. Other buildings collapsed in a heap, their floors stacking up like pancakes.

Telephone and electrical wires snapped, windows shattered, and huge chunks of concrete crashed to the sidewalks. Fires broke out, and the morning that was once sunny grew dark. Smoke and dust filled the air.

Three minutes later the earthquake, which measured 8.1 on the Richter scale, ended.

The greatest destruction occurred almost entirely in the low-lying center of Mexico City. Because this part of the land is an ancient lake bed, it is quite soft. Not only is the land unstable, but it also amplified the shock waves from the quake. As a result, the land in the center of Mexico City rocked back and forth like a huge bowl of jelly.

The destructive nature of the Mexico City earthquake is apparent in the statistics. More than 11,000 buildings were either destroyed or heavily damaged, 31,000 people were left homeless and had to find shelter in tent cities. Nearly 5,000 people lost their lives.

What steps might officials in Mexico City take to limit property damage during future earthquakes?

These rules, if followed quickly, may help prevent death, injury, and property damage.

Before an earthquake occurs, be prepared. Keep on hand a supply of ready-to-eat food, bottled water, flashlights, batteries, and a portable radio. Plan what you will do if an earthquake strikes while you are at home, in school, or in a car. Discuss these plans with your family and friends. Learn how to turn off the gas, water, and electricity in your home.

During an earthquake, stay calm. There are usually a few seconds between tremors, during which you can move to a safer posi-

Career Focus: Earth Forces

"I get . . . satisfaction when I can find out something that no one has discovered before . . . and come up with logical arguments to support my findings."

Tectonophysicist

Back in California after one month of fieldwork in the Soviet Union, Red China, and Japan, tectonophysicist Wayne Thatcher described the rewards of his work.

"When I can close the door, cut out the distractions, and think up new projects to understand how the earth works, I'm a very happy man," he said. "Sometimes I actually feel lucky to get paid for doing something I enjoy so much."

Thatcher, pictured left, studies the structure of the earth's crust and the forces that shape the crust. He is particularly interested in understanding the causes of earthquakes.

"I had a high school physical geography teacher who really got me started in the field. He was interested in theories about how the continents were formed and whether their drifting was important to how the earth's surface formed and changed."

Thatcher's work as chief of the tectonics branch of the United States Geological Survey includes administrative duties and research. He uses data that is gathered by

tion. If you are indoors, protect yourself from falling debris by standing in a doorway or crouching under a desk or table. Stay away from windows, heavy furniture, and other objects that might topple over. Do not run outside. If you are in school, follow the instructions given by your teacher or principal. If you are in an automobile, stop in a place away from buildings, tunnels, and bridges and remain in the car until the tremors cease.

After an earthquake, be cautious. Check for fire and fire hazards. Always wear shoes when walking among broken glass, and avoid downed powerlines and objects touched by downed wires.

observation and by seismic networks worldwide.

"Although we know earthquakes are almost exclusively the result of plate tectonics, more research is needed to define the actual processes involved."

Volcanologist

Volcanologists use modern equipment and techniques to monitor volcanoes and predict their activity. Using a seismograph, for example, a volcanologist, pictured lower left, can detect earthquakes, which often signal eruptions. Currently, most monitoring studies are conducted on volcanoes in Hawaii, Italy, Japan, New Zealand, and Russia.

Volcanologists also study rock formations and past lava flows to determine which way lava might flow in future eruptions. Their findings may help save hundreds of lives and homes from destruction.

Most volcanologists begin their career preparation with a bachelor's degree in geology or geophysics.

Exploration Geophysicist

Exploration geophysicists study the earth's subsurface. They may use their skills to locate sources of fresh water, predict earthquakes, or solve environmental problems. Most exploration geophysicists, however, search for sources of petroleum.

To help them in their work, these exploration geophysicists often rely on seismic surveys. In seismic surveys sound waves are used to detect rock layers where petroleum might be found.

After collecting data about a specific area, exploration geophysicists analyze the data, prepare maps and reports, and determine the best drilling locations.

Entry-level positions in exploration geophysics require a bachelor's degree in geophysics or geology.

For Further Information

For more information about careers in geophysics, write the American Geophysical Union, 2000 Florida Ave., NW, Washington, DC 20009.

Earthquake Warnings and Predictions

One of the earliest means of predicting earthquakes was to observe the behavior of animals. People knew that just before an earthquake, animals appeared nervous and restless, almost as if they could sense the coming vibrations. Today, strange animal behavior is still considered an earthquake warning sign. Unfortunately, the warning given by animals does not come soon enough to be of any practical value. Some areas experience major earthquakes at regular intervals. Using records of past earthquakes, scientists can make approximate predictions of future quakes. However, these predictions may be off by 25 years or more.

To make more accurate predictions, scientists try to detect changes in the earth's crust that can signal an approaching earthquake. Faults near many population centers have been located and mapped. Instruments placed along these faults measure small changes in rock movement around the fault and can detect an increase in strain. Along some faults, scientists have identified zones of immobile rock called **seismic gaps.** A seismic gap is a place where the fault is locked and unable to move. The strain in surrounding rock has increased, and no major earthquake has occurred in this location for at least 30 years. Scientists think seismic gaps are likely locations of future earthquakes. Several gaps that exist along the San Andreas Fault may be the sites of major earthquakes in the near future.

Figure 6–6. Earthquakes have periodically occurred along the San Andreas Fault throughout its life of about 15-20 million years. Seismologists predict that future great earthquakes along the fault will not come suddenly, but will probably be preceded by an increase in seismic activity.

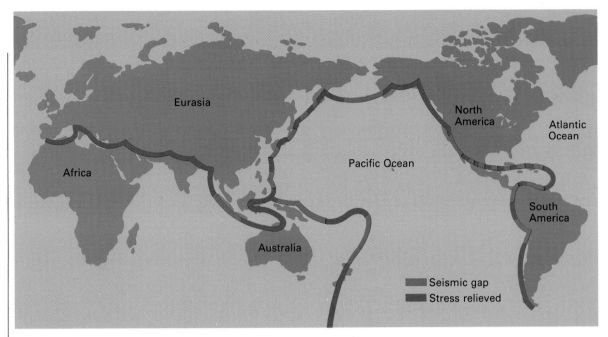

Figure 6–7. The purple areas represent fault zones in which earthquakes have relieved strain. The red areas represent seismic gaps in which strain is building.

Sometimes scientists can detect the slight tilting of the ground that occurs shortly before some earthquakes. They can also detect the strain and cracks in rocks caused by the stress that builds up just before an earthquake. The magnetic and electrical properties of rock change when these cracks fill with water. As a result, scientists may be able to detect small changes in the earth's magnetic field and in the way electricity is conducted by rocks prior to an earthquake. Scientists can also detect increased gas seepage from strained or fractured rocks.

An earthquake is often preceded by a decrease in the speed of local P waves. The P waves being measured have traveled from distant earthquakes. The decrease in the speed of local P waves may last for several days or several years preceding an earthquake. Just before the earthquake occurs, the speed of the local P waves suddenly returns to normal. Evidence even suggests that the longer the decrease in speed lasts, the stronger the earthquake will be.

Scientists would like to be able to control the force of earthquakes. A better understanding of earthquakes and their causes may enable them to develop a system of earthquake control. In fact, tests at Rangely, Colorado, showed that when water was injected along a fault, friction was reduced and earthquakes were smaller.

Section 6.3 Review

1. How do tall buildings usually respond during a major earthquake?
2. What causes tsunamis?
3. What should you do if an earthquake strikes while you are at home? In a car?
4. What are some early warning signs of earthquake activity?
5. What type of building construction and location regulations should be included in the building code of a city located near an active fault?

Reviewing Chapter 6

Key Concepts

- According to the elastic rebound theory, strain builds in rocks along a fault until they break and snap into new positions. See page 85.
- Most earthquakes occur near plate boundaries. See page 86.
- There are three types of seismic waves: P waves, S waves, and L waves. See page 89.
- The difference in times at which P waves and S waves arrive at a seismograph station helps scientists to locate the focus of an earthquake. See page 90.
- The Richter scale expresses the energy released by an earthquake. See page 90.
- Most earthquake damage is caused by the collapse of buildings and other structures. See page 92.
- Tsunamis often accompany ocean-floor earthquakes. See page 92.
- People in areas around active faults should know and follow safety rules in case of possible earthquakes. See page 93.
- Gaps in fault movement, tilting ground, and variations in seismic waves are among the changes in the earth's crust that scientists use in predicting earthquakes. See page 96.

Key Terms

aftershock (85)
deep-focus earthquake (86)
earthquake (85)
elastic rebound theory (85)
epicenter (85)
fault zone (87)
focus (85)
intensity (90)
intermediate-focus earthquake (86)
L wave (89)
Mercalli scale (90)
microquake (90)
P wave (89)
Richter scale (90)
Ring of Fire (86)
S wave (89)
seismic gap (96)
seismograph (89)
shallow-focus earthquake (86)
tsunami (92)

Review

On your paper, write the letter for the term that best completes each of the following statements.

1. Vibrations in the earth caused by the sudden movement of rock are called (a) epicenters (b) earthquakes (c) faults (d) gaps.
2. The elastic rebound theory states that, as a rock becomes strained, it first (a) deforms (b) melts (c) breaks (d) shifts position.
3. The point along a fault where an earthquake begins is called the (a) surface wave (b) epicenter (c) gap (d) focus.
4. The point on the earth's surface directly above the point where an earthquake begins is called the (a) focus (b) epicenter (c) surface wave (d) secondary wave.
5. A characteristic of earthquakes that cause the most damage is (a) a deep focus (b) an intermediate focus (c) a shallow focus (d) a deep epicenter.
6. Most severe earthquakes occur (a) in California (b) in rural areas (c) at plate boundaries (d) in the middle of a plate.
7. The boundary of the Pacific plate scrapes against that of the North American plate and forms (a) a single fault (b) folded mountains (c) a volcano (d) a fault zone.
8. P waves travel through (a) solids only (b) liquids and gases only (c) solids and liquids only (d) solids, liquids, and gases.

9. S waves cannot pass through (a) solids (b) the mantle (c) the earth's outer core (d) the asthenosphere.

10. L waves cause (a) rock particles to move in the direction in which the waves are traveling (b) the surface of the earth to rise and fall (c) rock particles to move horizontally at right angles to the direction in which the waves are traveling (d) little movement or damage.

11. By analyzing the difference in the time it takes for P waves and S waves to arrive at a seismograph station, scientists can determine an earthquake's (a) focus (b) L waves (c) fault zone (d) intensity.

12. The Richter scale expresses an earthquake's (a) magnitude (b) location (c) speed (d) depth.

13. Most injuries during earthquakes are caused by (a) the collapse of buildings (b) cracks in the earth's surface (c) the vibration of S waves (d) the vibration of P waves.

14. A large seismic sea wave is called (a) a tsunami (b) a P wave (c) an L wave (d) an S wave.

15. If an earthquake strikes while you are in a car, you should (a) continue driving (b) get out of the car (c) park the car under a bridge (d) stop the car in a clear space and remain in the car.

16. An earthquake is frequently preceded by a (a) temporary change in the speed of local P waves (b) temporary change in the speed of the surface waves (c) landslide (d) tsunamis.

Application

On your paper, write answers to the following questions.

1. Earthquakes with a very deep focus cannot be explained by the elastic rebound theory. Explain why.

2. Why do many earthquakes occur in the vicinity of mountain ranges?

3. Scientists have discovered that P waves slow down and S waves stop at a depth of 2,900 km. Why does this occur?

4. If a seismograph station measures P waves but no S waves from an earthquake, what can you conclude about the location of the earthquake?

5. Two cities are struck by earthquakes. The cities are the same size, are built on the same type of ground, and have the same types of buildings. The city in which the quake measured 4 on the Richter scale suffered $1 million in damage. The city in which the quake measured 6 on the scale suffered $50 million in damage. What might account for this difference?

6. You are going to choose a building site for a home. You would like a high place with a view, but you are concerned about earthquakes. What information do you need to make an informed decision about the site?

7. Would an earthquake in the interior of China be likely to form a tsunami? Explain why.

8. Why is it wise to stand in a doorway when an earthquake strikes?

9. Explain the relationship between the elastic rebound theory and the reason scientists monitor seismic gaps.

10. Imagine you are monitoring a seismograph station along the San Andreas Fault. An earthquake occurs in Mexico, and you notice that the P waves you are recording from that quake have a velocity that is less than normal. What does this tell you about the area around your seismograph station?

Extension

1. Research major earthquake activity along the San Andreas Fault over the past five years. Write a brief report on your findings.

2. Find out how and why the worldwide network of seismograph stations was formed. Also find out how all the stations in the network work together. Report your findings.

3. Find out if there is a building code designed to minimize earthquake damage in your city. Summarize the regulations within the code.

Chapter 7

Volcanoes

Mount Vesuvius, Krakatoa, Mount St. Helens, Olympus Mons—these are some of the best-known volcanoes in the solar system. Volcanoes are powerful and sometimes destructive reminders that the earth and other planetary bodies have been, and in some cases still are, geologically active.

In this chapter you will learn about the mechanics, the power, and the effects of volcanic activity.

Chapter Outline

In 1980 Mount St. Helens exploded in a volcanic eruption.

7.1 Volcanoes and Plate Tectonics

Section Objectives

- **Describe the formation and movement of magma.**
- **Define** *volcanism.*
- **List three locations where volcanism occurs.**

Scientists have no direct way to measure temperatures deep within the earth. However, analysis of seismic waves allows scientists to estimate those temperatures. Figure 7–1 shows estimates of the earth's inner temperatures and pressures, based on the behavior of seismic waves. As this graph shows, the combined temperature and pressure in the lower part of the mantle keep the rocks there below their melting point.

Despite the high temperature in the asthenosphere, most of this zone remains solid because of the great pressure of the surrounding rock. Sometimes, however, areas of the solid rock will melt, forming **magma,** or liquid rock. Geologists think that magma forms in areas where the surrounding rock exerts less-than-normal pressure. The lower pressure allows the particles of the rock to move more quickly; thus the rock becomes liquid.

Volcanism

Any activity that includes the movement of magma toward or onto the surface of the earth is called **volcanism.** Pockets of magma grow due to melting of some of the surrounding rock. As more rock melts, the magma pockets expand. The magma slowly pushes upward into the crust because magma is less dense than solid crustal rock. The magma slowly rises, forcing its way into cracks in the overlying rock. This process causes large blocks of overlying rock to break off and melt, adding still more material to the magma pockets.

Magma also forms at plate boundaries, where one lithospheric plate, usually of oceanic crust, is subducted beneath another plate, often continental crust. The subducting plate is forced deep into the asthenosphere, where parts of it melt to become magma.

Sometimes magma breaks through to the surface of the earth. Magma that erupts onto the earth's surface is called **lava.** The opening through which the molten rock flows onto the surface is called a **vent.** The vent and the volcanic material that builds up on the earth's surface around the vent is called a **volcano.**

Major Volcanic Zones

If you were to plot the location of the 600 or so volcanoes that have erupted within the past 50 years, you would see that they form a pattern across the earth. Most of these active volcanoes are found in zones near both convergent and divergent boundaries of the lithospheric plates.

Subduction Zones

Many volcanoes are located along subduction zones, where one plate is forced under another plate. When a plate with oceanic crust meets a plate with continental crust, the oceanic crust, which is more

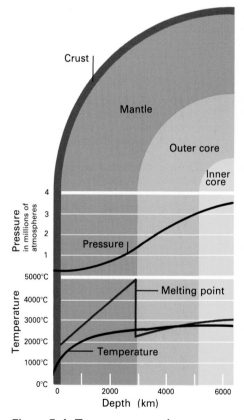

Figure 7–1. Temperature and pressure increase as the depth beneath the earth's surface increases.

dense, is forced beneath the continental crust. A deep trench forms on the ocean floor along the edge of the continent where the plate is being subducted. The plate with the continental crust buckles and folds, forming a line of mountains along the edge of the continent.

The subducted plate is forced deep into the hot mantle. There, heat from both the mantle and the friction between the moving plates begins to melt the subducted plate. Part of the melting plate becomes magma. Some of this magma erupts through the earth's surface, forming volcanic mountains along the edge of the continent.

A major zone of active volcanoes formed by subducting plates encircles the Pacific Ocean. This zone, called the *Ring of Fire*, results from plates subducting along the Pacific coasts of North America, South America, and Asia. As you read in Chapter 6, the Ring of Fire is also one of earth's three major earthquake zones.

If two plates with oceanic crust at their boundaries collide, one of the plates is subducted, forming a deep trench. As the plate descends into the mantle, some of it melts. The resulting magma breaks through to the surface along the trench. In time, a string of volcanic islands, called an *island arc,* forms along the trench. The early stages of this type of subduction produce a series of small volcanic islands. One example is the Aleutian Islands, which stretch across the north Pacific Ocean. As more magma surfaces, more islands appear, become larger, and join. Because of the way they are formed, island arcs are sites of frequent volcanic eruptions and earthquakes. The group of volcanic islands that make up Japan is an example of an older island arc.

Figure 7–2. When oceanic crust is subducted beneath continental crust (bottom, left), volcanoes often form along the edge of the continent. Similarly, when oceanic crust is subducted beneath oceanic crust (bottom, right), lava erupts from undersea volcanoes and forms islands, such as the Aleutians shown above.

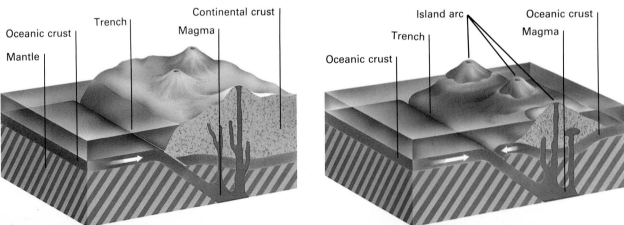

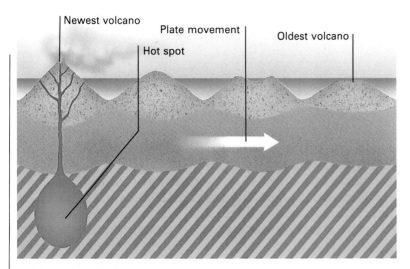

Newest volcano

Plate movement

Hot spot

Oldest volcano

Figure 7–3. Lithospheric plate movement over a hot spot produces a chain of volcanic islands. The islands farthest away froml the hot spot are the oldest. Eventually plate movement may carry these older islands beneath the surface of the ocean.

Mid-Ocean Ridges

The greatest amount of magma comes to the surface where plates are moving apart along mid-ocean ridges. Thus, a major zone of volcanoes is the interconnected mid-ocean ridges that circle the earth. Fractures between the spreading plates along the ridges reach down to the asthenosphere. Magma rises through the fractures and comes to the surface through long, narrow cracks called *rifts*. The emerging lava forms new ocean floor and adds material to the ridges.

Most volcanic eruptions along the mid-ocean ridges go unnoticed because they take place deep beneath the surface of the ocean. An exception is found on Iceland. Iceland is an exposed part of the Mid-Atlantic Ridge. One half of Iceland is on the North American plate and is moving westward. The other half is on the Eurasian plate and is moving eastward. Iceland is covered with large **fissures,** cracks through which lava flows. What do you think causes the fissures in Iceland?

Hot Spots

Not all volcanoes develop along plate boundaries. Sometimes magma works its way to the earth's surface within the interiors of lithospheric plates. These areas of volcanism within plates are called **hot spots.** Hot spots remain stationary. However, the lithospheric plate above a hot spot continues drifting slowly. As a result, the volcano on the surface is eventually carried beyond the hot spot. The activity of the volcano ceases because there is no longer any magma underneath to feed it. A new volcano then forms where new crust has moved over the hot spot, as shown in Figure 7–3. The Hawaiian Islands are an example of a chain of volcanic islands formed over a hot spot.

Section 7.1 Review

1. What must happen for volcanism to occur?
2. Explain how magma reaches the surface.
3. How does subduction produce magma?
4. What would be a likely explanation for the onset of volcanic activity in the central United States?

7.2 Volcanic Eruptions

Volcanoes can be thought of as windows to the interior of the earth. The lava that erupts from them provides an opportunity for scientists to study firsthand the materials that form deep within the earth's mantle. By analyzing the types of minerals found in hardened lava, geologists have concluded that there are three general types of lava. One type of lava is dark colored when hardened and rich in magnesium and iron. This type of lava is called **mafic lava.** The second type of lava contains much silica along with iron and magnesium. It has a lighter color when hardened and is called **felsic lava.** The third type of lava has a range of compositions that falls between the mafic and the felsic varieties.

Thin, mafic lava usually hardens with a wrinkled surface. This type of solidified lava is known as **pahoehoe** (puh-HOEEE-ʜoᴇᴇᴇ), which means "ropey" in Hawaiian. Rapid cooling on the surface of a lava flow can form a crust that breaks into jagged chunks as the liquid below continues to flow. The lava deposit that remains is described by its Hawaiian name **aa** (AH-ah). The term refers to the sharp, blocky shapes into which the hardened lava breaks. Sometimes the outer part of a mafic lava flow cools so rapidly that it forms a hardened shell around a liquid interior. This liquid material flows out, leaving tunnels in the hardened lava shell. When lava flows out of fissures on the ocean floor, it cools rapidly, often in rounded shapes. This formation is called **pillow lava.** Pillow lava is commonly found along mid-ocean ridges.

Kinds of Eruptions

The composition of the lava that reaches the surface largely determines the force with which a particular volcano will erupt. Why would lava that contains large amounts of dissolved gases usually produce a more explosive eruption than lava that contains few dissolved gases?

Figure 7–4. Pahoehoe lava has a wrinkled surface.

Oceanic volcanoes, both those that erupt beneath the ocean and those that erupt on islands, usually are produced by mafic lava. Mafic lava is thin and flows almost as easily as water. Because gases can easily escape from mafic lava, eruptions from oceanic volcanoes are usually quiet. That is, the lava flows out from the volcanic opening like a red-hot river.

In contrast to the fluid lavas produced by oceanic volcanoes, the felsic lavas of continental volcanoes tend to be thick. They also contain a large amount of gases, mostly water vapor and carbon dioxide. When a vent or fissure opens up, the dissolved gases within the lava boil out explosively, sending molten and solid particles shooting out of the opening.

Figure 7–5. Lava flows from a quiet eruption (left) like a red-hot river. During an explosive eruption, lava, steam, ash, and other volcanic material are ejected violently from the volcano (right).

Volcanic Rock Fragments

Unlike mafic lava, which tends to flow smoothly, felsic lava explodes, throwing **tephra** into the air. Tephra, sometimes called **pyroclastics,** is volcanic rock fragments ejected from a volcano. Some tephra forms when cooling magma breaks into fragments because of the rapidly expanding gases within it. Other tephra forms when a spray of lava cools and solidifies as it flies through the air.

Tephra particles less than 2 mm in diameter make up **volcanic ash.** Particles less than 0.25 mm in diameter are called **volcanic dust.** After a volcanic eruption, wind usually carries dust and ash away from the volcano. Most of the volcanic dust and ash settles on the land immediately surrounding the volcano. Some of the smallest particles, however, may travel completely around the earth in the upper atmosphere.

Larger tephra particles, less than 64 mm in diameter, are called **lapilli** (luh-PIL-ie), from a Latin word that means ''little stones.'' Lapilli generally fall near the vent.

Large clots of lava are thrown out of an erupting volcano while they are red-hot. As the clots spin through the air, they cool and develop a round or spindle shape. This tephra is called **volcanic bombs.** The largest tephra, formed from solid rock blasted from the fissure, is known as **volcanic blocks.** Some volcanic blocks are as big as houses.

Volcanic Features

Volcanic activity produces a variety of characteristic features on the earth's surface. These features are formed during both quiet and explosive eruptions. For example, the lava and tephra ejected during volcanic eruptions can create massive rock piles around the vent. These rock piles, known as *volcanic cones,* are classified into three main types.

Volcanic Cones

Volcanic cones that are broad at the base and have gently sloping sides are called **shield cones.** A shield cone covers a wide area and generally results from a quiet lava eruption. Layers of thin lava flow out around the vent, harden, and slowly build up to form the cone. The Hawaiian Islands are actually a cluster of shield cones built up from the ocean floor.

Explosive eruptions form different types of cones. Many explosive eruptions form **cinder cones.** A cinder cone is made of solid

Figure 7–6. Shield cones (top) have broad bases and gentle slopes. Cinder cones (middle) have narrow bases and very steep slopes. Composite cones (bottom) are both broad and steep. The photo shows a cinder cone at Lava Beds National Monument in California.

layers of lava

Shield cone

layers of cinders

Cinder cone

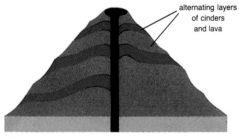

alternating layers of cinders and lava

Composite cone

106

Investigation

Volcanic Cones

A cinder cone is formed by material thrown out during an explosive volcanic eruption. A shield cone is made up of layers of lava that has flowed out during a quieter eruption. You can demonstrate that the very different shapes of these cones is the result of the different materials that forms them.

Science Process Skills Focus: constructing models, analyzing, inferring, comparing

Materials
bowl, teaspoon, plaster of Paris, measuring cup, water, mixing spoon, ruler, 2 paper plates, protractor, cereal flakes

Procedure
1. In a bowl, add 2 teaspoons of plaster of Paris to 60 mL of water. Stir until smooth.
2. Hold the measuring cup 2 cm over a paper plate. Pour the contents of the cup slowly and steadily onto one spot on the plate.
3. Wait 5–10 minutes for the cone to harden. Measure the angle of its slope with a protractor, as shown in the figure.
4. Pour dry cereal flakes slowly onto one spot on a clean plate until the mound is 5 cm high.

5. Without disturbing the mound, measure its slope with a protractor.

Analysis and Conclusions
1. Which cone that you formed represents a cinder cone? Which represents a shield cone? How do the angles formed by these cones compare?
2. How would the slope have been affected if the cereal was rounder? Thicker?
3. Suppose you formed a cone by pouring alternating layers of wet plaster of Paris and dry cereal. How would the shape and overall size of this cone differ from the two cones you have formed? Which type of volcano would such a cone look like?

---

fragments ejected from the volcano. Cinder cones have very steep slopes, often close to 40°. Because the fragments are loosely arranged and thus tend to roll down the slope easily, cinder cones are rarely more than a few hundred meters high.

Many volcanoes have both quiet eruptions and explosive eruptions. During a quiet eruption, the cone is formed mainly by lava flows, the type that usually result in shield cones. Then an explosive eruption occurs, depositing large amounts of tephra around the vent. The explosive eruption is followed again by quiet lava flows. Thus, the resulting cone is formed of alternating layers of hardened lava flows and tephra. These **composite cones,** also known as **stratovolcanoes,** often develop into high volcanic mountains. Some of the best-known volcanic mountains in the world are composite mountains. Among these are Mount Fuji in Japan and Mounts Rainier, Hood, Shasta, and St. Helens in the United States.

Figure 7–7. When calderas fill with water, they form scenic lakes such as the one at Crater Lake National Park, shown above.

Craters and Calderas

The funnel-shaped pit at the top of a volcanic vent is known as the **crater.** The crater is formed when material is blown out of the volcano by explosions. A crater usually becomes wider as magma melts and breaks down the walls of the crater, allowing loose materials to collapse back into the vent. Sometimes a smaller cone forms within a crater from materials erupting from the vent.

When the magma chamber below a volcano is emptied, the volcanic cone may collapse. Explosions may also completely destroy the upper part of the cone, leaving a large, basin-shaped depression called a **caldera** (cal-DER-uh). Krakatoa, a volcanic island group in Indonesia, is an example of a caldera. When the volcanic cone exploded in 1883, a caldera with a diameter of 6 km formed. This huge caldera changed the shape of the entire island.

Predicting Volcanic Eruptions

A volcanic eruption can be one of the earth's most destructive natural phenomena. When Mount St. Helens exploded in 1980, the entire surrounding area, including a great forest and large lake, were completely devastated. The eruption at Krakatoa exploded away more than half the island and produced tsunami waves that killed tens of thousands of people. The entire city of Pompeii was stilled by the power of a volcano. If volcanic activity could be predicted far enough in advance, many lives might be saved.

Scientists have made some progress in their efforts to predict the eruptions of active volcanoes. They use sensitive instruments to detect geological events that may signal the beginning of an eruption. For example, seismographs are used to monitor one of the most important warning signals—small earthquakes. These small earthquakes result from the growing pressure on the surrounding rocks as magma works its way upward. Temperature changes within the rock and the actual fracturing of the rock surrounding a volcano also contribute to the small earthquakes. The number of earthquakes often increases until they occur almost continuously just before an eruption. An increase in the strength of such earthquakes may also be a signal that an eruption is about to occur.

Another signal of volcanic activity scientists watch for is slight bulges in the sloping surface of a volcano. Before an eruption, the upward movement of magma beneath the surface may push out the surface of the volcano. These bulges slightly alter the distance between two marked points on the slope, just as spots on a balloon would move apart as the balloon is inflated. Special instruments can measure these small changes in distance as well as changes in the tilt of the ground surface. Another possible signal of an increase in volcanic activity that is being investigated is changes in the composition of the gases given off by volcanoes.

Accurately predicting the eruption of a particular volcano also requires knowledge of previous eruptions. To provide the best forecast, scientists compare the past behavior of the volcano with current

IMPACT

Krakatoa

For centuries, Krakatoa, a volcanic island group between Java and Sumatra, Indonesia, was geologically inactive. In late May of 1883, however, that calm was shattered by the first of a series of volcanic eruptions. The eruptions continued throughout the summer, growing increasingly violent. The climax came on August 27, when a tremendous explosion destroyed the islands, propelling ash and rock 80 km into the sky. The blast was heard 3,500 km away in Australia.

The eruption had a catastrophic impact. Twenty cubic kilometers of rock and great quantities of ash fell over an area of 800,000 km². A thick cloud of ash kept the area in total darkness for two and a half days. Volcanic ash buried all life on what was left of the islands.

The Krakatoa eruption triggered a series of tsunamis. One wave, estimated at 36 m high, took 36,000 lives in Java and Sumatra.

The effects of the blast continued long after the initial eruption. Fine dust and ash, which reflected sunlight, circled the earth, causing dramatic sunrises and sunsets for two years. This dust and ash also blocked some of the sun's rays for several years, causing global temperatures to drop by as much as 0.27°C. The climate worldwide remained below normal well into the 1890's, possibly as a result of the spectacular destruction of Krakatoa.

If you examined the remains of Krakatoa beneath the surface of the ocean, what volcanic feature would you expect to find?

daily measurements of earthquakes, surface bulges, and changes in gases. Unfortunately, only a few of the active volcanoes in the world have been scientifically studied long enough to establish any patterns in their activity. Also, volcanoes that have been dormant for long periods of time may, with little warning, suddenly become active.

Section 7.2 Review

1. Explain the difference between mafic and felsic lava.
2. Define *tephra* and list the major types.
3. Name and compare the three types of volcanic cones.
4. What event forms a caldera?
5. Name three events that may signal a volcanic eruption.
6. Would quiet eruptions or explosive eruptions be more likely to increase the height of a volcano? Why?

7.3 Extraterrestrial Volcanism

Many people think of the other planets in the solar system as geologically dead worlds. However, evidence gathered by human exploration and data sent back by various spacecraft now shows differently. Indications are that many of the planets and moons in the solar system, including the earth's moon, were volcanically active in the past. Some of these planets and moons remain active. At least one of them, Io, a moon of Jupiter, has far more volcanic activity than does the earth.

The Moon

The earth's moon is covered with basaltic lava flows. This is evidence that sometime in the past, active volcanoes dotted the surface of the moon. Most of the craters on the lunar surface result from meteorite bombardment. However, some of them have smooth, lava-flow interiors and gently sloping, channelled exteriors. These features suggest that the craters were once active volcanoes.

Questions yet to be resolved by scientists are how magma formed in the lunar interior and how it reached the surface. There is no evidence of plate tectonics or convection currents on the moon, so the magma must have been produced in other ways. Great heat would have been needed to produce both the magma in the upper layers of the moon and the cracks through which it flowed onto the lunar surface. Some scientists think this heat may have come from a long period of intense meteorite bombardment. How might meteorite bombardment have produced such great heat?

Mars

Spacecraft have landed on Mars and have sent back photographs that show numerous volcanoes and volcanic features on the Martian surface. The largest of these is the shield volcano Olympus Mons.

Figure 7–8. The structure of lunar craters, like the one shown on the left suggests past volcanism on the moon. The immense Martian volcano Olympus Mons (right) may still be active.

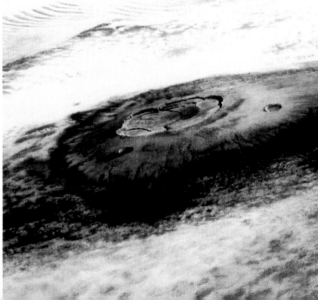

Olympus Mons rises nearly 28 km above the Martian surface. Its base measures 600 km across, and its caldera, 70 km across. Scientists think the volcano has grown to such a tremendous size because, unlike the earth's crust, the Martian crust does not drift. Consequently, Olympus Mons stayed over the lava source for perhaps millions of years with the lava continuously building the size of the volcano.

Whether Martian volcanoes are still active is a question scientists have yet to answer. However, Mars does seem to be seismically active. A Viking landing craft searching Mars for signs of life detected two geological events that produced waves similar to those of an earthquake. These "marsquakes" may also mean that the Red Planet is still volcanically active.

Io

Io, one of the many moons of Jupiter, is the first planetary body (other than earth) on which active volcanoes have been sighted. In 1979, the spacecrafts *Voyager 1* and *Voyager 2* flew by the moons of Jupiter. The photographs and other data they transmitted back to the earth showed nine volcanoes erupting on Io.

Io, about the size of the earth's moon, is probably the most volcanically active body in the solar system. Scientists calculate that volcanoes on Io eject several thousand metric tons of material each second. This means that every month the surface of Io is covered with volcanic material equal to that ejected by Mount St. Helens in May 1980.

Unlike volcanoes on earth, the material that erupts from the volcanoes on Io is neither mafic nor felsic. Because Io is colored a brilliant yellow-red, scientists think the volcanic material is primarily sulfur and sulfur dioxide.

Volcanoes on Io appear to be much more powerful than those on earth. Plumes of volcanic material reach heights of hundreds of kilometers. In general, the plumes are shaped like giant umbrellas.

Io moves inward and outward in its orbit around Jupiter because of the gravitational pull of the other moons of Jupiter. As Io is pulled back and forth, its surface also moves in and out. Heat from the friction caused by this surface movement probably results in the melting of the interior of Io and leads to volcanism.

Figure 7–9. This computer-enhanced photo provided some of the first visible evidence of active volcanism on Io, one of the moons of Jupiter.

Section 7.3 Review

1. What is responsible for most of the craters on the earth's moon?
2. What is the most likely explanation for the growth of Olympus Mons to its present size?
3. Why do scientists think Io, one of the moons of Jupiter, is yellowish-red in color?
4. What does the discovery of active volcanoes on other planets or moons suggest about the origin and development of most of the planets and moons in the solar system?

Reviewing Chapter 7

Key Concepts

- Magma develops in the asthenosphere. See page 101.
- Volcanism is the movement of magma toward the earth's surface. See page 101.
- Volcanism is common at the boundaries of lithospheric plates. See page 101.
- The collision of lithospheric plates may produce continental volcanoes or island arcs. See page 102.
- There are three types of lava, each with a different chemical composition. See page 104.
- Thin lava is associated with quiet eruptions; thick gas-containing lava is associated with explosive eruptions. See page 105.
- A volcano ejects a variety of tephra during an explosive eruption. See page 105.
- Volcanic cones are classified into three categories according to composition and form. See page 106.
- Events that may signal a volcanic eruption include changes in earthquake activity, surface changes in the volcano, and changes in the composition of gases given off by the volcano. See page 108.
- Evidence of volcanism is common throughout our solar system. See page 110.
- Volcanism on Io differs significantly from volcanism on earth. See page 111.

Key Terms

aa (104)	hot spot (103)	pyroclastics (105)	volcanic bomb (106)
caldera (108)	lapilli (105)	shield cone (106)	volcanic dust (105)
cinder cone (106)	lava (101)	stratovolcano (107)	volcanism (101)
composite cone (107)	mafic lava (104)	tephra (105)	volcano (101)
crater (108)	magma (101)	vent (101)	
felsic lava (104)	pahoehoe (104)	volcanic ash (105)	
fissure (103)	pillow lava (104)	volcanic block (106)	

Review

On your paper, write the letter of the term that best completes each of the following statements.

1. Factors that allow magma to push its way upward include temperature and (a) color (b) density (c) crust (d) thickness.
2. Activity caused by the movement of magma is called (a) extraterrestrial (b) tephra (c) volcanism (d) subduction.
3. The belt of volcanoes that encircles the Pacific Ocean is called (a) the subduction zone (b) an island arc (c) a hot spot (d) the Ring of Fire.
4. Island arcs are formed by the collision of (a) two plates with continental crust at their edges (b) two pyroclastics (c) two volcanic bombs (d) two plates with oceanic crust at their edges.
5. The greatest amount of magma reaches the surface of the earth along (a) mid-ocean ridges (b) subduction zones (c) island arcs (d) hot spots.
6. Magma comes to the earth's surface through long, narrow openings called (a) subduction zones (b) cones (c) calderas (d) fissures.

7. Areas of volcanism within plates are called (a) hot spots (b) pyroclastics (c) cones (d) fissures.
8. Lava that breaks into jagged chunks when it is subjected to rapid cooling is called (a) aa (b) pahoehoe (c) pillow lava (d) felsic lava.
9. Explosive volcanic eruptions result from (a) mafic lava (b) tephra lava (c) felsic lava (d) pahoehoe lava.
10. Tephra that forms into rounded or spindle shapes as it flies through the air is called (a) ash (b) lapilli (c) volcanic bombs (d) volcanic blocks.
11. The Hawaiian Islands are formed from (a) shield cones (b) cinder cones (c) composite cones (d) craters.
12. A cone formed only by solid fragments built up around a volcanic opening is a (a) shield cone (b) cinder cone (c) composite cone (d) stratovolcano.
13. The depression that results when a cone collapses into an empty magma chamber is a (a) crater (b) vent (c) caldera (d) fissure.
14. Shortly before a volcano erupts, magma may cause the surface of the volcano to (a) bulge out (b) cave in (c) get darker (d) melt.
15. Scientists have discovered that before an eruption earthquakes (a) completely stop (b) increase in number (c) bear no relation to volcanism (d) move to other locations.
16. Studying the behavior pattern of volcanoes helps scientists to (a) prevent eruptions (b) control eruptions (c) predict eruptions (d) channel eruptions.
17. Olympus Mons, the largest known volcano in the solar system, is found on (a) the moon (b) Io (c) Mars (d) Venus.
18. The material ejected from volcanoes on Io is probably (a) sulfur and sulfur dioxide (b) basalt (c) granite (d) lapilli.

Application

On your paper, write answers to the following questions.

1. Most magma that reaches the earth's surface goes unnoticed and unobserved. Why does this happen?
2. Can you assume that every mountain located along the edge of a continent is a volcano? Explain why.
3. Why would felsic lava not produce a natural tunnel within itself as it cools?
4. While exploring a volcano that has been dormant, you observe volcanic ash first and lapilli later. Are you more likely to be moving toward or away from the volcanic opening? Explain your answer.
5. If you see a steep volcanic cone that is only 300 m high, what can you assume about the type of cone and its composition?
6. To predict a volcanic eruption, what kinds of information would you seek?
7. If craters on a planetary body have channels in their sloping sides, were the craters more likely to be the result of meteorite bombardment or volcanism? Explain why.
8. There is no evidence of craters produced by meteorite bombardment on Io, even though the other moons of Jupiter are cratered. What might be the explanation for the craterless surface of Io?

Extension

1. Research major volcanic activity in the United States during the last five-year period. Write a short report on your findings. If possible, collect photographs to illustrate the report.
2. Draw a series of diagrams that depicts the usual order of events in the formation of a subduction-zone volcano. Start with lithospheric plates colliding and end with a cone that has been worn away.
3. Use modeling clay of various colors to build models of the three major types of volcanic cones. Label each part.
4. Prepare a report on the latest evidence for extraterrestrial volcanism.

Composition of the Earth

Capitol Reef National Park, Utah (background): Oil drilling rig. (inset)

Introducing Unit Three

S cientists at the U.S. Geological Survey were among the first to describe the minerals found in lunar samples collected during the Apollo 11 Mission. I will never forget the excitement at the first Lunar Science Conference in January 1970. Hundreds of dignitaries and scientists from around the world had come to hear what our findings would be.

When you learn something brand new about the earth's chemistry and composition, there's that same kind of excitement. Although scientists have been studying our planet for a long time, the possibilities for new discoveries remain endless. The earth still poses so many questions. The answer to one question often leads to a dozen more questions to take its place.

In this unit, you will explore some of the answers that have been discovered about the composition of the earth as well as explore some of the questions that remain.

Malcolm Ross

Malcolm Ross
Research Minerologist
U.S. Geological Survey

8 Earth Chemistry

9 Minerals of the Earth's Crust

10 Rocks

11 Resources and Energy

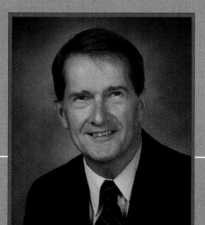

Chapter 8

Earth Chemistry

Scientists have long sought to understand the universe by identifying and examining increasingly smaller parts of it. Scientists have studied parts as large as the whole earth. They have classified the material that makes up the earth into solids, liquids, and gases. In their search for smaller and smaller parts, scientists have turned their attention to studying atoms and their interactions.

Chapter Outline

8.1 Matter
Atoms and Elements
Atomic Structure
Isotopes
Solids, Liquids, and Gases

8.2 Combinations of Atoms
Electron Energy Levels
Chemical Bonds
Chemical Formulas
Mixtures

This scene is from Geyser Basin in Yellowstone Park, Wyoming.

8.1 Matter

Every object in the universe is made up of particles, or basic units, of some kind of substance. Scientists use the word **matter** to describe the substance of which an object is made. Matter is anything that takes up space and has *mass*. The amount of matter in any object is the mass of that object. Scientists are able to identify the kind of matter that makes up a substance by observing the properties of that substance.

Each kind of matter has two types of properties. **Physical properties** are those characteristics that can be observed without changing the composition of the substance. For example, physical properties include color, hardness, freezing point, boiling point and the ability to conduct electricity. **Chemical properties** are those characteristics that describe how a substance interacts with other substances to produce different kinds of matter. For example, a chemical property of iron is that it interacts with oxygen to form rust. A chemical property of nitrogen is that nitrogen gas usually does not interact with other substances. Understanding the chemical properties of a substance requires some basic information about the particles that make up the substance.

Atoms and Elements

All matter is made up of *elements*. An element is a substance that cannot be broken down into a simpler form by ordinary chemical means. Figure 8–1 shows the most common elements in the earth's crust. Notice that there is a universally understood symbol of one or two letters that represents each element. Some 90 elements occur

Section Objectives

- State the distinguishing characteristic of an element.
- Describe the basic structure of an atom.
- Define atomic number and mass number.
- Explain what an isotope is.
- Compare solids, liquids, and gases.

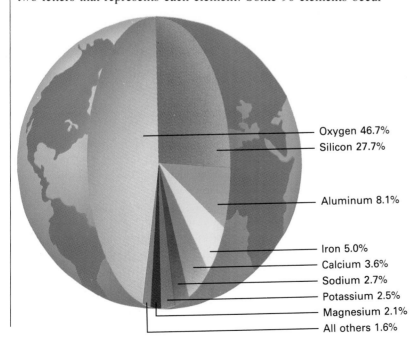

Figure 8–1 The earth's crust is made up primarily of these elements. The graph shows the percentage by weight of each element.

Oxygen 46.7%
Silicon 27.7%
Aluminum 8.1%
Iron 5.0%
Calcium 3.6%
Sodium 2.7%
Potassium 2.5%
Magnesium 2.1%
All others 1.6%

Figure 8–2 Quartz is made of silicon and oxygen, the two most abundant elements in the earth's crust.

naturally in the earth. More than a dozen others have been created in the laboratory. Of the natural elements, only a few make up most of the earth's crust. In fact, two elements—silicon and oxygen—make up almost 75 percent of the earth's crust.

Elements cannot be broken down into a simpler form because each element consists of only one basic kind of **atom.** An atom is the smallest unit of an element and has all the properties of that element. The word *atom* comes from a Greek word for "indivisible." You can identify an element by its atoms. The atoms of any one element are significantly different from the atoms of all other elements.

An atom is so small that the size of a single atom is difficult to imagine. To get an idea of how small an atom is, look at the thickness of this page. More than a million atoms lined up side by side would about equal that thickness.

Investigation

Electrically Charged Objects

Electrically charged objects behave in predictable ways. Objects with unlike charges attract each other; objects with like charges repel each other. Objects with a neutral charge neither attract nor repel each other. You can demonstrate this behavior with several common objects.

Science Process Skills Focus: demonstrating, observing, describing, explaining, inferring

Materials
balloon, thread, ruler, rubber or plastic comb, drinking glass, wool cloth, plastic bag

Procedure
1. Hang the inflated balloon from the ruler by a piece of thread, as shown in the figure. Hold the comb next to the balloon and observe.
2. Repeat Step 1 using the glass and not the comb.
3. Rub the balloon and the comb with the wool cloth. Hold the comb next to the balloon.
4. Rub the glass with the plastic bag. Hold the glass next to the balloon.

Analysis and Conclusions
1. Describe the behavior of the balloon, comb, and glass in Steps 1 and 2. What can you conclude about their electrical charges?

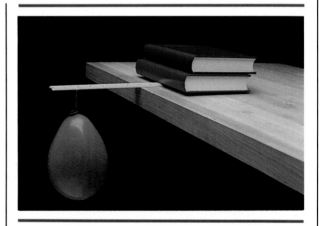

2. Describe the behavior of the balloon and comb in Step 3. What can you conclude about their electrical charges?
3. Describe the behavior of the balloon and glass in Step 4. What can you conclude about their electrical charges?
4. Rubbing the glass with the plastic bag produces a positive electrical charge on the glass. What is the charge of the balloon in Steps 3 and 4? What is the charge of the comb in Step 3?
5. Two sets of ions, such as sodium and chloride, may form a chemical bond. What can you infer about the charges of the ions?

Atomic Structure

As tiny as atoms are, they are made up of even tinier particles called subatomic particles. The three major kinds of subatomic particles are **electrons, protons,** and **neutrons.** Electrons carry a negative electrical charge, protons carry a positive charge, and neutrons have no electrical charge.

Subatomic particles are arranged in a similar way within all atoms. The protons and neutrons of an atom are packed close to one another. Together they form the **nucleus,** a small region in the center of an atom. Because neutrons do not carry an electrical charge, the protons give a nucleus a positive charge. Compared with the size of the whole atom, the nucleus takes up about the same percentage of space as a gumdrop takes up in a football stadium.

The electrons of an atom move in a certain region of space around the nucleus known as an **electron cloud.** Because unlike electrical charges attract, the negatively charged electrons are attracted to the positively charged nucleus. Consequently, the electrons tend to remain relatively close to the nucleus of an atom. However, the constant motion of the electrons keeps them from falling into the nucleus.

An atom of a specific element is distinguished from the atoms of all other kinds of elements by the number of its protons. For example, an atom of the element lithium contains three protons. No other element has three protons.

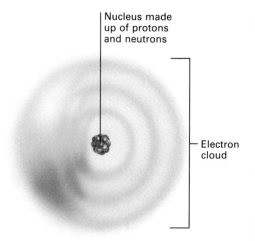

Nucleus made up of protons and neutrons

Electron cloud

Figure 8–3 The nucleus of the atom contains the protons and neutrons. The protons give the nucleus a positive charge. The negatively charged electrons are in the electron cloud around the nucleus.

Atomic Number

To help in the identification of elements, scientists have assigned an **atomic number** to each kind of atom. All atoms of any one element have the same atomic number. The atomic number is equal to the number of protons in the atom. An uncharged atom has an equal number of protons and electrons and so has neither a positive nor negative charge. Therefore, the atom is electrically neutral. Thus, the atomic number also equals the number of electrons in an uncharged atom of that element. For example, the atomic number of oxygen, which is 8, shows that an oxygen atom has eight protons and eight electrons. The atomic number of 1 for hydrogen shows that it has only one proton and one electron.

Atomic number is one of the components listed for each element in the periodic table. The periodic table is a system for classifying elements.

Mass Number

Each kind of atom has also been assigned a **mass number,** which is the sum of the number of protons and neutrons in the atom. For example, the mass number of oxygen is 16. Mass number is expressed by using a scale on which the mass of a proton is assigned the value 1. A neutron, which has nearly the same mass as a proton, also has a mass of 1. In contrast, electrons are much lighter than protons and neutrons. The combined mass of about 1,840 electrons

The Periodic Table

Metals

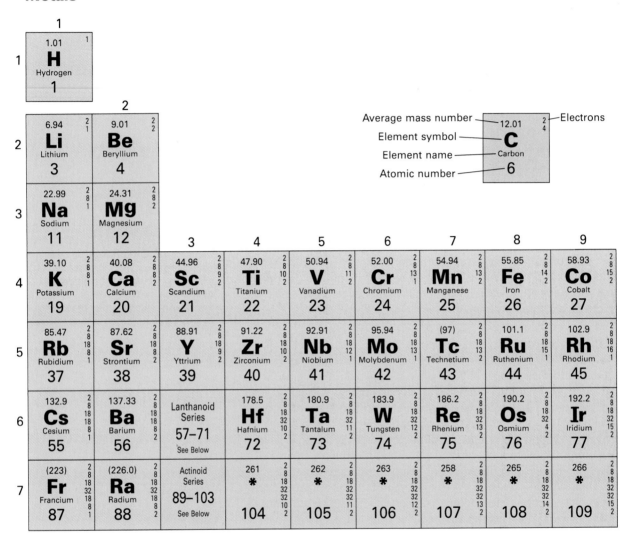

The Rare Earth Elements

Lanthanoid Series						
La Lanthanum 57	Ce Cerium 58	Pr Praseodymium 59	Nd Neodymium 60	Pm Promethium 61	Sm Samarium 62	Eu Europium 63

Actinoid Series						
Ac Actinium 89	Th Thorium 90	Pa Protactinium 91	U Uranium 92	Np Neptunium 93	Pu Plutonium 94	Am Americium 95

A value given in parentheses denotes the mass number of the isotope of longest known half-life.

* Elements synthesized, but not officially named.

Nonmetals

			18
			4.00 2 **He** Helium 2

13	14	15	16	17	
10.81 2/3 **B** Boron 5	12.01 2/4 **C** Carbon 6	14.01 2/5 **N** Nitrogen 7	16.00 2/6 **O** Oxygen 8	19.00 2/7 **F** Fluorine 9	20.18 2/8 **Ne** Neon 10
26.98 2/8/3 **Al** Aluminum 13	28.09 2/8/4 **Si** Silicon 14	30.97 2/8/5 **P** Phosphorus 15	32.06 2/8/6 **S** Sulfur 16	35.45 2/8/7 **Cl** Chlorine 17	39.95 2/8/8 **Ar** Argon 18

10	11	12						
58.71 2/8/16/2 **Ni** Nickel 28	63.55 2/8/18/1 **Cu** Copper 29	65.38 2/8/18/2 **Zn** Zinc 30	69.74 2/8/18/3 **Ga** Gallium 31	72.59 2/8/18/4 **Ge** Germanium 32	74.92 2/8/18/5 **As** Arsenic 33	78.96 2/8/18/6 **Se** Selenium 34	79.90 2/8/18/7 **Br** Bromine 35	83.80 2/8/18/8 **Kr** Krypton 36
106.4 2/8/18/18/0 **Pd** Palladium 46	107.9 2/8/18/18/1 **Ag** Silver 47	112.4 2/8/18/18/2 **Cd** Cadmium 48	114.8 2/8/18/18/3 **In** Indium 49	118.7 2/8/18/18/4 **Sn** Tin 50	121.8 2/8/18/18/5 **Sb** Antimony 51	127.6 2/8/18/18/6 **Te** Tellurium 52	126.9 2/8/18/18/7 **I** Iodine 53	131.3 2/8/18/18/8 **Xe** Xenon 54
195.1 2/8/18/32/17/1 **Pt** Platinum 78	197.0 2/8/18/32/18/1 **Au** Gold 79	200.6 2/8/18/32/18/2 **Hg** Mercury 80	204.4 2/8/18/32/18/3 **Tl** Thallium 81	207.2 2/8/18/32/18/4 **Pb** Lead 82	209.0 2/8/18/32/18/5 **Bi** Bismuth 83	(209) 2/8/18/32/18/6 **Po** Polonium 84	(210) 2/8/18/32/18/7 **At** Astatine 85	(222) 2/8/18/32/18/8 **Rn** Radon 86

157.3 2/8/18/25/9/2 **Gd** Gadolinium 64	158.9 2/8/18/27/8/2 **Tb** Terbium 65	162.5 2/8/18/28/8/2 **Dy** Dysprosium 66	164.9 2/8/18/29/8/2 **Ho** Holmium 67	167.3 2/8/18/30/8/2 **Er** Erbium 68	168.9 2/8/18/31/8/2 **Tm** Thulium 69	173.0 2/8/18/32/8/2 **Yb** Ytterbium 70	175.0 2/8/18/32/9/2 **Lu** Lutetium 71
(247) 2/8/18/32/25/9/2 **Cm** Curium 96	(247) 2/8/18/32/27/8/2 **Bk** Berkelium 97	(251) 2/8/18/32/28/8/2 **Cf** Californium 98	(252) 2/8/18/32/29/8/2 **Es** Einsteinium 99	(257) 2/8/18/32/30/8/2 **Fm** Fermium 100	(258) 2/8/18/32/31/8/2 **Md** Mendelevium 101	(259) 2/8/18/32/32/8/2 **No** Nobelium 102	(260) 2/8/18/32/32/9/2 **Lr** Lawrencium 103

equals the mass of 1 proton. Because electrons add so little to the total mass of an atom, their mass is usually not included in calculating the mass of an atom.

You can use an element's atomic number and mass number to determine the number of protons, neutrons, and electrons in an atom of the element. For instance, if you find sodium on the periodic table, you will see that its atomic number is 11. Since the atomic number equals the number of protons or electrons, you know that sodium has 11 protons and 11 electrons. Notice that sodium has a mass number of 22.99, which rounds to 23. Since the mass number equals the sum of the protons and neutrons, the mass number minus the atomic number equals the number of neutrons. Therefore, subtracting 11 from 23, you find that sodium has 12 neutrons. What is the number of protons, electrons, and neutrons for nitrogen?

Isotopes

Although all atoms of a given element contain the same number of protons, they do not always contain the same number of neutrons. Because the mass number is equal to the sum of the protons and

Horizons

Technology: The Smallest Particles

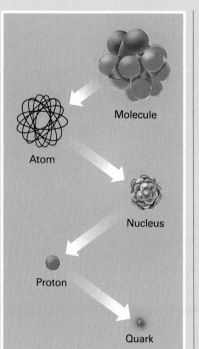

Molecule

Atom

Nucleus

Proton

Quark

Scientists at Fermi National Accelerator Laboratory near Chicago have applied the fundamental law of physics that states "the smaller the thing being looked at, the more energy you need to see it," on a grand scale. They have built Tevatron, a huge particle accelerator, to help them in their search for the smallest, and therefore, the most basic, particles of matter.

For a long time, scientists thought the most basic particles of matter were atoms. Then, when the protons, neutrons, and electrons found in atoms were discovered, these particles were thought to be the most basic. Now, the

honor goes to particles called *quarks* and *leptons*. The illustration to the left shows these increasingly smaller particles.

Because the magnitude of energy needed to see subatomic particles does not exist naturally, scientists have to build up enough energy to break apart these particles so

neutrons in a nucleus, each additional neutron increases the mass number. Atoms of the same element that differ from each other in mass number are called **isotopes** (IE-suh-topes). Isotopes of any given atom have the same number of protons but a different number of neutrons.

Hydrogen is an example of an element with several isotopes. All hydrogen atoms have the atomic number 1. That means a hydrogen atom has one proton in its nucleus and one electron moving around the nucleus. However, some hydrogen atoms also have one neutron in the nucleus. Because one neutron adds one unit to the mass, hydrogen atoms with one neutron in the nucleus have a mass number of 2. A very rare form of hydrogen has two neutrons in its nucleus. It has a mass number of 3. Figure 8–4 shows the three isotopes of hydrogen.

Solids, Liquids, and Gases

There are several ways to classify matter. One way is to classify it into one of three physical forms—**solid, liquid,** or **gas.** The particles that make up a solid are packed tightly together and are not free

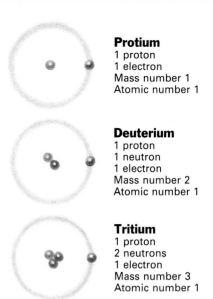

Protium
1 proton
1 electron
Mass number 1
Atomic number 1

Deuterium
1 proton
1 neutron
1 electron
Mass number 2
Atomic number 1

Tritium
1 proton
2 neutrons
1 electron
Mass number 3
Atomic number 1

Figure 8–4 Why does tritium, an isotope of hydrogen, have a mass number of 3?

individual parts can be isolated for study.

Tevatron, shown above, is an underground, ring-shaped tunnel, 6.4 km in circumference. It is built of 1,000 superconducting magnets, that move beams of tiny particles at increasingly higher speeds. As they gain speed, the particles build up energy.

When the energized particles are moving close to the speed of light—299,460 km/s— they are directed to hit either a fixed target or particles moving in an opposite direction. At impact, the particles split. The byproducts of the particles are separated and scattered. The smaller particles, quarks and leptons

among them, are isolated by a collider detector, shown in the photo far left.

The research being conducted with particle accelerators provides earth scientists with information about the structure of matter that makes up the earth's crust. In addition, because the matter studied is the same type of matter that formed galaxies and solar systems, the research also provides earth scientists with insight into the possible beginnings of the universe.

You have learned that the word atom *comes from the Greek word meaning "indivisible." Is an atom really indivisible? Explain your answer.*

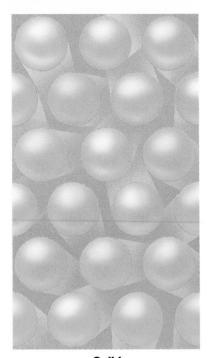

Solid

Liquid

Gas

Figure 8–5 The relative motion of particles in solids, liquids, and gases is shown in these diagrams.

to move very much. Therefore, a solid has a definite shape and volume. A liquid has a definite volume but does not have a definite shape. Instead a liquid takes the shape of the container that holds it. The particles that make up a liquid are less tightly packed than those of a solid and are more free to move than those in a solid. A gas does not have a definite volume or shape. The particles of a gas are loosely packed and are moving faster and more freely than those of a liquid. A gas is thus a formless collection of particles that tends to expand in all directions at the same time. If a gas is not confined, the space between its particles will continue to increase. Look at Figure 8–5, which shows the relative motions of atoms in solids, liquids, and gases.

To melt a solid material, heat must be added. Adding heat causes the particles in the solid to move faster. When enough energy has been added, the individual particles move so rapidly that they break away from each other. This results in the material becoming a liquid. Adding additional heat causes the particles in a liquid to move even faster. The result is the formation of a gas.

Section 8.1 Review

1. How do chemical properties differ from physical properties?
2. What is an element?
3. Name the three basic subatomic particles.
4. Compare atomic number and mass number.
5. Explain how isotopes of an element differ.
6. In what physical form—solid, liquid, or gas—are particles least tightly packed?
7. Which element has more electrons, iron (Fe) or carbon (C)? Explain your answer.

8.2 Combinations of Atoms

Section Objectives

- **Explain how atoms join together and form compounds.**
- **Describe two ways that electrons form chemical bonds between atoms.**
- **Read and interpret chemical formulas.**
- **Explain the difference between compounds and mixtures.**

Elements rarely occur in pure form in the earth's crust. Instead, they generally are found in combination with other kinds of elements. When the atoms of two or more elements are chemically united, the resulting substance is called a **compound.** A compound is a new substance with properties different from those of the elements that make it up. For example, water is a compound formed when atoms of hydrogen (H) and oxygen (O) combine. Sodium chloride is a compound formed from sodium (Na) and chlorine (Cl) atoms.

The smallest complete unit of a compound is called a **molecule.** In the compound water, for example, every individual water molecule is made up of two atoms of hydrogen and one atom of oxygen.

Some elements exist naturally as **diatomic** molecules, molecules made up of two atoms. For example, hydrogen occurs naturally only as a diatomic molecule. Free oxygen, oxygen that is not part of a compound, also occurs only as a diatomic molecule. The oxygen in the air you breathe is the diatomic molecule O_2. The O in this notation is the symbol for oxygen. The subscript $_2$ indicates the number of atoms of oxygen.

Electron Energy Levels

Different kinds of atoms join together and form compounds based upon the way their electrons are arranged. Remember that the electrons in an atom move in an electron cloud outside the nucleus. Within the electron cloud of an atom, electrons are arranged in **energy levels.** The electrons in each energy level have a specific amount of energy, and each energy level can hold only a certain maximum number of electrons. For example, the first energy level can hold a maximum of 2 electrons, and the second can hold up to 8 electrons.

An electron occupies only one energy level at a time, although it may go to higher or lower levels. Also, the electrons of an atom may fill several different levels. For example, the electrons of a large atom such as Radium may occupy seven levels. However many energy levels are occupied in an atom, the outermost level can never hold more than eight electrons. What is the number of electrons that an element with one energy level can have?

Not all atoms have the maximum number of electrons in their outermost energy level. However, the most stable atoms are those in which the outermost energy level is filled. In other words, the outermost level in these atoms holds the maximum number of electrons, or eight electrons. Atoms with filled outer energy levels do not easily lose or gain electrons, and so they do not easily form compounds with other elements.

Atoms of certain elements give up electrons more easily than atoms of certain other elements. Atoms with only one, two, or three electrons in the outermost level give up electrons easily. The

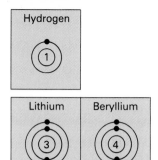

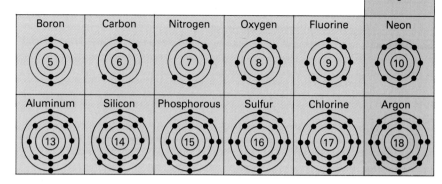

Figure 8–6 The diagrams above show the electron energy levels of the first 18 elements in the periodic table. The number at the center of each diagram indicates the number of protons in the nucleus.

elements that have these limited numbers of electrons in the outermost level are the same elements that have metallic properties and are classified as metals. Examples of metals are aluminum (Al), cobalt (Co), copper (Cu), gold (Au), iron, lead (Pb), and tin (Sn). Metals have a number of useful characteristics. Most metals are good conductors of heat and electricity. Many metals are malleable, which means they can be hammered into thin sheets. These metals are also ductile—they can be drawn out into wire.

Atoms with four, five, or six electrons in the outermost level do not lose electrons easily. These are elements without metallic properties and are classified as nonmetals. Carbon, nitrogen (N), and oxygen are examples.

Chemical Bonds

The atoms that make up a compound are held together by forces called **chemical bonds.** A chemical bond is produced by the interaction of electrons from the outermost energy levels of two or more atoms. Scientists can predict which kinds of atoms will form chemical bonds with each other. They compare the number of electrons in the outermost energy level of the atoms with the maximum number possible for that energy level. For example, a hydrogen atom has only one electron in its single energy level. Because this one energy level can hold two electrons, you can predict that hydrogen will accept another electron if it is available.

Atoms can form chemical bonds by sharing electrons or by transferring electrons from one atom to another. In both cases, the attraction that joins the atoms is the pull of an electrical charge. When electrons are transferred from one atom to another, the bond is an **ionic bond.**

Ionic Bonds

A compound formed through the transfer of electrons is called an **ionic compound.** Most ionic compounds are formed by the transfer of electrons from a metal to a nonmetal. A familiar example is sodium chloride, or common table salt. A sodium atom has 11 electrons, with 2 electrons in its first energy level, 8 electrons in its second energy level, and 1 electron in its outermost energy level. A chlorine atom has 17 electrons, with 2 electrons in its first energy level, 8 electrons in its second energy level, and 7 in its outer level. With only 7 electrons in its outer energy level, chlorine can accept 1 more electron. Because of the number of electrons in their respective outermost energy levels, sodium atoms have a strong tendency to lose 1 electron and chlorine atoms tend to gain 1 electron. Therefore, a sodium atom will give up an electron to a chlorine atom, as shown in Figure 8–7.

When an electron is transferred from one atom to another, both atoms become electrically charged. An **ion** is an atom or group of atoms that carries an electrical charge. Sodium becomes positively charged when it gives up an electron. The loss of an electron leaves a sodium atom with 11 protons in its nucleus but only 10 electrons. It has become a positively charged sodium ion (Na^+). In a similar way, chlorine becomes an ion when it gains an electron. The additional electron gives a chlorine atom 18 electrons, one more than its 17 protons. The extra electron changes the neutral chlorine atom into a negatively charged chloride ion (Cl^-).

As you have read, sodium and chloride ions combine and form the compound sodium chloride. The positively charged sodium ions and negatively charged chloride ions are held together because oppositely charged atoms attract one another. Each ion is strongly attracted to its oppositely charged neighbor and can fit into the ion arrangement of sodium chloride only in a definite position. Sodium chloride in the solid form has a cube shape.

An ionic bond creates an extremely strong attraction. Sodium chloride is a difficult compound to melt because its oppositely charged ions are held together by such strong bonds.

Figure 8–7 The compound sodium chloride is formed by ionic bonding.

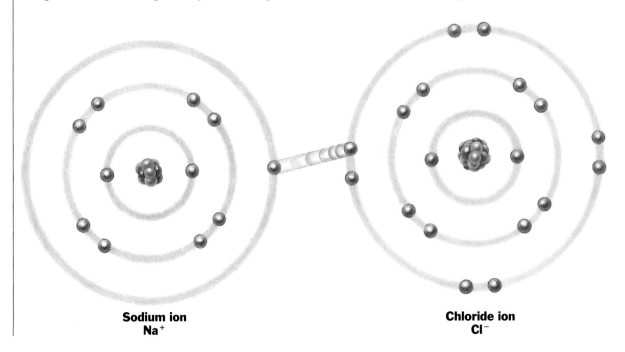

Sodium ion
Na^+

Chloride ion
Cl^-

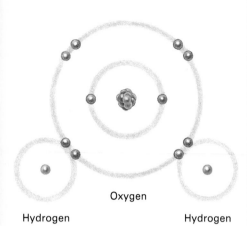

Oxygen

Hydrogen   Hydrogen

Figure 8–8 In a water molecule, hydrogen and oxygen atoms are joined by covalent bonds.

Covalent Bonds

A bond based on the attraction between atoms that share electrons is a **covalent bond.** Water is an example of a **covalent compound**– that is, a compound formed by the sharing of electrons. Two hydrogen atoms can share their single electrons with an oxygen atom, giving oxygen a stable number of eight electrons in the outermost energy level. At the same time, the oxygen atom shares two of its electrons, one with each hydrogen atom, giving each hydrogen atom a stable number of electrons in its outermost level of electrons. Thus, a water molecule consists of two atoms of hydrogen combined with one atom of oxygen. Each covalent compound consists of only one kind of molecule, just as each element consists of only one kind of atom.

When atoms share electrons, the positive nucleus of each atom is attracted to the negative electrons being shared. The pull between the positive and negative charges is the force that keeps the atoms joined. In diatomic gases such as nitrogen and oxygen, the most abundant gases in the earth's atmosphere, there are covalent bonds between the two atoms of each diatomic molecule. Because the nuclei are identical, they each pull with the same strength on the shared electrons. In a compound, the covalent bonds are between two different kinds of atoms. The nucleus of one atom pulls more strongly than the other. The electrons are shared but remain closer to the nucleus that pulls more strongly. In a water molecule, for example, the oxygen nucleus attracts the electrons more strongly than does either hydrogen nucleus. As a result, the molecule has a slightly negative charge at its oxygen end and a slightly positive charge at its hydrogen end.

The molecules that make up a covalent compound are usually held together by weak forces of attraction. These forces are far weaker than the powerful attraction between ions in an ionic compound such as sodium chloride. For this reason, ice, a covalent compound, melts at a much lower temperature than does sodium chloride, an ionic compound.

Chemical Formulas

In any given compound—either covalent or ionic—the elements that make up the compound are always found in the same proportion. Therefore scientists can represent any compound by using a **chemical formula.** A chemical formula indicates what elements the compound contains. It also indicates the relative number of atoms of each element present.

The chemical formula for water is H_2O, which indicates that each water molecule consists of two atoms of hydrogen and one atom of oxygen. In a chemical formula a subscript number is used after the symbol for an element to indicate how many atoms of that element are in a single molecule. For example, in the chemical formula for water, the number 2 after the H indicates that two atoms of hydrogen are in each molecule of water.

Mixtures

On the earth, elements and compounds are generally found mixed together. A **mixture** is material that contains two or more substances that are not chemically combined. The substances in a mixture keep their individual properties. Therefore, unlike a compound, a mixture can be separated into its parts by physical means. For example, you can use a magnet to separate a mixture of powdered sulfur and iron filings. The magnet will attract the iron and leave behind the sulfur, which is not magnetic.

Many substances studied in earth science are mixtures. Rocks and soil, for example, are mixtures of many different compounds. The atmosphere is a mixture of gases, including oxygen, nitrogen, carbon dixoide, and water vapor. Air pollution forms when dust and chemicals are added to this mixture. Dust and chemicals from industrial smokestacks and car exhausts form a mixture called **smog,** which can irritate people's eyes and impair their breathing.

Seawater is an example of a **solution,** a mixture in which one substance is uniformly dispersed in another substance. Sodium chloride and many other ionic compounds are dissolved in seawater. The ability of water to dissolve salt and other substances is related to what you have already read about the distribution of electrical charge in a water molecule. The positive end of a water molecule attracts negative chloride ions, and the negative end of the water molecule attracts positive sodium ions.

You probably think of solutions as liquids, but gases and solids may also be solutions. For example, an **alloy** is a solid solution of two or more metals. Common alloys include brass, a mixture of copper and zinc; and bronze, a mixture of copper and tin.

Figure 8–9 Dust and chemicals in the air form the unhealthful mixture known as smog shown left. Sandstone shown right is a mixture of sand grains, which are usually quartz, bound together by silica or calcium carbonate.

Section 8.2 Review

1. Define the term *compound*.
2. Which type of bonding is weaker—ionic or covalent?
3. Write out the words for this formula: $ZnCl_2$.
4. How does a mixture differ from a compound?

Reviewing Chapter 8

Key Concepts

- An element is a substance that cannot be broken down into a simpler form by ordinary chemical means. See page 117.
- An atom consists of electrons traveling around a nucleus made up of protons and neutrons. See page 119.
- The atomic number of an atom equals the number of protons in the atom. The mass number equals the sum of the protons and neutrons in the atom. See page 119.
- Isotopes are atoms of the same element that have different numbers of neutrons. See page 122.
- A solid holds its shape because it is made up of tightly locked particles. A liquid has no definite shape, and it flows freely. A gas is a formless collection of particles that tends to expand in all directions when not confined. See page 123.
- Different kinds of atoms join together and form compounds based upon the way their electrons are arranged in energy levels within the electron cloud. See page 125.
- Chemical bonds between atoms are formed by the sharing or transfer of electrons between atoms. See page 126.
- A chemical formula tells what elements a compound contains and the relative number of atoms of each element present. See page 128.
- A compound consists of atoms from two or more elements that are chemically united. A mixture consists of two or more substances that are not chemically united. See pages 125, 129.

Key Terms

alloy (129)	covalent bond (128)	ion (127)	molecule (125)
atom (118)	covalent compound	ionic bond (126)	neutron (119)
atomic number (119)	(128)	ionic compound (126)	nucleus (119)
chemical bond (126)	diatomic (125)	isotope (123)	physical property (117)
chemical formula (128)	electron (119)	liquid (123)	proton (119)
chemical property	electron cloud (119)	mass number (119)	smog (129)
(117)	energy level (125)	matter (117)	solid (123)
compound (125)	gas (123)	mixture (129)	solution (129)

Review

On your paper, write the letter of the term that best completes each of the following statements.

1. Color and hardness are examples of an element's (a) physical properties (b) chemical properties (c) atomic structure (d) molecular properties.
2. A substance that cannot be broken down into a simpler form by ordinary chemical means is (a) a mixture (b) a gas (c) an element (d) a compound.
3. The smallest unit of an element is (a) a molecule (b) an atom (c) an ion (d) an electron.
4. Particles in atoms that do not carry an electrical charge are called (a) neutrons (b) nuclei (c) protons (d) ions.
5. The number of protons in the nucleus indicates the atom's (a) mass number (b) electrical charges (c) isotope (d) atomic number.
6. The mass number of an atom is equal to its (a) number of protons (b) total number of

electrons and protons (c) total number of neutrons and protons (d) number of neutrons and electrons.

7. Atoms of the same element that differ in mass are (a) ions (b) isotopes (c) neutrons (d) molecules.

8. A material with a definite shape and tightly bonded particles is a (a) compound (b) liquid (c) gas (d) solid.

9. A liquid does not have a definite (a) shape (b) volume (c) chemical formula (d) bonding arrangement.

10. If a gas is not confined, the space between its particles will (a) decrease slowly (b) decrease rapidly (c) increase (d) not change.

11. Atoms of two or more elements that are chemically united form (a) a mixture (b) a nucleus (c) an ion (d) a compound.

12. An atom does not easily lose or gain electrons if it has (a) many protons (b) a filled outer energy level (c) many energy levels (d) few neutrons.

13. A molecule of water, or H_2O, has one atom of (a) hydrogen (b) helium (c) oxygen (d) osmium.

14. A material that contains two or more substances that are not chemically combined is (a) a mixture (b) a compound (c) an ion (d) a molecule.

Application

On your paper, write answers to the following questions.

1. Oxygen combines with hydrogen to form water. Is this process a result of the physical or chemical properties of oxygen?

2. How many different *kinds* of atoms make up the element helium?

3. What distinguishes a particular atom from all other kinds of atoms?

4. How many neutrons does a potassium atom have if its atomic number is 19 and its atomic mass is 39?

5. The atomic number of calcium is 20, and the atomic number of copper is 29. Which has more electrons, a calcium atom or a copper atom? How do you know?

6. Why do isotopes of an element have different mass numbers?

7. The mercury in a thermometer has a definite volume, but it takes the shape of the glass tube that holds it. Is the mercury in a thermometer a solid, a liquid, or a gas?

8. A helium atom has two electrons in its first and only energy level. Would you predict that helium easily forms compounds with other elements? Why or why not?

9. Calcium chloride is an ionic compound. Carbon dioxide is a covalent compound. Which of these compounds would you expect to have a lower melting point? Explain your answer.

10. Is a compound that is liquid at room temperature more likely to be a covalent compound or an ionic compound? Explain why you think this is so.

11. The chemical formula for carbon dioxide is CO_2. How many atoms of carbon and oxygen does a molecule of carbon dioxide contain?

12. What happens to the properties of a substance when it becomes part of a mixture?

Extension

1. Choose one of the elements that is abundant in the earth's crust. Prepare a report on the atomic structure of the element, its physical and chemical properties, and its economic importance. Illustrate the report with photographs, if possible.

2. List five common elements that you come across each day. Describe the form in which you find each. Write the symbol for each of these elements.

3. Use the library to research some common forms of air and water pollution. Find out the source of the pollution and what the pollution is made up of. Find out what lawmakers are doing to control these forms of pollution. Compile your findings in a chart.

Chapter 9
Minerals of the Earth's Crust

You probably would not recognize the blue substance in the rock pictured on the right as copper. Yet scientists have a number of methods for identifying copper and other minerals. Copper, like every other mineral, has its own chemical composition. Minerals can also be identified by properties such as structure, color, and hardness. These differences among minerals result from the way the atoms of each mineral are held together. This chapter explains how atoms combine to form the minerals of the earth's crust.

Chapter Outline

9.1 What Is a Mineral?
Kinds of Minerals
Crystalline Structure

9.2 Identifying Minerals
Characteristics of Minerals
Unusual Properties of
Minerals

This blue peacock copper ore is from Southern Arizona.

9.1 What Is a Mineral?

A ruby, a gold nugget, and a grain of salt look very different from one another, but they all have one thing in common. They are all **minerals,** the basic materials of the earth's crust. A mineral is a natural, **inorganic** solid formed in the earth. An inorganic substance is one that is not made up of living things or the remains of living things. Every mineral has a characteristic chemical composition and can be either an element or a compound. Most rocks that make up the crust are mixtures of various minerals.

To determine whether a substance is a mineral or a nonmineral, scientists ask four basic questions about the substance. If the answer is yes to all four questions, the substance is a mineral. First, scientists ask whether the substance is inorganic. Coal, for example is *organic*—it is composed of the remains of plants and animals. Thus it is not a mineral. Magnetite, composed of the inorganic substances iron (Fe) and oxygen (O), is a mineral.

Second, scientists ask whether the substance occurs naturally in the earth. The minerals quartz (SiO_2), silver (Ag), and sulfur (S), for example, all occur naturally in the earth. Manufactured substances, such as steel, are not minerals.

Third, scientists determine whether the substance is a solid. Petroleum and natural gas are naturally occurring substances in the earth's crust. They are not solid, however, and thus are not minerals. In addition, both are made up of the remains of plants and animals.

The fourth question is whether the substance has a definite chemical composition. The mineral gold (Au) is an element with only gold atoms. The mineral fluorite is a compound, made up of only calcium (Ca) and fluoride (F) ions in a specific pattern. A chunk of concrete, however, is made up of several substances. When concrete is made, the amounts of these substances vary according to the intended purpose of the concrete. Is concrete a mineral? Explain how you know.

Kinds of Minerals

The earth's crust contains more than 2,000 kinds of minerals, but fewer than 20 of them are common. The common minerals are called **rock-forming minerals** because they form the rocks of the earth's crust. Of the 20 rock-forming minerals, 10 are so common

Section Objectives

- **Define a mineral and distinguish between the two main mineral groups.**
- **Identify the elements found most abundantly in common minerals.**
- **Name six types of nonsilicate minerals.**
- **Distinguish among four main arrangements of silicon-oxygen tetrahedra found in silicate minerals.**

Figure 9–1. Plagioclase feldspar (left), muscovite mica feldspar (center), and orthoclase feldspar (right) are 3 of the 20 common rock-forming minerals.

Table 9—1: Major Groups of Nonsilicate Minerals

Carbonates

Compounds that contain a carbonate group (CO_3). Examples: dolomite [$CaMg(CO_3)_2$], shown left, and calcite ($CaCO_3$), shown right

Halides

Compounds that consist of chlorine or fluorine combined with sodium, potassium, or calcium. Examples: halite (NaCl), shown left, and fluorite (CaF_2), shown right

Native Elements

Elements uncombined with other elements. Examples: gold (Au), platinum (Pt), diamond (C), silver (Ag) shown left, and copper (Cu), shown right

that they make up 90 percent of the mass of the earth's crust. These minerals are quartz, orthoclase feldspar, plagioclase feldspar, mica, calcite, clay, dolomite, halite, gypsum, and ferromagnesian minerals, which include olivines, pyroxenes, and amphiboles. All minerals, however, can be classified into two main groups based on their chemical composition—**silicate minerals** and **nonsilicate minerals.**

Silicate Minerals

All silicate minerals contain atoms of silicon (Si) and oxygen. The common mineral quartz consists of only silicon and oxygen atoms. However, most silicate minerals also contain one or more other kinds of atoms. Feldspar is the most common silicate mineral. The type of feldspar that forms depends on which metal combines with the silicon and oxygen atoms. Orthoclase feldspar results when the metal is potassium (K). Plagioclase feldspar forms when the metal is sodium (Na), calcium, or both. Besides quartz and feldspar, common silicate minerals include talc, hornblende, and mica. Silicate minerals make up 96 percent of the earth's crust. Feldspar and quartz alone make up more than 50 percent of the crust.

Oxides

Compounds that contain oxygen and some element other than silicon. Examples: magnetite (Fe_3O_4), corundum (Al_2O_3) shown left, and hematite (Fe_2O_3), shown right

Sulfates

Compounds that contain a sulfate group (SO_4). Example: gypsum ($CaSO_4 \cdot 2H_2O$)

Sulfides

Compounds that consist of one or more elements combined with sulfur. Examples: sphalerite (ZnS), galena (PbS) shown left, and pyrite (FeS_2), shown right

Nonsilicate Minerals

Four percent of the earth's crust consists of nonsilicate minerals—that is, minerals that do not contain silicon. Based on their chemical composition, nonsilicate minerals are classified into six major groups: carbonates, halides, native elements, oxides, sulfates, and sulfides. Table 9–1 describes the characteristics of each group and lists representative examples.

Crystalline Structure

Almost all minerals in the earth's crust have a crystalline structure. Each type of crystalline material is characterized by a specific geometric arrangement of its atoms or ions. Crystalline material will sometimes form a **crystal,** a natural solid substance that has a definite shape. The conditions under which minerals are produced do not usually allow single crystals to grow. As a result, minerals more commonly consist of masses of fine crystalline grains that you can see only with a microscope. If a crystalline mineral forms unrestricted by surrounding material, however, the mineral will develop into a single, large crystal in one of six crystal shapes. The crystal shapes are helpful in identifying minerals. You will read about these crystal shapes later in this chapter.

Scientists use X rays to study the structure of crystals. X rays passing through a crystal and striking a photographic plate produce an image that shows the geometric arrangement of the atoms or ions that make up the crystal.

The Crystalline Structure of Silicates

Although there are many kinds of silicate minerals, their crystalline structure is made up of the same basic building blocks. Each of

Career Focus: *Mineral Resources*

"You can work for years to find one simple answer and only come up with more questions. I really enjoy that challenge."

Mining Engineer

Civil engineer Lani Boldt often talks to high school students about her work in mining research for the United States Bureau of Mines in Spokane, Washington. Each time her message is the same:

"You don't have to be a super brain to make it through an engineering curriculum. What you do need is interest, perseverance, and plain old-fashioned hard work," she said.

Like other engineers in mining, who must have at least a bachelor of science degree in civil engineering, Boldt's main objective is to make all kinds of mining safer, easier, and more profitable. In the future, this may include seabed and space mining. Currently Boldt, pictured left, is working to develop methods for using the wastes produced in mining.

The majority of Boldt's work involves both laboratory and field investigations using computers, soil and rock testing equipment, and environmental test chambers, devices used in the lab to simulate the natural environment.

"The most exciting thing about my work is its variability," she said. "I never know from one day to the next where I'll be or what I'll be doing. I just know I'll be learning something new and, I hope, helping someone in the meantime."

Gem Cutter

A gem cutter, shown below, evaluates and grades all colored gemstones in their rough state. The gem cutter will then determine

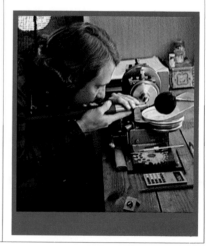

these building blocks consists of four oxygen atoms arranged in a pyramid with one silicon atom in the center. Figure 9–2 shows this four-sided structure. The structure is known as a **silicon-oxygen tetrahedron.**

The silicon-oxygen tetrahedra combine in different arrangements to produce the many diverse silicate minerals. The various arrangements are the result of the kinds of bonds formed between the oxygen atoms of the tetrahedra and other atoms. The bonds may form between the oxygen atoms and silicon atoms of neighboring tetrahedra, or they may form between the oxygen atoms in the tetrahedra

Figure 9–2. The silicon-oxygen tetrahedron is the basic structure of all silicate minerals.

how to cut and polish the stones to best reveal their color and luster.

The process begins when the gem cutter saws or slices jewelry-quality minerals into usable pieces. Those pieces are then ground or cut into one of two basic styles—cabochon or faceted. While cabochon cuts result in rounded surfaces, faceted cuts produce many small, flat surfaces organized in intricate patterns. In both cutting procedures, a knowledge of the form, structure, properties, and classification of crystals is necessary.

Gem cutters enter the field through a variety of on-the-job and technical training programs.

Mine Inspector

Mine inspectors evaluate various kinds of mines including coal, limestone, uranium and copper mines to ensure the health and safety of mine workers. Some mines are deep underground. Other mines, such as the copper mine shown above, are open.

Among the concerns of mine inspectors are explosive gases, poisonous fumes, unstable roof and ground structures, and harmful dust.

In addition to gathering information on health and safety conditions, a mine inspector enforces established health and safety regulations by notifying mine managers of any violations and then seeing that those conditions are corrected. Mine inspectors also investigate and report on mine accidents and may occasionally

become involved in directing rescue attempts.

Entry-level positions are available through apprenticeships, work experience, or four years of related education after high school.

For Further Information

For more information on careers in mining, write the Bureau of Mines, U.S. Department of Interior, 2401 E Street, NW, Washington, DC 20241.

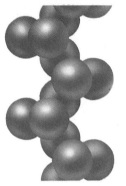

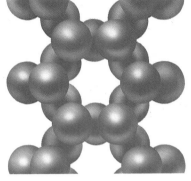

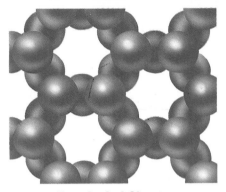

Single Chain   **Double Chain**   **Tetrahedral Sheet**

Figure 9–3. The silicon-oxygen tetrahedra shown above combine to form single chains (left), double chains (center), and sheets (right).

and other elements outside the tetrahedra. Figure 9–3 illustrates the kinds of arrangements formed by tetrahedra.

Ionic, Chain, and Sheet Silicates

Silicon-oxygen tetrahedra linked only by atoms of elements other than silicon and oxygen make up ionic silicate materials. For example, olivine is a mineral formed when the oxygen atoms of tetrahedra bond to magnesium (Mg) or iron atoms.

In the bonding arrangement called a single-chain silicate, each tetrahedron is bonded to two others by shared oxygen atoms. In double-chain silicates two single chains of tetrahedra bond to each other. Minerals made up of single chains are called pyroxenes, and those made up of double chains are amphiboles.

In tetrahedral sheets, each tetrahedron shares three oxygen atoms with other tetrahedra. The fourth oxygen atom bonds with an atom of potassium or aluminum (Al), which joins one sheet to another. These sheets separate easily because the three bonds among the tetrahedra are much stronger than the bond with potassium or aluminum. Mica is an example of tetrahedral sheets.

Network Silicates

In the more complex arrangement known as network silicate, each tetrahedron is bonded to four neighboring tetrahedra. Networks that contain only silicon-oxygen tetrahedra form the mineral quartz, which has the chemical formula SiO_2. Quartz is one of the hardest minerals because all of its atoms are tightly bonded together.

Feldspar is also a network silicate. Unlike quartz, however, some of the tetrahedra in feldspar have atoms of aluminum or other metals instead of silicon atoms. The bonds between these atoms are weaker than those between silicon and oxygen. Thus feldspar can be broken more easily than quartz.

Section 9.1 Review

1. What are the two main groups of minerals?
2. Which two elements are most commonly found in minerals?
3. Are carbonates and halides silicate or nonsilicate minerals?
4. How do sheet silicates differ from network silicates?
5. Silver is a naturally occurring inorganic substance that is formed in the earth's crust. It is a solid and has a definite chemical composition. Is silver a mineral? Explain your answer.

9.2 Identifying Minerals

Section Objectives
- **Describe some simple tests that help distinguish one mineral from another.**
- **List four unusual properties that may help identify certain minerals.**

If you have ever found an unfamiliar mineral, you may have wondered how to go about identifying it. Scientists called **mineralogists** conduct tests with special equipment to properly identify minerals. Mineralogists work in laboratories and in the field around the world, helping to identify minerals—from the most common to the most rare and precious.

Characteristics of Minerals

Each mineral has specific properties that are a result of its chemical composition. These properties provide useful clues for identifying minerals. You can identify some of these properties by simply looking at a sample of the mineral. You can determine other properties through simple tests.

Color

A property you easily observe is the color of a mineral. Some minerals have distinctive colors. For example, sulfur is bright yellow, and azurite is a deep blue. The mineral cinnabar is red, while serpentine is green. Color alone, however, is often not a reliable clue in identifying a mineral sample. Many minerals are similar in color, and very small amounts of certain elements may affect the color. For example, corundum is a colorless mineral composed of aluminum and oxygen atoms. However, corundum that contains traces of chromium (Cr) forms ruby, a rare red mineral. Sapphire, a rare blue mineral, consists of corundum with traces of iron and titanium (Ti). Amethyst is quartz that is purple because it contains a tiny amount of the element manganese (Mg).

Color is also an unreliable identification clue because weathered surfaces may hide the color of minerals. For example, iron pyrite is the color of gold, but it appears dark yellow when it is weathered. When you examine a mineral for color, be sure to inspect a freshly exposed surface.

Figure 9–4. Pure quartz (left) is colorless. Amethyst (above) is a form of quartz that looks purple because of the presence in it of manganese.

Streak

A more reliable clue to the identity of a mineral is the color of that mineral in powdered form, which is called its **streak.** The easiest way to observe the streak of a mineral is to rub some of the mineral against a piece of unglazed ceramic tile called a streak plate. Because the streak is the powdered form of the mineral, it may not be the same color as the larger piece of the mineral. For example, the streak of gold-colored pyrite is black. For most minerals, however, the streak is either colorless or a very light shade of the normal color of the mineral.

Luster

Light reflected from the surface of a mineral is called **luster.** Minerals that reflect light like polished metal are said to have a *metallic luster.* All other minerals have a *nonmetallic luster.* Mineralogists distinguish several types of nonmetallic luster. Transparent quartz and other minerals that look like glass have a glassy luster. Minerals with an appearance like the surface of candle wax have a waxy luster. Some minerals, such as mica, have a pearly luster. Diamond is an example of a mineral with a brilliant luster. A mineral that lacks any kind of shine has a dull luster. Some forms of the mineral cinnabar have a dull claylike luster.

Cleavage and Fracture

Some minerals tend to split easily along certain flat surfaces. This property, called **cleavage,** is related to the types of bonds in the internal structure of the mineral. The surface along which cleavage occurs runs parallel to the plane in the crystal where bonding is relatively weak. For example, minerals made of tetrahedral sheets tend to split parallel to the sheets because the bonds between tetrahedra in each sheet are much stronger than the bonds that hold different sheets together. Mineralogists can use cleavage to identify and describe some minerals. The mineral halite, for example, breaks into small cubes.

Most minerals, however, do not break along cleavage planes. Instead, they *fracture,* or break, unevenly into curved or irregular pieces. Mineralogists describe a fracture according to the appearance of the broken surfaces. For example, a rough surface is called *uneven* or *irregular.* A broken surface that looks like a piece of broken wood is called *splintery* or *fibrous.* A curved surface on a fractured mineral is called *conchoidal.*

Figure 9–5. All minerals have either a metallic luster like platinum (top) or a nonmetallic luster like muscovite mica (bottom).

Figure 9–6. The diagram (left) demonstrates cleavage in three directions. Galena is a mineral that cleaves in three directions. The diagram (right) demonstrates cleavage in one direction. Mica is a mineral that has cleavage in one direction.

Hardness

The measure of the ability of a mineral to resist scratches is called **hardness.** Hardness differs from resistance to cleavage or fracture. A diamond, for example, is extremely hard but it can be split along cleavage planes more easily than can iron, which is a softer mineral.

The hardness of an unknown mineral can be determined by scratching it against the minerals on **Mohs' scale,** shown in Table 9–2. This scale lists ten common minerals in order of increasing hardness. The softest mineral is talc with an assigned hardness of 1. The hardest mineral is diamond with an assigned hardness of 10. The difference in hardness between two consecutive minerals is about the same throughout the scale except for the difference between the two hardest minerals. Diamond is much harder than corundum, the mineral just before it on the scale. Care must be taken in testing hardness. For example, the mark usually left by talc on an unknown mineral may appear to be a scratch. Actually it is the streak made by talc, and it is easily rubbed off. A true scratch will remain when a harder mineral rubs a softer mineral.

To test an unknown mineral for hardness, you must determine which is the hardest mineral on the scale that it can scratch. For example, galena can scratch gypsum but not calcite. Between what two numbers on Mohs' scale does galena fall? If neither of two minerals scratches the other, they have the same hardness.

The hardness of a mineral is largely determined by the strength of the bonds between the atoms or ions that make up its internal structure. Both diamond and graphite consist exclusively of carbon atoms. Diamond has a hardness of 10, however, while the hardness of graphite varies between 1 and 2. The hardness of diamond results from a strong crystal structure in which each carbon atom is firmly bonded to four other carbon atoms. In contrast, the carbon atoms in graphite are arranged in layers with weak bonds between the layers.

Figure 9–7. Conchoidal fracture, characteristic of obsidian, is a type of fracture that helps to identify some minerals.

Table 9–2: Mohs' Scale of Mineral Hardness

Mineral	Hardness	Common Test
Talc	1	Easily scratched by fingernail
Gypsum	2	Can be scratched by fingernail
Calcite	3	Barely can be scratched by copper penny
Fluorite	4	Easily scratched with steel knife blade
Apatite	5	Can be scratched by steel knife blade
Feldspar	6	Easily scratches glass
Quartz	7	Easily scratches both glass and steel
Topaz	8	Scratches quartz
Corundum	9	No simple tests
Diamond	10	No simple tests

Table 9–3: The Six Basic Crystal Systems

Isometric or Cubic System

Three axes of equal length intersect at 90° angles. Examples: galena, halite, and pyrite

Orthorhombic System

Three axes of different lengths intersect at 90° angles. Examples: olivine, topaz, and staurolite

Triclinic System

The three axes are of unequal length and are oblique to one another. Examples: plagioclase feldspars, turquoise, and axinite

Monoclinic System

Of three axes of different lengths, two intersect at 90° angles. The third axis is oblique to the others. Examples: micas, augite, and gypsum

Hexagonal System

The three horizontal axes are the same length and intersect at 60° angles. The vertical axis is longer or shorter than the horizontal axes. Examples: beryl, calcite, apatite, hematite, and quartz

Tetragonal System

Three axes intersect at 90° angles. The two horizontal axes are of equal length. The vertical axis is longer or shorter than the horizontal axes. Examples: cassiterite, chalcopyrite, and zircon

Crystal Shape

As explained in Section 9.1, under the proper conditions, mineral crystals will form in one of six basic shapes. These shapes are illustrated in Table 9–3. Each kind of mineral is characterized by crystals of a specific shape. A certain mineral always has the same general shape because the atoms or ions that form its crystals combine in the same geometric pattern.

Density

When handling specimens of various minerals, you may notice that some feel heavier than others. For example, a piece of galena feels heavier than a piece of quartz of the same size. One way to compare these minerals is to lift, or heft, mineral samples of the same size. However, a more precise comparison can be made by measuring the

density of a sample. Density is the ratio of the mass of a substance to its volume. The units of density are grams per cubic centimeter (g/cm³). For example, if the mass (M) of a mineral sample is 85 g and its volume (V) is 34 cm³, its density (D) is found by solving the equation:

$$D = \frac{M}{V} = \frac{85 \text{ g}}{34 \text{ cm}^3} = 2.5 \text{ g/cm}^3$$

The density of a mineral depends on the kinds of atoms it contains and how closely they are packed. Most of the common minerals in the earth's crust have densities in the narrow range between 2 and 3 g/cm³. However, the densities of minerals containing such heavy metals as lead, uranium, gold, and silver range from 7 to 20 g/cm³. Therefore, density more readily helps identify the heavier minerals than it does the lighter ones.

Investigation

Mineral Identification

Most minerals can be identified by two or three physical properties. You can use some basic tests of mineral properties and a table of mineral characteristics to identify an unknown mineral.

Science Process Skills Focus: observing, recording data, interpreting data

Materials
mineral specimen, streak plate, Mohs' Scale of Hardness on page 141, copper penny, steel file, glass plate, mineral identification table on page 552-553.

Procedure
1. Copy Data Table onto a piece of paper.
2. Find a mineral specimen that seems to be the same color and makeup throughout. Record the color of your specimen.
3. If your specimen shines like polished metal, consider its luster metallic; otherwise, nonmetallic. Record your observations.
4. Rub your specimen across the streak plate. Record the color of the streak.
5. Study Mohs' Scale of Hardness. Use your fingernail, a penny, then a steel file, and finally a glass plate to scratch your specimen. Match the item that scratched your specimen with

Data Table

Color	
Luster	
Streak	
Hardness	

those listed beside the scale. Record your results.
6. Compare your results with the mineral identification table. List all those minerals that fit the description of your specimen.
7. Refer to the mineral identification chart again and make a list of five minerals that clearly do not fit the description of your specimen.

Analysis and Conclusions
1. Which tests were the most useful in selecting the minerals that fit the description of your specimen? Tentatively identify the type of mineral you might have.
2. Suggest other tests that would positively identify your specimen.

Figure 9–8. Notice the change in color of the fluorescent mineral willemite calcite as it goes from ordinary light (left) to ultraviolet light (right).

Unusual Properties of Minerals

All minerals exhibit the properties described earlier in this section. In addition to those properties, however, a few minerals have some unusual properties that can aid in their identification.

Magnetism

A magnet passed through some sand or loose soil will often attract small particles of iron-containing minerals. Magnetite is the most common among this group of magnetic minerals. Lodestone is a form of magnetite that acts as a magnet. When lodestone is cut into an elongated shape, it automatically becomes polarized, with a north pole at one end and a south pole at the other, just like any other bar magnet. The needles of the first magnetic compasses used in navigation were made of lodestone.

Fluorescence

The mineral calcite is greenish white in ordinary light, but under ultraviolet light it often appears red. This ability to glow under ultraviolet light is referred to as **fluorescence.** Fluorescent minerals absorb ultraviolet light and then produce visible light of various colors. For example, willemite is light brown in ordinary light, but under ultraviolet light it appears green. You can see the effect of fluorescence on minerals in Figure 9–8.

Some minerals subjected to ultraviolet light will continue to glow after the ultraviolet light is cut off. Minerals that continue to glow have the property called **phosphorescence.**

Figure 9–9. The mineral calcite exhibits double refraction.

Double Refraction

Light rays bend as they pass through transparent minerals. This bending of light rays as they pass from one substance, such as air, to another, such as a mineral, is called **refraction.** Crystals of calcite and some other transparent minerals bend light in such a way that they produce a double image of any object viewed through them. This property is called **double refraction.** Double refraction occurs because the light ray is split into two parts as it enters the crystal.

Radioactivity

Some minerals have a property known as *radioactivity.* You learned in Chapter 8 that certain atoms have unstable electron arrangements. Some atoms also have unstable arrangements of protons and neutrons in their nuclei. Radioactivity results as unstable nuclei decay

IMPACT

Mineral Resources

As far back as two and a half million years ago rocks that littered the earth's surface were fashioned into crude tools by primitive people. Since then the minerals on and within our planet have served as the raw materials for building an evermore sophisticated world.

For example, the towering skyscrapers that punctuate urban skylines would be impossible to build without the high-strength steel girders that support them. Steel is a manufactured substance made up mostly of the mineral iron. The stone facings on these buildings and even the glass that makes up their windows are also products of the earth's crust.

Automobiles, which have made our society the most mobile in history, are stamped out of corrosion-resistant steel. The aircraft that crisscross the planet depend on lightweight, high-strength metals such as alloys of magnesium and aluminum to form their skins. The high-performance jet engines that propel them require metals like titanium and molybdenum to operate at high temperatures for long periods of time.

Gemstones, too, play a role in modern technology. Diamonds, the hardest of the minerals, are used to coat cutting and grinding tools. Diamonds are also used to coat high-resolution optical lenses on cameras and telescopes.

Steel, which is essential to the building of automobiles and skyscrapers, is not a mineral. How do you know it is not a mineral? What mineral makes up much of our steel, however?

over time into stable nuclei by releasing particles and energy. Uranium (U) and radium (Ra) are examples of radioactive elements that occur in mineral deposits. Pitchblende is the most common mineral containing uranium. Other uranium-bearing minerals include carnotite, uraninite, and autunite.

Section 9.2 Review

1. What are the two main types of luster?
2. How would you determine the hardness of a mineral sample?
3. Is color or streak a more reliable clue to the identity of a mineral? Explain your answer.
4. What is fluorescence? Name one mineral that has this property.
5. Gypsum has a ranking of 2 on Mohs' scale, while topaz has a ranking of 8. Can gypsum scratch topaz? Explain your answer.

Reviewing Chapter 9

Key Concepts

- A mineral is a natural inorganic solid substance with a characteristic chemical composition. See page 133.
- Oxygen and silicon are the most abundant elements in common minerals. See page 134.
- The six major groups of nonsilicate minerals are composed of carbonates, halides, native elements, oxides, sulfides, and sulfates. See page 135.
- Clues to identifying minerals are revealed by simple tests for streak, hardness, density, and other properties. See pages 140–142.
- Unusual properties such as magnetism, fluorescence, double refraction, and radioactivity can

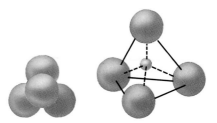

- **Silicate minerals consist of pyramid-shaped units called silicon-oxygen tetrahedra, which may be arranged in a variety of ways. See page 137.**

aid in the identification of certain minerals. See page 144.

Key Terms

cleavage (140)	inorganic (133)	phosphorescence (144)	silicon-oxygen
crystal (135)	luster (140)	refraction (144)	tetrahedron (137)
density (143)	mineral (133)	rock-forming mineral	streak (140)
double refraction (144)	mineralogist (139)	(133)	
fluorescence (144)	Mohs' scale (141)	silicate mineral (134)	
hardness (141)	nonsilicate mineral (134)		

Review

On your paper, write the letter of the term that best completes each of the following statements.

1. A natural inorganic solid substance with a characteristic chemical composition is called (a) an atom (b) a gemstone (c) a mineral (d) a tetrahedron.
2. Minerals that contain silicon and oxygen are called (a) sulfide minerals (b) sulfate minerals (c) ores (d) silicate minerals.
3. The most common silicate minerals are the (a) feldspars (b) halides (c) carbonates (d) sulfates.
4. Four percent of the earth's crust is made up of (a) sulfur and lead (b) silicate minerals (c) copper and aluminum (d) nonsilicate minerals.
5. The basic structural units of all silicate minerals consist of (a) tetrahedral networks (b) silicon-oxygen tetrahedra (c) single chains (d) double chains.
6. An example of a mineral with a basic structure consisting of single tetrahedra linked by atoms of other elements is (a) mica (b) olivine (c) quartz (d) feldspar.
7. When two single chains of tetrahedra bond to each other, the result is called a (a) single-chain silicate (b) sheet silicate (c) network silicate (d) double-chain silicate.
8. The appearance of the light reflected from the surface of a mineral is called (a) color

(b) streak (c) luster (d) fluorescence.
9. The words *waxy, pearly,* and *dull* describe a mineral's (a) luster (b) hardness (c) streak (d) fluorescence.
10. The words *uneven* and *splintery* describe a mineral's (a) cleavage (b) fracture (c) hardness (d) luster.
11. Mohs' scale is used in measuring a mineral's (a) hardness (b) specific gravity (c) color (d) luster.
12. The ratio of the mass of a mineral to its volume is the mineral's (a) atomic weight (b) density (c) mass (d) weight.
13. The needles of the first magnetic compasses used in navigation were made of the magnetic mineral (a) iron pyrite (b) silver (c) cinnabar (d) lodestone.
14. When calcite absorbs ultraviolet light and gives off red light, it is displaying the property of (a) radioactivity (b) double refraction (c) magnetism (d) fluorescence.
15. A mineral that is radioactive probably contains the element (a) uranium (b) silicon (c) fluorine (d) calcium.
16. Double refraction is a distinctive property of crystals of (a) mica (b) feldspar (c) calcite (d) galena.

Application

On your paper, write answers to the following questions.
1. Natural gas is a substance that occurs naturally in the earth's crust. It is a gas and is made up of the remains of dead plants and animals. Is it a mineral? Explain how you know.
2. Which of the following are you most likely to find in the earth's crust: the silicates feldspar and quartz or the nonsilicates copper and iron? Explain your answer.
3. Which, if any, of the following mineral groups contain silicon: carbonates, halides, or sulphates? Explain how you know.
4. Describe the tetrahedral arrangement of olivine, an ionic silicate material.
5. Why is it difficult to identify a mineral simply by its color?
6. Iron pyrite (FeS_2) is called *fool's gold* because it looks very much like gold. What simple test could you use to determine whether a mineral sample is gold or pyrite? Explain what it would show.
7. A mineral sample has a mass of 51 g and a volume of 15 cm^3. What is the density of the mineral sample?
8. Can you determine conclusively that an unknown substance contains magnetite using only a magnet? Explain.

Extension

1. Use an encyclopedia and other reference sources to research properties of ten common rock-forming minerals. Make a chart showing the name of each mineral, its chemical formula, hardness, luster, color, and density. If possible, include its streak, how it breaks, and what special properties it has.
2. Use the atlas in your school library to find a mineral map of the United States. Find out what minerals are most common in the United States. List these minerals and the states in which they are found. Present your findings to the class.
3. Research several minerals that are mined in or near your area. Include descriptions of how each mineral is extracted from the earth and how people use the mineral. Present your findings to the class.
4. Collect several different types of rock samples. Using a magnifying glass, examine the samples carefully for any visible crystals. Make a sketch of any crystals you find and use the table on page 142 to identify the crystal type.
5. Use your fingernail, a penny, a steel knife blade, and a piece of glass to determine the hardness of various mineral samples. Be extremely careful when handling the steel knife and the piece of glass.

Chapter 10
Rocks

Have you ever collected rocks? Sooner or later, as the rock collection grows, even a casual collector finds a need to organize the collection. Rock samples may be grouped by color, shape, or size. They may also be organized into scientific categories used by geologists. This chapter describes these scientific categories and explains how each kind of rock forms and changes.

Chapter Outline

Mount Whitney is in the Sierra Nevada Mountains in California.

148   Unit 3   Composition of the Earth

10.1 Rocks and the Rock Cycle

Section Objectives

- **Identify the three main types of rock and explain how each is formed.**
- **Summarize the steps in the rock cycle.**

Hot, molten rock, or *magma,* from the earth's interior is the parent material for all rocks. From the time magma cools and hardens at or near the surface of the earth, the resulting rock begins to change. In time, the rock formed from the original magma is altered many times. Geologists study not only the crust of the earth but also the forces and processes that act upon the rocks of the crust. Based upon these studies, geologists have concluded that rocks form in one of three ways. Thus, they have classified rocks into three types based on the way the rocks are formed.

Three Types of Rock

Studies of volcanic activity provide information about the formation of one rock type—**igneous rock.** The word *igneous* means "from fire." Igneous rock develops when magma cuts its way through the crust. When the magma cools and hardens, it forms igneous rock.

Forces such as wind and waves break down igneous and other types of rock into small fragments. Rock, minerals, and organic matter that have been broken into fragments are known as **sediment.** Sediment is carried away and deposited by water, ice, and wind. When sediment deposits are pressed and cemented together and harden, they form a second type of rock. This type of rock is called **sedimentary rock.**

Certain forces and processes, including tremendous pressure, extreme heat, and chemical reaction, also can change the form of existing rock. The existing rock can thus be changed into a third

Figure 10-1. When hot molten rock, or magma, reaches the surface of the earth, it is called *lava.*

type of rock that is called **metamorphic rock.** The word *metamorphic* means "changed form." Figure 10–2 shows an example of each type of rock.

Any of the three types of rock can be changed into another type. Various geological forces and processes cause rock to change from one type to another and back again. This series of changes is called the **rock cycle.**

The Rock Cycle

As you have read, cooled and hardened magma forms igneous rock. Igneous rock thus provides a good beginning for an examination of the rock cycle. Study Figure 10–3, which shows the steps in the rock cycle.

Once a body of igneous rock has formed, a number of processes on the earth's surface break down the igneous rock into sediments. When the sediments from the igneous rocks are compacted and hardened, they form sedimentary rocks. If the resulting sedimentary rocks are subjected to extremely high temperature and great pressure within the earth, they become metamorphic rocks. If the heat and pressure become even more intense, the metamorphic rock will melt and form magma. In turn, this magma may then cool and form new rock. What kind of rock—igneous, sedimentary, or metamorphic— will this new rock be?

All the rocks in the earth's crust have probably passed through the rock cycle many times during the earth's history. However, as Figure 10–3 shows, a particular body of rock does not always pass through each stage of the complete rock cycle. For example, igneous rock may never be exposed at the earth's surface where it would be changed into sediments. Instead, the igneous rock may be changed

Figure 10-2. Rhyolite (left) is an igneous rock. Chert (center) is a sedimentary rock. Marble (right) is a metamorphic rock.

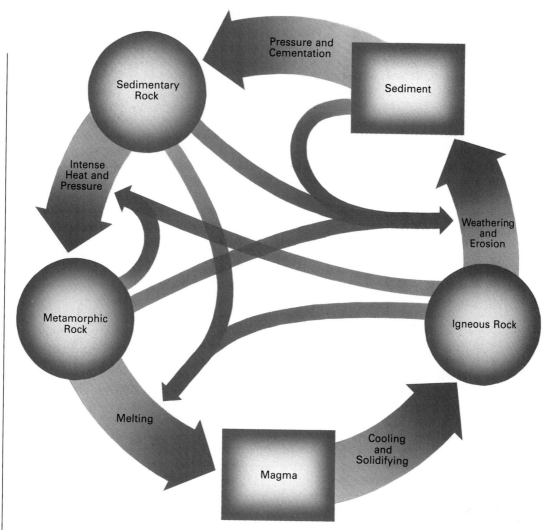

directly into metamorphic rock while still beneath the earth's surface. Igneous and sedimentary rock may melt and become magma without first becoming metamorphic rock. All three of these rock types may form sediments when they are exposed to processes at the surface of the earth.

Figure 10-3. The rock cycle illustrates the changes that sedimentary, igneous, and metamorphic rocks undergo.

Section 10.1 Review

1. Which type of rock—igneous, sedimentary, or metamorphic— forms from magma that cools and hardens?
2. Which type of rock is composed of fragments of rock, minerals, or organic matter?
3. Which type of rock forms from other rocks as a result of intense heat, pressure, or chemical reactions?
4. What is the rock cycle?
5. Does every rock go through the complete rock cycle, from igneous rock to sedimentary rock to metamorphic rock and back again to igneous rock, each time around? Explain your answer.
6. Limestone is changed into marble, a type of metamorphic rock. Name the three processes that cause this change.

- **Describe how the cooling rate of magma and lava affects the structure of igneous rocks.**
- **Classify igneous rocks according to their mineral composition.**
- **Describe a number of identifiable igneous rock structures.**

10.2 Igneous Rock

As you have read, when molten rock cools and hardens it is called igneous rock. There are two types of igneous rocks. They are classified according to where the molten rock cools and hardens. The cooling of magma deep below the crust produces **intrusive igneous rocks.** These rocks are so named because the magma that forms them intrudes, or enters, into other rock masses beneath the earth's surface. The magma then cools and hardens to form intrusive igneous rocks. The rapid cooling of lava, or melted rock on the earth's surface, produces **extrusive igneous rocks.** The many types of intrusive and extrusive igneous rocks differ in the minerals they contain and in the size of their crystalline mineral masses or grains. The size of the crystalline grains in igneous rock is called its *texture*. Both the mineral composition and the texture of igneous rocks are largely determined by the cooling rate of the magma that formed the rock.

Texture of Igneous Rock

Intrusive igneous rocks form when magma cools and hardens slowly deep underground. The slow loss of heat allows time for the minerals in the cooling magma to form large, well-developed crystalline grains. Intrusive igneous rocks composed of these large mineral grains have a coarse-grained texture. An example of a coarse-grained rock is gabbro, shown in Figure 10–4.

Extrusive igneous rocks form when lava cools rapidly on the earth's surface. Surface temperatures are cooler than underground temperatures. The rapid loss of heat does not allow time for large crystalline grains to form and, thus, produces fine-grained rock. Most extrusive igneous rocks have small mineral grains that cannot be seen by the unaided eye. An example of a common fine-grained igneous rock is basalt. Figure 10–5 shows Devil's Tower, which is composed of basalt.

Some intrusive igneous rocks form from magma that cools slowly at first and then more rapidly as it comes to the surface. This type of cooling produces large grains embedded within a mass of smaller ones. An igneous rock with a mixture of large and small grains is called a **porphyry** (POR-fuh-ree).

Extremely rapid cooling produces a kind of igneous rock in which crystals were unable to form. The obsidian, or volcanic glass, shown in Figure 10–5, illustrates the glassy appearance of this type of rock. During extremely rapid cooling, gases escaping from the molten material may become trapped in the rock and form many small bubbles. These are evident in the pumice in Figure 10–5.

Composition of Igneous Rock

The mineral composition of an igneous rock is determined by the chemical composition of the magma from which the rock develops. Different types of igneous rocks have similar mineral compositions.

Figure 10-4. Notice the coarse-grained texture of this sample of gabbro, an intrusive igneous rock.

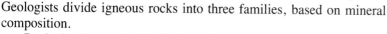

Geologists divide igneous rocks into three families, based on mineral composition.

Rocks in the granite family form from magmas called *felsic*, which are high in silica. Rocks in this family have the light coloring of their main mineral components, orthoclase feldspar and quartz. These rocks may also contain plagioclase feldspar, hornblende, and mica. In addition to the coarse-grained rock known as granite, several other rocks are part of the granite family. The fine-grained rock rhyolite also has the mineral composition that makes it a member of the granite family. Obsidian, a glassy rock of the granite family, may be bluish, black, or red, depending on its mineral composition.

Rocks in the basalt family form from magmas called *mafic*, which are low in silica. The main mineral components of rocks in this family are plagioclase feldspar and augite. These rocks may also include dark-colored minerals such as olivine, biotite, and hornblende. These three components, along with augite, give basaltic rocks a dark coloring. The dark gray or black rock known as basalt is a fine-grained member of this family. Coarse-grained gabbro is also a member of the basalt family.

The medium-colored rocks of the diorite family are made up of the minerals plagioclase feldspar, hornblende, augite, and biotite. Components of rocks in this family, however, include little or no quartz. Rocks in the diorite family include the coarse-grained diorite and the fine-grained andesite. If you know only that an igneous rock has coarse grains, can you identify the family it is in? Why?

Figure 10-5. Devil's Tower in Wyoming (left) consists of basalt, a fine-grained igneous rock. Obsidian (top, right) has no crystalline structure and, thus, has a glassy texture. Gases trapped inside magma as it cools give pumice (bottom, right) its spongelike appearance.

Igneous Rock Structures

A number of identifiable rock formations are formed of igneous rock. The underground rock masses made up of intrusive igneous rocks are called **intrusions.** The surface rock masses made up of extrusive igneous rocks are called **extrusions.**

Intrusions

The largest of all intrusions are called **batholiths.** Batholiths are very large masses of igneous rock that cover hundreds of square kilometers. The word *batholith* means "deep rock." Batholiths were once thought to extend to great depths. However, recent studies have shown that many batholiths do have lower boundaries, and they are several thousand meters thick. Batholiths form the cores of many major mountain ranges. The largest batholith in North America forms the core of the Coast Range in British Columbia. A **stock,**

Investigation

Crystal Formation

The rate of cooling affects the size of the crystals of minerals found in igneous rock. You can demonstrate this relationship by cooling crystals of Epsom salts at three different rates.

Science Process Skills Focus: observing, identifying cause-and-effect relationships

Materials
3 glasses, water, ice cubes, 3 large test tubes, small pan, spoon, gram scale or measuring cup, Epsom salts, stove, tongs, clock or watch

Procedure
1. Add the following until each glass is ⅔ full: glass 1, water and ice cubes; glass 2, warm water; glass 3, water at room temperature. Place one large test tube in each glass.
2. In a small pan, mix 350 g, about 2 cups, of Epsom salts in 500 mL water. Heat over low heat. Do not let the mixture boil. Stir until all the crystals have dissolved.
3. Carefully pour equal amounts of the Epsom salts mixture into the 3 test tubes. Use the tongs to steady the test tubes as you pour.
4. Observe crystal formation in the test tubes at 10-minute intervals for a total of 30 minutes.

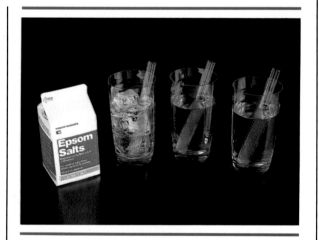

Analysis and Conclusions
1. In which test tube are the crystals the largest? The smallest?
2. How does the rate of cooling affect the size of crystals formed? Explain your answer.
3. How would you change the procedure to obtain even larger crystals of Epsom salts? Why?
4. Some igneous rocks that are thrown out of a volcano contain large crystals surrounded by very small ones. Based on your observations in this activity, explain why the crystals are different sizes.

illustrated in Figure 10–6, is an intrusion similar to a batholith that covers less than 100 km².

When magma flows between rock layers and spreads upward, it sometimes pushes the overlying rock layers into an arc. The floor of the intrusion is parallel to the rock layer beneath it. This type of intrusion is called a **laccolith.** *Laccolith* means "lake of rock." Laccoliths are frequently found in groups. You can sometimes identify them by the small dome-shaped mountains they push up on the earth's surface, like the one shown in Figure 10–6. Many laccoliths are located beneath the Black Hills of South Dakota.

When a sheet of magma flows between the layers of rock and hardens, a **sill** is formed. A sill lies parallel to the rock layers surrounding it, even if the layers are tilted. Sills vary in thickness from a few centimeters to hundreds of meters and can extend laterally for several kilometers. Big Bend National Park in Texas has examples of sills.

Magma sometimes forces its way through rock layers by following existing vertical fractures or by creating new ones. When the magma solidifies, a **dike** is formed. Dikes differ from sills in that they cut across rock layers rather than lie parallel to the rock layer. Dikes are common in areas of past volcanic activity.

Extrusions

When lava erupts onto the earth's surface, it often forms a *volcano.* A volcano is a cone of extrusive rock particles surrounding a central shaft through which the lava flows. When the eruption stops, the lava in the shaft cools and solidifies. When a volcano stops erupting for a long period, its cone gradually wears away. Eventually the softer parts of the cone are carried away by wind and water, and only the hard, solidified rock in the shaft remains. The solidified central shaft is called a **volcanic neck.** Narrow dikes that sometimes radiate out from the neck may also be exposed. A dramatic example of a volcanic neck, called *Shiprock,* is located in New Mexico.

Extrusive rock masses may also take other forms. Many extrusions are simply erratically shaped masses of extrusive rock called **lava flows.** Some extrusions, however, take the form known as a **lava plateau.** A lava plateau develops from lava that streams out of long cracks in the earth's surface. The lava then spreads over a vast area, filling in valleys and covering hills. When the lava hardens, it forms a raised plateau.

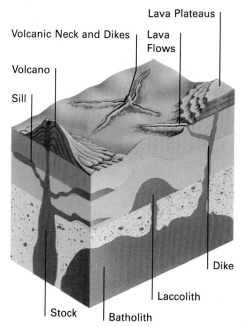

Figure 10-6. Intrusions and extrusions formed from igneous rocks create a number of characteristic landforms (top). A laccolith forms a small dome-shaped hill when exposed at the surface (bottom).

Section 10.2 Review

1. What determines whether an igneous rock will have large crystals or small crystals?
2. Name the three families of igneous rocks.
3. What is a batholith?
4. An unidentified light-colored igneous rock is made up of orthoclase feldspar and quartz. To what family of igneous rocks does it belong? Explain your answer.

10.3 Sedimentary Rock

All sedimentary rock is formed from sediments. **Cementation** and **compaction** are the processes that form sedimentary rock. In cementation, water carries dissolved minerals through the sediments. These minerals are left between the fragments of sediment and provide a cement to hold the fragments together. During compaction, extreme pressure pushes the fragments together and squeezes out air and water from between the fragments. Geologists classify sedimentary rocks according to the kind of sediments that form them.

Formation of Sedimentary Rock

One class of sedimentary rock is made up of rock fragments carried away from their source by water, wind, or ice and left as deposits elsewhere. Over time, the separate fragments may become cemented and compacted into solid rock. The rock formed from these deposits is called **clastic sedimentary rock.** A second class of sedimentary rock, called **chemical sedimentary rock,** forms from minerals that have been dissolved in water. A third class, called **organic sedimentary rock,** forms from the remains of decaying organisms.

Clastic Sedimentary Rock

Clastic sedimentary rocks are classified by the size of the sediments from which they are formed. One group consists of gravel-sized fragments that are cemented together by minerals. Rock composed of rounded gravel-sized fragments, or pebbles, is called a **conglomerate.** If the fragments are angular with sharp corners, the rock is called a **breccia** (BRECH-ee-uh). In conglomerates, as shown in Figure 10–7, and breccias the individual pieces of sediment can be seen.

Figure 10-7. Conglomerate rock is composed of rounded fragments held together by a cement (left). The formation on the right is made up of conglomerate rock.

Figure 10-8. Sandstone (top) is made up of small quartz grains cemented together by the mineral calcite. The flaky clay particles in shale (bottom) compress into flat layers.

Another group of clastic sedimentary rocks are made up of sand-sized grains that have been cemented together. These rocks are the sandstones. Since most sediments of sand-grain size are made of quartz, quartz is the major component of sandstone. Many sandstones have pores between the sand grains through which liquids move easily. An example of sandstone is shown in Figure 10–8.

A third group of clastic sedimentary rock is shale, which consists of clay-sized particles cemented and compacted under pressure. The flaky clay particles are usually pressed into flat layers that will easily split apart. Figure 10–8 shows these characteristic layers.

Chemical Sedimentary Rock

Some sedimentary rock is not made up of rock fragments but is chemical in origin. These rocks form from minerals that were once dissolved in water. Some form from dissolved minerals that precipitate, or settle, out of the water as a result of a change in temperature. For example, a certain type of limestone forms when cool currents lower the temperature of warm ocean water, and calcite precipitates and solidifies on the ocean floor.

Another type of sedimentary rock results when water evaporates and leaves behind the minerals dissolved in the water. The dissolved minerals left behind form solid rock called **evaporite.** Gypsum and halite, or rock salt, are two examples of sedimentary rock formed by evaporation. Examples of large evaporites can be found in Great Salt Lake in Utah and in Mono Lake in California.

Organic Sedimentary Rock

The third class of sedimentary rock is *organic,* which means "formed from the remains of living things." Coal and limestone are both examples of organic sedimentary rock. Coal forms from decayed plant remains that are buried and compacted into matter that is mostly carbon. You have learned that some limestone can be precipitated as a chemical sediment. However, the formation of most limestones begin when the mineral calcite is removed from sea water by animals, such as clams, oysters, and mussels. These animals use the calcite to make their shells. When the animals die, their shells and bones become limestone. Chalk is a type of limestone made up of the shells of tiny, one-celled organisms. The chalk originally formed as mud at the floor of an ancient sea. The white cliffs of Dover in England are made up of chalk.

Sedimentary Rock Features

Sedimentary rocks have a number of easily identifiable features. These include stratification, ripple marks, mud cracks, fossils, concretions, and geodes.

Horizons

Technology: SPOT

On February 21, 1986, a European Ariane rocket, shown left, launched the most powerful civilian earth-observation satellite in the world, *SPOT.* *SPOT* is an acronym for *Système Probatoire d'Observation de la Terre,* or Probative System for the Observation of Earth. The French-made satellite was designed to make multi-spectral images of surface details on the earth's crust as small as 9.9 m across. *SPOT* images can be processed so that different types of rock are highlighted. These images are available to anyone who needs them. Real estate developers might want to know the types of rock that make up a tract of land. Oil company executives might use *SPOT* images to determine whether a certain region contains enough oil-bearing shale to make drilling profitable. A mining company might be looking for new sites for granite or marble quarries.

Just two days after the successful launch of *SPOT* from Kourou, French Guiana, its electronic scanner recorded the picture shown at the right of the Atlas Mountains in Northern Algeria. The image, made possible by the sophisticated microchip technology of *SPOT,* clearly shows three major features of the earth's surface composition.

The dark blue-colored

Stratification

Layering, or **stratification,** of sedimentary rock occurs when there is a change in the kind of sediment being deposited. The type of deposit varies for a number of reasons. Changes in river currents or sea level, for example, may result in a different kind of sediment. The stratified layers, or beds, vary in thickness. Although most strata are deposited horizontally, some sediments are characterized by cross-bedding. Cross-bedding is an inclined type of stratification, as shown in Figure 10–9. When sediment deposition occurs in curved slopes, the layers will be cross-bedded.

When various sizes and kinds of materials are deposited within a layer, graded bedding will occur. Graded bedding occurs as different sizes and shapes of sediment settle to different levels. Where in each layer do you think the coarsest, heaviest, and most spherical grains of sediment settle?

Ripple Marks and Mud Cracks

Some sedimentary rocks clearly display **ripple marks.** Ripple marks are formed by the action of wind or water on sand. When the sand becomes sandstone, the ripple marks may be preserved.

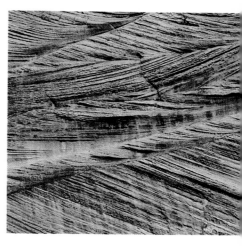

Figure 10-9. The deposition of this sedimentary rock formation occurred in slopes. Notice the cross-bedding in the rock layers.

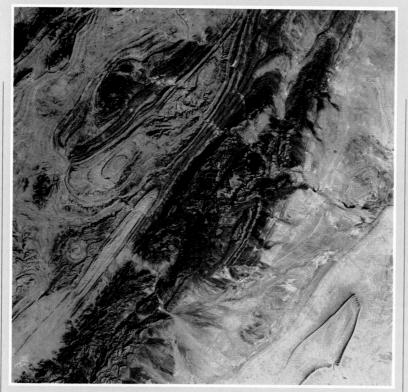

ridges in the upper left portion of the photo are hard

sandstone, and the yellow fan-shaped areas are composed of

alluvial sands that have washed down from the mountains. The small red patches running along the rivers and streams indicate there is some type of vegetation in those areas.

The *SPOT* images are available in both visible light and in a group of three spectral bands. They cover areas as large as 2,080 km². The images are said to be approximately three times as sharp as any satellite images previously available to civilians.

How can SPOT help a geologist create a map showing the distribution of the three rock types across North America?

Figure 10-10. Features of sedimentary rock may include ripple marks (top, left), mud cracks (top, right), and geodes (bottom).

The ground in Figure 10–10 shows **mud cracks,** another feature of sedimentary rock. Mud cracks result when muddy deposits dry and shrink. The shrinking causes the dried mud to crack. A river flood plain is a likely place to find mud cracks. Once the area is again flooded, new deposits fill in the cracks and preserve them as the mud hardens to solid rock.

Fossils

Remains or traces of ancient plants and animals that are usually preserved in sedimentary rock are called **fossils.** As sedimentary deposits pile up, plant and animal remains are buried. Harder parts of these remains may stay in the rock. More often, even the harder parts dissolve, leaving only an impression in the rock. Fossils are valuable in helping earth scientists learn about the development of the earth's crust. Fossils are discussed in detail in Unit 5.

Concretions and Geodes

Sedimentary rocks sometimes contain lumps, or nodules, of rock with a composition different from that of the main rock body. These nodules are known as **concretions.** Concretions form when minerals precipitated from solutions build up around an existing rock particle.

Groundwater sometimes deposits dissolved material inside cavities in sedimentary rock. These materials crystallize to form quartz or calcite, which fills the cavities, as shown in Figure 10–10. The crystal-filled cavities are called **geodes.**

Section 10.3 Review

1. How does clastic sedimentary rock differ from chemical sedimentary rock?
2. What kind of sedimentary rock forms from the remains of decaying organisms?
3. What term describes the remains or impressions of plants and animals in sedimentary rock?
4. You suspect that a rock you have found is sedimentary rock. What features would you look for to confirm your identification?

10.4 Metamorphic Rock

The changing of one type of rock to another by heat, pressure, and chemical processes is called **metamorphism.** Most metamorphic rock forms deep beneath the surface of the earth. All metamorphic rock is formed from existing igneous, sedimentary, or metamorphic rock.

Section Objectives

- **Distinguish between regional and contact metamorphism.**
- **Distinguish between foliated and unfoliated metamorphic rocks and give an example of each.**

Formation of Metamorphic Rock

During metamorphism, heat and pressure can cause certain minerals to change into other minerals. Minerals may also change in size or shape or separate into parallel layers that give the rock a banded or layered appearance. Hot liquids or gases from magma may circulate through the rock, changing the mineral composition by dissolving some materials and adding others. All of these changes are part of metamorphism. Two types of metamorphism occur in the crust of the earth. One type of metamorphism occurs when rocks come into direct contact with magma. The other type occurs due to the heat and pressure created by tectonic activity.

When hot magma pushes through existing rock, the heat from the magma can change the structure and mineral composition of the surrounding rock. This type of metamorphism is called **contact metamorphism** because only rocks near or actually touching the hot magma are metamorphosed by its heat. Chemicals and gases working through fractures may also cause changes in the surrounding rock during contact metamorphism.

Metamorphism sometimes occurs over an area of thousands of square kilometers during periods of tectonic activity. This type of metamorphism is called **regional metamorphism.** The movement of one tectonic plate against another creates tremendous heat and pressure in the rocks at the plate edges. This heat and pressure causes chemical changes in the minerals of the rock. Most metamorphic rock is formed by regional metamorphism. However, volcanism often accompanies tectonic activity. Thus, rocks formed by contact metamorphism often are found where regional metamorphism has occurred.

Classification of Metamorphic Rocks

Metamorphic rocks are classified according to their structure. Metamorphic rocks have either a **foliated** structure or an **unfoliated** structure. Rocks with a foliated structure have visible layers, or bands. Rocks without visible layers are unfoliated.

Foliated Rocks

Foliated rocks can form in one of two ways. Extreme pressure may flatten the mineral crystals in the original rock and push them into parallel layers. Foliation also occurs as minerals of different densities separate into layers, producing a series of alternating dark and light bands.

Common metamorphic rocks with foliated structure include slate, schist, and gneiss. Slate, shown in Figure 10–11, is formed by great pressure acting on the sedimentary rock shale, which contains clay minerals of a flaky consistency. The fine-grained minerals in slate are compressed into thin layers, which split easily into flat sheets.

A great amount of heat and pressure change slate into phyllite, another form of metamorphic rock. Additional heat and pressure cause the phyllite to become the coarser-grained metamorphic rock known as *schist,* shown in Figure 10–11. Deep underground, intense heat and pressure cause the elements in schist to change to very coarse-grained minerals separated into layers of different densities. This greatly metamorphosed rock with bands of light and dark minerals is gneiss, which you can also see in Figure 10–11.

Unfoliated Rocks

Unfoliated metamorphic rocks do not have layers of crystals. One common unfoliated rock is quartzite, the result of metamorphism of sandstone. During metamorphism, the sandstone is compacted so tightly that the spaces between the particles disappear. Because

Figure 10-11. The pictures below illustrate metamorphism of slate (top, left) to schist (top, right) to gneiss (bottom) as each rock is subjected to increasingly greater heat and pressure.

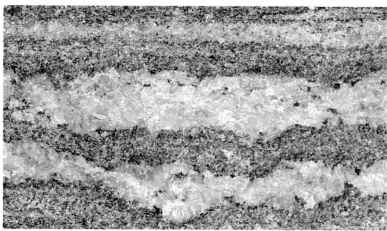

Science Notebook

Moon Rocks

NASA's Apollo moon missions have brought almost 400 kg of lunar rocks back to earth. Soon after the Apollo 11 astronauts set foot on the moon in 1969, they began to fill two boxes with brown and gray moon rocks (below). About 2,000 specimens have now been collected.

Geologists discovered that moon rocks were similar to earth rocks in composition and in the way they were formed. Geologists used their knowledge of earth rocks to analyze the moon rocks and learn about the geological history of the moon (right).

Much of the rock matter brought back from the moon was in a powdery form. This pulverized rock, called *regolith,* covers much of the surface of the moon. Rock-dating methods proved regolith to be the oldest of the moon rocks. Geologists concluded that during the first billion years of the moon, a shower of meteorites pulverized most of the then-existing moon rocks.

The solid rocks on the moon are of two types—highland rocks and mare rocks. The highland rocks are igneous rocks high in plagioclase feldspar.

Mare refers to dark areas on the moon. Mare rock formed after meteor showers dug craters on the moon. Lava poured onto the crater floors, covering wide areas of the lunar surface. The lava then cooled and hardened into basalt.

Is mare rock igneous, metamorphic, or sedimentary? Explain your answer.

quartzite is very hard and durable, it remains to form a large part of many hills and mountains after weaker rocks have been worn away. Do you think that quartzite will break into layers? Explain how you know.

Section 10.4 Review

1. Which kind of metamorphism affects only those rocks near or actually touching the hot magma?
2. What is a foliated structure? In what two ways do rocks get a foliated structure?
3. What is an unfoliated structure? Explain how quartzite gets its unfoliated structure.
4. The metamorphic rock phyllite breaks into flat sheets. Is phyllite foliated or unfoliated? Explain your answer.

Reviewing Chapter 10

Key Concepts

- Based on how they form, rocks are classified into three types. These types are igneous rock, sedimentary rock, and metamorphic rock. See page 149.

- In the rock cycle, rocks change from one type into another. See page 150.

- The rate at which magma cools determines the composition and crystal size of igneous rock. See page 152.

- Igneous rocks are classified into three families based on their mineral composition. These families are granite, basalt, and diorite. See page 153.

- Igneous rock structures take two basic forms. They are intrusions and extrusions. See page 154.

- Sedimentary rock forms in one of three ways. It may form from rock fragments, minerals once dissolved in water, or the remains of decaying organisms. See page 156.

- Sedimentary rocks have a number of identifiable features, including stratification, ripple marks, mud cracks, fossils, concretions, and geodes. See page 158.

- Metamorphic rock is formed by heat and pressure caused by hot magma or plate movement. See page 161.

- Metamorphic rocks can have a foliated or unfoliated structure. See page 161.

Key Terms

batholith (154)	dike (155)	laccolith (155)	ripple mark (159)
breccia (156)	evaporite (157)	lava flow (155)	rock cycle (150)
cementation (156)	extrusion (154)	lava plateau (155)	sediment (149)
chemical sedimentary rock (156)	extrusive igneous rock (152)	metamorphic rock (150)	sedimentary rock (149) sill (155)
clastic sedimentary rock (156)	foliated (161)	metamorphism (161)	stock (154)
compaction (156)	fossil (160)	mud crack (160)	stratification (159)
concretion (160)	geode (160)	organic sedimentary rock (156)	unfoliated (161)
conglomerate (156)	igneous rock (149)		volcanic neck (155)
contact metamorphism (161)	intrusion (154) intrusive igneous rock (152)	porphyry (152) regional metamorphism (161)	

Review

On your paper, write the letter of the term that best completes each of the following statements.

1. Rock that is formed from magma is called
   (a) igneous (b) metamorphic (c) sedimentary
   (d) clastic.

2. The changes during which rock changes from one type to another and back again is called
   (a) a rock family (b) the rock cycle (c) contact metamorphism (d) foliation.

3. Intrusive igneous rocks are characterized by a coarse-grained texture because they contain
   (a) heavy elements (b) small crystals
   (c) large crystals (d) fragments of different sizes and shapes.

4. Light-colored igneous rocks are part of the family called (a) basalt (b) hornblende
   (c) granite (d) diorite.

5. Magma that solidifies underground forms rock masses that are known as (a) extrusions (b) volcanic cones (c) lava plateaus (d) intrusions.
6. One example of an extrusion is a (a) stock (b) dike (c) batholith (d) lava plateau.
7. Sedimentary rock formed from rock fragments is called (a) organic (b) chemical (c) clastic (d) granite.
8. One example of a chemical sedimentary rock is (a) evaporite (b) coal (c) gneiss (d) breccia.
9. Hollow sedimentary rocks that become filled with quartz or calcite are called (a) dikes (b) fossils (c) sills (d) geodes.
10. Contact metamorphism is a result of (a) plate movement (b) hot magma (c) sedimentation (d) lava flows.
11. Regional metamorphism is a result of (a) plate movement (b) hot magma (c) cementation (d) compaction.
12. The splitting of slate into flat layers illustrates its (a) contact metamorphism (b) formation (c) sedimentation (d) foliation.

Application

On your paper, write answers to the following questions.
1. What type of rock will be formed from a sedimentary rock that comes under extreme pressure and heat but does not melt? Explain your answer.
2. Explain how metamorphic rock can change into either of the other two types of rocks in the rock cycle.
3. Suppose you found an igneous rock with a coarse texture. Would the magma that formed the rock have cooled slowly or quickly? Explain how you know.
4. A certain rock is made up mostly of plagioclase feldspar and augite. It also includes olivine, biotite, and hornblende. Will the rock have a light or dark coloring? Explain your answer.
5. There is a huge batholith in the northwestern part of Idaho. What can you say about the landscape in that area? Explain your answer.
6. If you know that a certain area in South Dakota has a number of laccoliths, what can you say about the landscape of that area? Explain how you know.
7. Some of the powdery rock found on the moon serves as the cementing agent for sedimentary moon rocks. What type of sedimentary rocks are these? How do you know?
8. Imagine that you have found a piece of limestone, a sedimentary rock, with strangely shaped lumps on it. Will the lumps have the same composition as the limestone? Explain your answer.
9. The western part of California is located on a boundary of two tectonic plates. Would most of the metamorphic rock in that area occur in small patches or wide regions? How do you know?
10. Which would be easier to break, the foliated rock slate or the unfoliated rock quartzite? Explain how you know.

Extension

1. Imagine that you have decided to start a rock collection. List the first 20 kinds of rocks you would acquire. Use the encyclopedia entry "Rocks" for help. Include on your list a method for classifying the rocks in your collection. Explain your system of classification to the class.
2. Find out what types of rock are common in your state. Draw a chart of the rock cycle and indicate where on the chart the rocks fall. Present your findings to the class.
3. Make a clay cross-sectional model showing a batholith, a laccolith, a dike, a stock, and a sill. Present your model to the class.
4. Concrete is an artificial rocklike material similar to natural conglomerate rock. Look at a piece of concrete and compare its makeup with that of the conglomerate shown in Figure 10–7. Draw a diagram comparing the two. Explain your diagram to the class.

Chapter 11
Resources and Energy

Supplies of valuable resources are diminishing rapidly. The world supply of
essential minerals, for example, is limited. Energy
sources, such as coal, petroleum, and natural gas are
also limited. As these resources dwindle, scientists
search for new and renewable alternate resources.

Chapter Outline

11.1 Mineral Resources
   Formation of Ores
   Uses of Mineral Resources
   Mineral Conservation

11.2 Fossil Fuels
   Coal
   Petroleum and Natural Gas
   Uses of Fossil Fuels
   Fossil Fuel Supplies
   Fossil Fuels and the
      Environment

11.3 Nuclear Energy
   Fission
   Fusion

11.4 Alternate Energy Sources
   Solar Energy
   Geothermal Energy
   Energy from Water and
      Wind

This hydroelectric plant is on the Colorado River in Arizona.

11.1 Mineral Resources

Section Objectives

• **Explain what ores are and how they form.**

• **Discuss the wide variety of uses for mineral resources.**

The earth's crust contains a wealth of useful mineral resources. Scientists have identified about two thousand minerals so far. These minerals, however, are **nonrenewable resources**—their supply is limited and they cannot be replaced once they are used. The processes that form these resources may take billions of years. **Renewable resources,** such as air, water, and plants, can be replaced as they are used. Trees, for example, are a renewable resource because more trees can be planted to replace those cut down.

Mineral resources may be *metals,* such as gold (Au), silver (Ag), and aluminum (Al), or *nonmetals,* such as sulfur (S) and quartz (SiO_2). Metals can be identified by their shiny surfaces, ability to conduct heat and electricity, and tendency to bend easily when in thin sheets. Nonmetals are dull and are poor conductors of heat and electricity. Which would be a better insulator for a hot water pipe, a metal or a nonmetal? Why?

Formation of Ores

Silver, copper (Cu), and a number of other metals are found in the earth's crust as native, or uncombined, elements. However, most metal and nonmetal elements are found in chemically combined forms as minerals in the crust. Deposits of minerals from which metals and nonmetals can be removed profitably are called **ores.** For example, iron (Fe) can be obtained from magnetite and hematite ores. Mercury (Hg) can be separated from cinnabar, and aluminum can be separated from bauxite.

Ores and Cooling Magma

Ores form within the earth's crust in a variety of ways. Chromium (Cr), nickel (Ni), and iron ores commonly form within cooling magma. As magma cools, dense metallic minerals sink to the bottom of the body of magma. Layers of these minerals accumulate and form ore deposits within the hardened magma.

Ores and Contact Metamorphism

Some lead (Pb), copper, and zinc (Zn) ores form through contact metamorphism. In contact metamorphism, hot magma comes into contact with existing rock. Heat and fluids from the magma change the surrounding rock. The amount of contact metamorphism that takes place depends upon such things as temperature of the magma, pressure, and composition and permeability of the existing rock. Often a band of mineral ore carried by the fluids forms in the new metamorphic rock that surrounds the cooled magma.

A similar but far more extensive process forms a different type of ore deposit. Hot mineral solutions spread through many small cracks in a large mass of rock and deposit minerals in narrow finger-like bands called **veins.** Figure 11–1 shows mineral veins. A large number of thick mineral veins form a deposit called a **lode.**

Figure 11–1. Hot mineral solutions have left veins of quartz in this rock. Lead, silver, gold, and copper are also found in veins.

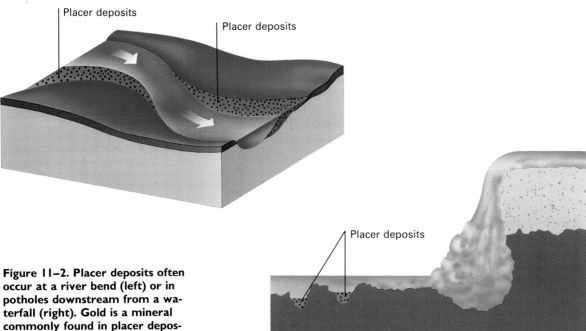

Figure 11–2. Placer deposits often occur at a river bend (left) or in potholes downstream from a waterfall (right). Gold is a mineral commonly found in placer deposits. A stream carries gold grains and nuggets dropping them where the current is weak.

Ores and Moving Water

The movement of water helps to form ore deposits in two ways. In one process, fragments of native metals such as gold are released as rock breaks down due to natural processes. Streams carry the fragments until, because of their density, they are deposited where the currents are weak. The fragments become concentrated at the bottom of stream beds in layers called **placer deposits,** as shown in Figure 11–2. In the second process, water dissolves minerals as it flows through cracks in rocks on the earth's surface. These dissolved minerals accumulate in placers, forming ore deposits. Ores of iron, lead, copper, and aluminum often form in this way.

Uses of Mineral Resources

Some metallic ores, such as gold, platinum, and silver, are prized for their beauty and rarity. Other metallic ores are sources of valuable elements. Some nonmetallic minerals are **gemstones,** rare minerals that display extraordinary brilliance and color when specially cut. Other nonmetallic minerals such as limestone and clay are used as building materials. Table 11—1 shows various uses for metallic and nonmetallic minerals.

Mineral Conservation

Most minerals are being consumed at an increasing rate each year. Every person in the growing population of the world represents a need for additional mineral resources. One source of untapped mineral resources lies beneath the ocean floors. However, these deposits are difficult to locate and even more difficult to remove. The time may come when even these mineral deposits are used up.

Table 11-1: Minerals and Their Uses

Metallic Minerals	Uses
Hematite and magnetite (iron)	in making steel
Galena (lead)	as plumbing, in "tin cans"
Gold, silver, and platinum	in electronics, in dental work, as objects such as coins, jewelry, eating utensils, and bowls
Chalcopyrite (copper)	as wiring, in coins and jewelry, as building ornaments
Sphalerite (zinc)	in making brass
Bauxite (aluminum)	as wiring, in cans, in aircraft and spacecraft

Nonmetallic Minerals	Uses
Kimberlite (diamond)	in drill bits and saws, as phonograph needles, in jewelry
Limestone (calcite)	in cement, as building stone
Halite (salt)	in food preparation
Kaolinite (clay)	in ceramics, cement, and bricks
Quartz (sand)	as glass
Sulfur	in gunpowder, medicines, and rubber
Gypsum	in plaster

The only sure way to preserve mineral resources is through conservation. One way to conserve minerals is to use other, more abundant materials in place of minerals. One such substitute is *plastics*. Another way to conserve minerals is by recycling them. Recycling is using materials over again. Many metals, such as iron, copper, and aluminum, can be recycled. Many building materials can also be used over and over. In fact, many people prefer the appearance of used, rather than new, bricks on their buildings.

Section 11.1 Review

1. What is an ore?
2. Explain how some ores form through the process of contact metamorphism.
3. What are some uses of the metals gold, silver, platinum, and copper?
4. Explain what is meant by renewable and nonrenewable resources.

Section Objectives

- **Explain why coal is a fossil fuel.**

- **Describe how petroleum and natural gas are formed and how they are removed from the earth.**

- **Discuss the importance of fossil fuels as a source of energy and of petrochemical products.**

- **Explain that fossil fuels are nonrenewable resources that must be used wisely.**

- **Describe some of the effects that the use of fossil fuels has on the environment.**

11.2 Fossil Fuels

Buried deep within the earth's crust lie the sources of much of the energy you use every day. These natural resources—coal, petroleum, and natural gas—formed from the remains of living things. They have developed and accumulated over long periods of time earlier in the earth's history. Because of their origin, coal, petroleum, and natural gas, are called **fossil fuels.** Fossil fuels consist primarily of compounds of carbon and hydrogen called **hydrocarbons.** These compounds contain energy originally obtained from sunlight by plants and animals that lived millions of years ago. When hydrocarbons are used, energy is released in the forms of heat and light.

Coal

Coal is a dark-colored, organic rock. Complex chemical and physical changes produced coal from the remains of plants such as giant ferns that flourished in prehistoric swamps millions of years ago. Usually dead plants and other organisms are decomposed by microorganisms. However, if oxygen is limited, microorganisms can only partially decompose the remains and coal may form.

Formation of Coal

The vast coal deposits of today are the remains of plants that have undergone **carbonization.** Carbonization may occur when partially decomposed trees and other plants are buried in swamp water. Bacteria consume some of the plant material. These bacteria then release marsh gas, which includes methane (CH_4), water vapor (H_2O), and carbon dioxide (CO_2). As the gas escapes, the original complex chemical compounds present in the plants gradually change, and only carbon remains.

Types of Coal

The partial decomposition of plant remains produces a brownish-black material called **peat.** Over time, peat deposits are covered by layers of sediments. The weight of these overlying sediments

Figure 11–3. Peat deposits are still forming today. Some people in Ireland and Scotland heat their houses with peat. In Ireland and the Soviet Union peat is used to fuel some electric power plants.

Figure 11–4. Coal that is deposited in horizontal beds, or layers, near the surface is strip mined, as shown on the left. Large earth-moving machines strip away soil and rock to expose the coal.

squeezes out water and gases from the peat, which then becomes a denser material called **lignite,** or brown coal. The pressure of more deposited sediments further compresses the lignite and forms **bituminous coal,** or soft coal. Bituminous is the most abundant type of coal. Where the folding of the earth's crust produces extremely high temperatures and pressure, bituminous coal is changed into **anthracite,** the hardest of all forms of coal. Bituminous coal and anthracite consist of 80 to 90 percent carbon and produce a large amount of heat when they burn. Is anthracite an igneous, sedimentary, or metamorphic rock? Why?

Petroleum and Natural Gas

Petroleum and natural gas are mixtures of hydrocarbons. The hydrocarbons formed largely from microorganisms that lived in oceans or large lakes millions of years ago. Petroleum, also called oil, consists of liquid hydrocarbons. Natural gas is made up of hydrocarbons in gaseous form.

Formation of Petroleum and Natural Gas

When microorganisms die in shallow prehistoric oceans and lakes, their remains accumulate on the ocean floor and lake bottoms and are buried by sediments. As in coal formation, these sediments limit the available oxygen supply and prevent the remains from decomposing completely. As more and more sediments accumulate, the heat and pressure on the buried organisms increases. When the heat and pressure become great enough, chemical changes occur that convert the remains into petroleum and natural gas.

Petroleum and Natural Gas Deposits

The sedimentary rocks in which petroleum is found have many interconnected spaces between rock particles. Liquids can flow through these spaces. Rock through which liquids can easily flow is called *permeable rock*. As sedimentary rock becomes deeply buried under

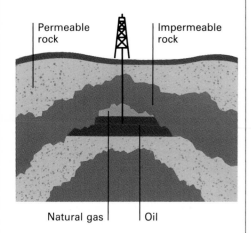

Permeable rock | Impermeable rock

Natural gas | Oil

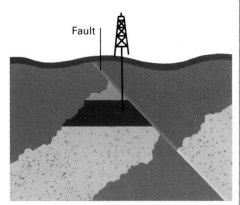

Fault

Figure 11–5. Many oil traps are anticlines, or upward-arching folds, in rock layers (top). Another common type of oil trap is a fault, or fracture, in the earth's crust (bottom).

overlying sediments, pressure increases. It forces water and hydrocarbons out of the pores and up through the layers of permeable rock. The petroleum and water continue to move until they meet a layer of impermeable rock, or rock through which liquids cannot flow. This impermeable rock layer, most commonly shale, is called *cap rock*. Petroleum that accumulates beneath the cap rock forms an oil pool. Because petroleum is less dense than water, it floats on top of the trapped water. The water beneath holds the oil pool in place. If natural gas is present, it is on top of the petroleum because natural gas is less dense than both oil and water.

Geologists explore the earth's crust to discover the kinds of rock structures, like the anticline in Figure 11–5, that are likely to contain petroleum traps. When a well is drilled into an oil pool, the petroleum and natural gas often flow to the surface. The pressure that causes water and hydrocarbons to move up through the permeable rock forces the petroleum up and out through the well.

Uses of Fossil Fuels

Fossil fuels are the main sources of energy for industries, such as transportation, farming, and construction. **Crude oil,** or unrefined petroleum, is also the source of many useful products, as shown in the graph in Figure 11–6. Notice the term **petrochemical** on the graph. Petrochemicals are chemicals derived from petroleum. Petrochemicals are the essential components of over 3,000 products. These products include plastics, synthetic fabrics, medicine, building materials, synthetic rubber, insecticides, chemical fertilizers, detergents, and shampoos.

Fossil Fuel Supplies

Fossil fuels, like minerals, are nonrenewable resources. Coal is the most abundant fossil fuel in the world. Every continent has coal but almost two thirds of the known deposits are found in three countries—the United States, the Soviet Union, and China. At the present rate of use, scientists estimate that the worldwide coal reserves will last only about 200 years.

The United States has been explored for petroleum more completely than has any other part of the world. Scientists estimate that at least 75 percent of all the petroleum in the United States has already been discovered. Much of the undiscovered supply is thought to lie on the ocean floor along the edges of North America.

Petroleum production in the United States has almost certainly reached its peak and will decline in the future. In other areas of the world, however, more than 90 percent of the petroleum known to exist is still in the ground. Scientists also think that much natural gas remains to be discovered, but it lies more than 4,600 m below the earth's surface. Oil shale also contains petroleum. However, the cost of finding and recovering the oil from shale is far greater than the present cost of recovering petroleum and gas.

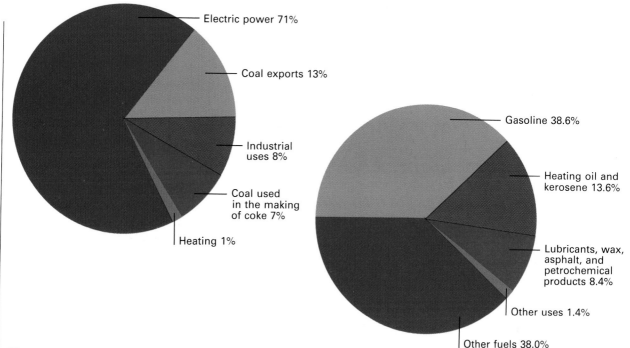

Electric power 71%

Coal exports 13%

Industrial uses 8%

Coal used in the making of coke 7%

Heating 1%

Gasoline 38.6%

Heating oil and kerosene 13.6%

Lubricants, wax, asphalt, and petrochemical products 8.4%

Other uses 1.4%

Other fuels 38.0%

Figure 11–6. The pie graphs show the uses of coal in the United States (left) and the products derived from petroleum (right).

Fossil Fuels and the Environment

The use of any fossil fuel has an impact on the environment. Strip mining of coal, shown in Figure 11–4, leaves deep ditches where coal is removed. Rocks and topsoil that are displaced to expose the coal are left in steep slopes. Without plants and topsoil to protect it, the exposed land will wear away. When wet, rocks exposed during mining give off acids. Rain may carry the acids into nearby rivers and streams, causing harm to living things there. As a result, United States legislators have passed laws to limit the damage done by strip mining.

Fossil fuels can also cause air pollution. The burning of coal with a high sulfur content releases large amounts of sulfur dioxide into the atmosphere. Although petroleum and natural gas are cleaner-burning fuels than coal, they too can damage the environment. The burning of gasoline in automobiles, for example, is a major contributor to air pollution. Other parts of the environment are also susceptible to pollution from fossil fuels. You may have heard of spills from oil wells, tankers, or pipelines that polluted the ocean and harmed wildlife.

Section 11.2 Review

1. List the four types of coal.
2. Describe the kind of rock structures in which pools of petroleum are formed.
3. Are fossil fuels renewable or nonrenewable resources?
4. Name two environmental problems associated with the mining and use of coal.
5. Why would manufacturing less plastic help conserve petroleum?

- **Summarize the process of nuclear fission.**
- **Summarize the process of nuclear fusion.**

11.3 Nuclear Energy

You read about the structure of atoms in Chapter 8. Remember that only the outermost electrons of atoms are involved in the formation of chemical bonds. The chemical changes that bond elements into molecules and compounds have no effect on the nucleus of the atom. The atomic nuclei, however, can undergo changes that release great amounts of energy.

Fission

Scientists and engineers have developed a technology that uses nuclear reactions to produce energy for commercial use. This technology is based upon **nuclear fission,** the splitting of the nucleus of a large atom into two or more smaller nuclei. Fission involves a release of a tremendous amount of heat energy, which can be used to generate electricity.

Only one kind of naturally occurring element can be used for nuclear fission. It is a rare isotope of the element uranium called *uranium-235* (U-235). Before U-235 can be used for nuclear fission, it must be separated from deposits of natural uranium. Next U-235 is mixed with radioactive U-238 and formed into pellets. This uranium is then shaped into rods called *fuel rods*. Bundles of these fuel rods are bombarded by neutrons. When struck by a neutron, the U-235 nuclei in the fuel rods split. When one U-235 nucleus in these fuel rods splits, it releases neutrons and more energy. A chain reaction occurs as these neutrons strike neighboring U-235 atoms and cause their nuclei to undergo fission. In turn, the neutrons from these nuclei strike other U-235 atoms, and the chain reaction continues. The nuclear chain reaction causes the fuel rods to become very hot. What do you suppose accounts for this heat?

A liquid, usually water, is pumped around the fuel rods to absorb and carry away the heat. The resulting hot liquid or steam becomes a source of energy for running electric generators. Once the liquid gives off its heat, it is again pumped around the fuel rods to absorb more heat.

The chain reaction that occurs during nuclear fission can be controlled. The neutron flow can be regulated so that it is slowed down, speeded up, or stopped. The equipment in which controlled nuclear fission is carried out is called a *nuclear reactor*. A reactor is shown in Figure 11–7.

The production of nuclear power has a number of drawbacks. The waste products of nuclear fission give off dangerous radiation. The radiation can destroy plant and animal cells and cause harmful changes in the genetic material within the cells of living things. In the past, nuclear waste has been stored in the ocean, but it has proved harmful to marine life. The wastes may also be harmful to the human beings who eat this marine life. A method used currently to dispose of nuclear waste is to store it in salt mines located deep

Figure 11–7. The chain reaction that produces energy by nuclear fission can be harnessed and controlled in reactors like this one.

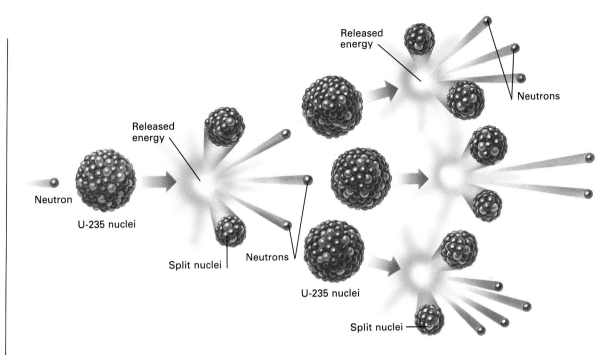

Neutron

U-235 nuclei

Released
energy

Split nuclei

Neutrons

U-235 nuclei

Split nuclei

Released
energy

Neutrons

underground. However, the radiation can pass through many types of barriers, and some radioactive waste remains dangerous for thousands of years.

Figure 11–8. The splitting of nuclei increases rapidly in a chain reaction during nuclear fission.

Fusion

All of the energy that reaches the earth from the sun is produced by another kind of nuclear reaction, which is called **nuclear fusion.** In nuclear fusion, the nuclei of smaller atoms combine and form larger nuclei. For example, nuclear fusion in the sun releases energy when hydrogen nuclei combine and form helium nuclei. These reactions can take place only at temperatures of more than 15 million degrees Celsius.

Scientists are trying to harness nuclear fusion for the production of large amounts of electricity. Much more research is needed before a commercial fusion reactor can be built. If such a reactor is built in the future, hydrogen atoms from ocean water could be used as the fuel. Thus, the amount of energy available from nuclear fusion on earth would be almost limitless. The wastes from fusion are also less harmful than those from fission.

Section 11.3 Review

1. What is the naturally occurring element that must be used for nuclear fission? From where is this element obtained?
2. What occurs during a fission chain reaction?
3. Through which kind of nuclear reaction does the sun produce heat and light?
4. Can the waste products of nuclear fission be safely disposed of in rivers and streams? Explain why.

11.4 Alternate Energy Sources

As energy needs increase, the world supply of fossil fuel will continue to be used at a much faster rate. Nuclear energy provides an alternative to fossil fuels. In addition, nations are looking into the use of other alternate energy sources in order to find a safe and renewable energy source.

Solar Energy

Every 15 minutes the earth receives enough energy from the sun to meet the energy needs of the entire world for one year. If scientists could capture and use even a small part of this solar energy, they could solve the energy problems of the world. There are two general ways of converting solar energy into useful forms. One method is called a *passive system* because it requires no mechanical setup. An example of a passive solar energy system is a house with windows facing the warming rays of the sun. Sunlight enters the building and warms the building material, which may be stone or brick, for example. The warmed material heats the building and stores some heat for the evening.

The other method of capturing solar energy is called an *active system*. Some buildings have **solar collectors** on their roofs. A solar collector is a box with a glass or plastic top. Beneath the box are tubes that circulate water through the building. The sun heats the water as it moves through the tubes, providing heat and hot water. On cloudy days, however, there may not be enough sunlight to heat the water. On these days the system must use the heat it has stored.

Geothermal Energy

In many locations, water flows within the earth's crust far beneath the surface. This water may flow through rock that is heated by a nearby body of magma or by hot gases released by magma. The

Figure 11–9. Geothermal steam fields produce electricity in New Zealand, shown right, Italy, Japan, Mexico, the Soviet Union, Iceland, and the United States.

Investigation

Solar Collector

You can demonstrate the principles of solar collection with a simple model.

Science Process Skills Focus: constructing models, recording data, identifying variables

Materials

small pan, sheet of black plastic, thermometer, adhesive tape, water at room temperature, plastic wrap, rubber bands, clock or watch

Procedure

1. Line the inside of the pan with black plastic. Tape the thermometer to the inside of the pan as shown. Fill the pan three fourths full with water at room temperature. Fasten plastic wrap over the pan with a rubber band. Be sure you can read the thermometer.
2. Place the pan in a sunny area. Record the temperature every minute until it remains constant for 3 minutes. Discard the water.
3. Repeat the procedure in Steps 1 and 2 but do not cover the pan with plastic wrap.
4. Repeat the procedure in Steps 1 and 2 but do not line the pan with black plastic.
5. Repeat the procedure once again but do not line the pan with black plastic and do not

cover the pan with plastic wrap.
6. Calculate the rate of temperature change for each trial. To do so, subtract the beginning temperature from the ending temperature. Divide this number by the number of minutes it took for the temperature to remain constant.

Analysis and Conclusions

1. What are the variables in this investigation? Which model had the greatest rate of temperature change? The least?
2. Which variable that you tested has the more significant effect on temperature change?
3. What materials would you use to make an efficient solar collector of your own design?

---

water is heated as it flows through the rock. The hot water, or the steam that results, is the source of a huge supply of energy. This energy is called **geothermal energy,** which means "energy from the heat of the earth's interior."

Energy experts have harnessed some of this energy by drilling wells to reach the hot water. A steady stream of very hot water or steam is pumped up through the wells. Sometimes water must first be pumped down into the hot rocks if water does not already flow through them. The resulting steam and hot water can be used as a direct source of heat. The steam and hot water also serve as sources of power to drive turbines, which generate electricity. The city of San Francisco, for example, gets some of its electricity from a geothermal power plant located in nearby mountains. In Iceland, 80 percent of the homes are heated by geothermal energy, and Italy and Japan have developed power plants using geothermal energy.

Energy from Water and Wind

Two additional renewable resources can help satisfy the growing energy needs of the world. These resources are wind and water. Energy from water can come from the running water of rivers and streams or from ocean tides.

Energy from Running Water

A great proportion of the energy needs can be met by **hydroelectric power,** the power produced by running water. Today, 11 percent of the electricity in the United States comes from hydroelectric power plants. At a hydroelectric plant, massive dams hold back running water and channel it through the plant. Inside the plant, the force of the water spins turbines of electrical generators, which, in turn, produce electricity.

Energy from Tides

Twice each day, water in the oceans moves toward and then away from the shore. These movements are called *tides*. High tide occurs when the water reaches its highest point along the shore, low tide, when it reaches its lowest point. To make use of this tidal power,

Horizons

Technology: Hot Dry Rock

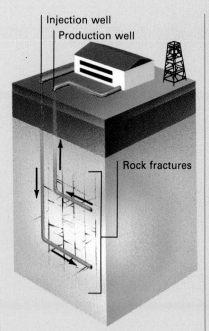

Injection well
Production well

Rock fractures

I n 1986 scientists in Los Alamos, New Mexico, unveiled new technology that made geothermal energy an option for people in many areas.

Until recently the use of geothermal energy had been restricted to people living near natural underground hot water reservoirs and springs. The scientists at the Los Alamos National Laboratory, however, used hot rocks deep in the earth's interior to produce energy. Their project, named *Hot Dry Rock*, generated enough energy during a monthlong trial to power a town of 2,000 people.

The key to a successful hot dry rock installation is the availability of hot underground rock formations. In order to be most useful as a source of geothermal energy, the hot rock formations should have temperatures of 100°C or more and be less than 10 km below the surface. Most such rock deposits are

dams trap the water at high tide and release it at low tide. Based on what you have read about hydroelectric power, how does the channeled tide water generate electricity?

Energy from Wind

Wind energy is now being used worldwide to produce electricity. Small wind-driven generators are used to meet some energy needs of individual homes. Larger devices produce power to meet most electric needs in certain locations. There are only a few places, however, where the wind generators are practical. Even in the most favorable locations, the wind does not always blow. Consequently, wind-driven generators work best if some electrical energy is stored. Because wind-driven generators are practical only in areas with suitable winds, they can supply very little of the energy needs worldwide.

Section 11.4 Review

1. How does a solar collector work?
2. How do most of the people in Iceland heat their homes?
3. Is tidal power renewable or nonrenewable? Why?

found where magma has invaded the crust.

For the hot dry rock installation, a drill bit, like that shown in the photo at the left, was used to drill two wells. The wells, like those shown in the diagram far left, were each about 4 km deep. Then the scientists pumped water under extreme pressure into one well to fracture the rock structure. After the rock was fractured, cold water was forced down the same well. The cold water passed through the fractured, naturally hot rock. It became heated to more than 100°C, and it then moved up the second well to the surface. At the surface, the hot water turned to steam. This steam could drive turbines to generate electricity, or it could be used directly to heat homes and businesses.

Scientists estimate that a hot dry rock installation could remain an effective energy producer for 10 to 40 years. The continual pumping of cold water, however, eventually causes an installation to become ineffective.

What do you think causes the installation to lose its effectiveness?

Reviewing Chapter 11

Key Concepts

- Ores are mineral deposits from which metals and nonmetallic minerals can be removed. See page 167.
- Minerals are important sources of many useful and valuable materials. See page 168.
- Millions of years ago, chemical and physical changes turned the remains of plants into coal. See page 170.
- Petroleum and natural gas formed from the remains of ancient microorganisms. Both are found beneath impermeable layers of rock. See page 171.
- Coal, petroleum, and natural gas provide much of the energy for developed societies. Numerous additional products are made of chemicals derived from petroleum. See page 172.
- Fossil fuels are nonrenewable resources, which must be used wisely. See page 172.
- The extraction and use of fossil fuels may damage the environment. See page 173.
- Nuclear fission can produce energy to generate electricity. Nuclear fission, however, gives off dangerous radiation. See page 174.
- If scientists could produce nuclear fusion, the energy available would be limitless. See page 175.
- Energy from the sun can be harnessed by both passive and active methods. See page 176.
- Experts can utilize geothermal energy—energy from the heat of the earth's interior. See page 176.
- Alternative forms of energy include hydroelectric power and wind power. See page 178.

Key Terms

anthracite (171)	geothermal energy (177)	nonrenewable resource (167)	petrochemical (172)
bituminous coal (171)	hydrocarbon (170)	nuclear fission (174)	placer deposit (168)
carbonization (170)	hydroelectric power (178)	nuclear fusion (175)	renewable resource (167)
crude oil (172)	lignite (171)	ore (167)	solar collector (176)
fossil fuel (170)	lode (167)	peat (170)	vein (167)
gemstone (168)			

Review

On your paper, write the letter of the term that best completes each of the following statements.

1. Metals are known to (a) have a dull surface (b) provide fuel (c) conduct heat and electricity (d) be found in shale.
2. Aluminum can be taken out of bauxite, which is (a) an ore (b) an energy source (c) a renewable resource (d) a fossil fuel.
3. Hot mineral solutions that spread through cracks in rock, form bands called (a) placer deposits (b) crystal (c) veins (d) silicates.
4. Gemstones include rubies and (a) gold (b) diamonds (c) silver (d) platinum.
5. Energy resources that have formed from the remains of living things are called (a) minerals (b) metals (c) gemstones (d) fossil fuels.
6. At the top of an oil pool is a layer of (a) coal (b) cap rock (c) peat (d) water.
7. Plastics, synthetic fabrics, and synthetic rubber are composed of chemicals derived from (a) anthracite (b) peat (c) petroleum (d) shale.
8. The most abundant fossil fuel is (a) coal (b) petroleum (c) natural gas (d) shale.

9. Scientists estimate that 75 percent of the petroleum deposits in the United States (a) have not yet been discovered (b) have been used up (c) cannot be used for fuel (d) have already been discovered.
10. One problem caused by the strip mining of coal is (a) increased rainfall (b) loss of soil (c) the release of sulfur dioxide into the air (d) the drying up of rivers.
11. The splitting of the nucleus of an atom to produce energy is called (a) geothermal energy (b) nuclear fission (c) nuclear fusion (d) hydroelectric power.
12. The energy that reaches the earth from the sun is produced by (a) nuclear fission (b) geothermal power (c) fossil fuels (d) nuclear fusion.
13. Hydrogen atoms may someday provide fuel for (a) nuclear fission (b) hydroelectric power (c) geothermal energy (d) nuclear fusion.
14. Most solar collectors require (a) coal (b) water (c) fission (d) wind.
15. Energy experts have harnessed geothermal energy by (a) building dams (b) building wind generators (c) drilling wells (d) burning coal.
16. In a hydroelectric plant, running water produces energy by spinning a (a) turbine (b) fan (c) windmill (d) reactor.

Application

On your paper, write answers to the following questions.
1. What might an engineer from a mining company look for in rock masses to identify possible copper deposits? Explain your answer.
2. Do you think it would be profitable for a mining company to mine hematite? Explain your answer.
3. Explain the term *fossil fuel,* using coal as an example.
4. A certain area has extensive deposits of shale. Why might a petroleum geologist be interested in the area?
5. List five products commonly found in households composed of petrochemicals.
6. If the United States continues using petroleum in vast amounts, it will eventually have to depend upon foreign sources for this resource. Why?
7. Imagine that you are on a committee to reduce air pollution in a crowded city. Would you recommend the use of high-sulfur coal as a fuel? Why?
8. A certain company in your area produces U235 pellets and fuel rods. With which energy source is the company involved? How do you know?
9. Imagine that your senator is thinking of proposing a bill to cut off funding for nuclear fusion research. What might you say in a letter to change the senator's mind? What might you say to agree with the senator?
10. If your school was converting to solar energy, how would you decide the best place on the roof for a solar collector?
11. Why might hot dry rock technology increase the use of geothermal energy?
12. Is hydroelectric power considered a renewable resource or a nonrenewable resource? Explain your answer.

Extension

1. Choose one important mineral and research its various uses. Present your findings to the class in a chart. Indicate the original ore, the mineral extracted from it, and a number of uses of the mineral.
2. Using Figure 11–5 as a reference, make a clay model of an oil trap. Use clay of various colors to show the different substances in the oil trap. Present your model to the class.
3. Research the effects of automobile fuel on the quality of air. Report your findings to the class.
4. Research a typical nuclear power plant. Draw a large diagram showing the major parts of this type of power plant. Present your diagram to the class.

UNIT
4
Reshaping the Crust

Capitol Reef National Park, Utah (background); Yellowstone Falls, Yellowstone National Park, Wyoming (inset)

Introducing Unit Four

As a park ranger, I know that protecting and preserving our natural environment isn't something done just for future generations. It's something we do for ourselves—for today—as well.

In addition to supporting the delicate balance of nature, parks also play a vital role in rejuvenating the visiting public. Somehow, being surrounded by unaltered nature puts the details of our busy lives into perspective. In these respects, what I and others are doing to preserve our environment for the future also has a direct, positive effect on those things that are important to us today.

This unit discusses the forces that have shaped and continue to shape the earth. Understanding these forces may help you find ways to help preserve our natural environment—for today as well as tomorrow.

Pat Buccello

Patricia Buccello
Park Ranger
Sequoia National Park

12 Weathering and Erosion

13 Water and Erosion

14 Groundwater and Erosion

15 Glaciers and Erosion

16 Erosion by Wind and Waves

Chapter 12

Weathering and Erosion

The face of our planet is constantly changing. A massive landslide may make dramatic changes in an instant. Other changes, such as those that create sculptured canyons, take place gradually over thousands of years. In this chapter, you will study how natural agents such as wind and rain have helped reshape the surface of the earth.

Chapter Outline

12.1 Weathering Processes
Mechanical Weathering
Chemical Weathering

12.2 Rates of Weathering
Rock Composition
Amount of Exposure
Climate
Topography

12.3 Weathering and Soil
Soil Composition
Soil Profile
Soil and Climate
Soil and Topography

12.4 Erosion
Soil Erosion
Accelerated Soil Erosion
Soil Conservation
Gravity and Erosion
Erosion and Landforms

These formations are found in Bryce Canyon National Park, Utah.

12.1 Weathering Processes

Most rocks deep within the earth's crust were formed under high temperature and pressure. When these rocks are uplifted to the surface, they are exposed to much lower temperature and pressure and to the gases and water in the air. As a result, rocks on the earth's surface undergo changes in their appearance and composition. The change in the physical form or chemical composition of rock materials exposed at the earth's surface is called **weathering.**

There are two types of weathering processes. **Mechanical weathering** processes break rock into smaller pieces but do not change its chemical composition. **Chemical weathering** processes break down rock by changing its chemical composition.

Mechanical Weathering

Mechanical weathering is strictly a physical process. Common agents of mechanical weathering are ice, plants and animals, gravity, running water, and wind. Physical changes within the rock itself aid mechanical weathering. For example, granite is formed deep beneath the earth's surface. As overlying rocks are removed, the pressure on the granite decreases. As a result, the granite expands, and long, curved cracks parallel to the surface, or **joints,** develop in the rock. When joints develop on the surface of the rock, the rock breaks into curved sheets that peel away from the underlying rock in a process called **joint sheeting.**

Ice Wedging

A common kind of mechanical weathering is called **ice wedging.** Ice wedging occurs when water seeps into cracks or joints in rock and freezes. When the water freezes, its volume expands by about 10 percent, creating pressure on the rock. Every time the ice thaws and refreezes, it wedges farther into the rock, and the crack widens and deepens. This process eventually will split the rock apart, as shown in Figure 12–1.

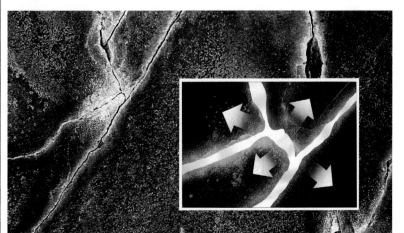

Figure 12-1. Ice acts as a wedge that breaks rock apart, as shown in the inset. Ice wedging is responsible for much of the shattering seen in the photograph.

Ice wedging commonly occurs at high altitudes. It also occurs in places where the temperature regularly varies above and below the freezing point, such as in the northern United States.

Organic Activity

Plants and animals are important agents of mechanical weathering. The roots of plants can work their way into cracks in rock. As the roots grow and expand, they create pressure that wedges the rock apart. The digging activities of burrowing animals constantly expose new rock surfaces to weathering. These processes may appear trivial, but together they can be effective in weathering rocks over a long period of time.

Abrasion

The collision of rocks with one another, resulting in the breaking and wearing away of the rocks, is a form of mechanical weathering

Investigation

Mechanical Weathering

You can demonstrate the effects of mechanical weathering by water by placing rocks in a container of water and shaking the container.

Science Process Skills Focus: constructing models, observing, recording data, interpreting data

Materials

silicate rocks; hand lens; 2 plastic containers, one with tight-fitting lid; water; strainer; clear glass jar

Procedure

1. Examine the rocks with a hand lens, noting the shape and surface texture.
2. Fill the plastic container with the tight-fitting lid about half full of rocks. Add water to barely cover the rocks. Tighten the lid and shake the container 100 times.
3. Hold the strainer over the other container. Pour the water and rocks into the strainer.
4. Run your finger around the inside of the empty container. Write down what you feel.
5. Use the hand lens to observe the rocks.
6. Pour the water into the glass jar and examine the water with a hand lens.
7. Put the rocks and water back into the container with the tight-fitting lid. Tighten the lid and shake the container 100 times.
8. Repeat Steps 4 through 7 three more times.

Analysis and Conclusions

1. Has the amount and particle size of the residue left in the container changed during the course of the investigation? Explain your answer.
2. How has the appearance of the rocks changed? The appearance of the water?
3. If the water from a small stream ran over a ledge of rock into a pool below, what would you expect to find at the bottom of the pool?

Figure 12-2. The roots of plants act like wedges to split rock.

called **abrasion.** Agents of abrasion are gravity, running water, and wind. Gravity causes loose soil and rocks to move down the slope of a hill or mountain. Rocks break into smaller pieces as they fall and collide. Running water or wind also can carry particles of sand or rock. These particles scrape against each other and against other stationary rocks, thus abrading the exposed surfaces. What evidence of abrasion have you seen at the beach?

Chemical Weathering

Chemical weathering, or decomposition, occurs when chemical reactions take place between the minerals in the rock and water, carbon dioxide, oxygen, and acids. These chemical reactions alter the internal structure of the original mineral and lead to the formation of new minerals. As a result, both the chemical composition and the physical appearance of the rock undergo changes.

Hydrolysis
Water plays a crucial role in chemical weathering. The change in the composition of minerals when they react chemically with water is called **hydrolysis.** For example, a type of feldspar combines with water and produces a common clay called *kaolin.* In this reaction, hydrogen ions from the water displace such elements as potassium or calcium in the feldspar. The feldspar thus is changed into clay by a chemical reaction.

The minerals affected by hydrolysis may dissolve in water. The water then carries the dissolved minerals to lower layers of rock in a process called **leaching.** Mineral ore deposits may form when leaching causes a mineral to concentrate in a thin layer beneath the earth's surface.

Carbonation
When carbon dioxide (CO_2) from the air dissolves in water (H_2O), a weak acid solution called *carbonic acid* (H_2CO_3) is produced:

$$H_2O + CO_2 \longrightarrow H_2CO_3$$

Figure 12-3. Carbonic acid dissolved the calcite in limestone to produce this underground cavern.

The acid has a higher concentration of hydrogen ions than does pure water. These ions speed up the process of hydrolysis.

When some minerals come in contact with carbonic acid, they combine chemically with the H_2CO_3 and form a new product. This process is called **carbonation.** For example, carbonic acid reacts with calcite, which is a major component of limestone, and converts it to calcium bicarbonate. Calcium bicarbonate dissolves easily in water, so the limestone eventually is eaten away. The dissolving action of carbonic acid on limestone sometimes produces underground caverns such as that shown in Figure 12–3.

Oxidation

When metallic elements combine with oxygen, **oxidation** occurs. Oxidation often attacks rock with iron-bearing minerals. Iron (Fe) combines quickly with oxygen (O_2) dissolved in water and forms rust, or iron oxide (Fe_2O_3):

$$4Fe \ + \ 3O_2 \longrightarrow 2Fe_2O_3$$

The red color of much of the soil in the southeastern states is due mainly to the presence of iron oxide produced by oxidation.

Exfoliation

When the minerals of granite have been combined with water, the physical structure of the rock is weakened. In time, thin sheets of the weakened rock flake off the surface. This peeling action is called **exfoliation.** Exfoliation gives the weathered rock a rounded appearance, as shown in Figure 12–4.

Figure 12-4. Notice the peeling action of exfoliation shown below.

Acid Rain

Rainwater is naturally slightly acidic because it contains dissolved carbon dioxide. However, in industrial and heavily populated areas, waste gases containing oxides of nitrogen and sulfur are released into the air. These compounds combine with water in the atmosphere and produce **acid rain.** Acid rain has a greater ability to weather rock than does ordinary rain. Thus, the chemical weathering processes are greatly accelerated.

IMPACT

The Price of Corrosion

Corrosion is the natural chemical weathering of metals. However, when corrosion occurs in metals used in building or manufacturing, it poses a serious problem.

The cost of replacing or treating metals damaged by corrosion in the United States alone is estimated at $170 billion a year. However, the drain on natural resources also is enormous. For example, producing a single ton of steel requires 4 tons of coal.

Rust, an oxide that affects iron and steel, is by far the most common form of corrosion. Rust not only discolors but also weakens metal. It attacks the metal in buildings, bridges, culverts, and highways. Left unchecked, rust will eventually destroy these structures.

Rust can also pose a threat within the home, where it attacks pipes and pipe fittings, furnace exhaust ducts, and electrical conduits.

Rust may ruin the appearance of an automobile, but it can also threaten the lives of drivers and passengers. Rust can damage the floor of the passenger compartment, allowing carbon monoxide fumes to enter. It can also weaken the suspension, steering, and brake systems.

Trying to stop rust is, at best, difficult. Steel can be galvanized, a process that covers the metal with a zinc coating, but this only delays the inevitable. Wherever moisture, heat, and steel exist together, you are likely to find rust.

What chemical reaction produces both corrosion and iron oxide in soil?

Plant Acids

Acids are produced naturally by some plants. Lichens and mosses grow on rocks and produce weak acids that can dissolve the surface of the rock. The acids also can seep into the rock and produce cracks that eventually cause it to break apart.

Section 12.1 Review

1. In what kind of climate does ice wedging usually occur?
2. How do plants and animals help weather rocks?
3. What agents promote abrasion?
4. What chemical weathering process occurs when minerals in the rock react with carbon dioxide? When minerals react with water?
5. Automobile exhaust contains nitrogen oxides. How might these pollutants affect chemical weathering processes?

- **Explain how rock composition affects the rate of weathering.**
- **Discuss how the amount of exposure determines the rate at which rock weathers.**
- **Describe the effects of climate on the rate of weathering.**

12.2 Rates of Weathering

The processes of weathering generally work very slowly. For example, carbonation dissolves limestone at an average rate of only about one twentieth of a centimeter every 100 years. At this rate, it would take 30 million years to dissolve a 150-m layer of limestone from the earth's surface.

Rocks do not weather at the same rate. Different rates of weathering produce different formations, such as those shown in Figure 12–5. The rate at which rock weathers depends on a number of factors.

Rock Composition

The composition of rocks is a major factor in their rate of weathering. Often, igneous and metamorphic rocks on the earth's surface remain almost unchanged after all the surrounding sedimentary rock has weathered away. Of the minerals in igneous and metamorphic rock, quartz is the least affected by chemical weathering. Because

Figure 12-5. Different rock composition and structures account for the different rates of weathering in this rock formation.

quartz is one of the hardest minerals, it also resists mechanical weathering and retains its structure in tiny grains of sand.

Among sedimentary rocks, limestone and other rocks containing calcite are most rapidly weathered. Although these rocks resist mechanical weathering, they are easily weathered by carbonation.

Many other sedimentary rocks are affected mainly by mechanical weathering processes. The rate at which these rocks weather depends on the material that holds the fragments of the sediments together. For example, shales and sandstones that are not firmly cemented together gradually break up into clay and sand particles. However, conglomerates and those sandstones that are strongly cemented by silicates may resist weathering processes even longer than most igneous rocks.

Amount of Exposure

The more exposure a rock receives to weathering processes, the faster it will weather. The amount of time the rock is exposed and the amount of surface area available for weathering are important factors.

Most rocks at the surface of the earth are broken by fractures and joints. Fractures also form natural channels through which water flows. Water penetrates the rock through these channels and breaks the rock by ice wedging. These processes usually split the rocks into a number of smaller blocks. Fractures and joints increase the surface area of a rock and allow weathering to take place more rapidly.

For example, picture a block of rock as a cube with six sides exposed. When the block is broken once, two more surfaces are created. When each of those blocks is broken, four more surfaces are created. Splitting the original block into eight smaller blocks, as shown in Figure 12–6, doubles the total surface area available for weathering.

Climate

In general, the freezes and thaws produced by alternating hot and cold weather and rainfall have the greatest effect on the rate of weathering. What chemical and mechanical weathering processes do you think are affected most by these conditions?

In hot, dry climates, weathering takes place slowly. The lack of water limits the rate of the many chemical and mechanical weathering processes associated with water, such as carbonation and ice wedging. Weathering is also slow in very cold climates. In warm, humid climates, chemical weathering is fairly rapid. The constant moisture is highly destructive to exposed surfaces.

Climates in which variable weather conditions exist can cause ice wedging fractures, which help to expose new rock surfaces. Chemical weathering then can attack the fractured rock more quickly. When temperatures rise daily or seasonally, the rate at which chemical reactions occur also accelerates.

Figure 12-6. When fractures bisect every edge of a smooth cube of rock, the surface area exposed to weathering doubles.

Figure 12-7. The photograph on the right shows Cleopatra's Needle after only one century in New York City. Compare the effects of weathering on the carvings. The top inset shows the carvings just after the obelisk was moved to New York. The bottom inset shows the carvings after one century in New York.

The effect of climate on weathering rates is seen on Cleopatra's Needle, an Egyptian obelisk made of granite. For 3,000 years, the obelisk stood in Egypt, where the hot, dry climate scarcely changed its surface. Then, in 1880, the United States government received Cleopatra's Needle and moved it to Central Park in New York City. After only 100 years of exposure to moisture and industrial and automobile pollution, Cleopatra's Needle has been damaged severely. Ice wedging and chemical weathering due to acid rain did more harm in a single century than had been caused in the preceding 30 centuries.

Topography

The topography of the land, or altitude and slope of the surface where rock is located, also influences the rate of weathering. Because temperatures are lower at higher elevations, ice wedging often increases as the altitude increases. On steep slopes, such as mountainsides, weathered rock fragments are pulled downhill by gravity and washed along by heavy rains. New surfaces of the mountain are thus exposed more quickly to weathering.

Section 12.2 Review

1. Which mineral in igneous and metamorphic rocks is least affected by chemical weathering?
2. Why does fractured rock weather more rapidly than smooth rock?
3. What climatic factors influence weathering rates?
4. How does the topography of a region affect weathering rates?
5. How would Cleopatra's Needle probably have been affected if it had been in the cold, dry climate of Siberia for 100 years?

12.3 Weathering and Soil

Section Objectives

- **Explain how the composition of parent rock affects soil composition.**

- **Predict the type of soil produced in various climates.**

Weathering processes break and alter almost all the rocks on the surface of the earth. A result of weathering is the formation of **regolith,** the layer of weathered rock fragments covering much of the earth's surface. **Bedrock** is the solid, unweathered rock beneath the regolith. The relation between bedrock and regolith can be seen in Figure 12–8.

The lower regions of regolith are partly protected by those above and do not weather as rapidly as the upper regions. Eventually the uppermost rock fragments weather and form a layer of very fine particles. This layer of small rock particles provides the basic components of **soil.** Soil is a complex mixture of minerals, water, gases, and the remains of dead organisms. As plants and animals die, their remains decay and produce **humus,** a dark, organic material. Humus is part of any fully developed soil.

Soil Composition

The rock material in soil consists of three main types: clay, sand, and silt. These materials are classified by the size of their particles. Clay particles are less than 0.002 mm in diameter, silt particles range from 0.002 to 0.05 mm, and sand particles from 0.05 to 2 mm. The proportion of these three materials in a particular soil depends mainly on the rock from which the soil was weathered, called its *parent rock.*

Soils containing large amounts of clay are weathered from parent rock that is rich in feldspar. When feldspar weathers, fine grains of silicate material containing aluminum and water are formed. The individual grains of clay are too small to be felt or seen by the unaided eye. Other weathered minerals containing aluminum also may form clay.

Weathered granite and other rocks containing large amounts of quartz form sandy soils. Generally, sand consists of small quartz

Figure 12-8. A blanket of regolith overlies the solid bedrock below.

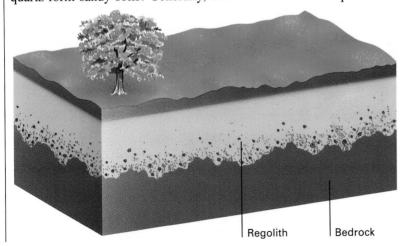

Regolith | Bedrock

grains, which are large enough to be both felt and seen. Quartz rarely is weathered chemically. So quartz grains remain behind after other minerals in rock have broken down.

Silt particles are too small to be seen easily, but they give a gritty feel to the silt. Currents often carry silt along riverbeds and deposit it on soils near river banks.

The weathered mineral grains that form soil may be carried away from the location of the parent rock by water, wind, or glaciers. The soil resulting from the deposit of this material, called *transported soil,* may have a different composition than the bedrock on which it rests.

Soil Profile

Soil that rests on top of its parent rock, called *residual soil,* often develops distinct layers over a long period of time. To determine the composition of these soil layers, scientists study a **soil profile.** A soil profile is a cross section in which layers of the soil and the bedrock beneath the soil can be seen. Figure 12–9 shows the layers of a soil profile. The layers are called **horizons.** Fully developed, or

Figure 12-9. Notice the A horizon (black), B horizon (gray), and C horizon (brown) in this soil profile.

mature, residual soils generally consist of three principal horizons: the A horizon, B horizon, and C horizon.

The A horizon of a mature soil is topsoil. Topsoil is a mixture of organic and small rock materials. Almost all the living things that live in soil inhabit the A horizon. Therefore, the A horizon contains humus and other organic materials. The A horizon is also the zone from which surface water leaches the minerals.

The B horizon is immediately beneath the A horizon. The B horizon is the subsoil, which contains the minerals leached from the topsoil, clay, and, sometimes, humus. In dry climates, the B horizon also may contain minerals that accumulate as the water in the soil evaporates rapidly.

The C horizon—the bottom layer—consists of bedrock that has been partially weathered. This bottom layer is in the first stages of mechanical and chemical change. In time, weathering will cause this horizon to develop into a B horizon and, finally, an A horizon.

Transported soils do not have horizons. Instead, they are deposited in sorted layers or in unsorted masses.

Soil and Climate

Climate is the most important factor influencing soil formation. Climate determines the weathering processes that occur in a region. These processes partly determine the composition of the soil.

In tropical climates, where there is much rain and temperatures are high, chemical weathering causes thick soils to develop rapidly. These soils are called **laterites** (LAT-UH-rites). This type of soil contains iron and aluminum minerals that do not dissolve easily in water. Rain leaches other minerals from the A horizon of the soil and the minerals sometimes collect in the B horizon. The heavy rains also wash away much of the topsoil, thus keeping the A horizon thin. However, because of the dense vegetation in these climates, organic material is continuously added to the soil. As a result, a thin layer of humus usually covers the B horizon. Why would tropical soil not be good for farming?

In desert regions, where rainfall is minimal, chemical weathering processes do not occur rapidly. As a result, the soil is thin and consists mostly of regolith—evidence that mechanical processes are occurring. The soil in arctic climates also is formed mainly by mechanical weathering. Because chemical weathering takes place slowly in arctic regions, the soil also is thin and is made up of rock fragments.

In temperate regions, where temperatures range between cool and warm and rainfall is not excessive, both mechanical and chemical weathering processes take place. All three soil horizons develop and reach depths of several meters. In some regions, alternate freezing and thawing causes the soil to be somewhat denser than that found in areas with more moderate climates.

Two main soil types are found in temperate regions. In areas receiving more than 65 cm of rain a year, a type of soil called

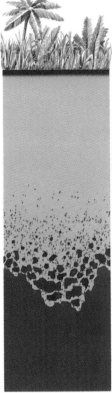

Arctic soil

Tropical soil

Figure 12-10. Tropical climates produce thick, infertile soils. Arctic climates produce thin soils.

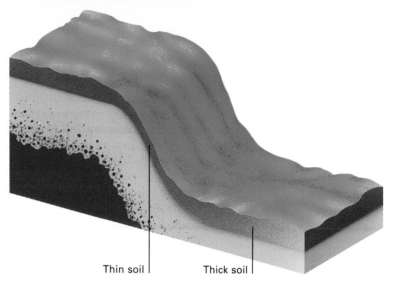

Figure 12-11. Soil is thick at the top and bottom of a slope; soil is thin along the slope itself.

Thin soil   Thick soil

pedalfer (PUH-DAL-fur) is found. Pedalfer soils contain clay, quartz, and iron compounds. The Gulf states and states east of the Mississippi River have pedalfer soils. In areas receiving less than 65 cm of rain a year, a soil called **pedocal** (PED-uh-KAL) is found. The southwestern states and states west of the Mississippi River have pedocal soils. Pedocal soils contain large amounts of calcium carbonate, which combines with excess hydrogen in the soil. This process makes the soil less acidic and very fertile.

Soil and Topography

Topography, or configuration of the land, also plays a role in soil formation. Because rainwater tends to run down steep slopes, much of the topsoil of the slope is washed away. Therefore, as shown in Figure 12–11, the soil at the top and bottom of a slope tends to be thicker than that on the slope itself. A study of soils in Manitoba, Canada, showed that the A horizon of soils on flat areas was more than twice as thick as that on 10° slopes.

The topsoil that does remain on a slope is often too dry and thin to support dense plant growth. The resulting lack of vegetation contributes to the development of a poor-quality soil. A thick humus layer cannot form because little organic matter is added to the soil. The soils on the sides of mountains are commonly thin and rocky, with few nutrients. Thus, cultivation of mountain soils is quite difficult. In contrast, lowlands that retain water tend to have heavy, wet soils with a high concentration of organic matter, which forms humus. A fairly flat area with good drainage provides the best surface for formation of thick fertile layers of residual soil.

Section 12.3 Review

1. Why do some soils contain large amounts of clay?
2. Describe the differences between residual and transported soils.
3. Describe the three horizons of a mature residual soil.
4. Although desert and arctic climates are extremely different, their soils may be somewhat similar. Explain why.

12.4 Erosion

Various forces may move weathered fragments of rock away from where the weathering occurred. The process by which the products of weathering are transported is called **erosion.** The agents of erosion are gravity, wind, glaciers, and water in the form of ocean waves and currents, streams, and rainfall.

Soil Erosion

Soil erosion occurs worldwide and is normally a slow process. Ordinarily, new residual soil forms about as fast as the existing soil erodes. The eroded soil is often deposited elsewhere as transported soil. However, unwise use of the land and unusual climatic conditions can upset this natural balance. Once the balance is upset, soil erosion accelerates.

Accelerated Soil Erosion

Unwise farming and ranching methods increase soil erosion. For example, clearing trees and small plants and allowing animals to overgraze destroy the soil protection provided by plants. The upper soil layers are then exposed to the full effects of erosion.

Furrows in plowed land, especially those that are plowed along slopes, allow water to run swiftly over the soil. As soil is washed away with each rainfall, a furrow becomes larger, forming a small gully, or ditch. Eventually, the land is covered with deep ditches. This type of accelerated soil erosion is called **gullying.** The farmland shown in Figure 12–12 has been ruined by gullying. What two agents of erosion are at work in gullying?

Figure 12-12. This field is useless for farming because of heavy erosion due to gullying.

Another type of soil erosion strips away parallel layers of topsoil and eventually exposes the surface of the subsoil or the partially weathered bedrock. This process is called **sheet erosion.**

Sheet erosion may occur when continuous rainfall evenly washes away the topsoil. Carried to streams by flowing water, the eroded topsoil may clog the stream or even change its course. Wind also can cause sheet erosion during unusually dry periods. The soil, made dry and loose by a lack of moisture, is carried away by the wind as clouds of dust and drifting sand. These wind-borne particles may move as dust storms that cover a wide area.

Constant erosion reduces the fertility of the soil by removing the A horizon, which contains the rich humus. The B horizon, which does not contain organic matter, is thereby exposed. The soil in this horizon is difficult to cultivate because it resists plant growth. Without plants, the B horizon has nothing to protect it from further erosion. As a result, within a few years, all of the soil layers could be removed by continuous erosion.

Soil Conservation

Rapid, destructive soil erosion can be prevented by soil conservation methods. Cultivating cover plants to hold the topsoil in place can protect topsoil in some areas. Farmers also can practice various crop planting methods designed to reduce erosion.

In a method called **contour plowing,** the soil is plowed in circular bands that follow the contour, or shape, of the land. This method of planting, shown in Figure 12–13, prevents water from flowing directly down slopes and so prevents gullying.

In **strip-cropping,** crops are planted in alternate bands. For example, a band of a crop planted in rows, such as corn, may be planted in one band. Next to it may be planted a crop that fully covers the surface of the land, such as alfalfa. The cover crop protects the soil by absorbing and holding rainwater. Strip-cropping is often combined with contour plowing. The combination of these methods can reduce soil erosion by 75 percent.

Figure 12-13. Contour plowing helps prevent gullying by causing water to flow along plowed furrows (left). Terracing builds up low ridges that slow movement of water downhill (right).

The construction of steplike ridges that follow the contours of a sloped field is called **terracing.** Terraces, such as those shown in Figure 12–13, prevent or slow the downslope movement of water, thus preventing rapid erosion.

In **crop rotation,** farmers plant one type of crop one year and a different type of crop the next. For example, crops that expose the soil to the full effects of erosion may be planted one year, and a cover crop the next year. Crop rotation stops erosion in its early stages, allowing small gullies formed during one growing season to fill with soil during the next.

In developed areas, clearing vegetation and removing soil to build housing and roads also contributes to erosion. It is important that people recognize the impact of these activities and that city planners take soil conservation measures.

Gravity and Erosion

Gravity, through its downward pull, causes rock fragments to move down inclines. This movement of fragments down a slope is called **mass movement.** Mass movement can be either rapid or slow.

Rock breaks into fragments that move downslope. The rock fragments accumulate at the base of the slope in piles called **talus** (TAY-lus). As you can see in Figure 12–14, talus forms a slope that covers the base of a cliff.

Mechanical and chemical weathering may reduce talus to smaller fragments, which can move farther down the slope. Some fragments wash into gullies. Carried into successively larger waterways, the fragments eventually flow into rivers.

Figure 12-14. Talus accumulates at the base of a slope.

Rapid Mass Movements

The most dramatic mass movements occur rapidly. You have probably heard news reports of devastating, sudden slides of earth and rock. The fall of rock from a steep cliff is called a **rockfall.** A rockfall is the most rapid kind of mass movement. Rocks in rockfalls may range in size from tiny fragments to giant boulders.

The sudden movement of masses of loose rock and soil down the slope of a hill, mountain, or cliff is a **landslide.** Large landslides involving loosened blocks of bedrock generally occur on very steep slopes. You may be familiar with small landslides on cliffs and steep hills overlooking highways. Heavy rainfall, spring thaws, volcanic eruptions, and earthquakes can trigger landslides.

The rapid movement of a large mass of mud creates a **mudflow.** Mudflows occur in dry, mountainous regions during sudden, heavy rainfall or as a result of volcanic eruptions. Masses of mud move down slopes and through valleys, frequently spreading out in a large fan shape at the base of the slope. Many of the disastrous mass movements that occur in hillside residential communities of southern California are referred to as landslides but are actually mudflows.

Sometimes a large block of soil and rock becomes unstable and moves downhill under the influence of gravity. It then slides along the curved slope of the surface in one piece. This type of movement is called a **slump.** Slumping occurs along oversteepened slopes. Saturation by water and loss of friction with underlying rock causes weak soil to slip downhill over the more resistant rock.

Slow Mass Movements

Although most slopes appear to be stable, some slow movement actually is occurring. Catastrophic landslides are considered the most hazardous downslope movement. However, more material is ultimately moved by the greater number of slow mass movements.

Career Focus: *Natural Resources*

"I get a great sense of satisfaction from knowing the work I'm doing isn't just for today. It's going to matter in the future, too."

Soil Conservationist

For Lewis Nichols, the career path to soil conservation began in a high school greenhouse.

"I had always been interested in plants and flowers. Then, when I was in high school, I became involved in a program in vocational agriculture. I worked in the greenhouse at the school during the year and had a job with the Soil Conservation Service during the summers," Nichols said.

Nichols, pictured left, was determined to turn his interests into a profession. This East St. Louis, Illinois, native completed a bachelor's degree program in agronomy, the study of crop production and soil management, at Western Illinois University.

Like other soil conservationists, Nichols uses his technical training and on-the-job experience in many ways. He assists government agencies, industries, and private individuals in the areas of soil and water conservation, insect control, and land use. Though most of

their work is done in the field, soil conservationists return to their offices to develop specific conservation practices for the problems being studied.

"Because your ability to provide practical assistance depends so much on work experience, I was somewhat frustrated in the beginning," he said. "Now that I have that experience, I find it very rewarding to know I'm able to help

One form of slow mass movement is called **solifluction.** Solifluction, which means "soil flow," occurs in arctic regions, where the subsoil is permanently frozen. In spring and summer, only the top layer of soil thaws. The moisture from this layer becomes trapped and cannot penetrate the frozen layers beneath. As a result, the surface layer becomes muddy and slowly flows downslope. Solifluction can also occur in warmer regions where the subsoil consists of hard clay. The clay layer acts like the permanently frozen subsoil in arctic regions, forming a waterproof barrier that prevents penetration of water deep into the soil.

society by looking out for the future and providing a better world for our children—a world where the words *natural resources* will still have meaning."

Hydrologist

Hydrologists study the earth's surface and underground waters to locate adequate supplies of fresh water. They analyze the chemical and physical properties of water sources both on-site and in the laboratory.

A hydrologist, shown lower left, works for government agencies and private industry in a variety of fields. Among these are soil and water conservation, pollution control, hazardous waste management, irrigation, development of hydroelectric power, and flood control. Irrigation projects such as the one shown above benefit from the hydrologist's knowledge of drainage and crop production.

Entry into the field of hydrology requires a bachelor's degree with course work in engineering, geology, geography, chemistry, mathematics, and physics.

Water Purification Technician

Every time you take a drink of water, your health depends on the work being done by water purification technicians at local water treatment plants.

Water purification technicians make sure that water is fit to drink by monitoring the chemical processes and equipment involved in filtering and purifying water. Water purification technicians also know how to repair all of the water purification equipment. This

equipment can include systems for removing impurities in the water by various methods such as filtration and disinfection.

Technicians learn through on-the-job training or in programs available in two-year colleges.

For Further Information

For more information on a career in resource management, write the Bureau of Land Management, Denver Service Center, Federal Center Bldg. 50, Denver, CO 80255.

The extremely slow downhill movement of weathered rock material is known as **creep.** Soil creep is the most effective of all mass movements. However, it usually goes unnoticed unless buildings, fences, or other objects on the surface are moved along with it.

Many factors contribute to soil creep. Water separates and lubricates rock particles, allowing them to move freely. Growing plants produce a wedgelike pressure forcing particles apart. The burrowing of animals and repeated freezing and thawing also loosen rock particles. Gravity then slowly pulls the particles downhill.

Erosion and Landforms

Through weathering and erosion, the earth's surface is shaped into different physical features, or **landforms.** There are three major landforms: mountains, plains, and plateaus. Mountains are landforms of very high elevations. Plains are large, flat or gently sloped surfaces not far above sea level. Plateaus are high, flat surfaces. Minor landforms include hills, valleys, and dunes. The shape of a landform is influenced by rock composition and structure.

All landforms are the result of two opposing processes. One process bends, breaks, and lifts the earth's crust, generally creating elevated landforms. The other process is the wearing action of weathering and erosion, reducing the land surface to sea level.

Erosion of Mountains

During the early stages in the history of a mountain, it undergoes uplift. As long as tectonic forces continue to uplift the mountain, it usually rises faster than it is eroded. Mountains that are being lifted are said to be youthful. Youthful mountains are rugged and have sharp peaks and deep, narrow valleys. During the intermediate, or mature, stage, a mountain is no longer rising. Weathering and erosion wear down the rugged peaks to rounded peaks and gentle slopes. The formations in Figure 12–15 illustrate the difference between youthful and mature mountains.

In its old stage, a mountain is reduced to a low, almost featureless surface near sea level. This surface is called a **peneplain** (PEEN-ih-PLANE), which means "almost flat." A peneplain usually has low, rolling hills, as seen in southern New England. Knobs of rock may resist erosion and protrude above the peneplain. These knobs are called **monadnocks** (muh-NAD-NAHCKS).

Figure 12-15. The youthful mountains on the left contrast with the mature mountains on the right.

A peneplain can be mistaken for a true plain. However, the rocks beneath a peneplain have been folded and tilted by tectonic forces, while the rocks beneath a true plain lie in horizontal layers.

Erosion of Plains and Plateaus

A plain is a broad, flat landform not far above sea level. There is no real distinction among the stages in the life cycle of a plain, because very little erosion must occur before this landform reaches sea level. However, a youthful plain has narrow, fast streams, whereas a mature or old plain has slow-moving, curving streams.

Like a plain, a plateau is a broad, flat landform. However, a plateau is higher, usually more than 1,500 m above sea level. Thus, a plateau is subject to much more erosion than is a plain. A young plateau usually has deep stream valleys that separate broad, flat regions. Mature plateaus, such as those found in the Catskill region in New York State, can be eroded into rugged hills and valleys similar to eroded mountains. In old age, plateaus are worn down to form a peneplain near sea level.

The effect of weathering and erosion on a plateau depends on the climate and the composition and structure of the rock. In dry climates, resistant rock produces plateaus with nearly flat tops. As a plateau enters old age, erosion may dissect the plateau into smaller, tablelike areas called **mesas** (MAY-suhz). Mesas ultimately erode down to small, narrow-topped formations called **buttes** (BYOOTS). Mesas and buttes are shown in Figure 12–16. In dry regions, these landforms have steep walls with flat tops. In humid climates, the landforms are more rounded.

Figure 12-16. A mesa (left) is a small tablelike area. A butte (right) is a smaller, narrow-topped formation.

Section 12.4 Review

1. List six agents of erosion.
2. What natural phenomena can trigger a landslide?
3. What three farming methods increase the rate of soil erosion?
4. Could you determine if a landform is a peneplain or a true plain by flying over it in an airplane? Explain your answer.

Reviewing Chapter 12

Key Concepts

- Agents of mechanical weathering break rock into smaller pieces but do not change its chemical composition. See page 185.
- Chemical weathering changes the composition of the minerals within a rock. See page 187.
- Rock weathers at different rates depending partly on its mineral composition. See page 190.
- The greater the amount of exposure, the faster a rock weathers. See page 191.
- Rock weathers more rapidly in regions where rainfall is abundant and where alternating freezes and thaws occur. See page 191.
- The parent rock from which soil is formed is the major factor determining the composition of the soil. See page 193.
- Thick soils develop in tropical climates. Thin soils develop in climates where rainfall is minimal. See page 195.
- The movement of the products of weathering away from where the weathering occurred is caused by natural agents. See page 197.
- Farming methods that result in loss of topsoil accelerate soil erosion. Plowing fields and planting crops wisely help conserve soil. See page 197.
- Slow and rapid mass movements of rock, soil, and mud are the major causes of erosion. See page 199.
- Erosion wears away landforms as it levels the earth's surface. See page 202.

Key Terms

abrasion (187)	exfoliation (188)	leaching (187)	regolith (193)
acid rain (188)	gullying (197)	mass movement (199)	rockfall (199)
bedrock (193)	horizon (194)	mechanical weathering	sheet erosion (198)
butte (203)	humus (193)	(185)	slump (200)
carbonation (188)	hydrolysis (187)	mesa (203)	soil (193)
chemical weathering	ice wedging (185)	monadnock (202)	soil profile (194)
(185)	joint sheeting (185)	mudflow (199)	solifluction (201)
contour plowing (198)	joints (185)	oxidation (188)	strip-cropping (198)
creep (202)	landform (202)	pedalfer (196)	talus (199)
crop rotation (199)	landslide (199)	pedocal (196)	terracing (199)
erosion (197)	laterite (195)	peneplain (202)	weathering (185)

Review

On your paper, write the letter of the term that best completes each of the following statements.

1. Physical stress on exposed surfaces cause rock to (a) decompose (b) compress (c) melt (d) become buried.
2. A common kind of mechanical weathering is called (a) oxidation (b) carbonation (c) ice wedging (d) leaching.
3. Oxides of sulfur and nitrogen that combine with water vapor cause (a) iron rain (b) acid rain (c) mechanical weathering (d) carbonation.
4. The mineral in igneous rock that is least affected by chemical weathering is (a) calcite (b) quartz (c) feldspar (d) mica.
5. The surface area of rocks exposed to weathering is increased by (a) burial (b) leaching

(c) quartz grains (d) joints.
6. Chemical weathering is most rapid in
   (a) hot, dry climates (b) cold, dry climates
   (c) cold, wet climates (d) hot, wet climates.
7. The chemical composition of soil depends to
   a large extent on (a) topography (b) its A
   horizon (c) the parent material (d) its B horizon.
8. The soil in tropical climates is often
   (a) thick (b) dry (c) thin (d) fertile.
9. The transport of weathered materials by a
   moving natural agent is called (a) mass

movement (b) weathering (c) erosion
(d) creep.
10. The most effective of all mass movements is
    (a) a landslide (b) a rockfall (c) a mudflow
    (d) creep.
11. All of the following farming methods prevent
    gullying, except (a) terracing (b) contour
    plowing (c) strip-cropping (d) irrigation.
12. When a mountain reaches old age, it is
    eroded to an almost featureless surface called
    (a) peneplain (b) monadnock (c) talus slope
    (d) mesa.

Application

On your paper, write answers to the following
questions.
1. Compare the weathering processes that affect
   a rock on top of a mountain and a rock buried
   beneath the ground.
2. What might be the result when rocks containing iron-bearing minerals undergo hydrolysis?
   Explain why.
3. Do you think a limestone building or a
   quartzite building will withstand the effects of
   acid rain better? Explain why.
4. Which do you think would weather faster, a
   sculptured marble statue or a smooth marble
   column? Explain your answer.
5. Compare the appearance of a 50-year-old
   highway that runs through a desert with that
   of a highway of the same age that runs
   through New York City.
6. If the A horizon and B horizon of a soil are
   relatively red, what do you think the underlying C horizon is composed of? Why?
7. If you were a farmer and could choose the

ideal climate in which to grow your crops of
deep-rooted plants, what climate would you
choose? Explain your reasons.
8. Mudflows in the southern California hills are
   usually preceded by a dry summer and widespread fires, followed by torrential rainfall.
   Explain why.
9. After the top layer of soil on a slope in the
   Arctic moves downhill, what do you think
   happens to the unweathered rock beneath it?
10. Suppose you wanted to cultivate grapevines
    on a hillside in Italy. What farming methods
    would you use? Explain why.
11. How could you determine if a plateau in an
    advanced stage of evolution is made of sand-
    stone? Explain.
12. Suppose that a mountain has been wearing
    down at the rate of about 2 cm per year for
    ten years. In the eleventh year, scientists find
    that the mountain is no longer being eroded.
    What do you think has happened?

Extension

1. Rocks on the moon are subjected to alternating very high and very low temperatures. The
   wide variation in temperatures does not produce rain or snow, however. Find out how
   these alternating temperatures affect moon
   rocks. Summarize your findings.
2. The Dust Bowl created in the 1930's in the
   southwestern United States was partially due
   to poor farming practices. Find out in what

ways farmers damaged the soil. Find out, too,
what farmers in that region are doing now to
prevent a recurrence of the Dust Bowl. Report your findings.
3. Find out the evolutionary stages of the Andes
   Mountains, the Rocky Mountains, and the
   Appalachian Mountains and make a relief
   map that illustrates the differences.

Chapter 13

Water and Erosion

*T*hundering in a mighty falls or seeping silently through the ground, water is always at work. It carries away more soil and moves more of the rocks that make up the earth's crust than all the other forces of erosion combined. In this chapter you will learn about the water supply of the earth and how running water helps to reshape the surface of the land.

Chapter Outline

13.1 The Water Cycle
Water Budget
Water Conservation

13.2 River Systems
Headward Erosion
Stream Erosion
Stages of a River System

13.3 Stream Deposition
Deltas and Alluvial Fans
Flood Deposits
Flood Control

The Goosenecks in Utah were formed by the Colorado River.

13.1 The Water Cycle

Section Objectives

- **Outline the stages of the water cycle.**
- **Explain the components of a water budget.**
- **List two approaches to water conservation.**

The origin of the earth's water supply has puzzled people for centuries. Aristotle and other ancient Greek philosophers believed that vast rivers, such as the Nile and the Danube, could be supplied by rain and snow alone. It was not until the middle of the seventeenth century that scientists precisely measured the amount of water received on the earth and the amount flowing in rivers. These measurements showed that the earth actually may receive up to five times as much water as rivers carry off. So, a more puzzling question than ''Where does the earth's water come from?'' is ''Where does the water go?''

There is a continuous movement of water from the air to the earth's surface and back to the air again. This cycle of water movement is called the hydrologic cycle, or **water cycle,** and is illustrated in Figure 13–1.

The process by which liquid water changes into water vapor is called **evaporation.** Each year, about 500,000 km^3 of water evaporate into the air. About 86 percent of this water evaporates from the ocean. The remaining 14 percent of the water evaporates from other sources, such as lakes, streams, and the soil. Water vapor also enters the air by **transpiration,** a process in which plants give off water vapor into the atmosphere. Together, the processes of evaporation and transpiration are called **evapotranspiration.** Evapotranspiration is one of the three processes of the water cycle. The continents lose about 70,000 km^3 of water each year through evapotranspiration. Another 40,000 km^3 flows over the land into streams and rivers as **runoff** or soaks deep into the soil and rock underground, becoming **groundwater.**

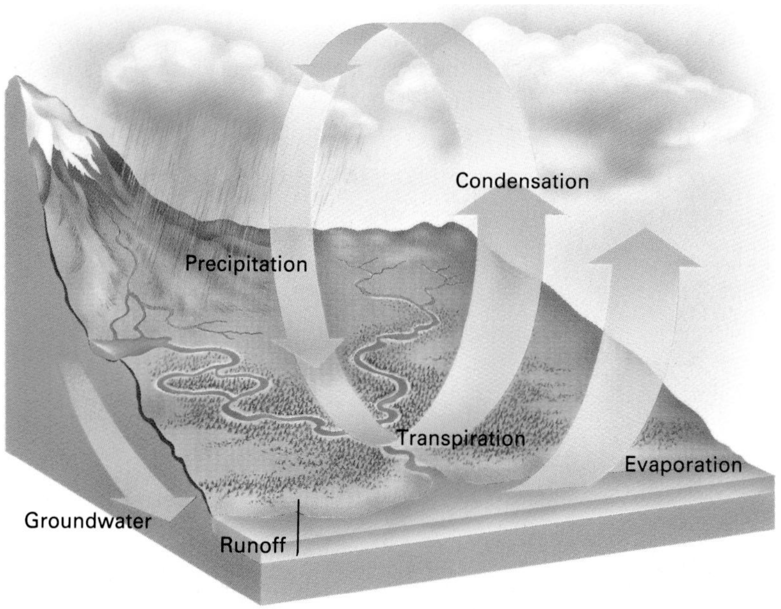

Figure 13-1. Evapotranspiration, condensation, and precipitation make up the continuous process called the *water cycle*.

Another process of the water cycle is **condensation.** When water vapor rises in the atmosphere, it expands and cools. As the vapor becomes cooler, some of it condenses, or changes into liquid water, and forms clouds.

The third process of the water cycle is **precipitation.** Precipitation is the process by which water falls from clouds to the earth as rain, snow, sleet, and hail. About 75 percent of all precipitation falls on the earth's oceans. The remaining 25 percent flows over the surface as runoff or into it as groundwater. Essentially all of this water eventually returns to the air by evapotranspiration, condenses, and falls to earth to begin the cycle again.

Water Budget

The continuous cycle of evapotranspiration, condensation, and precipitation gives the earth its **water budget.** A financial budget is a statement of expected income—money coming in—and expenses—money going out. When the money coming in equals the money going out, the budget is balanced. In the earth's water budget, precipitation is the income. Evapotranspiration and runoff are the expenses. The water budget of the earth as a whole is balanced because the amount of precipitation is equal to the amount of evapotranspiration and runoff. However, the water budget of a particular area, called the *local water budget,* usually is not balanced.

Factors that affect the local water budget include temperature, the presence of vegetation, wind, and the amount and duration of rainfall. When the precipitation exceeds the evapotranspiration and runoff of an area, the result is moist soil and possible flooding. When evapotranspiration exceeds precipitation, the soil becomes dry and irrigation may be necessary. Vegetation reduces runoff in an area. Winds, which can carry away great masses of humid air, can increase the rate of evapotranspiration.

The factors that affect the local water budget vary geographically. For example, the Sahara Desert receives very little precipitation compared with that received in the tropical rain forests of

Figure 13-2. Tropical vegetation (left) requires large amounts of rainfall yearly. Deserts (right) have low yearly rainfalls.

Science Notebook

Life Cycle of a Lake

All precipitation does not either evaporate or immediately flow from the land to the ocean. Sometimes this water collects in a depression in the land and creates some of the most beautiful formations of water on the earth—lakes.

Lakes occur most frequently in high latitudes and in mountainous areas. However, they also are common in low areas near the ocean, such as Florida, and near rivers with low gradients. Most of the water in lakes comes from precipitation and melting ice and snow. Springs, rivers, and runoff directly from the land also are sources of lake water.

Like rivers, lakes have a life cycle. Unlike rivers, most lakes are relatively short-lived. Many lakes eventually disappear because too much of their water drains away or evaporates. The most common

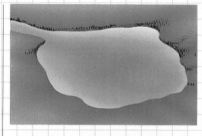

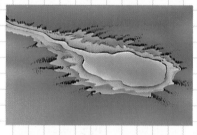

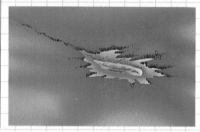

cause of excess drainage is an outflowing stream that cuts its bed below the lake basin. A lake also may lose its water if the climate becomes drier and evaporation exceeds precipitation.

Lake basins may be destroyed if they are filled with sediments. Streams that feed a lake carry and deposit sediments in the lake. Sediments also may be eroded from the lake basin itself by waves cutting into its sides. These sediments build up on the shores of the lake and gradually fill it in, as shown in the illustrations.

Organic deposits from vegetation also may accumulate in the basins of shallow lakes. As these deposits grow denser, a bog or swamp may form. The lake basin eventually may become dry land.

What process of stream deposition is similar to the filling of a lake basin with sediments?

Brazil. The rate of evapotranspiration in Kenya, which is close to the equator, is generally much greater than the rate in Finland, which is near the North Pole.

The local water budget also changes with the seasons in most areas of the earth. Generally, cooler temperatures slow the rate of evapotranspiration. During the warmer months, evapotranspiration increases. Would you expect that streams generally would transport more water in warmer or in cooler months?

Water Conservation

According to recent estimates, each person in the United States uses about 95 m^3 of water each year. Water is used for bathing, washing clothes and dishes, watering lawns, carrying away wastes, and, of course, drinking. Agriculture and industry use very large amounts of

Figure 13-3. As water becomes polluted, the supply available for human consumption is reduced.

water. Each year, millions of cubic meters of water are used to irrigate crops. As the population of the United States increases, so, too, does the demand for water.

On the average, 90 percent of the water used by cities and industry is returned to rivers or to the oceans as waste water. Much of this waste water contains harmful materials, such as toxic chemicals and metals. These toxic materials destroy plants and animals in the water and create polluted rivers, such as the one shown in Figure 13–3. Polluted rivers cannot support life.

Scientists have identified two approaches that can be used to ensure that enough fresh water is available today and in the future. The first approach is conservation. Many people favor enacting and strictly enforcing antipollution laws. Many concerned people also support programs that educate the public about the limits of water resources and the danger of wasting them. The second approach is finding other supplies of fresh water. Scientists are experimenting with several methods of **desalination,** the process of removing salt from ocean water. Someday desalination may provide the earth with nearly all of its fresh water. The best way of maintaining an adequate supply of fresh water at present, however, is the wise use and conservation of the fresh water that is now available.

Section 13.1 Review

1. Name the major processes of the water cycle.
2. What is evapotranspiration?
3. Explain why many local water budgets vary.
4. What is desalination?
5. Regions that have much vegetation generally receive much rainfall. How do these factors affect the local water budget?

13.2 River Systems

A river system is made up of a main stream and all the feeder streams, called **tributaries**, that flow into it. The land from which water runs off into these streams is called the drainage basin, or **watershed**, of the river system. The ridges or elevated regions of high ground that separate watersheds are called **divides**.

A river system begins to form when local precipitation exceeds evapotranspiration. The soil in the area soaks up as much water as it can hold. Gravity causes water that cannot soak into the soil to move downslope as runoff. When runoff moves across the earth's surface, it picks up and carries away weathered rock.

The force of the water running across surface sediments erodes a narrow ditch, called a *gully*. The runoff from nearby slopes tends to flow into the gully. Each time rain falls, the water moving through the gully erodes the gully further, making it wider and deeper. Eventually, these processes—precipitation, downslope movement of water and sediments, and erosion of the land—will form a fully developed valley with a permanent stream.

Headward Erosion

The path that a stream follows is called its **channel**. The main stream and its tributaries form a network of channels that drain the watershed of the river system. These channels lengthen and branch out at their upper ends, where runoff first enters the streams. The process of lengthening and branching of a stream is called **headward erosion**. Headward erosion carries away sediments from the slopes of a watershed at the upper end of a stream. Erosion of the watershed slopes extends a river system which, in turn, extends the area of land drained.

Headward erosion also can enlarge a river system through **stream piracy.** Stream piracy is the capture of a stream in one watershed by a stream with a higher rate of erosion in another

Figure 13-4. The tributaries feeding into this river are fed by runoff from slopes. Together the slopes make up the watershed of the river.

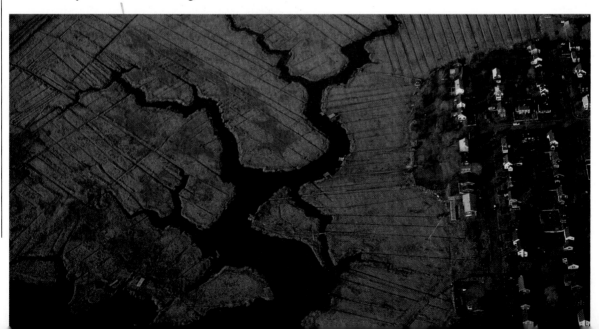

watershed. In this process, the captured stream drains into the other river system. Stream piracy extends the area of land drained by the capturing stream.

Stream Erosion

The edges of a stream channel that are above water level are its banks. The part of the stream channel that is below the water level is its bed. A stream gradually becomes wider and deeper as it erodes its banks and bed. Various factors affect the rate at which a stream erodes its channel.

Stream Loads

As a stream flows, it carries more than just water. The stream also transports various sediments, including soil, loose rock fragments,

Investigation

Soil Erosion

The uncontrolled runoff of surface water is a major cause of soil erosion. You can demonstrate soil erosion by making a model of a hillside.

Science Process Skills Focus: constructing models, observing, describing, comparing and contrasting

Materials

rectangular pan about 23 × 33 cm, sand, sink lined with paper towels, brick, water faucet, container to catch water and sand, clock or watch

Procedure

1. Fill the pan about half full with moist sand as shown in the figure.
2. Place the pan in the sink so that the end resting on the brick is under the water faucet.
3. Place the second container so it catches sand and water that flow out of the pan.
4. Open the faucet until a slow, gentle stream of water falls onto the sand in the pan. Let the water run for 15 seconds.
5. Draw the pattern of water flow over the sand.
6. Turn off the water and press the sand back into place. Turn on the water so that the rate of flow is greater than it was in Step 4 but

does not splash the sand. Draw the pattern of water flow.
7. Repeat Steps 4 and 5 two more times with different rates of water flow. For each trial, observe and draw the pattern of water flow.

Analysis and Conclusions

1. Compare the rate of erosion to the rate of water flow.
2. Compare the patterns of erosion to the rate of water flow.
3. How does the rate of water flow affect gullies?
4. Without changing the rate of water flow, how could the rate and effects of erosion be reduced on an actual hillside?

and dissolved minerals. The sediments carried by a stream are called the **stream load.** The stream load has three forms—the **suspended load,** the **bed load,** and the **dissolved load.** The suspended load consists of particles of sand, silt, and mud. The velocity, or speed, of the water keeps these particles suspended, and they do not sink to the stream bed. The bed load is made of larger, coarser materials, such as gravel and rocks. This material moves by sliding, by rolling, and by **saltation,** or short jumps. The dissolved load is mineral matter transported in the form of ions or molecules.

Although all three forms of stream load contribute to stream erosion, the bed load has the greatest effect. As rock fragments are carried along, they abrade the bottom and sides of the stream channel, wearing it away. Therefore, streams with large bed loads erode their channels more quickly than those with small bed loads. Large rocks are sometimes scraped over one area of the stream bed in a whirlpool motion. This scraping creates a bowl-shaped cavity called a *pothole,* like the one shown in Figure 13–5.

Figure 13-5. This photograph shows a pothole formed by the whirlpool action of rocks carried by a stream.

Discharge and Gradient

The **discharge** and velocity of a stream, as well as its load, affect how a stream cuts down and widens its channel. The volume of water moved by a stream in a given time is its discharge. The faster a stream flows, the higher is its discharge and the greater is the load it can carry. A swift stream carries more sediment and larger particles than a slow-moving stream carries. Hence, swift streams erode their channels more quickly than slow-moving streams do.

The velocity of a stream depends mainly on its **gradient,** or the steepness of its slope. Gradient is the change in elevation of the stream over a given horizontal distance. The gradient of a stream varies along the stream channel. Near the **headwaters,** or the beginning of a stream, the gradient is steep. This area of the stream generally has a high velocity, which causes rapid channel erosion. As the stream nears its mouth, where the stream enters a larger body of water, its gradient often becomes less steep. Its velocity and erosive power also decrease. The stream channel eventually is eroded to a nearly flat gradient.

Water and Wind Gaps

Movements of the earth's crust can raise or lower the surface of the land. When the land surface is uplifted slowly by geologic forces, the existing stream channels usually are eroded downward at the same rate as the land around them is elevated. When this happens, a deep notch is formed where the stream has eroded its channel through the raised mountains. This notch is called a **water gap.** An example is the Delaware Water Gap. In some cases, the land is uplifted faster than the stream can erode its channel. Because streams cannot flow uphill, the water gap is abandoned. The notch through which water no longer flows is called a **wind gap.** A water gap also is changed into a wind gap if the stream flowing through the gap is captured by another stream. The Cumberland Gap in the Appalachian Mountains is an example of a wind gap.

Stages of a River System

As a river erodes its banks and bed, it changes the landforms it passes through and alters its own course. The development of a river system is divided into three stages—youthful, mature, and old. These stages are not based on the actual age of the river but on its shape and how it erodes the land. The time required for a river to pass through each of these stages depends on the composition and structure of the rock through which the river flows.

Youthful Rivers

In its youthful, or early stage, a stream usually erodes its bed more rapidly than it erodes its banks. This produces a V-shaped valley with steep sides, like the one shown in Figure 13–6. Waterfalls and rapids are common features of youthful streams. These features are especially common in stream channels cut into hard rock, because the rock resists erosion.

Youthful rivers usually have relatively few tributaries. For this reason, a youthful river usually carries a small volume of water. Much of the precipitation falling on the watershed of a youthful river system does not reach the main stream because so few tributaries have developed. Instead, the precipitation remains in lakes and swampy areas at high elevations.

Mature Rivers

A mature river, by comparison, has well-established tributaries. It drains its watershed effectively. Because of good drainage and many tributaries, a mature river can carry a larger volume of water than a youthful river can carry. A mature river, however, tends not to deepen its channel as much as a youthful stream does. Instead, erosion occurs mostly along the valley walls when the river overflows

Figure 13-6. A young river has a narrow valley and fast-moving water (left). An old river forms a broad, shallow plain (right).

its banks and covers the valley floor. A mature-river channel usually occupies only a small part of the wide and relatively flat valley floor that it produces. Most of the waterfalls and rapids that existed during the youthful state of a mature river have disappeared. The gradient also has become less steep.

A mature stream with a low gradient tends to curve back and forth across the flat valley floor. A slight bend in the stream channel usually becomes a wider curve, because the water flows fastest around the outside edge of the curve. The faster-flowing water erodes the outside bank of the curve more quickly than the slower-moving water erodes the inner bank. The slower-moving water often deposits sediments along the inner bank. This process enlarges the curve and shifts the stream channel toward the outside bank. Generally, a series of these wide curves, called **meanders,** forms across the valley floor.

Frequently, a meander becomes so curved that it almost forms a loop, separated by only a narrow neck of land. When the river eventually cuts across this neck, it deposits sediments at both ends of the meander. The meander is thus isolated from the river, as shown in Figure 13–7. If the water remains in the isolated meander, an **oxbow lake** is formed.

Old Rivers

As a river continues to age, its gradient and velocity decrease. The stream no longer erodes the land; instead, it begins to deposit its sediments in its own channel and on its banks. A broad, shallow plain is formed. More meanders develop, and there are fewer tributaries, as smaller tributaries merge and become larger. How do you think the drainage of an old river compares with the drainage of a mature river?

Rejuvenated Rivers

Any movement of the earth's crust that increases the slope of the land will change the gradient of existing streams. A **rejuvenated** river is one whose gradient has become steeper in this way. The increased gradient of a rejuvenated river allows the river to cut more deeply into the valley floor. Rejuvenation often results in the formation of steplike terraces on both sides of a stream valley. These terraces provide evidence that the valley floor has been uplifted and a new floor has been cut through. There are many terraces along the Mississippi River.

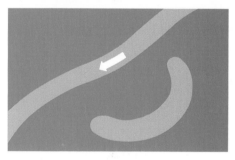

Figure 13-7. An oxbow lake results when part of a meander is cut off.

Section 13.2 Review

1. How does a river system begin?
2. Define *watershed* and *divide*.
3. What process results in the lengthening and branching of a river channel?
4. Name the three types of stream loads.
5. What feature might help you distinguish a rejuvenated river from a young river?

13.3 Stream Deposition

The total load a stream can carry is greatest when a large volume of water is flowing swiftly. When the velocity of the water decreases, the ability of the stream to carry its load generally also decreases. As a result, part of the stream load is deposited as sediment. When the velocity increases again, some or all of the deposited sediment is carried away. In some cases, stream deposits remain in place for only a brief time. In other cases, these deposits can become permanent features of the land.

Deltas and Alluvial Fans

Most of the load carried by a stream is deposited when the stream reaches a large body of water. As a stream empties into a large body of water such as an ocean, a gulf, or a lake, the velocity of the stream decreases greatly. The load then is deposited at the mouth of the stream in a fan shape with its tip facing upstream, as illustrated in Figure 13–8. This fan-shaped deposit at the mouth of a stream is called a **delta.** The exact shape and size of a delta are determined by the local waves and tides.

When a stream descending a steep slope reaches a flat plain, the speed of the stream is suddenly reduced. As the stream slows down rapidly, it deposits some of its load on the level plain at the base of the slope. This deposit forms a fan-shaped heap with its tip pointing upstream. This deposit is called an **alluvial fan.** In desert and semi-desert regions, temporary streams often form alluvial fans. An alluvial fan differs from a delta in the following three ways. First, the sediment that forms an alluvial fan is deposited on dry ground, whereas the sediment that forms a delta is deposited in water. Second, an alluvial fan is made up of coarse, angular sand and gravel. A delta, though, usually is made up of fine sand, silt, and mud. Third, the surface of an alluvial fan is sloping, while the surface of a delta is relatively flat.

Section Objectives

- List two types of stream deposition and explain the differences between them.

- Describe the change in a stream that causes flooding.

- Identify direct and indirect methods of flood control.

Figure 13-8. When a stream meets an ocean or a gulf, the sediment carried by the stream is deposited in a delta (below). Runoff from slopes in desert regions deposits sediment in an alluvial fan (inset).

IMPACT

A River and a City

At the beginning of the fifteenth century, Bruges, Belgium was one of the great commercial and cultural centers of Europe. Waterways, such as the canal shown below, brought the city wealth through trade. Located at the base of the Zwin estuary, Bruges was easily accessible from the North Sea. Merchant ships from all over Europe and the Far East regularly docked at the harbor on the Zwin. In turn, traders from Bruges sailed to other parts of the world, carrying fine fabrics from its famous textile mills.

However, the prosperity of Bruges was short-lived. By the end of the fifteenth century, the last of the textile mills were closing, the shipping trade was virtually nonexistent, and the once powerful and populous city had been nicknamed *Bruges-la-Morte,* Bruges-the-Dead.

The decline of Bruges can be traced to stream deposition, the process whereby a stream or river deposits its sediment when it enters a larger body of water. The Reie River, which flows through Bruges, broadened at the city of Damme, just northeast of Bruges, to form the Zwin estuary. The Zwin then emptied into the North Sea. As the Reie flowed into the Zwin, it deposited some sediment. In turn, as the Zwin flowed into the North Sea, it deposited its sediment. Thus, the Zwin estuary was becoming blocked at both ends. Soon, access to Bruges was cut off, and the golden age of the city ended.

What natural process in the development of a river system contributed to the stream deposition at Bruges?

Flood Deposits

The size of a stream channel is determined by the average volume of water that flows in the stream. Some channels are small enough to step across; others are hundreds of meters wide. If the volume of water in a stream remains constant, its channel changes very little. However, the volume of water in nearly all streams changes continually. A dramatic increase in volume can cause a stream to flood, or overflow its banks and wash over the valley floor. The part of the valley floor usually covered in a flood is called a **floodplain.**

Spring floods are common near headwaters in areas where the winters are harsh. During the winter, little evapotranspiration occurs and snow and ice remain on the land. The water released by melting snow cannot be absorbed by the frozen ground and runs off into surface streams, increasing the volume of water in those streams.

Figure 13-9. The regularly flooded area along a stream is called the *floodplain.*

Ice jams, too, increase the chance of spring flooding when ice blocks the stream channels.

Human activity also contributes to the size and number of floods in many areas. Sometimes the natural ground cover of plants and trees, which protects the surface from heavy runoff, is removed. Forest fires, some logging practices, and the clearing of land for cultivation or housing development can increase the volume of runoff. What would have to happen to balance the local water budget in such an area?

When a stream overflows its banks and spreads out over the nearby land, the stream quickly deposits the coarser part of its load along the banks of the channel. The accumulation of these deposits along the banks eventually produces raised banks, which are called **natural levees.** Natural levees may be quite high and prominent along mature river channels, such as the lower Mississippi River, where repeated flooding has occurred.

Not all of the load deposited by a stream in a flood will form levees. Finer sediments are carried over the floodplain by the flood waters and are deposited there. These fine sediments form a fairly even layer on the floodplain. A series of floods produces a thick layer of this deposited material. Swampy areas are common on floodplains because drainage is usually poor in the area between the levees and the outer walls of the valley.

Flood Control

Flooding is a natural stage in the development of a stream and will continue to occur as the river system matures. However, flooding can become a problem when cities and farms are located on the fertile floodplains of rivers. The best safety measure is to leave such

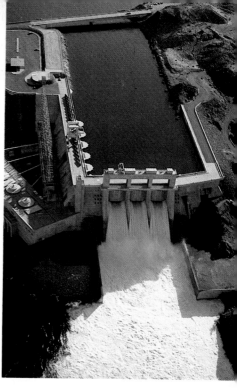

Figure 13-10. Artificial levees (left) and dams (right) are often built to control floods.

areas in their natural state. However, since cities, towns, industrial complexes, and other construction is done in these areas, flood control is often necessary to minimize the loss of life and property.

There are indirect and direct methods of flood control. Indirect methods include forest and soil conservation measures that prevent excess runoff during periods of heavy rainfall. The most common method of direct flood control is the building of a dam. The artificial lakes that form behind dams act as reservoirs for excess runoff. The stored water can be used to generate electric power, to supply fresh water to populated areas, to irrigate farmland during dry periods, and for recreation.

Another direct method of flood control is the building of artificial levees. However, artificial levees offer only temporary protection. As a river deposits sediment along its bed, the height of the levee must be raised. Artificial levees also must be protected against erosion by the river.

In some cases, a permanent overflow channel, or **floodway,** can be an effective means of flood control. When the volume of water in a river increases, a floodway carries away excess water and prevents the main stream from overflowing.

Section 13.3 Review

1. Describe the differences between a delta and an alluvial fan.
2. What is a floodplain?
3. Why are spring floods common where the headwaters of a river are in an area of harsh winters?
4. What advantage other then flood control does a dam provide?
5. If you were picking a material to make an artificial levee, what major characteristic would you look for?

Reviewing Chapter 13

Key Concepts

- The continuous cycle of water movement on earth is called the water cycle. See page 207.
- The earth's water budget is the amount of water received and the amount lost. See page 208.
- Careful conservation of the earth's limited supply of fresh water is important. See page 209.
- The development of river systems illustrates the erosive power of water. See page 211.
- The velocity of its water and the abrasiveness of its loads cause a stream to erode its channel. See page 212.
- Youthful river systems tend to erode the land much faster than do mature or old river systems. See page 214.
- A stream can deposit its sediments on land or in the water. See page 216.
- Flooding is often the result of a large and sudden increase in the volume of water carried by a stream. See page 217.
- Floods can be controlled through natural conservation and by building various flood-control structures. See page 218.

Key Terms

alluvial fan (216)	evaporation (207)	headwater (213)	stream piracy (211)
bed load (213)	evapotranspiration	meander (215)	suspended load (213)
channel (211)	(207)	natural levee (218)	transpiration (207)
condensation (208)	floodplain (217)	oxbow lake (215)	tributary (211)
delta (216)	floodway (219)	precipitation (208)	water budget (208)
desalination (210)	gradient (213)	rejuvenated (215)	water cycle (207)
discharge (213)	groundwater (207)	runoff (207)	water gap (213)
dissolved load (213)	headward erosion	saltation (213)	watershed (211)
divide (211)	(211)	stream load (213)	wind gap (213)

Review

On your paper, write the letter of the term that best completes each of the following statements.

1. The change of water vapor into liquid water is called (a) runoff (b) evaporation (c) desalination (d) condensation.
2. Vegetation gives off water vapor into the atmosphere through a process called (a) condensation (b) rejuvenation (c) saltation (d) transpiration.
3. In a water budget, the income is precipitation and the expense is (a) evapotranspiration and runoff (b) condensation and saltation (c) erosion and conservation (d) rejuvenation and sedimentation.
4. The process that turns seawater into fresh water is (a) desalination (b) transpiration (c) conservation (d) rejuvenation.
5. The slopes from which water runs off into a stream are called its (a) tributaries (b) divides (c) watershed (d) gullies.
6. Tributaries branch out and lengthen as a river system develops through (a) headward erosion (b) condensation (c) saltation (d) runoff.
7. The stream load that includes gravel and large rocks is the (a) suspended load (b) dissolved

load (c) runoff load (d) bed load.
8. When a young river deepens its channel faster than it can cut into its sides, the result is (a) a gradient (b) a V-shaped valley (c) a floodway (d) an oxbow lake.
9. A stream whose gradient has been increased by movement of the earth's crust is said to be (a) rejuvenated (b) meandering (c) eroded (d) suspended.
10. The triangular formation that occurs when a stream deposits its sediment on land is called (a) a delta (b) a meander (c) an oxbow lake (d) an alluvial fan.
11. The part of a valley floor that is usually covered in a flood becomes the (a) floodway (b) groundwater (c) floodplain (d) artificial levee.
12. One indirect method of flood control is (a) soil conservation (b) dams (c) floodways (d) artificial levees.

Application

On your paper write answers to the following questions.
1. How would the earth's water cycle be affected if a significant portion of the sun's rays was blocked by dust or other contaminants in the atmosphere?
2. How might the local water budget of Singapore differ from that of Stockholm? Use an atlas to determine the geographic location of these two cities before thinking about your answer.
3. Desalination someday may provide an almost endless supply of fresh water. What other problem must be solved before the desalinated water can be used?
4. In the desert areas of the southwestern United States, there are many shallow, narrow ditches that cut through the landscape. What do you suppose these ditches are? What was the most likely cause of their formation? If these ditches were located anywhere else, what might happen to them? Why does this not happen in the desert?
5. The Colorado River is usually grayish-brown as it flows through the Grand Canyon. What causes this color?
6. Assume that you decided to test the color of the water of the Colorado River over a period of years and found that it was becoming clearer. What would you conclude was happening to the river?
7. If you were trying to locate a mature river for geologic exploration, what characteristics would you look for?
8. Why do you think the surface of an alluvial fan is sloping and that of a delta is flat?
9. Developers are planning to build retirement communities on the floodplain of a river, but away from the banks. Considering only the safety aspect, would you argue for or against building these communities? Support your argument.
10. What steps might the developers in Question 9 take to protect the people and property in the communities?

Extension

1. Do some research in your community to find how much precipitation your area receives, how much water is lost to evapotranspiration, how much runs off into streams and rivers, and how much becomes groundwater. How does the water budget for your community change with the seasons? Report your findings to the class.
2. Select a major river system in your area of the country. Prepare a map that shows the main stream channel, major tributaries, watershed, divides, waterfalls or major rapids, delta, alluvial fans, and oxbow lakes.
3. Choose a part of a river system in your area that is prone to flooding. Find out what caused each flood in the past. Were lives and property lost? How did the people respond to the floods? Have the flooded communities taken any steps to control future flooding? If so, what has been done?

Chapter 14

Groundwater and Erosion

Powerful agents of erosion are at work underground as well as on the earth's surface. Old Faithful Geyser in Yellowstone National Park and Mammoth Cave in Kentucky are evidence of those underground processes. In this chapter, you will learn how water moves beneath the ground and how it erodes the crust.

Chapter Outline

Old Faithful is located in Yellowstone National Park, Wyoming.

14.1 Water Beneath the Surface

Section Objectives

- **Distinguish between porosity and permeability.**
- **Identify the two moisture zones below the earth's surface.**
- **Relate the contour of the water table to the contour of the land.**
- **Describe how groundwater can be polluted.**

Precipitation that does not run off into streams and rivers usually seeps down through the soil. As the water seeps into the upper layers of the earth's crust, it occupies the air-filled openings, or pores, between rock particles. This underground water that fills almost all the pores in rock and sediment is called *groundwater.*

Groundwater is a plentiful source of fresh water on the earth. An estimated 90 percent of the earth's fresh water supply is stored beneath the surface. In the United States alone, groundwater supplies about a fifth of the fresh water needs. The amount of groundwater is possibly 50 times as great as the total amount of water in all the earth's rivers and lakes.

Rock Properties That Affect Groundwater

A body of rock through which large amounts of water can flow and in which much water can be stored is called an **aquifer.** Some aquifers are composed of porous layers of sediments, such as sand and gravel. Rocks with large pores, such as sandstone, and highly fractured rock, such as limestone, also may be aquifers. Although a rock may be able to hold much water, if its pores or fractures are not connected, the water cannot flow freely through it. Consequently, it is not an efficient aquifer.

Porosity

The amount of water a rock can hold is determined by the **porosity** of the rock. Porosity refers to the percentage of open spaces in a given volume of a rock or sediment. These spaces can be fractures, cavities caused by erosion, or pores between rock particles.

Porosity is influenced by several factors. One factor, **sorting,** is illustrated in Figure 14–1. Sorting is the amount of uniformity in the size of the particles, or grains, in rock or sediment. When a sediment is well sorted, its particles are all about the same size. This

Figure 14–1. Well-sorted, large-grained rock (left) has high porosity. Well-sorted, small-grained rock (center) has equally high porosity. Rock with grains of many different sizes (right) has low porosity.

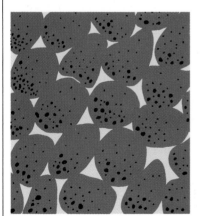
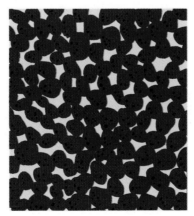

uniformity means that few, if any, smaller particles fill the spaces between the larger grains. The size of the grains is unimportant. A sediment composed of all large particles can have the same porosity as a sediment composed of all tiny particles. A poorly sorted sediment contains particles of many sizes. Smaller particles fill the spaces between larger ones, making the rock less porous.

Another factor that influences porosity is the way particles are packed together. If they are packed loosely, there are many open spaces for water storage, and the rock has high porosity. Tightly packed rocks have few open spaces and, thus, low porosity.

Permeability

The **permeability** of a rock or sediment indicates how freely water passes through the open spaces in it. For a rock to be permeable, the open spaces must be connected. A rock can have high porosity, but

Investigation

Permeability

You can demonstrate permeability and calculate rate of drainage with a simple model.

Science Process Skills Focus: constructing models, observing, interpreting data

Materials

sharpened pencil, 3 large polystyrene cups, cheesecloth, rubber bands, ruler, sand, soil, gravel, 3 small thread spools, saucer, measuring cup, water, clock or watch

Procedure

1. With the pencil point, make 7 small holes in the bottom of each cup. Cover the holes with cheesecloth secured by a rubber band.
2. Mark a line 2 cm from the top on each cup. Fill the first cup to the line with sand, the second with soil, and the third with gravel.
3. Place the first cup on the spools in a saucer, as shown in the figure. Pour 120 mL of water into the cup. Time and record how long it takes for the water to drain through the cup.
4. Pour the water from the saucer into the measuring cup. Record the amount of water.
5. Repeat Steps 3 and 4 with the other two cups.
6. Calculate the rate of drainage for each cup by

dividing the amount of water that drained by the time it took the water to drain.

7. Calculate the percentage of water retained in each cup by subtracting the amount of water that drained into the saucer from the original 120 mL. Divide the difference by 120.

Analysis and Conclusions

1. Which cup had the highest drainage rate?
2. Which cup retained the least water?
3. Consider the sample with the highest drainage rate and lowest percentage of water retained to be the most permeable. Which sample is the most permeable? The least permeable?

if the pores are not connected, the rock is not permeable. Permeability is affected by the size and sorting of the particles that make up a rock or sediment. The larger and more consistently sorted the particles are, the more permeable the rock or sediment tends to be. The most permeable rocks are those composed of coarse particles, such as sandstone. Other rocks, such as limestone, are permeable because they have interconnected cracks. Clay is composed of fine-grained rock particles. Because of this, clay is **impermeable,** meaning water cannot flow through it.

Regions of Groundwater

Gravity pulls water down through soil and rock until it reaches impermeable rock. Water then begins to fill, or saturate, the spaces in the rock above the impermeable layer. As more water soaks into the ground, its level rises underground, creating the lower of the two zones of groundwater—the **zone of saturation.** The zone of saturation, as shown in Figure 14–3, is the layer of ground where all the pores are filled with water. The upper surface of the zone of saturation is called the **water table.**

The upper zone of groundwater lies between the water table and the earth's surface and is called the **zone of aeration.** The zone of aeration is composed of three regions. The uppermost region holds soil water—water that forms a film around grains of topsoil. The bottom region, just above the water table, is the **capillary fringe.** Water is drawn up from the zone of saturation into the capillary fringe by capillary action. Capillary action is the attraction of water molecules to other materials, such as soil. When you use a paper towel to soak up a spill, you are relying on capillary action to draw moisture into the towel. Between the region of soil water and the capillary fringe is a middle region that normally remains dry except during rainfalls. If this middle region remains wet for an extended time, would you expect capillary action in the capillary fringe to increase or decrease?

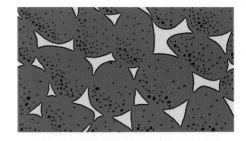

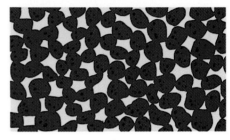

Figure 14–2. Friction slows the flow of water through rock. A large-grained rock (top) has less surface area to cause friction, shown as dark areas around particles, than does a small-grained rock (bottom).

Figure 14–3. The diagram on the left illustrates the moisture zones beneath the earth's surface.

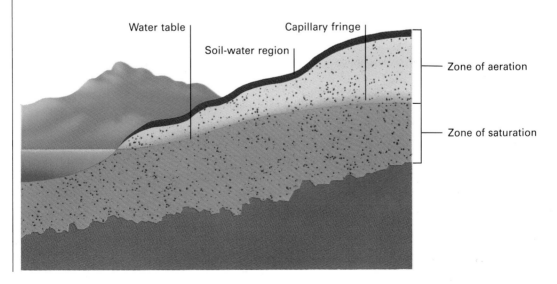

Water table

Soil-water region

Capillary fringe

Zone of aeration

Zone of saturation

Movement of Groundwater

The permeability of the rocks in the zone of saturation influences the flow of groundwater. Like water on the earth's surface, groundwater flows downward in response to gravity. For example, it flows away from divides toward the lower-lying stream valleys.

The rate at which groundwater flows through the zone of saturation depends on the permeability of the aquifer and the **hydraulic gradient** of its water table. The hydraulic gradient is the steepness of the slope. Water passes quickly through highly permeable rock but slows as the permeability of the rock decreases. The velocity of a stream increases as its gradient increases. In like manner, the velocity at which groundwater moves increases as the gradient of the water table increases.

Topography and the Water Table

The water table is found at different levels. The level depends on the topography of the land, the permeability of the rock, and the amount of rainfall. Generally, the water table parallels the contours of the land, rising beneath hills and dipping under valleys. During periods of prolonged rainfall, the water table rises. During periods of drought, the water table flattens as water flows toward stream valleys and is not replaced.

Under most surface areas, there is only one water table. In some areas, however, a layer of impermeable rock lies near the surface above the main water table. This rock layer prevents water from seeping down into the zone of saturation. Water collects on top of this rock layer, creating a second zone of saturation and, thus, a second water table. These secondary water tables, called **perched water tables,** are illustrated in Figure 14–4.

Conserving Groundwater

Groundwater is the principal source of water for irrigation in the United States. In many communities, it is the only source of fresh water. Although groundwater is renewable, the supply is limited.

Figure 14–4. A perched water table lies above the main water table.

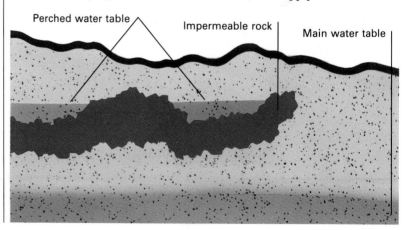

Perched water table   Impermeable rock   Main water table

Groundwater collects and moves slowly, and the water taken from aquifers may not be replenished completely for thousands of years. Like other natural resources that are limited, groundwater must be conserved.

Groundwater can be polluted in many ways. Waste dumps for toxic chemicals and other hazardous materials are a common source of groundwater pollution. Agricultural fertilizers and pesticides are other groundwater pollutants. These materials seep into the ground and mix with the groundwater. Also, if too much groundwater is pumped from an aquifer near the ocean, the water table may drop below sea level. The lowered water table allows salt water from the ocean to flow through the aquifer, contaminating the water supply.

Communities can regulate their use of groundwater to help conserve this valuable resource. They can monitor the level of the local water table and discourage excess pumping when the water table falls too low. After water is used, it generally drains into sewers or sewage canals. Because groundwater replenishes itself slowly, some communities recycle used water. Used water is purified and then is pumped back into the ground to replenish the groundwater supply.

Figure 14–5. Polluted groundwater can be rid of toxic material in water purification plants such as this one in San José, California.

Section 14.1 Review

1. What is the difference between the porosity and the permeability of rock?
2. Name and describe the two zones of groundwater.
3. How does the contour of the water table compare with the local topography?
4. Pumping too much water from an aquifer can deplete the supply of groundwater. In what type of location might this practice lead to tainting of the groundwater supply? How?

14.2 Wells and Springs

Two common ways through which groundwater comes to the earth's surface are wells and springs. A well is a hole that is dug below the water table and that fills with groundwater. A spring is a natural flow of groundwater to the earth's surface that is found where the ground dips below the water table. Wells and springs are classified into two groups: ordinary and artesian.

Ordinary Wells and Springs

For an **ordinary well** to function properly, it must penetrate deep into highly permeable sediment or rock. If the rock is not sufficiently permeable, groundwater cannot flow into the well quickly enough to replenish the water that is withdrawn. If the well is not deep enough, it will dry up when the water table falls below the bottom of the well.

Pumping water from a well creates a cone-shaped depression in the water table around the well, as illustrated in Figure 14–6. This lowered area of the water table is called a **cone of depression.** If a large amount of water is taken from a well, the cone of depression may drop to the bottom of the well, causing the well to go dry. The lowered water table around a well can extend several kilometers from the well, causing surrounding wells to go dry also.

Ordinary springs are commonly located in steep hills because the surface of the slope often drops below the water table. However, such springs usually do not flow continuously because the water table may drop well below the slope of the hill during dry periods.

However, a spring that is formed where a perched water table intersects the surface of a hill is likely to flow continuously. Water filters down through permeable rock until it is stopped by the impermeable rock that forms the perch. At this point the water comes to the surface of the hill as a spring. Because of the impermeable rock beneath, the level of the zone of saturation can never fall below

Figure 14–6. A cone of depression develops around a pumping well. Notice that the lowered water table has caused a neighboring well to go dry.

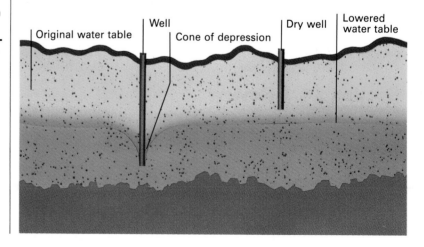

Original water table | Well | Cone of depression | Dry well | Lowered water table

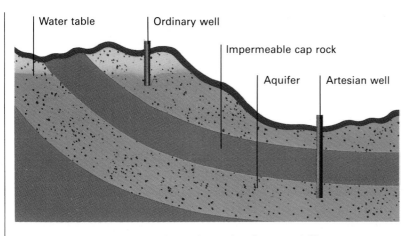

Figure 14-7. The aquifer in this artesian formation dips under the impermeable cap rock. An artesian well must penetrate the cap rock to reach the groundwater in the aquifer.

the slope of the hill. Therefore, the spring is essentially permanent. The spring will disappear only if there is a severe drought that causes the perched water table to dry up completely.

Artesian Wells and Springs

The groundwater that supplies many wells comes from local precipitation. However, one type of well, called an **artesian well,** gets its water from far, possibly hundreds of kilometers, away. An artesian well is one through which water flows freely with no pumping necessary.

The source of water for an artesian well is an arrangement of permeable and impermeable rock called an **artesian formation.** An artesian formation is a sloping layer of permeable rock sandwiched between two layers of impermeable rock and exposed at the surface. The permeable rock is the aquifer, and the top layer of impermeable rock is called the **cap rock,** as illustrated in Figure 14-7. Precipitation drains down into the exposed area of the aquifer. As the water flows downward, pressure builds because of the weight of the overlying water. An artesian well is dug through the cap rock to reach the water in the aquifer. Because the water in the aquifer is under pressure, it quickly flows up through the well. The level of the water in an artesian well usually is much higher than the level of the surrounding water table.

In addition to supplying artesian wells, artesian formations are the source of water for some springs. When cracks occur naturally in the cap rock, water under pressure in the aquifer flows through the cracks. This flow forms **artesian springs,** also known as *fissure springs.* An artesian spring is often the source of water in a desert oasis. Why do you think ordinary springs are not the usual source of water in desert oases?

Hot Springs and Geysers

Groundwater is sometimes heated beneath the earth's surface. The water can be heated as it passes through areas where there has been recent volcanic activity or near pockets of molten rock. Hot groundwater that rises to the surface before cooling produces a **hot spring.**

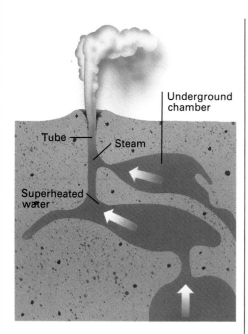

Figure 14–8. The underground chambers and tubes of a geyser enable water to become superheated and eventually erupt to the surface.

The labels in the figure read: Underground chamber, Tube, Steam, Superheated water.

To be called a *hot spring,* the water must be at least as warm as human body temperature—37°C.

Because greater quantities of minerals dissolve in hot water than in cold water, hot springs often contain large concentrations of minerals. Many of these dissolved minerals are deposited when the water reaches the surface and cools. They frequently accumulate around the mouths of hot springs, forming steplike layers, or terraces. Most of these terraces are made of **travertine,** a form of calcite. Travertine is usually white when freshly deposited but turns gray as it weathers. It also may be red, brown, or yellow if iron compounds or other minerals are present in the spring.

Hot springs that form in areas of recent volcanic activity may become **mud pots.** A mud pot forms when the rock surrounding a hot spring is chemically weathered by volcanic gases dissolved in the water. The weathered rock mixes with the hot water to form a sticky, liquid clay that bubbles at the surface. Mud pots sometimes are called *paint pots* because the clay may be highly colored by the minerals or organic material it contains.

Hot springs that erupt periodically are called **geysers.** Some geysers erupt from open pools and shoot up sheets of water and steam. Others erupt through small surface openings, sending narrow columns of water and steam high into the air. Both types of geysers tend to build deposits of the silicate mineral geyserite. The water erupted from the geyser evaporates, leaving behind the minerals.

The underground structure of a geyser consists of a crooked, narrow tube that leads to the surface and connects with one or more underground chambers, as illustrated in Figure 14–8. Heated groundwater fills both the tube and the chambers. The water at the bottom of the tube is superheated by the hot rocks around it and is under pressure from the water above. Because water boils at a higher temperature when it is under high pressure, the water at the bottom of the tube does not boil. Eventually, the water higher in the tube, which is under less pressure, begins to boil. The boiling water produces steam that pushes the water above it to the surface. Release of the water near the top of the tube relieves the pressure on the superheated water farther down. With the sudden release of pressure, the superheated water changes almost immediately into steam and explodes toward the surface. The eruption continues until most of the water and steam are emptied from the tube and storage chambers. After the eruption, groundwater begins to collect again and the process is repeated, often at regular intervals.

Section 14.2 Review

1. Why are ordinary springs commonly found on steep hills?
2. Describe the rock layers in an artesian formation.
3. In what two ways can heated groundwater reach the surface of the earth?
4. What features of geysers cause them to erupt rather than flow to the surface like hot springs? Explain why.

14.3 Groundwater and Chemical Weathering

Section Objectives

- **Explain how caverns and sink-holes form.**
- **Identify the features of karst topography.**

As groundwater passes through permeable rock, it dissolves minerals in the rock. The warmer the rock is and the longer the water travels through the rock, the more minerals the water will dissolve. Water that contains relatively large amounts of dissolved minerals, especially ions of calcium, magnesium, and iron, is called **hard water.** Water that contains few minerals is called **soft water.**

Dissolved minerals make hard water unacceptable for many common uses. For example, soap added to hard water will not produce suds. Also, many people prefer not to drink hard water because of its salty, metallic taste. What are some household appliances or fixtures that might be damaged by the buildup of mineral deposits from hard water?

Results of Weathering by Groundwater

One way that minerals become dissolved in groundwater is through carbonation. As water moves through rock and soil, it leaches carbon dioxide from the rock and combines with it to form carbonic acid. This weak acid chemically weathers the rock it passes through, breaking down and dissolving the minerals in the rock.

Caverns

Rocks rich in the mineral calcite, such as limestone, are especially vulnerable to carbonation. Although limestone is not a porous rock, its layers are usually fractured by vertical and horizontal cracks. As groundwater flows through these openings, carbonic acid slowly dissolves the limestone and enlarges the cracks. When this chemical weathering process enlarges a number of connected cracks and cavities, a cavern forms, as shown in Figure 14–9. A cavern is a large tunnel or cave, often containing many smaller, connecting chambers. Caverns are common in areas with extensive limestone deposits. The Carlsbad Caverns in New Mexico and Mammoth Cave in Kentucky are examples of limestone caverns.

Figure 14–9. The formations in the Carlsbad Caverns in New Mexico are made of calcite, a mineral dissolved from limestone.

Sinkholes

During dry periods, the water table is low and caverns are not completely filled with water. When there is no longer support from the water beneath, the roof of a cavern near the surface may collapse. This process forms a circular depression called a **sinkhole,** or sink. When sinkholes later fill with water, they become sinkhole ponds or sinkhole lakes.

Stalactites and Stalagmites

A cavern that lies above the water table does not fill with water; however, water still passes through the surrounding rock. When water containing dissolved calcite drips from the ceiling of a limestone cavern, some of the calcite solidifies on the ceiling. As this calcite builds up, it forms a suspended, cone-shaped deposit called a **stalactite** (stuh-LAK-TITE). When drops of water fall on the cavern

IMPACT

Disappearing Land

One day in May 1981, the ground beneath a house in Winter Park, Florida, collapsed, creating a hole 300 m deep. That hole eventually expanded to 100 m across—the size of a football field. It swallowed up several houses, part of a four-lane highway, a swimming pool, a parking lot, five cars, and a truck. The cost of the damage from this cave-in reached about $2 million.

The catastrophe that struck Winter Park actually started thousands of years earlier. The groundwater that flows beneath central Florida began to dissolve the surface layer of limestone. Over the years, so much limestone eroded away that giant underground caverns were created.

When groundwater fills a cavern, it takes the place of rock that has dissolved. Groundwater provides the pressure that helps to support the limestone above the cavern. However, if the water level is lowered, the earth's surface above the cavern loses its support and caves in, forming a sinkhole.

This is what happened in Winter Park after a two-year drought. With no precipitation to replace the groundwater used by Winter Park residents, the caverns beneath the town remained empty and the surface layers collapsed.

During the same week, eight more sinkholes appeared in Winter Park.

What measures might Winter Park residents take to help prevent more sinkholes from forming?

floor, deposited calcite builds up to form a cone called a **stalagmite** (stuh-LAG-MITE). Often a stalactite will grow downward and a stalagmite will grow upward until they meet, forming a continuous column of calcite.

Natural Bridges

When the roof of a cavern collapses in several places, a relatively straight line of sinkholes is created. The uncollapsed rock between each pair of sinkholes forms an arch called a **natural bridge.**

A natural bridge also can form when a surface river enters a crack in a rock formation, runs underground, then reemerges. As shown in Figure 14–10, the river weathers the rock it passes through, enlarging the opening, and eventually forms a natural bridge.

Karst Topography

Regions where the effects of the chemical weathering due to groundwater are clearly visible are said to have **karst topography.** This topography is named after the Karst region of Yugoslavia, where this type of topography is well developed. The features of these areas include many closely spaced sinkholes, caverns, and streams that disappear into fissures in the rock. These streams then emerge in caves or through other fissures many kilometers away. In the United States, karst topography is found in Kentucky, Tennessee, southern Indiana, and northern Florida.

Perhaps the most spectacular karst topography on earth is on the Vogelkop peninsula of New Guinea. Sinks more than 300 m deep have formed so close together that they are separated only by razor-sharp ridges. The terrain forms a pattern like that left in dough cut out by a biscuit cutter.

Features characteristic of karst topography also can be found in some arid regions. In such areas, the water table is far below the surface, and groundwater must pass through a thick zone of aeration. As it seeps downward, the groundwater may wash out fine grains from weakly cemented sediments. Small pipes form and extend outward from drainage gulleys. In some areas of the Badlands of South Dakota, for example, these pipes extend for many meters.

When the pipes collapse, sinkholes and caverns like those of areas with karst topography are formed. However, the pipes never extend below the water table, as do many sinkholes in limestone regions with true karst topography.

Figure 14–10. The weathering effect of an underground river produces a natural bridge.

Section 14.3 Review

1. What is the effect of carbonic acid on rocks?
2. How are caverns formed?
3. What is the difference between a stalactite and a stalagmite?
4. Why might you expect to find springs in regions with karst topography?

Reviewing Chapter 14

Key Concepts

- The porosity and permeability of a rock or sediment determine whether it is an efficient aquifer. See page 223.
- The contour of the water table is influenced by the local topography. See page 226.
- Pollution can make the use of groundwater dangerous to living organisms. See page 227.
- Artesian formations are the source of artesian wells and springs. See page 229.
- Hot springs and geysers result when groundwater is heated. See page 229.
- Caverns and sinkholes form as a result of chemical weathering. See page 231.

- **There are two zones of moisture beneath the ground. See page 225.**
- Karst topography shows the erosive effects of groundwater. See page 233.

Key Terms

aquifer (223)
artesian formation (229)
artesian spring (229)
artesian well (229)
cap rock (229)
capillary fringe (225)
cone of depression (228)

geyser (230)
hard water (231)
hot spring (229)
hydraulic gradient (226)
impermeable (225)
karst topography (233)
mud pot (230)
natural bridge (233)

ordinary spring (228)
ordinary well (228)
perched water table (226)
permeability (224)
porosity (223)
sinkhole (232)
soft water (231)
sorting (223)

stalactite (232)
stalagmite (233)
travertine (230)
water table (225)
zone of aeration (225)
zone of saturation (225)

Review

On your paper, write the letter of the term that best completes each of the following statements.

1. Any body of rock through which water can flow and which can store enough water for domestic or industrial use is called (a) a well (b) an aquifer (c) a sinkhole (d) an artesian formation.
2. The total volume of the open spaces in a rock is called its (a) viscosity (b) permeability (c) hydraulic gradient (d) porosity.
3. When all the particles in a sediment are about the same size, the sediment is said to be

(a) fractured (b) well sorted (c) permeable (d) porous.
4. The ease with which water can pass through a rock or sediment is called (a) permeability (b) porosity (c) carbonation (d) velocity.
5. The underground layer of rock where all of the open spaces are filled with water is called the (a) zone of aeration (b) cap rock (c) water table (d) zone of saturation.
6. The upper surface of the zone of saturation is called the (a) capillary fringe (b) water table (c) hydraulic gradient (d) travertine.

7. The slope of a water table is its (a) hydraulic gradient (b) permeability (c) porosity (d) aquifer.
8. It takes a long time for polluted water to become pure again because (a) groundwater can only be replaced artificially (b) groundwater is replenished slowly (c) groundwater travels very quickly (d) groundwater can never be replaced.
9. A natural flow of groundwater that has reached the surface is (a) a spring (b) an aquifer (c) a well (d) a travertine.
10. Pumping water from a well causes a lowering of the water table known as a (a) cone of depression (b) horizontal fissure (c) hot spring (d) sinkhole.
11. Hot springs that are formed in areas of recent volcanic activity may become (a) caverns (b) natural bridges (c) geysers (d) mud pots.
12. Travertine usually forms (a) terraces (b) natural bridges (c) sinkholes (d) caverns.
13. When part of the roof of a cavern collapses, the result is a (a) sinkhole (b) natural bridge (c) horizontal fissure (d) geyser.
14. Calcite formations suspended from the ceiling of a cavern are called (a) stalagmites (b) stalactites (c) sinks (d) aquifers.
15. Regions where the results of chemical weathering by groundwater are clearly visible are said to have (a) sink topography (b) karst topography (c) limestone topography (d) artesian formations.

Application

On your paper, write answers to the following questions.

1. A rock can be porous, yet impermeable. Explain how.
2. What effect would you expect an unusually dry season to have on the capillary action in the soil?
3. In areas where the water table is along the surface of the land, what type of terrain would you expect to find?
4. How is groundwater replenished artificially? Why is this an important process?
5. Describe an artesian formation and explain how the water in an artesian well may have entered the ground many hundreds of kilometers away.
6. Explain how a mud pot forms. Why are mud pots sometimes called paint pots?
7. Explain the process that results in the formation of stalactites and stalagmites. Can you think of another process in nature that produces shapes similar to stalactites?
8. Do you think an area with karst topography would have many or few surface streams? Explain your answer.

Extension

1. Research the source of the fresh water used in your community. How much of it is groundwater? Are there times of the year when the water table in your area drops low enough that compulsory water conservation measures are enacted? Write a report that summarizes your findings.
2. Garbage, trash, and other solid waste products often are buried in sanitary landfills. Water that seeps into the wastes can carry soluble pollutants into the groundwater. Find out how solid wastes are disposed of in your community. What measures are taken to protect the groundwater supply from contamination? How is the groundwater tested to verify its purity? Write a report on your findings.
3. When a sinkhole develops in a populated community, the consequences can be quite serious. Use the *Reader's Guide to Periodical Literature* to look up articles on recent sinkhole activity and the damage caused. Report to the class on your findings.
4. Make a relief map of an area with karst topography out of clay. Try starting with a solid block and carving out caverns, sinkholes, and natural bridges as they would be sculpted by an underground stream.

Chapter 15

Glaciers and Erosion

Massive sheets of moving ice once covered much of the earth. The ice advanced and retreated many times during the earth's geologic history. Scientists are not sure why this occurred, but they have proposed several theories.

Although glaciers still exist in some areas of the world, evidence of their past movements exists worldwide. In this chapter, you will learn to recognize the signatures left by these powerful reshapers of the earth's surface.

Chapter Outline

The Columbia glacier is in Prince William Sound, Alaska.

15.1 Glaciers: Moving Ice

Section Objectives

• **Describe how glaciers form.**

• **Compare the two kinds of glaciers.**

• **Explain two processes by which glaciers move.**

A single snowflake is lighter and more delicate than a feather. However, you can gather a handful of soft snow and squeeze it to make a firm snowball. In a process similar to the compacting of snow to make your snowball, natural forces compact snow to create enormous masses of moving ice. These masses of moving ice, called **glaciers,** are powerful agents of erosion. In fact, they have carved some of the most spectacular features on the earth's surface.

Formation of Glaciers

Water that reaches the earth's surface as rain quickly runs off and eventually flows to the ocean or seeps into the earth as groundwater. However, at high altitudes and in polar regions, moisture often reaches the surface as snow and remains on the ground all year. The **snowline** is the elevation above which ice and snow remain throughout the year. The elevation of the snowline varies from place to place. At the equator, it occurs at about 5,500 m above sea level; near the poles, the snowline is at about sea level. Figure 15–1 shows the snowline in the Rocky Mountains.

A **snowfield,** or ice field, is an almost motionless mass of permanent snow and ice. Snowfields are formed by an accumulation of ice and snow above the snowline. Snowfields cover most of the land near the poles and the tops of some mountains at lower latitudes.

Because the average temperature at high altitudes and in polar regions is always near or below the freezing point of water, the snow accumulates in thick layers year after year. The partial melting and refreezing that occur periodically changes the snow crystals into small grains of ice. This grainy ice is called **firn,** from a German word that means ''of last year.''

In the deepest layers of accumulated snow, the pressure of the overlying layers becomes so great that the air between the ice grains is squeezed out. The firn then loses its white color and becomes a

Figure 15–1. The snowline in the United States Rocky Mountains is at about 2,800 m.

Figure 15–2. A valley glacier descends from a snowfield on this Alaskan mountain (left). An enormous snowfield in Denali National Park, Alaska covers most of the mountain top (right).

bright steel-blue color that is characteristic of solid ice. Each year more snow is added on top of the ice layers. Eventually the weight of the ice becomes great enough to cause the body of ice to move, and a glacier is formed. Relatively few places on earth have average temperatures that are low enough and snowfall that is great enough for glaciers to form.

The growth of a glacier depends on the balance between snowfall received and ice lost by melting and evaporation. As long as new snow is added faster than it melts and evaporates, the glacier will continue to increase in size. When the ice melts faster than snow is added, the glacier will decrease in size. Very slight differences in average yearly temperatures and snowfall may upset the balance between new snow and melting ice. Thus, the increase or decrease in the size of a glacier can be a relatively sensitive indicator of annual climatic change.

Types of Glaciers

There are two types of glaciers; these are distinguished by the way they are formed and by their size. One type of glacier is formed in the snowfields of mountainous areas. As the ice moves down the mountain, it produces a **valley glacier.** A valley glacier is a long, narrow, wedge-shaped mass of ice that usually moves through a mountain valley. The other type of glacier covers large land areas. These massive ice sheets, called **continental glaciers,** or *ice caps,* occupy millions of square kilometers. Today, continental glaciers are found only in Greenland and Antarctica.

Antarctica is covered by the largest continental glacier in the world. This glacier is one and a half times as large as the area of the mainland United States. In most places, the Antarctic ice cap is more than 3,000 m thick.

About 80 percent of Greenland also is buried under a continental glacier. The Greenland glacier is about 3,000 m deep at its thickest point. The amount of water contained in the two great glaciers of Greenland and Antarctica is enormous. Scientists estimate that if melted, the ice would release enough water to raise worldwide sea level by more than 60 m. What effect would this rise in sea level have on continental coastlines?

Movement of Glaciers

Glaciers are sometimes called *rivers of ice*. A glacier, however, moves very differently from the water in a river. Water in a stream and ice in a glacier both move downward in response to gravity. However, the ice in glaciers cannot move rapidly or flow easily around barriers as can water in a stream. On the average, a glacier moves about 100 m per year. Some glaciers may travel only a few centimeters per year, while others may suddenly start to move forward relatively quickly.

Glacial motion has been a subject of much scientific study. Most scientists agree that glaciers move by two basic processes—**sliding** and **internal flow.**

Sliding

The weight of the ice in a glacier eventually exerts enough pressure to melt the ice where it comes in contact with the ground. The water from the melted ice acts as a lubricant between the ice of the glacier and the underlying rock. All glaciers move by sliding.

A glacier may work its way over small barriers in its path by melting and then refreezing. For example, if the ice pushes against a stone, the pressure causes a little of the ice to melt. The water from the melted ice then flows around the barrier and freezes again as the pressure is removed.

Internal Flow

Glaciers also move by internal flow, in which ice crystals slip over each other, causing a slow forward motion. In studying glacial motion, scientists have driven stakes across a valley glacier, as shown in Figure 15–3. They found that the speed of internal flow is not the same at all parts of the glacier. The rate of motion at a given point is determined by the slope and by the thickness and temperature of the

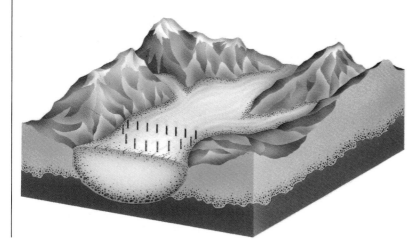

Figure 15–3. A line of blue stakes driven into a valley glacier moves to the position of the red stakes as the glacier flows. Notice that the center of the glacier moves faster than the sides.

ice. The internal flow is faster at the surface of a glacier than at its base. Friction caused by contact of the glacier with the valley slows the base. The center of the glacier also moves faster than its sides. Why do you think this occurs?

Features of Glaciers

Friction along the base and sides of a glacier slows the movement of the glacier. However, it also causes the ice to melt and refreeze, enabling those parts of the glacier to flow plastically. The surface ice, meanwhile, remains brittle. Because the glacier flows unevenly beneath the surface, points of pressure build on the surface. The brittle glacier surface then buckles, forming arches of ice called **pressure ridges.**

Large cracks, called **crevasses** (krih-VASS-uz), form between the pressure ridges on the surface of a glacier. Crevasses may be more than 30 m deep. They may be hidden by a thin crust of snow that breaks under the slightest weight. Traveling over the top of a glacier is therefore very dangerous and should be attempted only by experienced people.

Horizons

Technology: Computer Predictions

Visitors to the Alaskan coastal glacier Columbia in the year 2030 will take home very different photographs than those taken today. The glacier that you see in a recent photograph at the left will be only about half as long as it is now.

Scientists at the U.S. Geological Survey made this prediction with the aid of a new computer program. The computer program analyzes the position of the glacier, the number of icebergs detaching, and the rate of ice flow near the tip of the glacier. The program can predict how quickly the glacier will retreat and when huge blocks of ice, such as those shown right, may break loose.

Although no one can yet test the accuracy of predictions concerning the year 2030, the original 1980 computer forecast was on target. During 1981, the Columbia glacier made its largest retreat—450 m—exactly as predicted. The 1980 and 1981 boundaries are shown at the right.

A continental glacier usually has centers where the accumulation of snow and the resulting pressure build up. Glaciers move outward in all directions from these centers. For example, the ice sheets that now cover Antarctica and Greenland move outward toward the shores of their landmasses. Along some parts of the coast of Antarctica, the glacier has moved out over the ocean, forming ice shelves. The largest of these shelves, the Ross Ice Shelf, is equal to the area of Texas. The rise and fall of the tides breaks off large pieces of the leading edge of this ice. These large blocks of ice, called *icebergs*, drift into the ocean. Because the major portion of an iceberg is below the surface of the water, icebergs pose a hazard to ships. One of the largest icebergs ever observed was found in the Antarctic and was twice the size of Connecticut.

Section 15.1 Review

1. In what regions does snow accumulate year after year?
2. What type of glacier covers Greenland?
3. Explain the type of glacial movement known as *internal flow*.
4. Compare and contrast valley glaciers and continental glaciers.

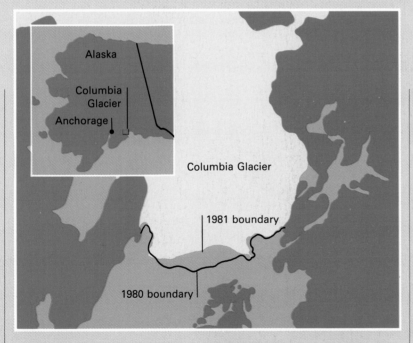

Computer-generated predictions about glacial movements directly benefit many groups, including the U.S. Coast Guard. These predictions indicate when potentially damaging icebergs are likely to break off from glaciers and fall into the Alaskan waters. Using this knowledge, the Coast Guard can plan to keep shipping lanes clear for tankers transporting oil from the Alaskan pipeline. One plan calls for tugboats to tow the icebergs out of the shipping lanes.

The Columbia glacier is the last great Alaskan coastal glacier to begin its rapid recession from the sea. Nevertheless, scientists feel that the ability to make computer predictions of glacial movements will be important in the future. The continuing rise in global temperatures and the simultaneous reduction in snowfall are causing other glaciers to begin to retreat.

In what period of an ice age would climatic conditions like these occur?

15.2 Landforms Created by Glaciers

Both valley glaciers and continental glaciers are powerful agents of erosion. Valley glaciers carve out rugged features in the mountains through which they flow. Massive continental glaciers produce a smoother landscape as they plane off all existing land features except the largest mountains.

Glacial Erosion

The glacial processes that change the shape of mountains begin in the upper end of the valley where a valley glacier forms. As a glacier wedges its way through a narrow valley, it breaks off rock from the valley walls, causing the walls to become steeper. The moving glacier also pulls blocks of rock from the floor of the upper valley. These actions create a bowl-shaped depression called a **cirque** (SURK), illustrated in Figure 15–4. This French word, meaning "circus," refers to the resemblance of the depression to a round circus theater. The ridges between cirques that form close together become sharp and jagged. These ridges are called **arêtes** (uh-RATES), which means "spines" in French. Sometimes, several arêtes are joined and form a sharp, curved peak called a **horn.**

Figure 15–4. This diagram illustrates the features created by glacial erosion. Notice in the top left diagram that the glacier is eroding the mountain to form cirques, arêtes, and horns. In the diagram at the bottom right, small lakes, or tarns, fill the cirques.

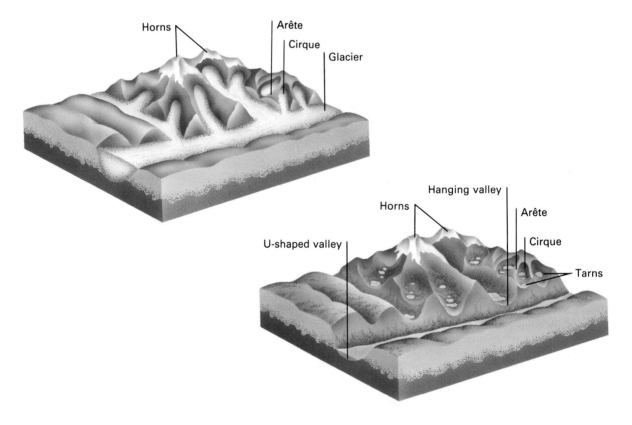

As the valley glacier flows down through an existing valley, it continues to pick up large amounts of rock from the valley. These rock fragments, which range in size from microscopic pieces to large blocks, become embedded in the glacier and help weather the rock over which the ice moves.

Solid rock over which a glacier has moved often is polished by the scraping action of the tiny rock particles embedded in the ice. Larger rocks carried by the ice may gouge out deep scratches or grooves in the bedrock. Large rock projections may be rounded by a passing glacier. These rounded projections usually have a smooth, gently sloping side facing the direction from which the glacier came. The other side of the projection is steeper and jagged because rock is pulled away as the ice passes over. The resulting rounded knobs of rock are called **roches moutonnees** (RAWSH-MOOT-un-AZE), which means ''sheep rocks,'' because they resemble the backs of sheep.

As the valley walls and floor are scraped away by a glacier, the original V-shape of the valley is changed into a U-shape. Because glacial action is the only means by which a valley can acquire a U-shape, scientists can easily tell if a glacier has passed through a particular valley.

Small glaciers in adjacent valleys often flow into the main glacier. As the small glaciers melt and disappear, their valleys are no longer eroded. The valley of the main glacier, however, continues to be eroded until it is much lower than the abandoned valley of the small glacier. The abandoned valley is left suspended on the mountain high above the main valley floor and is called a **hanging valley.** What natural feature do you think will appear when a stream flows through a hanging valley?

The landscape created by massive continental glaciers is quite different from the sharp, rugged features shown in Figure 15–5, which were produced by valley glaciers. Continental glaciers level most landforms, producing relatively smooth, rounded landscapes.

Figure 15–5. Notice the rugged landscape of this typical U-shaped glaciated valley in the Lamoille Canyon, Nevada.

Investigation

Glacial Erosion

Many of the earth's natural features have been shaped by the movement of glaciers. You can demonstrate the effects of glacial erosion with a simple model.

Science Process Skills Focus: constructing models, observing, describing, predicting

Materials

plastic container, about 10 × 5 × 5 cm; mixture of sand, gravel, and small rocks; water; freezer; modeling clay; hand towel; sand; flat box, about 30 × 20 cm; soft wood board, about 15 cm long

Procedure

1. Put the sand, gravel, and rocks in the bottom of the container. Fill the container with water to a depth of about 4 cm. Freeze the container until the water is solid. Remove the ice block from the container.
2. Flatten the clay into a rectangle about 10 × 10 × 1 cm. Grasp the ice block firmly with the hand towel. Place the block with the gravel-and-rock side down at one end of the clay. Press down lightly on the ice block and move it along the clay. Sketch the pattern made in the clay by the ice block.

3. Press damp sand into the rectangular box. Repeat Step 2, substituting first the sand and then the board for the clay.

Analysis and Conclusions

1. Describe the effects of the ice block on the clay, sand, and board.
2. Did any material from any surface become mixed with material from the ice block? Did the ice deposit material on any surface?
3. What glacial land features are represented by the features of your clay model? Your sand model? Your wood model?
4. Based on your observations, predict the results of glacial erosion on rock.

Existing valleys may be gouged out and deepened. Exposed rock surfaces are flattened by continental glaciers, just as bulldozers flatten landscapes. Surfaces also are scratched by coarse sand and pebbles. These scratches, or striae, run parallel to the direction of glacial movement.

Glacial Deposition

The ice of a glacier may melt when a valley glacier reaches lower altitudes or when a change in climate melts the ice sheets of continental glaciers. When a glacier melts, all of the material accumulated in the ice is deposited. Glacial deposits usually can be identified easily. For example, large boulders, called **erratics,** are transported by glaciers. The composition of erratics usually differs from that of the bedrock over which they lie.

Various land features are formed by **glacial drift**—sediments deposited by a glacier—or by the melted ice from a glacier. One type of glacial drift, called **till,** is unsorted deposits of rock material. Till either is deposited from sediments scraped off the bottom of the glacier or is left behind when glacial ice melts. Another type of glacial drift is called **layered drift.** Layered drift is material that has been sorted and deposited in layers by the action of streams of melted ice, or **meltwater,** flowing from the glacier. Many transported soils originate as glacial drift.

Till Deposits

Till often forms a moraine, which is a ridge or mound of unsorted rock material on the ground or on the glacier itself. There are several types of moraines, as illustrated in Figure 15–7. A **lateral moraine** is one that is deposited along the sides of a valley glacier, usually as a long ridge. When two or more valley glaciers join, their adjacent lateral moraines are joined in the middle of the combined glacier. These two moraines combine in the center of the new glacier to form a **medial moraine.**

Sometimes the bottom of a glacier becomes overloaded with rock material. Part of this load then is deposited, and the ice moves over it. All the unsorted material left beneath the glacier and the material deposited when the ice melts make up a **ground moraine.** After the glacier has melted completely, the surface of the ground moraine forms gentle hills and depressions. Much of the landscape from eastern Ohio west to the Rockies and north to Canada is made up of ground moraine. The soil of ground moraines is often very fertile. A subsequent advance of a glacier may mold ground moraine into **drumlins.** Drumlins are long, low, tear-shaped mounds of till, often found in clusters. The long axes of the drumlins are parallel to the direction of glacial movement.

Figure 15–6. Both continental glaciers and valley glaciers deposit unsorted rock material called till. Note the till on the floor of this valley in Jasper National Park; Alberta, Canada.

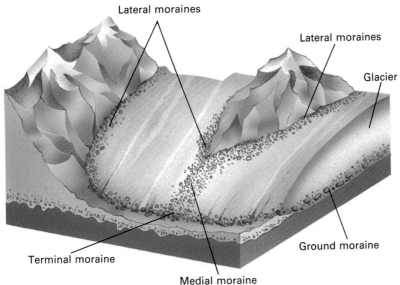

Figure 15–7. This diagram shows the types of moraines formed by glacial deposition.

Lateral moraines

Lateral moraines

Glacier

Terminal moraine

Medial moraine

Ground moraine

Till deposited at the leading edge of a melting glacier produces a **terminal moraine.** Terminal moraines are belts of small ridges of till with many depressions that contain lakes or ponds. Large terminal moraines, some more than 100 km long, can be seen in Minnesota, the Dakotas, Wisconsin, northern Illinois, Indiana, and Ohio. All of Long Island in New York is a terminal moraine.

Layered Drift Deposition

Melting occurs almost constantly in a glacier. Streams of meltwater flow from the edges, the surface, and beneath the glacier, particularly during warm weather. The glacial meltwater usually appears milky due to the presence of very fine rock particles carried from the glacier. Along with the small rock particles, the meltwater carries drift that it deposits as a large **outwash plain** in front of the glacier. An outwash plain is a fan-shaped deposit of layered drift. The plain usually lies in front of a terminal moraine and is crossed by many meltwater streams.

Most outwash plains are pitted with depressions called **kettles.** A kettle forms when a portion of glacial ice is buried by layered drift. As the piece of ice melts, it leaves a cavity in the drift. The layered drift sinks to the bottom of the cavity and produces a depression. Kettles often fill with water, forming kettle lakes.

When continental glaciers recede, long, winding ridges of gravel and coarse sand may be left behind. These ridges, called **eskers** (ESS-kurz), consist of layered drift deposited by streams of meltwater flowing in tunnels within the glaciers. Eskers may extend for tens of kilometers like raised, winding roadways.

Figure 15–8. This kettle lake in Saskatchewan, Canada was formed by glacial erosion.

Glacial Lakes

Glaciers often form lake basins by scooping out surface areas, leaving depressions and deepening existing valleys. Many lake basins in New England and New York State were gouged from solid rock by a continental glacier.

However, most glacial lakes are formed by deposition rather than by erosion. Many lake basins were left in the uneven surface of ground moraine deposited by glaciers. Lake basins such as these are found in many areas of upper North America and northern Europe. These basins filled with water from melting snow, rainfall, or groundwater.

Other glacial lakes formed when terminal and lateral moraines blocked existing streams. Minnesota, the Dakotas, Wisconsin, Indiana, Ohio, and northern Illinois have belts of moraines and associated lakes.

History of the Great Lakes

The Great Lakes of North America are the result of a combination of erosion and deposition by continental glaciers. Glacial erosion widened and deepened broad river valleys covered by the glaciers. Moraines to the south blocked off the ends of these valleys. As the glaciers melted, the meltwater flowed into the valleys and was held there by the moraines to form lakes. In their early stages, the lakes had outlets only to the south through the Wabash and Illinois rivers, which flowed into the Mississippi River. As more ice melted, the lakes grew larger and also began to drain into the Atlantic Ocean through the Susquehanna River. At a later stage, the lakes also drained through the Mohawk and Hudson River valleys.

The Great Lakes expanded until they were slightly larger than they are now. Then, as the glaciers retreated, the pressure of the ice on the land surface was reduced. As a result, the surface was uplifted. The lake beds were also uplifted and slowly shrank to their present size. The upward shift in the land also caused the lakes to drain to the north through the St. Lawrence River. As Lake Erie flowed north into Lake Ontario, Niagara Falls also were created. Figure 15–9 illustrates the history of the Great Lakes.

Salt Lakes

Many of the lake basins found in the southwestern part of the United States were large lakes during the periods when glaciers were present. As the glaciers retreated, meltwater and rainfall filled these low-lying areas. However, many of these lakes had no outlet streams, so water could leave them only by evaporation. When the water evaporated, salt that was dissolved in the water was left behind. This made the water increasingly salty. For example, the Great Salt Lake in Utah is the remains of the glacial Lake Bonneville. A lake that has no natural outlet or is deprived of its outlet may become a salt lake. However, the lake must be located in a region where evaporation is rapid and precipitation is low.

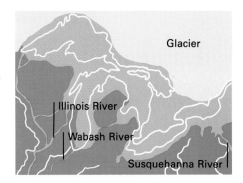

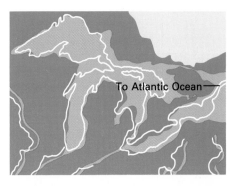

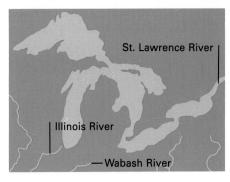

Figure 15–9. The glacier that covered the northern United States formed enormous lakes that drained to the south. As the glacier retreated, the lakes became smaller and the drainage pattern changed (center). Uplifting of the land reduced the Great Lakes to their present size and established current drainage north through the St. Lawrence River (bottom).

Science Notebook

Lake Agassiz

As the ice continued to retreat after the formation of the Great Lakes, a tremendous glacial lake was formed to the northwest of the Great Lakes. This was Lake Agassiz, named for Louis Agassiz, a nineteenth-century Swiss scientist. Agassiz became an early and active proponent of the idea of widespread glaciation, based on his extensive observations of land features in the Swiss alpine valleys.

Lake Agassiz developed when an ice dam blocked the northward flow of the glacial meltwater in the Red River Valley. The lake covered parts of North Dakota, Minnesota, and the Canadian province of Manitoba. The water of Lake Agassiz spilled southward and eventually flowed into the Mississippi River where St. Paul is now located.

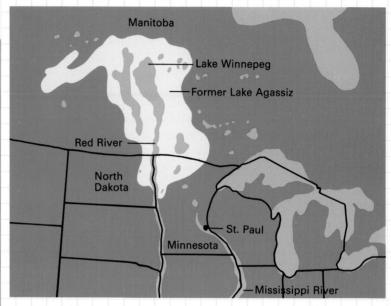

When the ice block melted, the lake quickly drained, leaving a great fertile plain in its place. Depressions in the lake floor remained filled with water and formed remnant lakes. Lake Winnipeg in Canada is the chief remnant of the former vast Lake Agassiz.

Today, the Red River flows for nearly 950 km through farms of wheat that now cover this ancient lake bed.

What type of glacial deposit formed the lake bed of Lake Agassiz?

The composition of the water in a salt lake depends upon the dissolved minerals it contains. The minerals either are brought in by streams or result from the chemical and biologic processes occurring in the lake. The salt deposits left in the dry beds of ancient salt lakes often contain valuable minerals. For example, borax, which is used as a cleaning material and in the manufacture of glass, is mined from such deposits.

Section 15.2 Review

1. How is a cirque produced?
2. How does a kettle form on an outwash plain?
3. How can terminal and lateral moraines form glacial lakes?
4. Compare land features formed by till with those formed by deposits from glacial meltwater.

15.3 Ice Ages

A long period of climatic cooling during which ice sheets cover large areas of the earth's surface is known as an **ice age.** Scientists have found evidence of at least four major ice ages during the earth's geologic history. These ice ages lasted an average of 50 thousand years. The earliest occurred more than 600 million years ago. The most recent began more than 2 million years ago. Its massive ice sheets started to retreat only 18 thousand years ago.

Section Objectives

- **Describe the climatic cycles that exist during an ice age.**
- **Identify and summarize the theory that best accounts for the ice ages.**

Climate During Ice Ages

An ice age probably begins with a long-term decrease in the earth's average temperatures. A drop in average temperatures of only about 5°C, combined with an increase in snowfall, sets the stage for an ice age. The greater snowfall increases the size of the glaciers. Pressure from the accumulating snow starts the glacier moving.

Scientists know that climates were generally colder during the last ice age. However, studies of landforms show at least four cycles of advance and retreat of glaciers during this ice age. The glaciers advanced during cold periods and retreated during warmer periods. The time when a glacier advances during an ice age is a **glacial period.** The times of warmer temperatures between glacial periods are called **interglacial periods.**

Glacial Periods

As Figure 15–10 shows, glaciers covered nearly one third of the earth's land surface during the last glacial period. In some parts of

Figure 15–10. At its peak, the last great glacial period covered about 30 percent of the earth's land area, shown here in light blue.

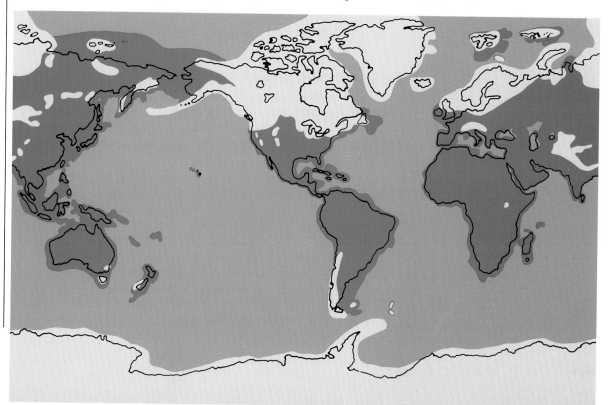

Figure 15–11. During the ice ages, many valley glaciers had a geologic impact on the land. In Alaska, valley glaciers such as this one in Glacier Bay, continue to shape the land.

North America, the ice was several kilometers thick. So much water was locked up in ice that the sea level may have been as much as 100 m lower than it is today.

Canada and the mountainous regions of Alaska were buried under ice. In the mountains of the western United States, valley glaciers joined to form several large valley glaciers. They flowed westward toward the Pacific and eastward toward the foothills of the Rocky Mountains. A great continental glacier with its center in the Hudson Bay region of Canada spread as far south as the Missouri and Ohio rivers. A large glacier, centered over the Baltic Sea, spread south to Germany, Belgium, and the Netherlands and west to the British Isles. Continental glaciers reached Poland and Russia, and valley glaciers formed in the Alps and in Siberia. Large glaciers also formed in the southern hemisphere. The Andes Mountains of South America and most of New Zealand were covered by large ice sheets. Many land features that formed during the last glacial period are still easily recognizable. What features would you expect to see?

Causes of Ice Ages

Scientists have proposed a number of theories to account for ice ages. Each theory explains why the earth experienced the gradual cooling and the increase in precipitation that brought on the expansion of the glaciers. The theories also explain why the glaciers retreated during the interglacial periods.

The theory that most scientists now accept is called the **Milankovitch theory.** Milutin Milankovitch, a Yugoslavian scientist, proposed that small, regular changes in the earth's orbit and in the tilt of the earth's axis caused the ice ages.

Three kinds of periodic changes occur in the way the earth moves around the sun. One change is in the shape of the earth's orbit. The orbit varies from nearly circular to more elongated and back to circular about every 100 thousand years. Figure 15–12 illustrates this change. A second change is in the tilt of the earth's axis. Over a period of about 41 thousand years, the tilt of the axis varies

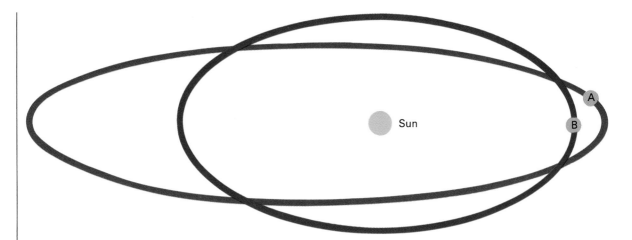

Figure 15–12. The distance of the earth from the sun changes as the earth's orbit varies between exaggerated pattern A and pattern B

between 21.5° and 24.5°. A third periodic change results from the circular motion, or precession, of the earth's axis. Precession causes the axis to change its position. The axis of the earth traces a complete circle in space every 26 thousand years. See Chapter 2 for a more complete discussion of precession. Milankovitch calculated how these three changes periodically decrease the amount of solar energy reaching the earth.

Evidence from the ocean floor—in the shells of dead marine animals—has provided support for the Milankovitch theory. Scientists have observed that the ratio of two isotopes of oxygen in the shells is related to the temperature of the water when the animals lived. The shells are found in layers of sediment on the ocean floor. They can show if a sediment was formed during an ice age, when the water temperature was low. Scientists have found that the record of ice ages left in the oceans closely follows the cycle of cooling predicted by the Milankovitch theory.

There also are other theories to explain the causes of ice ages. Like the Milankovitch theory, several of these theories state that the ice ages resulted from decreased solar energy reaching the earth's surface. However, the explanations for the decrease vary. Some theories propose that the decrease in solar energy might have been caused by a change in the amount of heat produced by the sun. Other theories propose that volcanic dust blocked the sun's rays. Another theory proposes that ice ages resulted from the changing positions of the continents, which prevented warm ocean currents from reaching regions near the poles.

Section 15.3 Review

1. What happens to the sea level during an ice age?
2. What is a glacial period?
3. What must a theory of the ice ages explain?
4. Explain the theory of the ice ages that most scientists accept at the present time.
5. Briefly describe three other theories of the cause of ice ages.

Reviewing Chapter 15

Key Concepts

- Formation of glaciers requires adequate snowfall and low temperatures. See page 237.
- The two major types of glaciers are valley glaciers and continental glaciers. See page 238.
- Glaciers move by sliding and by internal flow. See page 239.
- Glaciers erode the valleys through which they flow, producing characteristic landforms. See page 242.

- When glaciers melt, they deposit their sediments, creating features that can be readily identified. See page 244.
- Glaciers may form lake basins by eroding the land or by depositing sediments. See page 247.
- During an ice age, glacial advance alternates with glacial retreat. See page 249.
- The Milankovitch theory is the most commonly accepted theory of the cause of ice ages. See page 250.

Key Terms

arête (242)
cirque (242)
continental glacier (238)
crevasse (240)
drumlin (245)
erratic (244)
esker (246)
firn (237)

glacial drift (245)
glacial period (249)
glacier (237)
ground moraine (245)
hanging valley (243)
horn (242)
ice age (249)
interglacial period (249)

internal flow (239)
kettle (246)
lateral moraine (245)
layered drift (245)
medial moraine (245)
meltwater (245)
Milankovitch theory (250)
outwash plain (246)

pressure ridge (240)
roche moutonnée (243)
sliding (239)
snowfield (237)
snowline (237)
terminal moraine (246)
till (245)
valley glacier (238)

Review

On your paper, write the letter of the term that best completes each of the following statements.

1. Snow accumulates year after year in polar regions and (a) in southern Canada (b) in the United States (c) below the snowline (d) at high altitudes.
2. Glaciers formed in the snowfields of mountainous areas are called (a) continental glaciers (b) valley glaciers (c) icebergs (d) ice shelves.
3. Antarctica is covered by the earth's largest (a) iceberg (b) valley glacier (c) continental glacier (d) outlet glacier.
4. A glacier will move by sliding when the ice and rock are separated by a thin layer of (a) water (b) snow (c) pebbles (d) drift.

5. When a glacier moves by internal flow, its (a) center moves fastest (b) bottom moves fastest (c) edges move fastest (d) parts all move at the same speed.
6. Glacial erosion may produce a bowl-shaped depression known as (a) a moraine (b) an esker (c) a cirque (d) a horn.
7. As a glacier moves through a valley, it carves out (a) a U-shape (b) a V-shape (c) an esker (d) a horn.
8. Unsorted glacial deposits are called (a) layered drift (b) outwash plains (c) eskers (d) till.
9. Glacial drift that has been sorted and deposited by meltwater is (a) till (b) layered drift (c) a moraine (d) a drumlin.

10. Long, winding ridges of gravel and sand left behind by continental glaciers are called (a) eskers (b) drumlins (c) outwash plains (d) medial moraines.
11. A kettle is a (a) hill (b) depression (c) ridge (d) mound.
12. A fan-shaped deposit of layered drift is called (a) a drumlin (b) an outwash plain (c) ground moraine (d) a roche moutonnée.
13. Most glacial lakes are formed by (a) valley glaciers (b) pressure ridges (c) erosion (d) deposition.
14. During the last glacial period, the average temperature was about (a) 5° lower than today (b) 15° lower than today (c) 35° lower than today (d) 50° lower than today.
15. One component of the Milankovitch theory is (a) the circular motion of the earth's axis (b) continental drift (c) volcanic activity (d) landslide activity.
16. A proposed cause of the ice ages is decreased solar energy reaching the earth due to (a) a lunar eclipse (b) blockage by volcanic dust (c) sinking of the land (d) increased storm activity.

Application

On your paper, write answers to the following questions.
1. Imagine that there is a village in Greenland located at the edge of the ice cap. During the year there is an unusually large amount of snowfall. How might this snowfall affect the ice sheet? What danger might this pose for the inhabitants of the village?
2. Why is it important for scientists to monitor and study the continental glaciers that cover Greenland and Antarctica?
3. Antarctic explorers need special training to travel safely over the ice cap. Besides the cold, what structural aspects of the glaciers might be dangerous?
4. List some features caused by glacial erosion that you might see on a car trip.
5. What evidence of glacial deposition might you see on the trip in Question 4?
6. In what ways might past glacial action in New England and New York State affect tourism and recreation in those regions today?
7. In addition to decreasing temperature and increasing snowfall, what other phenomenon might signal an impending ice age?
8. A group of scientists is trying to find evidence to support Milankovitch's theory of the earth's ice ages. List some kinds of research they might need to do.

Extension

1. Many tourists visit Kettle Moraine Park in Wisconsin to see its glaciated landscape. Find out about other areas in the United States and Canada where you can see glacial features. Send for information about one of these places. You might try writing to the local tourist bureau. Once you have obtained your information, put together a travel brochure that would persuade people to visit the place.
2. Imagining the enormous scale of the work done by the ice age glaciers may be difficult. Use an encyclopedia, an atlas, or your local library to gather material for a profile of each of the Great Lakes. Find out the width and depth of each of these lakes. Draw a diagram to scale comparing these lake profiles with each other and with other familiar bodies of water. How does the depth of the deepest of the Great Lakes compare with the height of Mt. Everest? Include Mount Everest on your diagram to illustrate this relationship.
3. Antarctica and Greenland are among the few places on earth where you can experience ice age conditions. Imagine that you are a scientist studying the ice sheet on Antarctica. Write a diary of your activities during one day, including observations of the scenery and the research projects you are working on.

Chapter 16
Erosion by Wind and Waves

Imagine that you are standing on a beach. Suddenly you see swirling sheets of sand. All you can hear is the howl of the wind and the thundering of waves against the shore.

In reading this chapter, you will learn how the wind leaves its mark on the land. You will also learn how the wind and water work together to constantly change features of the earth where the land and oceans meet.

Chapter Outline

This rugged coastline is in Big Sur, California.

16.1 Wind Erosion

Section Objectives

• **Describe two ways that the wind erodes the land.**

• **Compare the two types of wind deposits.**

There is energy in the wind. Some of the energy can be used to move a ship or turn a wind turbine. However, this energy can also erode the land. The wind erodes dry land much more effectively than it erodes moist land. Moisture makes soil heavier and causes some soil particles to stick together. Moist particles are more resistant to the force of the wind than are dry particles.

As the wind erodes the land, it carries along with it rock particles of various sizes. These particles are categorized as either sand or dust, usually silt and clay. Sand is loose fragments of weathered rocks and minerals. Most grains of sand are made up of quartz. Other common minerals in sand grains are mica, feldspar, and magnetite. Sand grains range in diameter from 0.07 mm to 5 mm. The smallest grain of sand is barely visible to the unaided eye; the largest grain of sand is about the size of a pinhead.

Dust consists of particles smaller than the smallest sand grain, or less than 0.07 mm in diameter. Most dust—silt and clay—is microscopic fragments of rock and minerals that come from the soil or from volcanic eruptions. Other sources of dust are plants, animals, and bacteria. Dust may also occur as a by-product of the burning of fuels and of certain manufacturing processes.

Wind moves sand and dust in different ways. Wind cannot keep aloft even the smallest particle of sand. Instead, sand grains are moved along by a series of jumps and bounces, much the way pebbles are moved by streams. Such movements are called *saltation*. Saltation occurs when the wind speed becomes great enough to roll sand grains along the ground. When rolling sand grains collide with one another, some sand grains bounce up, as illustrated in Figure 16–1. Once in the air, a sand grain moves ahead a short distance, then falls. As a sand grain falls, it strikes other sand grains. These sand grains also may be thrown into the air or rolled ahead by the impact. Saltating sand grains move in the same direction that the wind is blowing. However, the grains do not rise more than 1 m above the ground, even in very strong winds.

In contrast to sand grains, dust particles can be lifted by the wind and carried high into the air. Even gentle air currents can keep dust particles suspended in the air. Dust from volcanic eruptions may

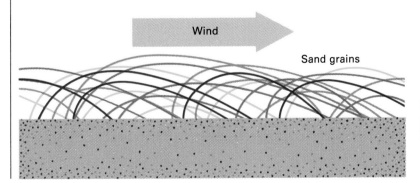

Wind

Sand grains

Figure 16–1. Sand grains move by saltation, making low, arcing jumps when blown by the wind.

remain in the atmosphere for several years before falling to the ground. Strong winds may lift enormous amounts of dust into the air and create large dust storms. Some dust storms cover hundreds of square kilometers and move more than 100 billion kilograms of dust.

Effects of Wind Erosion

The force of the wind erodes the land in a number of ways. The most common form of wind erosion is **deflation.** The term *deflation* is from the Latin word *deflare,* which means "to blow away." During the process of deflation, wind removes the top layer of fine, very dry soil particles. Left behind are rock fragments that are too large to be lifted by the wind. These remaining gravel particles often form a surface of closely packed small rocks called **desert pavement,** which protects the underlying land from erosion. Desert pavement is shown in Figure 16–2.

Deflation is a serious problem for farmers because it blows away the best soil for growing crops. Deflation may cause shallow depressions to form in areas where the natural plant cover has been removed. As an area of dry, bare soil is exposed to wind, the wind strips off a layer of the soil. Once the soil is gone, a shallow depression called a **deflation hollow** remains. A deflation hollow may be eroded further by wind and water, expanding to a width of several kilometers and a depth of 5 m to 20 m.

Desert pavements and deflation hollows are very common in desert areas because wind erosion occurs more rapidly in deserts than it does in humid regions. In the desert, soil layers are thin and contain little moisture. Consequently, desert soils may be swept away easily by the wind.

Sand grains carried by the wind increase its erosive power, just as the stream load increases the erosive power of running water. Abrasion is the weathering of rock particles by the impact of other rock particles. Abrasion by wind-blown sand results in much erosion in areas where there are strong, steady winds, large amounts of loose sand, and relatively soft rocks.

Pebbles and small stones in deserts and along some beaches are exposed to wind abrasion. As a result, the surfaces of the rocks become flattened and polished on two or three sides. This polishing produces facets on the stones. Stones thus smoothed by wind abrasion are called **ventifacts,** from the Latin word *ventus,* meaning "wind." The direction of the wind that formed a ventifact can be determined by the appearance of the ventifact. How would you determine wind direction by studying ventifacts in a certain area?

Scientists once thought that large rock structures, such as desert basins, natural bridges, rock pinnacles, and rocks perched on pedestals, were caused by wind erosion. However, it is more likely that such large features were produced by erosion due to running water. Erosion of large masses of rock by wind-blown sand takes place very slowly and is limited to the area close to the ground, where saltation occurs.

Figure 16–2. The formation of desert pavement prevents the erosion of the material beneath it.

Wind Deposition

All the material eroded by the wind is eventually deposited. The size of the particles carried by the wind depends on the wind speed. As its speed increases, the wind is able to carry larger particles. When the speed of the wind decreases, the heavier part of the load drops, forming a deposit. If the deposited fragments are not carried away by another strong wind, they are covered by future deposits. Eventually the weight of the top layers of built-up deposits compresses the covered fragments together. Pressure and the cementing action of minerals carried in water can bind the fragments together, thus forming sedimentary rock.

Dunes

The best-known wind deposits are **dunes,** which are mounds of wind-blown sand. Dunes form where the soil is dry and unprotected and the wind is strong. Dunes are common in some deserts and along the shores of oceans and large lakes.

A dune begins to form when a barrier slows the speed of the wind. This reduction in wind speed causes sand to accumulate on the sheltered side of the barrier, as illustrated in Figure 16–3. Sand also accumulates at the base of the exposed side. As more sand is added, the deposit acts as a barrier itself, and the dune continues to grow, eventually burying the original barrier.

Usually the gentlest slope of a dune is the side facing the wind, or the windward side, because the wind is constantly reshaping that side. The sand that is blown over the crest, or peak, of the dune tumbles down the opposite side, called the *slipface*. The slipface has a steeper slope than the windward side. Wind sweeping around the ends of a dune often causes two long, pointed extensions to form.

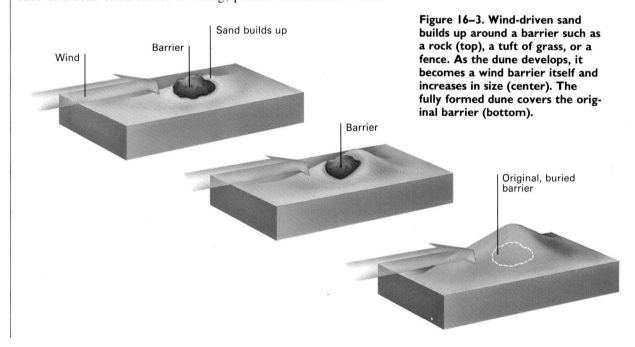

Wind | Barrier | Sand builds up

Barrier

Original, buried barrier

Figure 16–3. Wind-driven sand builds up around a barrier such as a rock (top), a tuft of grass, or a fence. As the dune develops, it becomes a wind barrier itself and increases in size (center). The fully formed dune covers the original barrier (bottom).

These give the dune a crescent shape. A crescent-shaped dune is called a **barchan** (BAHR-kan). The open side of the crescent faces away from the wind. Barchans a most common type of dune and are found in most deserts. Another type of dune, called a *parabolic dune,* is also crescent-shaped. However, the open side of a parabolic dune faces into the wind. These dunes often form as sand is blown from a deflation hollow and collects around its rim.

In desert areas, where the amount of sand is great, a series of ridges of sand form in long, wavelike patterns. These ridges are called *transverse dunes*. This type of dune forms at a right angle to the wind direction. A similar type of dune, called a *longitudinal dune,* also forms in the shape of a ridge. However, longitudinal dunes lie parallel to the direction in which the wind blows.

The complicated shapes of most dunes result from the force and direction of the wind. If the wind usually blows toward the same

Investigation

Dune Migration

You can demonstrate how dunes migrate by making a model dune and simulating natural events.

Science Process Skills Focus: constructing models, demonstrating, observing, predicting outcomes

Materials

pen, ruler, flat cardboard box, paper bag large enough to hold one end of box, dry sand, safety goggles, hair dryer, clock or watch

Procedure

1. Use the pen and ruler to mark the side of the box at 5-cm intervals, as shown in the figure.
2. Place the box halfway inside the paper bag so the bag will catch any blowing sand.
3. Fill the box about half full of dry sand. Make a dune in the sand. Look at the side of the box and record to the nearest centimeter the location of the peak of the dune.
4. Put the safety goggles on. Hold the hair dryer level with the top of the dune about 90 cm from the open end of the box.
5. Turn the hair dryer to low speed for 1 minute. Identify and record the new location of the peak of the dune to the nearest centimeter.

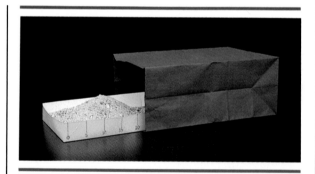

6. Repeat Step 5 three times, first running the hair dryer for 2 minutes, then 3 minutes, and then 5 minutes. After each trial, record the location of the peak of the dune.
7. Flatten the sand. Place a barrier such as a rock in the sand. Position the hair dryer as in Step 5. Run the dryer for 3 minutes.

Analysis and Conclusions

1. In what direction did the dune migrate?
2. How far overall did the dune migrate?
3. What was the average distance the dune migrated per minute?
4. In Step 7, where does the dune form? What steps might be taken to slow down the process of dune migration? Explain your answer.

Figure 16–4. These loess deposits are in Gila National Forest, New Mexico.

direction, dunes will move, or migrate, in that direction. Migration occurs as sand is blown off the windward side, over the crest, and is built up on the slipface. In fairly level areas, dune migration continues until a barrier is reached. To prevent dunes from drifting over highways, buildings, and farmland, the planting of grasses, trees, and shrubs or the building of fences is often necessary.

Loess

The wind carries dust higher and much farther than it carries sand. Fine dust may be deposited in such thin layers that it is not noticed. However, thick, unlayered deposits of yellowish, fine-grained sediment can be formed by the accumulation of wind-blown dust. This material is known by its German name, **loess** (LESS), meaning ''loose.'' Although loess is soft and easily eroded, it can break into vertical slabs. Loess sometimes forms steep bluffs, such as those shown in Figure 16–4.

A large area in northern China is covered entirely with loess. The material in this deposit came from the Gobi desert of central Asia. Large deposits of loess are also found in central Europe. In North America, loess is found in the north-central states, along the eastern border of the Mississippi River valley, and in eastern Oregon and Washington State. These deposits probably were built up by dust from dried beds of glacial lakes and streams. Loess deposits are extremely fertile and provide most of the cropland in these regions of the United States.

Section 16.1 Review

1. Define *deflation*.
2. Describe and compare barchans and parabolic dunes.
3. Describe how loess deposits are formed.
4. Compare the composition and shape of dunes and loess deposits.

Section Objectives

- **Compare the formation of six features produced by wave erosion.**
- **Define a beach and discuss the way in which it is formed.**
- **Describe the movement of sand along a shore and the features it produces.**

16.2 Wave Erosion

As wind moves over the ocean, it produces waves and currents that erode the bordering land. Wave erosion changes the shape of the earth's **shorelines,** places where the ocean and the land meet. Shorelines are temporary and unstable boundaries.

Shoreline Erosion

The power of waves striking rock along a shoreline may shake the ground like a small earthquake. Seismographs often record such vibrations. The great force of waves may break off pieces of rock and throw them back against the shore. The rock fragments grind together in the tumbling water. This abrasive action eventually reduces most of the rock fragments to small pebbles and sand grains.

Chemical weathering also attacks the rock along a shoreline. The waves force salt water and air into small cracks in the rock. Substances in the air and water produce a chemical action that may enlarge the cracks. The enlarged cracks, in turn, provide an increased surface area for physical and chemical weathering.

Much of the erosion along a shoreline takes place during storms. Large waves release tremendous amounts of energy against the shore. Huge blocks of rock, like those shown in Figure 16–5, can be broken off and eroded by such waves. A severe storm can noticeably change the appearance of a shoreline in a single day.

Sea Cliffs

In places where waves strike directly against rock, erosion usually produces a steep structure called a **sea cliff,** like the one shown in Figure 16–5. The waves slowly notch the base of the cliff. The notch cuts under the overhanging rock, until the rock eventually falls. The cliff is gradually worn back and made steeper. Many shorelines consist of such high, nearly vertical sea cliffs.

Figure 16–5. Sea cliffs develop where waves strike directly against rock along a shoreline. Point Bonita Lighthouse in Golden Gate, California is built on a rock projection that extends from a sea cliff.

The rate at which a sea cliff is eroded by waves is partially dependent upon the nature of the rock along the shoreline. Soft rock is eroded very rapidly. For example, cliffs made up of loose glacial deposits along the shore of Cape Cod are being worn away at the rate of about 1 m/yr. Old maps of parts of the shoreline of England show that cliffs made of soft sedimentary rock have been worn back several kilometers during the last 2,000 years. In contrast, shorelines made up of hard rock, such as the granite of the Seychelles Islands in the Indian Ocean, show little change over hundreds of years.

Sea Caves, Arches, and Stacks

A sea cliff is seldom eroded evenly. Fractured rock is more easily eroded than nonfractured rock, and projections of hard rock may be left on the cliff. Waves often cut deeply into fractures and weak rock along the base of the cliff, forming a large hole, or a **sea cave.** In a rock projection, continued wave action can enlarge a sea cave producing a **sea arch** when the waves cut completely through the projection. Continued erosion of a sea arch may cause its middle to collapse. The remaining isolated columns of rock, which had once been the sides of the arch, are called **sea stacks.** In time, the sea stacks, too, are eroded so that they stand no higher than the water.

Terraces

Most erosion of a sea cliff takes place above water. However, as a sea cliff is worn back, a nearly level platform usually remains beneath the water at the base of the cliff. This platform is called a **wave-cut terrace.** As the waves cause the cliff to retreat, some of the rock from the base of the cliff scrapes the wave-cut terrace until it is almost flat. Other eroded material may be deposited some distance from the shore, creating an extension to the wave-cut terrace called a **wave-built terrace.**

Figure 16–6. Wave action causes the slow erosion of a sea cave into first a sea arch and eventually into sea stacks shown below.

The water above a terrace is shallow. Waves lose much of their energy as they pass through shallow water. As the energy of the waves lessens over a terrace, the rate of erosion of the cliff is greatly reduced. However, if the terraces erode completely or if the sea level becomes higher, waves may begin to erode the cliff again.

Beaches

Waves create various features along shorelines by eroding the land and depositing the rock fragments. A **beach** is one such shoreline feature. A beach is a deposit of sand size or larger rock fragments along an ocean shore or a lakefront. As waves wash up on the shore, they move sand and small rock fragments forward. In retreating, the water moves some rock fragments away from the shore. Beaches form where the amount of rock fragments moving toward the shore is greater than the amount moving away from the shore. After the beach has formed, the rates at which fragments move toward and away from shore tend to equalize.

Composition of Beaches

Most people think of beaches as areas of light-colored, small-sized sand grains along the water. Actually, the sizes and kinds of materials found on beaches vary widely, as indicated in Table 16–1. Many beaches are covered with pebbles or large rock fragments and have no sand.

The composition of beach materials depends on the source rock. For example, granite yields light-colored fragments that are mostly quartz and feldspar. Beaches of this type of sand are common along the North American shorelines, where granite is abundant. Beaches of black sand, as shown in Figure 16–7, are found on Hawaii and on other volcanic islands. The black sand comes from the volcanic rocks common to these islands. Some beaches along the Oregon and Washington shorelines are also made up of dark volcanic sand.

Table 16–1: Beach Materials

Material	Diameter (mm)
Boulders	More than 200
Cobbles	76 to 200
Gravel	
Coarse	19 to 76
Fine	5 to 19
Sand	
Coarse	2 to 5
Medium	0.4 to 2
Fine	0.07 to 0.4
Silt	Less than 0.07

Figure 16–7. This beach in Hawaii is made up of black sand eroded from the volcanic rock that makes up the island.

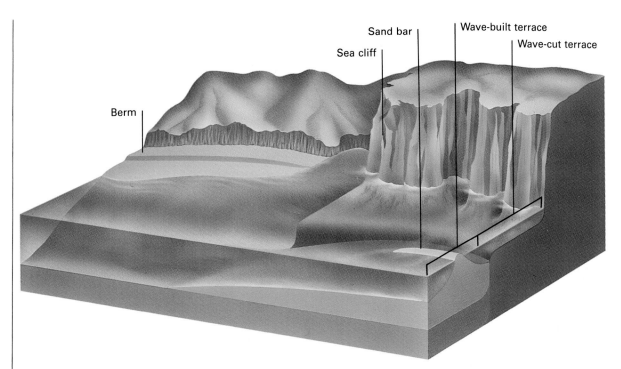

Berm

Sea cliff

Sand bar

Wave-built terrace

Wave-cut terrace

In many locations sand is carried to the shore by rivers. Thus the source rock of much of the sand found along a beach may be in the valley of a nearby river. Some beaches, such as those in Florida, may consist of fragments of shells and coral that are washed ashore. In other locations, such as Cape Cod in Massachusetts, beach sand has been deposited by glaciers.

Figure 16–8. Notice that erosion and deposition may create a variety of features on the beach and along the coastline.

The Berm

Most rock fragments, from fine sand to large cobbles, which are rounded fragments larger than pebbles, are moved toward and away from the shore by waves. The smaller and lighter fragments are moved most easily. Each wave reaching the shore moves individual sand grains forward slightly. Although each advance is small, the total action of several thousand waves each day may move sand grains a great distance. The sand piles up on the shore, producing a sloping surface. During high tides or when large waves come in, sand is deposited at the back of this sloping beach. As a result, most beaches have a raised section called the **berm.** The berm is the part of the beach most often used by people for recreation.

The appearance of most beaches changes seasonally. During winter storms, large waves remove sand from the beach on the seaward side of the berm. Thus the berm is higher and steeper during the winter. Outgoing waves move the beach sand seaward. The sand carried away from the berm is deposited offshore. These sand deposits form a long underwater ridge called a **sand bar.** Sand bars are usually visible only at low tides, if at all. In summer, waves return the sand from offshore to widen the beach on the seaward side of the berm.

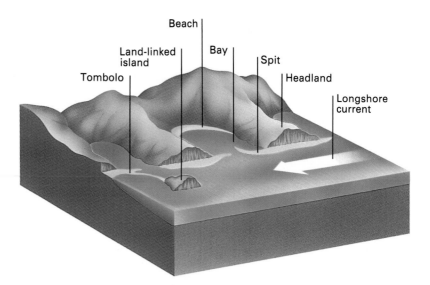

Beach

Land-linked
island

Bay

Spit

Tombolo

Headland

Longshore
current

Figure 16–9. Spits and tombolos form as longshore currents deposit sand at headlands or at openings to a bay.

Longshore-Current Deposits

The way a wave approaches the shore determines how it will move the sand grains and other materials found on the beach. Most waves approach the beach at a slight angle and retreat in a direction that is more perpendicular to the shore. As a result, individual sand grains are moved by the waves in a zig-zag motion. The general movement of sand along the beach is in the direction in which the waves strike the shore. The effects of waves approaching the shore at a slight angle are shown in Figure 16–9.

Waves moving at an angle to the shoreline push water along the shore, creating *longshore currents*. A longshore current is a movement of water parallel to and near the shoreline. These currents occur where the shore is uniformly sloped and straight. Longshore currents transport sand in a direction parallel to the shoreline.

Sand moving along a relatively straight shore keeps moving until the shoreline changes direction. Shoreline direction changes at openings to a bay or at projections of the shore called **headlands.** The longshore current slows and sand is deposited at the far end of the headland. A long, narrow deposit of sand connected at one end to the shore is called a **spit.** Currents or waves often curve the end of a spit into a hook shape. A chain of spits may develop between islands and connect them. Such connecting ridges of sand, called **tombolos,** may link an island to the mainland. What process of stream deposition is similar to the way a spit forms?

Section 16.2 Review

1. Name six features produced by wave erosion.
2. What determines the composition of beach materials?
3. Define *longshore current*.
4. What are the similarities in the way a barchan forms and the way a hooked spit forms?

16.3 Coastal Erosion and Deposition

Section Objectives

- **Explain how changes in sea level relative to the land affect coastlines.**
- **Describe the features of a barrier island.**
- **Compare the types of coral reef.**
- **Analyze the effect of human activity on coastal land.**

Coastlines, the boundaries between the land and the ocean, are among the most rapidly changing parts of the earth's surface. The coastal area extends from relatively shallow water to several kilometers inland. The study of a coastline can help scientists understand its history and predict future changes in it.

Processes That Affect Coastlines

Most coastlines are the result of various processes working together. However, scientists have found that many coastline features are formed by a change in sea level relative to the land. Coastlines are affected by the long term rising or falling of sea level and by the long term uplifting or sinking of the land that borders the water. These and other more rapid processes, including wave erosion and deposition, are constantly changing the appearance of coastlines.

Coastlines and Sea-Level Changes

A change in the amount of ocean water causes sea level to rise or fall, covering or exposing coastlines. During the last glacial period which ended about 11,000 years ago, some of the water that is now in the ocean existed as ice on land. Scientists estimate that the glaciers then held about 70 million km^3 of ice. Now glaciers hold only about 25 million km^3 of ice. During the last glacial period, the water that made up the additional 45 million km^3 of ice must have come from the oceans. As a result, sea level was probably as much as 100 m lower at that time than it is today. Since the last glacial period, the glaciers have been melting and sea level has been rising, as indicated in Figure 16–10. Sea level is now rising at an average rate of about 1 mm/yr. If all the glaciers were to melt, the oceans would rise about 65 m. Low-lying coastal regions would be submerged. Many of the largest cities, including New York, Tokyo, and London, are located in such coastal areas. Consequently, they would be endangered by a substantial rise in sea level.

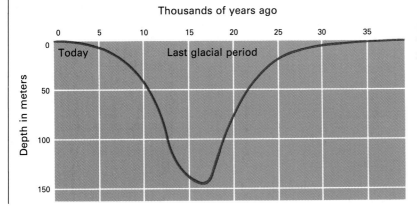

Figure 16–10. This graph shows how sea level has changed during the past 35,000 years. 0 on the graph represents sea level.

Coastlines and Land Changes

An exposed coastline can become covered with water if the land sinks. A coastline that is covered with water can be uncovered if the land rises. Sinking or rising of the land is sometimes part of the isostatic adjustment to weight changes of the earth's crust. The lithosphere that becomes heavier will sink lower on the asthenosphere; the lithosphere that becomes lighter will rise.

In some parts of Europe, particularly in Scandinavia, the land is rising. Docks have actually been raised completely out of the water over the course of the past several centuries. Apparently a heavy ice sheet that covered this region during the last glacial period caused the land to sink, or subside. As the ice began to melt and the ice sheet retreated, this weight was removed and the land began to rise. Parts of North America near Hudson Bay are also rising, probably for the same reason.

Large-scale movements of the earth's crust can cause coastlines to sink or to rise. For example, much of the eastern coast of the United States is sinking. Some scientists think that the side of a continent near the trailing edge of a moving crustal plate tends to sink. This hypothesis may help explain the sinking of the east coast of the United States. The east coast is located toward the trailing edge of the North American plate, which is moving westward.

Coastlines of continents near a plate boundary may also be exposed to forces created as the plates come together. Parts of such coasts may rise and others may sink. This seems to be the case on the west coast of the United States, which shows evidence of rising in some places and sinking in others. For example, much of the southern coast of California has risen over the past few million years. In contrast, most of the west coast from northern Oregon through Washington state has sunk.

Submergent Coastlines

When sea level rises or the land sinks, a **submergent coastline** is formed. Divides between neighboring valleys become headlands separated by bays and inlets. Beaches are usually short, narrow, and rocky because they have just begun to form. The highest parts of the

Figure 16–11. Notice how the features of a submergent coastline are eroded from youth (left) to old age (right) as sea level rises relative to the land.

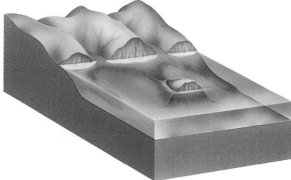

submerged land may form offshore islands. Irregular coastlines in many parts of the world, such as the northeast coast of the United States, seem to have been formed by submergence. However, submergent coastlines are eventually changed by wave action. The headlands are eroded, shorelines become straighter, and beaches become longer and wider.

The mouth of a river valley that is submerged by ocean water may become a wide, shallow bay that extends far inland. This type of bay is called an **estuary** (ESH-chuh-WER-ee). The mouths of most rivers along the Eastern United States coast are estuaries. The slightly salty water of estuaries is an important source of such seafood as shrimp and oysters. Pollution of the rivers flowing into an estuary can destroy this valuable food source.

The U-shaped glacial valleys along the coasts of Alaska, Chile, Greenland, Norway, and western Canada became flooded with ocean

IMPACT

The Flooding of Venice

The city of Venice, Italy, called "the Jewel of the Adriatic," is built on about 120 islands in the Adriatic Sea. Venice is famous not only for its canals but also for its priceless art treasures.

However, the waters of the Adriatic Sea have become a serious threat to the city. At 2:30 A.M. on February 1, 1986, the third of a series of recent floods hit Venice with unexpected fury. By dawn the water had risen 1.5 m above its normal level. Residents were evacuated, but art masterpieces were submerged under water and mud.

These floods have been caused by three factors—erosion, sinking land, and rising sea level. Erosion of barrier beaches allows the high tides to reach further into Venice. The weight of added sediments is causing the city to sink about 1 mm each year. Withdrawal of too much groundwater and the weight of the buildings within the city have caused additional sinkage. Also, like oceans worldwide, the Adriatic Sea is rising. Venice was almost 13 cm lower in 1987 than it was 50 years earlier.

To prevent future and possibly permanent damage to Venice, the city is constructing massive floodgates 800 m wide in the Adriatic. These tubular gates will rest on the seafloor until a high tide threatens the city. Then they will be filled with air and will surface. Thus they will block the incoming waters. Also, Venice has restricted the pumping of groundwater.

What will be the effect of the restriction on pumping of groundwater?

water as sea level rose. These flooded valleys, called **fiords** (fee-ORDZ), form very narrow, deep bays with steep walls. In Britain these inlets are called *firths* or *sea lochs*.

Emergent Coastlines

When the land rises or sea level falls, an **emergent coastline** results. If an emergent coastline has a steep slope and is exposed rapidly, it will become jagged and dotted with sea cliffs, narrow inlets, and bays. A series of wave-cut terraces may be exposed as well. Along some emergent coastlines, a gentle slope is formed when part of the continental slope is lifted and exposed. A relatively smooth coastal plain with few bays or headlands and many long, wide beaches results. Much of the Florida coast is an emergent coastline.

Barrier Islands

Most of the coastline of the United States from New England to Texas is a relatively flat coastal plain. As sea level has risen over the past 15,000 years, the shoreline has moved inland as much as 150 km across this gently sloping coast. Some geologists think that this movement isolated former dunes from the new shoreline, forming **barrier islands.** Barrier islands are ridges of sand, typically 2 km to 5 km wide and 10 km to 100 km long. They lie nearly parallel to the shoreline 3 km to 30 km offshore. There are nearly 300 barrier islands between Cape Cod, Massachusetts, and Padre Island, Texas. Other barrier islands may have formed when sand spits were separated from the land by storms. Waves also may pile up ridges of sand scraped from the sea bottom. These deposits then may be moved toward the shore, or shoreward, by waves, currents, and winds. Winds blowing toward the land often create a line of dunes on the shoreward side of the islands. The dunes are usually 3 m to 6 m high, although they may reach a height of 30 m.

Between barrier islands and the shoreline is a narrow region of shallow water called a **lagoon.** Lagoons are often nearly filled with mud brought in by streams. At high tide, water may cut channels through a barrier island to form a passageway, or tidal inlet, into the lagoon. As a result, large areas of some lagoons become **tidal flats.** A tidal flat is a muddy or sandy part of the shoreline that is visible at low tide and is under water at high tide. Most tidal flats are covered with plants that can grow in the salty water. What offshore feature is visible only at low tide, if at all?

Barrier islands have become popular recreation and resort areas, but they can be dangerous places to live. Because of their low elevation, barrier islands are severely eroded by high tides and storms. During a storm, sand is washed from the ocean side toward the inland side of the island, causing most barrier islands to migrate toward the shoreline. The beaches on some barrier islands are being eroded at an average rate of 20 m/yr. Many houses built along these beaches have collapsed as the beaches have disappeared.

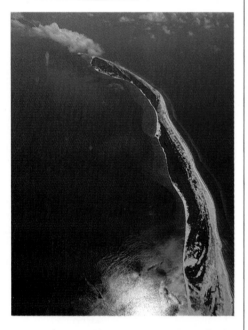

Figure 16–12. The beach on Petit Bois Island, a classic barrier island is being eroded rapidly.

Coral Reefs

Another coastal feature is a **coral reef.** Corals are small animals that live in warm, shallow waters. A coral extracts calcium carbonate from ocean water and uses it to build a hard outer skeleton. Corals attach to each other to form large colonies. New corals grow on top of dead ones, forming a coral reef, which is a ridge made up of millions of coral skeletons.

Some coral reefs form around volcanic islands. The coral colony continues to grow in the shallow water near the shore even when the volcano becomes dormant. This type of coral reef around the borders of the island is called a **fringing reef.** As the ocean lithosphere bends under the weight of the volcano, both the volcano and reef sink. The coral reef, however, builds higher because the animals can only live near the surface of the water. The coral thus forms a **barrier reef** around the remnant of the volcanic island. Finally, the island disappears completely under water, leaving only a nearly circular coral reef, called an **atoll,** surrounding a shallow lagoon.

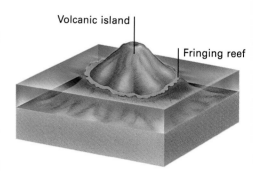

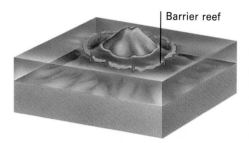

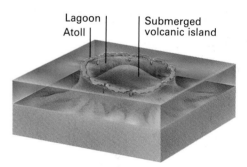

Figure 16–13. A coral reef changes from a fringing reef (top) to a barrier reef (center) to an atoll surrounding a lagoon (bottom). These changes occur as a volcanic island slowly sinks and the reef builds higher.

Preserving the Coastline

Coastal lands are used for commercial fishing, shipping, industrial and residential development, and recreation. However, coastlines are fragile, and they can be easily damaged or destroyed.

As the earth's population increases, the use of shorelines will also increase. For example, about 5 percent of the land in the United States is coastline. But, population experts estimate, by the year 2000, one half of its people will live on that land. New harbors will be dug along the Atlantic and Pacific oceans, the Gulf of Mexico, and the Great Lakes. Pollution, which has already destroyed one tenth of the shellfish-producing waters of the country, will probably continue to increase rapidly. Oil spills are a constant threat as tankers travel near the shorelines and oil wells are drilled offshore. If shoreline use continues as in the past, much of the existing coastal land will be unlivable.

If the coastal zone is to be used at all, private owners and government agencies must work to preserve it. Some efforts have been made. For example, some New Jersey shoreline towns have brought sand from inland to rebuild beaches eroded by severe winter storms. The ocean floor just offshore also can be a source of sand for such coastline-preservation projects.

Section 16.3 Review

1. How does the formation of a submergent coastline differ from the formation of an emergent coastline?
2. Why are barrier islands particularly subject to erosion?
3. How does a fringing reef become an atoll?
4. Predict the effect that cutting into the shoreline to create a new harbor will have on the longshore current.

Reviewing Chapter 16

Key Concepts

- The wind erodes the land by weathering and by removing sediments. See page 255.
- The two types of wind deposits are made up of particles of different sizes. See page 257.
- Beaches result from the deposition of sediments by waves. See page 262.
- Longshore currents move sediments along a shoreline. See page 264.
- Coastlines are exposed or submerged as sea level changes. See page 265.
- Barrier islands are unstable coastline features. See page 268.
- Coral reefs that surround volcanic islands undergo

- **Waves weather and erode the shoreline, producing characteristic features. See page 263.**

changes as the islands sink. See page 269.
- Human activities affect land along the coasts. See page 269.

Key Terms

atoll (269)	desert pavement (256)	loess (259)	submergent coastline
barchan (258)	dune (257)	sand bar (263)	(266)
barrier island (268)	emergent coastline	sea arch (261)	tidal flat (268)
barrier reef (269)	(268)	sea cave (261)	tombolo (264)
beach (262)	estuary (266)	sea cliff (260)	ventifact (256)
berm (263)	fiord (268)	sea stack (261)	wave-built terrace
coral reef (269)	fringing reef (269)	shoreline (260)	(261)
deflation (256)	headland (264)	spit (264)	wave-cut terrace (261)
deflation hollow (256)	lagoon (268)		

Review

On your paper, write the letter of the term that best completes each of the following statements.

1. Loose fragments of rock and minerals measuring between 0.07 mm and 5 mm are referred to as (a) pollen (b) dust (c) sand (d) desert pavement.
2. Sand moves by (a) saltation (b) abrasion (c) emergence (d) depression.
3. The most common form of wind erosion is (a) migration (b) abrasion (c) saltation (d) deflation.
4. Dunes move primarily by (a) abrasion (b) deflation (c) migration (d) submergence.
5. Unlayered, yellowish, fine-grained deposits are called (a) beaches (b) loess (c) dunes (d) desert pavement.
6. The most common type of dune is a (a) barchan (b) transverse dune (c) deflation hollow (d) parabolic dune.
7. All of the following shoreline features are produced by the erosion of sea cliffs except (a) dunes (b) sea stacks (c) wave-cut terraces (d) sea arches.
8. The composition of beach deposits depends

on (a) the climate (b) the source rock (c) the time of year (d) wave action.

9. Deposition of sand at the end of a headland produces a (a) sand bar (b) dune (c) spit (d) sea cliff.
10. Sea level is now (a) stationary (b) falling about 1 mm/yr (c) rising about 1 cm/yr (d) rising about 1 mm/yr.
11. A coastline resulting from a rise in sea level or subsidence of the land is (a) emergent (b) submergent (c) glaciated (d) volcanic.
12. Barrier islands tend to migrate (a) seaward (b) along the shore (c) in the summer (d) toward the shore.
13. Coral reefs are produced by living organisms that (a) swim in circles (b) extract minerals from seawater to form hard skeletons (c) attach themselves to sand bars (d) build nests of sand.
14. If shoreline resources continue to be used as in the past, (a) coastal land will be greatly improved (b) the damage can be repaired (c) sea level will fall (d) much of the existing coastal land will be unlivable.

Application

On your paper, write answers to the following questions.

1. The deserts of the southwestern United States contain many tall, sculpted rock formations that are the result of weathering and erosion. Was wind or water the more likely agent responsible for these formations? Explain your answer.
2. Beautifully colored sunsets and sunrises are the result of dust in the atmosphere. Such sunsets and sunrises were visible around the world for two years after Krakatoa, a volcanic island in the Indonesian chain, erupted in 1883. Explain how this is possible.
3. Suppose that once each month for a year a satellite orbiting the earth takes a photograph of the same 1-km^2 area of the Sahara. Would the surface features shown in these 12 photographs remain essentially the same, or would they be very different? Explain your answer.
4. What effect does the development of a wave-built terrace have on erosion of the shoreline? What phenomenon might counteract this effect?
5. Suppose that scientists have observed that the size of the sand bars in a particular area has been decreasing steadily over the past five years. What does this observation suggest about the climate of the area? Explain why.
6. As you approach a large landmass by ship, you notice an island connected to the shore by tombolos. What type of shoreline do you predict the mainland will have? Explain your answer.
7. Describe two ways that the melting of continental glaciers affects the relative level of the land and the sea.
8. Why do you think barrier islands form parallel to the shoreline?
9. Why are atolls nearly circular?
10. How might the creation of landfills for various purposes alter coastal land?

Extension

1. Imagine that you are a newspaper reporter who has traveled to another planet. Scientists know that, at one time, both wind and water eroded the surface of the planet. Prepare a newscast describing the landscape you see and explaining the processes that produced it. Present the newscast to the class.
2. Suppose that sea level continuously rises at the rate of 1 mm/yr and that other factors affecting the coastlines do not change. Draw a map showing what the Massachusetts coast might look like in 10,000 years.
3. Imagine that you are a geologist. You have been asked by the local government of a coastal city to prepare a plan for preserving their coastal land. Your plan should consider both the necessary uses as well as the unnecessary abuses of those resources. Present your plan to the class and, following a discussion, have them vote for or against it.

Glen Canyon, Arizona (background): Dawn Redwood fossil leaf, Republic, Washington (inset)

Introducing Unit Five

Today's world emphasizes making immediate records of what is happening around us, but today's news is always a part of tomorrow's history. Knowing where we are in the field of earth science is important, but it's just as helpful to know how we got there—especially if we're looking at our current situation in terms of future implications.

The complex process of earth science is just that—a process that encompasses yesterday, today, and tomorrow. To understand that process, it is necessary to have the basic details recorded somewhere. Fortunately, even before there were science writers to help make those records, the earth recorded—and continues to "write"—its own detailed history.

In the following unit, you will study the earth's history through rock and fossil records that are as important to us today as they will be in the future.

Richard A. Kerr

Richard Kerr
Senior Writer, Research News Staff
Science

17 The Rock Record

18 A View of the Earth's Past

19 The History of the Continents

Chapter 17
The Rock Record

Rocks contain clues to the earth's past, including evidence of ancient life forms. For example, the piece of rock shown here was once a living tree. Earth scientists can use the information found in the rock record to help them understand the geologic processes that have shaped the earth's crust.

In this chapter you will read about the methods earth scientists use to uncover the geologic history of the earth.

Chapter Outline

17.1 Determining Relative Age
Law of Superposition
Unconformities
Crosscutting Relationships

17.2 Determining Absolute Age
Rates of Erosion and Deposition
Varve Count
Radioactive Decay

17.3 The Fossil Record
Kinds of Fossils
Interpreting the Fossil Record

A petrified log lies in a petrified forest at Blue Mesa, Arizona

17.1 Determining Relative Age

Section Objectives

- **State the theory of uniformitarianism.**
- **Explain how the law of superposition can be used to determine the relative age of rocks.**
- **Compare three types of unconformity.**
- **Explain how the law of crosscutting relationships can be used to determine the relative age of rocks.**

Geologists estimate that the earth is about 4.6 billion years old. The idea that the earth is billions of years old originated with the work of James Hutton, an eighteenth-century Scottish physician and gentleman farmer. Hutton was a keen observer of the geologic changes taking place on his farm. Using scientific methods, Hutton drew conclusions based on his observations.

Hutton theorized that the forces changing the landscape of his farm were the same forces that had changed the earth's surface in the past. He thought that by studying the present, people could learn about the earth's past. Hutton's theory, called **uniformitarianism,** states that current geologic processes, such as volcanism and erosion, are the same processes that were at work in the past. This theory is one of the foundations of the science of geology. Geologists later refined Hutton's theory by pointing out that although the processes are the same, the rates at which those processes occur vary over time.

Before Hutton's work, most people thought that the earth was only about six thousand years old. They also thought that all geologic features had been formed at the same time. Hutton's theory of uniformitarianism, however, raised some serious questions about the earth's age. Hutton observed that the forces changing the land on his farm operated very slowly. He reasoned that millions of years must have been needed for those same forces to create the complicated rock structures found in the earth's crust. Hence, he concluded, the earth must be much older than previously thought.

Hutton's observations and conclusions about the age of the earth encouraged others to learn more about the earth's history. One way to learn about the earth's past is to determine the order in which rock layers and structures were formed.

Figure 17–1. The layers of sedimentary rock that make up Canyon de Chelly were deposited over millions of years.

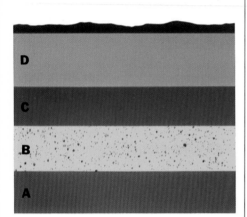

Figure 17–2. By applying the law of superposition, geologists are able to determine that layer *D* is the youngest layer in this rock bed.

Layers of rocks, called *strata,* are like the pages in a history book, detailing the sequence of events that took place in the past. Using a few basic principles, scientists can determine the order in which rock layers were formed. Once the order of formation is known, a **relative age** can be determined for each rock layer. Relative age indicates that one layer is older or younger than another layer. It does not indicate the exact age of the rock.

Various kinds of rocks form layers. Igneous rocks form layers when successive lava flows stack up. In some metamorphic rock, such as marble, layers are also visible. To determine the relative age of rocks, however, scientists commonly study the layers in sedimentary rocks.

Law of Superposition

The formation of sedimentary rock begins when sediments are deposited on a flat surface. As the sediments accumulate, they are compressed and harden into sedimentary rock layers, or beds. The boundary between two beds is called a **bedding plane.**

Scientists use a simple principle, called the **law of superposition,** in attempting to determine the relative age of a layer of sedimentary rock. This principle states that each sedimentary rock layer is older than the layers above it and younger than the layers below it. According to the law of superposition, layer *A* shown in Figure 17–2 was the first layer deposited, and, thus, it is the oldest. The last layer deposited was layer *D*. Is layer *B* older or younger than layer *C?*

Scientists also know that sedimentary rock generally forms in horizontal layers. Therefore, they can assume that most sedimentary rock layers that are not horizontal have been tilted by crustal movements after the layers were formed. Sometimes violent movements in the crust push older layers on top of younger ones, or the movements may overturn a group of rock layers. In such cases the law of superposition cannot be easily applied. Scientists must look for clues to the original arrangement of layers and then apply the law of superposition.

One clue to the original arrangement of rock layers lies in the size of the particles found in a layer of sedimentary rock. As you may recall from Chapter 10, the largest particles of sediment are usually deposited near the bottom of a layer. Therefore, the study of particle sizes in a sedimentary rock layer may reveal whether the layer has been overturned.

Another clue to the original position of rock layers is found in the shape of the bedding planes. In cross-bedded layers of sedimentary rocks, for example, sediments are deposited in curved sheets at an angle to the bedding plane. The tops of cross-bedded layers are often eroded before new layers are deposited. Therefore, the cross-beds are still curved at the bottom but flat across the top. By studying the shape of the bedding planes, scientists can determine the original position of cross-bedded layers.

Ripple marks can also be helpful in determining the order of rock layers. Ripple marks are small waves formed on the surface of sand by the action of water or wind. When the sand becomes sandstone, the ripple marks may be preserved. In undisturbed sedimentary rock layers, the peaks of the ripple marks point upwards. Thus, by examining ripple marks, scientists can establish the original arrangement of the rock layers. The relative ages of the rocks can then be determined using the law of superposition.

Unconformities

Sometimes movements of the earth's crust lift up rock layers that were buried and expose these layers to erosion. Later, if the eroded surface is lowered or the sea level rises, sediments will again be deposited, forming new rock layers. The rock layers lost to erosion create a break in the geologic record, making it difficult to determine the relative ages of layers above and below the break. This break in the geologic record is called an **unconformity.** An unconformity indicates that for a period of time deposition stopped, rock was removed by erosion, and then deposition resumed. As Figure 17–3 shows, there are three types of unconformities.

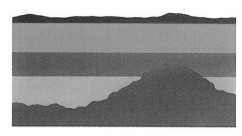

Noncomformity

Nonconformity

Sedimentary rock is stratified, or deposited in layers. Metamorphic and igneous rocks are usually unstratified. An unconformity in which stratified rock rests upon unstratified rock is called a **nonconformity.** For example, unstratified rock such as granite forms deep within the earth. The granite may be lifted to the earth's surface by crustal movements. Once exposed, the granite begins eroding. Sediments may then be deposited on the eroded surface. The boundary between the sandstone and the granite layers is a nonconformity. It represents an unknown period of time during which the granite was eroded.

Angular unconformity

Angular Unconformity

Another type of unconformity results when rocks deposited in horizontal layers are folded or tilted and then eroded. When erosion stops, a new horizontal layer is deposited on a tilted layer. The boundary between the tilted layers and the horizontal layer is called an **angular unconformity.** As Figure 17–3 shows, the bedding planes of the older rock layers are not parallel to those of the younger rock layers deposited above the angular unconformity.

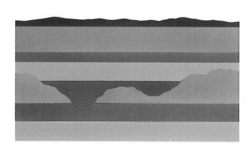

Disconformity

Figure 17–3. The three types of unconformity represent gaps in the geologic record.

Disconformity

Sometimes layers of sediments on the ocean floor are lifted above sea level without folding or tilting. Exposed to wind and running water, the surface layers are eroded. Eventually, the area again falls below sea level and deposition resumes. The boundary between the older, eroded surface and the younger, overlying layers is nearly horizontal and is called a **disconformity.** Although the rock layers look as if they were deposited continuously, a large gap exists in the ages of the upper and lower layers.

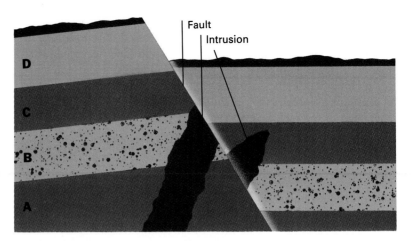

Figure 17–4. The law of crosscutting relationships can be used to determine the relative ages of rock layers and the faults and intrusions within them.

Crosscutting Relationships

When crustal movements have disturbed rock layers, determining relative age using only the law of superposition may be difficult. In such cases, scientists may also apply the **law of crosscutting relationships.** The law of crosscutting relationships states that a fault or an intrusion is always younger than the rock layers it cuts through. A fault is a break or crack in the earth's crust along which rocks shift their position. An intrusion is a mass of igneous rock formed when magma cools below the earth's surface. Crosscutting relationships can be extremely complex, and careful observation is necessary to accurately date rocks. Figure 17–4 shows a series of rock layers that contains both a fault and an intrusion. As you can see, an intrusion cuts across layers *A* and *B*. According to the law of crosscutting relationships then, the intrusion is younger than layers *A* and *B*. Look at the fault. What is the relative age of the fault compared to the rock layers?

Both the law of superposition and the law of crosscutting relationships can be applied to unconformities. According to the law of superposition, all rocks beneath an unconformity are older than those rocks above the unconformity. If a fault or intrusion cuts through the unconformity, the fault or intrusion is younger than all the rocks it cuts through above and below the unconformity.

Section 17.1 Review

1. What is the theory of uniformitarianism?
2. How can the law of superposition be used to determine the relative age of sedimentary rock?
3. Compare an angular unconformity with a disconformity. Which would be more difficult to recognize?
4. Suppose you find a series of rock layers in which a fault is cut off below an unconformity. Explain how you would apply the law of crosscutting relationships to determine the relative age of the fault and of the rock layers that were deposited above the unconformity.

17.2 Determining Absolute Age

Relative age indicates only that one rock layer is younger or older than another rock layer. In order to learn more about the earth's history, scientists often need to determine the specific age, or **absolute age,** of a rock layer.

Rates of Erosion and Deposition

One method of determining absolute age involves studying rates of erosion. If scientists can measure the rate at which a stream erodes its bed, they can determine the approximate age of the stream. For example, geologists have studied Niagara Falls and found that the edge of the falls is eroded at an average rate of 1.3 m per year. Based on the average rate of erosion, scientists then determined that the falls probably formed about 9,900 years ago, when glaciers of the last ice age started to melt.

Determining absolute age using the rate of erosion is practical only with geologic features that formed within the past 10,000 to 20,000 years. For older surface features, like the Grand Canyon, the method is much less dependable. The Grand Canyon developed over millions of years. During that period, the rates of erosion have varied greatly.

Section Objectives

- **Summarize the limitations of using the rates of erosion and deposition to determine the absolute age of rocks.**
- **Describe the formation of varves.**
- **Explain how the process of radioactive decay can be used to determine the absolute age of rocks.**

Figure 17–5. The rocky ledge above Niagara Falls has been eroding for nearly 9,900 years.

Another method of determining absolute age involves calculating the rate of sediment deposition. Based on data collected over a long period of time, geologists have estimated the average rates of deposition for common sedimentary rocks such as limestone, shale, and sandstone. On the average, 30 cm of sedimentary rock are deposited over a period of 1,000 years. However, any specific sedimentary layer may not have been deposited at an average rate. For example, a flood can deposit many meters of sediment in just one day. In addition, the rate of deposition has probably changed over time. Therefore, this means of determining absolute age is not always reliable.

Varve Count

You may have heard of people estimating the age of a tree by counting the growth rings in its trunk. Scientists have devised a similar method for estimating the age of certain sedimentary deposits. Some sedimentary deposits show definite annual layers. The layers, called **varves,** consist of a light-colored band of coarse particles and a darker band of fine particles.

Varves are usually formed in glacial lakes. During the summer, when snow and ice melt rapidly, a rush of water can carry large amounts of sediment into a lake. Most of the coarse particles settle quickly to form a layer on the bottom of the lake. With the coming of winter the surface of the lake begins to freeze. Fine clay particles, still suspended in the water, settle slowly to form a thin layer on top of the coarser sediments. A coarse summer layer and the overlying fine winter layer make up one varve. Thus, each varve represents one year of deposition. By counting the varves, scientists can estimate the age of the sediments.

Figure 17–6. Counting varves is one method of determining the absolute age of certain sediments.

Radioactive Decay

The discovery of radioactivity provided scientists with an accurate method for finding the absolute age of rocks. As you read in Chapter 8, the nuclei of some elements emit particles and energy at a relatively constant rate. Elements that emit particles and energy are called *radioactive*. As radioactive elements emit particles and energy, they eventually form new, nonradioactive elements. Because the rate of emission is measurable and is not affected by external factors, radioactive elements function as natural clocks. Scientists determine the amounts of the old, radioactive element and the newly formed, nonradioactive element in a rock. They then compare the relative amounts of the radioactive and nonradioactive elements to determine the absolute age of the rock.

Uranium is a radioactive element found in some rocks. One form of uranium, U-238, is particularly useful in establishing the absolute ages of rocks. U-238 has a mass number of 238 and an atomic number of 92. Remember, the atomic number is the number of protons in the nucleus. The mass number is the sum of the num-

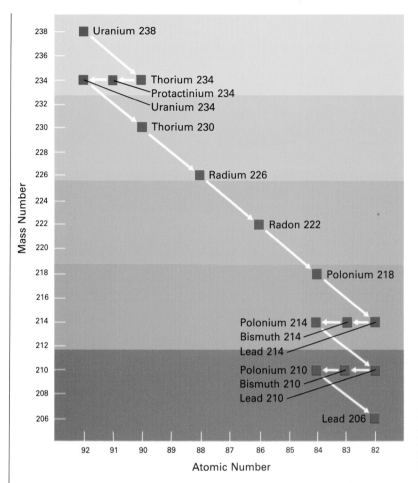

Figure 17–7. Through radioactive decay, the radioactive element U-238 eventually becomes the nonradioactive daughter element Pb-206.

ber of protons and neutrons in the nucleus. The highly radioactive nucleus of U-238 spontaneously emits two protons and two neutrons. This process, called *radioactive decay,* is illustrated in Figure 17–7. The loss of protons and neutrons from the nucleus decreases both the atomic number and the mass number. The result is a new element. In the first decay of U-238, the new nucleus has a mass number of 234 and an atomic number of 90. The new element is called *thorium* (Th-234).

The Th-234 nucleus is also radioactive, and one of its neutrons splits into a proton and an electron. The electron is emitted, leaving the proton in the Th-234 nucleus. This proton increases the atomic number by one, and thorium thus becomes the new element protactinium (Pa-234).

Protactinium is also radioactive, and the cycle of radioactive decay continues until a stable, or nonradioactive, form of the element lead (Pb) is produced. Pb-206, a form of lead with a mass number of 206, is the final product of this radioactive-decay chain. In this particular chain of decay, U-238 is known as the *parent element* and Pb-206 is known as a *daughter element.* Figure 17–7 shows the many daughter elements that form during the radioactive decay of U-238.

Half-Life

The decay of radioactive U-238 into the stable element Pb-206 occurs at a very slow rate. The rate of radioactive decay is relatively constant. It is not affected by changes in temperature, pressure, or other environmental conditions. Scientists have determined that the time required for half of any given amount of U-238 to decay into Pb-206 is 4.5 billion years. If you were to begin with 10 g of U-238, after 4.5 billion years you would have 5 g of U-238. Because half of the U-238 decays over 4.5 billion years, that period of time is called the **half-life** of U-238. A half-life is the time it takes for half the mass of a given amount of a radioactive element to decay into its daughter elements. At the end of another 4.5 billion years, or a total of two half-lives, one fourth, or 2.5 g, of the original U-238 would remain. Three fourths would now be daughter elements. After another 4.5 billion years, or three half-lives, how much of the U-238 would remain?

By comparing the amounts of U-238 and Pb-206 in some rock samples, scientists can determine the age of the sample. The greater the percentage of lead present in the sample, the older is the rock. Scientists know that from a million grams of U-238, 1/7,600 g of

Horizons

Technology: Laser Dating

Using laser beams, scientists at Simon Fraser University in Burnaby, British Columbia, are developing techniques for more accurately determining the age of sediments.

The research of physicists D. J. Huntley and M.L.W. Thewalt and archeologist D. I. Godfrey-Smith concerns the buildup and release of electrons in sediments. Electrons become trapped in sediments, such as those shown right, when the sediments are exposed to ionizing radiation from the sun. The sun's ionizing radiation is present even when the sun is not visible. The trapped electrons are freed when the sediments are exposed to sunlight. Once sediments are covered by other sediments, the effects of sunlight are blocked. The electrons continue to be trapped, but none can be released.

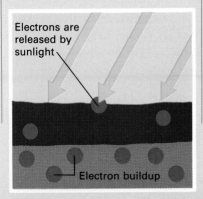

Ionizing radiation from the sun

Trapped electrons

Electrons are released by sunlight

Electron buildup

Pb-206 per year will be produced by decay. The U-Pb ratio can be used only when all the lead in the rock is known to have come from the decay of uranium. Because U-238 has a half-life of 4.5 billion years, it is most useful for dating geologic samples more than 10 million years old.

In addition to U-238, several other radioactive elements are used to date rock samples. Among these are potassium-40 (K-40), with a half-life of 1.3 billion years, and rubidium-87 (Rb-87), with a half-life of 48 billion years. Potassium is found in mica, clay, and feldspar. Potassium-40 is used to date rocks between 10 thousand and 4.6 billion years old. Dating with K-40 showed that the oldest Hawaiian islands are farthest from the hot spot where they originated. Often found in minerals with K-40, Rb-87 can be used to verify the age of rocks previously dated with K-40.

The amount of time that has passed since a rock was formed determines which radioactive element will give the more accurate age measurement. If too little time has passed since the radioactive decay process has begun, there may be too little of the daughter element for accurate dating. If too much time has passed, there may not be enough of the parent element left to obtain an accurate age measurement.

To estimate how old a sediment is, scientists need to determine how long the electrons have been building up in the sediment. Scientists free the trapped electrons by using laser beams, shown far left, as a substitute for sunlight. The electrons freed by the laser beams produce light. Scientists then use photomultipliers, instruments that measure light, to determine the exact amount of light being produced. The number of trapped electrons increases with the age of the sediment. Therefore, the older a sediment is, the greater is the amount of light produced when it is exposed to the laser beams.

Researchers published the initial research findings on this technique in January 1985. At that time, they estimated that laser dating could be used to determine the ages of sediments up to 700,000 years old.

Currently, more research is being done on the usefulness of this new, highly sensitive dating technique. The laser method may become particularly important in the dating of river deposits and glacial sediments, which are covered soon after they are deposited.

Would a scientist use laser dating or carbon dating to date the remains of a prehistoric fern? Explain why.

Investigation

Radioactive Decay

Radioactive elements decay at a constant, measurable rate. The time it takes for half of any given amount of a radioactive element to change into a nonradioactive element is called a *half-life*. You can demonstrate the principle of radioactive decay with a simple model.

Science Process Skills Focus: constructing models, interpreting data, hypothesizing

Materials
clock or watch with a second hand; sheet of ruled notebook paper, about 28 × 22 cm; scissors

Procedure
1. Record the time.
2. Wait 20 seconds, then carefully cut the sheet of paper in half. Select one piece, set the other piece aside.
3. Wait 20 seconds, then cut the selected piece of paper in half. Select one piece, set the other piece aside.
4. Repeat Step 3 until nine 20-second intervals have elapsed.

Analysis and Conclusions
1. In terms of radioactive decay, what does the

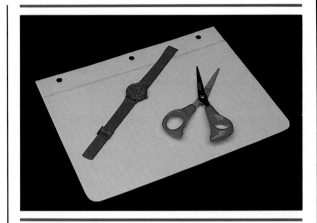

whole piece of paper used in this investigation represent?
2. What does the piece of paper that you set aside in each step represent?
3. What is the half-life of your "element"?
4. How much of your paper "element" was left after the first three intervals? After six intervals? After nine intervals? Express your answers as percentages.
5. What two factors in your model must remain constant for your model to be accurate? Explain your answer.

Carbon Dating

Some more recently formed geologic samples contain organic materials, the remains of once-living things. To determine the age of these samples, scientists use a form of radioactive carbon called *carbon-14* (C-14). Carbon-14 decays to form the daughter element nitrogen-14 (N-14).

Carbon-14 occurs naturally in the atmosphere combined with oxygen as carbon dioxide (CO_2). Most CO_2 in the atmosphere is formed with nonradioactive carbon-12 (C-12). Only a small amount of CO_2 is formed with C-14. The proportion of C-12 to C-14 in the atmosphere remains relatively constant.

All living plants and animals take in both C-12 and C-14. Plants absorb C-12 and C-14 in the form of CO_2 during photosynthesis. Carbon-12 and C-14 move through the food chain as animals eat plants. The C-12 and C-14 enter the tissues of plants and animals.

Figure 17–8. When organisms, such as this water beetle and grasshopper, die, the amount of C-14 in their tissues begins to decrease at a measurable average rate.

Living organism

C-14 absorbed

Organism dies

No new C-14 absorbed

After 5,800 years

50% of original C-14 remains

After 11,600 years

25% of original C-14 remains

While those organisms are alive, the ratio of the two elements remains relatively constant. When a plant or an animal dies, however, the ratio begins to change. The organism no longer takes in C-12 and C-14. The amount of C-14 in an organism's tissues decreases as the radioactive carbon atoms decay to nonradioactive nitrogen-14 (N-14).

The half-life of C-14 is only about 5,730 years. To establish the age of a small amount of organic material, scientists first determine the proportion of C-14 to C-12 in the sample. They then compare that proportion with the proportion of C-14 to C-12 known to exist in a living organism.

C-14 is often used to establish the age of wood, bones, shells, and the organic remains of early humans. Improved techniques for detecting C-14 now make it possible to use carbon dating to establish the age of samples up to 50 thousand years old.

Section 17.2 Review

1. Why are calculations of absolute age based on rates of deposition not always accurate?
2. What are varves?
3. Explain how radioactive dating is used to determine absolute age.
4. Suppose you have a shark's tooth that you suspect is about 15,000 years old. Would you use U-238 or C-14 to date the tooth? Explain your answer.

Section Objectives

- **Describe four ways in which entire organisms can be preserved as fossils.**
- **List four examples of fossilized traces of organisms.**
- **Describe how index fossils can be used to determine the relative age of rocks.**

17.3 The Fossil Record

Scientists called **paleontologists** (PAY-lee-ON-TOL-uh-justs) study *fossils* to learn about the earth's past. Fossils are the remains or traces of animals or plants from a previous geologic time. Fossils are an important source of information for establishing both the relative and absolute ages of rocks. Fossils also provide clues to past geologic events and **evolution,** or the gradual change of living things over time. The study of fossils is called **paleontology.**

Almost all fossils are found in sedimentary rock. Sediments, generally small and lightweight, cover but do not damage a dead organism. Sediments also protect dead organisms from being destroyed by other animals. The covering sediments may also slow down or stop the decaying process.

Fossils are almost never found in igneous rock or metamorphic rock. Igneous rocks are formed from hardened magma, or hot, molten rock. Therefore, living things covered by igneous rock usually burn up, leaving no remains to become fossilized. Metamorphic rocks undergo changes. Because the changes are caused by intense heat, pressure, and chemical reactions, any fossils in metamorphic rock usually are destroyed.

Kinds of Fossils

Fossils form in many different ways. Usually only the hard parts of organisms, such as bones, shells, and teeth, are preserved. In rare cases, an entire organism may be preserved. In some fossils, only a replica of the original organism remains. Other fossils merely provide evidence that life existed.

Figure 17–9. Trace fossils, such as these fern imprints, contain clues to the earth's past.

Preservation of Organisms

Normally, dead plants and animals are eaten by other animals or decomposed by bacteria. Left unprotected, even hard parts such as bones decay, leaving no trace of the organism. Only dead organisms that are buried quickly or protected from decay can become fossils.

Mummification   One means through which an organism may be preserved is **mummification,** or drying. Mummified remains are often found in desert caves or buried beneath desert sand. Because most bacteria cannot survive without water, the mummified organism does not decay.

Amber   Many insects have been found preserved in **amber,** or hardened tree sap. These insects became trapped in the sticky sap and were preserved when the sap hardened. In many cases, even the delicate wings and antennae of insects have been preserved, as shown in Figure 17–10.

Figure 17–10. This insect is perfectly preserved in amber.

Tar Beds   The remains of animals have also been found preserved in tar beds. Tar beds are formed by thick petroleum oozing to the earth's surface. The tar beds were often covered by water. Animals that came to drink the water became trapped in the sticky tar. Other animals preyed upon the trapped animals and became trapped themselves. The remains of the animals were covered by the tar and preserved. For example, the bones of thousands of animals that lived more than 15,000 years ago have been found in the La Brea Tar Pits in southern California. In tar beds in Poland, entire furry rhinoceroses have been found with their flesh and fur mostly intact.

Freezing   The low temperatures of frozen soil and ice also protect and preserve organisms. Because most bacteria cannot survive freezing temperatures, organisms buried in frozen soil or ice do not decay. The frozen bodies of mastodons and woolly mammoths, two types of extinct elephantlike animals, have been found in Siberia and Alaska. Furry rhinoceroses have also been found preserved in frozen arctic soil. These animals lived 8,000 to 20,000 years ago.

Petrification

Sometimes organisms are preserved through **petrification.** In this process, mineral solutions such as groundwater remove the original organic materials and replace them with minerals. Some common petrifying minerals are silica, calcite, and pyrite.

The replacement of the original materials is generally a very slow process, probably taking place molecule by molecule. The substitution of mineral for organic material often results in the formation of a nearly perfect mineral replica of the original organism. The petrified wood shown in Figure 17–11 is one example of this type of fossil. In the United States, petrified logs of cone-bearing trees can be seen in Petrified Forest National Park in Arizona. The minerals agate and chalcedony give these logs an over-all gray color. However, some bright colored streaks of other minerals are also visible. When some petrified fossils are viewed under a microscope, the de-

Figure 17–11. Look carefully at this photograph of petrified logs in Petrified Forest National Park, Arizona. Notice that the structure of the bark and the wood grain have been preserved.

tailed cell structure of the original tissues is clearly visible. In other petrified fossils, however, only the general outline of the organism remains.

Traces of Organisms

When no part of the original organism survives in fossil form, other types of fossils, called *trace fossils,* can still provide information about prehistoric life. The fossilized traces of ancient organisms are often remarkably clear and detailed.

Imprints   A fossil may be an imprint, such as the footprint of an animal or the outline of a leaf. When scientists discover an imprint, such as that shown in Figure 17–12, they try to trace its history. Suppose a giant dinosaur left deep footprints in soft mud. Sand or silt may have blown or washed into the footprints so gently that the footprints remained intact. Then more sediment may have been deposited over the prints. As time passed, the mud containing the footprints hardened into sedimentary rock. The footprints were thus preserved. Scientists have discovered footprints of ancient reptiles, amphibians, birds, and mammals. Imprints of leaves, stems, flowers, and fish made in soft mud or clay have also been found preserved in a similar way.

Figure 17–12. By studying the size and depth of these dinosaur footprints, scientists can learn about the size and weight of the dinosaur that made them.

Molds and Casts   Shells of snails, parts of trees and plants, and similar organic remains were often buried in sediments that later formed sedimentary rock. Eventually these remains decayed or dissolved. In place of the organism, an empty cavity, called a **mold,** remained. A mold retains the shape and surface markings of the

Figure 17–13. Over time, the mold of this *cryptolithus* (left) filled with sediments. The sediments then hardened and formed a cast of the original organism (right).

original organism; however, it gives little information about the internal structure of the organism.

Sometimes sand or mud fills a mold and hardens, forming a natural **cast** such as that shown in Figure 17–13. A cast is a replica of the outer surface of the original organism. An artificial cast can be made by filling a mold with plaster.

Coprolites   Some ancient snails, sponges, and crabs are known to have existed because fossil remains of their waste material have been found. For instance, as ancient sea worms fed on small organisms on the ocean floor, they swallowed sand or mud as well. After the food was digested, the sand or mud was excreted as castings, or waste materials. These castings are found as fossils in some marine sediments.

Fossilized masses of waste material from animals are called **coprolites.** They can be cut into thin sections and observed through a microscope. The materials identified in these sections reveal the feeding habits of ancient animals.

Gastroliths   Some dinosaurs had stones in their digestive systems to help grind their food. In many cases, these stones, called **gastroliths,** survive as fossils. Gastroliths can often be recognized by their smooth, rounded, and polished surfaces. However, identification of these stones is certain only if they are found within the remains of a dinosaur.

Interpreting the Fossil Record

The fossil record provides scientists with many clues to the geologic history of the earth. For instance, scientists can use fossils to find the relative ages of rocks. Fossils also reveal the ways in which environments have changed and how the changes have affected organisms throughout the geologic past of the earth.

Figure 17–14. Trilobite fossils make good index fossils. If a trilobite fossil, such as this one, is found in a rock layer, the rock layer was probably formed 500 to 600 million years ago.

Index Fossils

Certain fossils are found in rock layers of only one geologic age. These fossils are called **index fossils** or *guide fossils*. To be considered index fossils, fossils must meet certain requirements. First, they must be present in rocks scattered over a wide area of the earth's surface. Second, index fossils must have features that clearly distinguish them from all other fossil organisms. Third, the organisms from which the index fossils formed must have lived during a relatively short span of geologic time. Fourth, they must occur in fairly large numbers within the rock layers.

Paleontologists can use index fossils to establish the relative ages of the rock layers in which the fossils are found. For example, a type of marine organism called a *graptolite* lived 350 to 450 million years ago. Another marine creature, called a *trilobite,* lived 500 to 600 million years ago. Therefore, in dating rock layers, scientists know that any rock layer containing trilobite fossils is older than a layer containing graptolite fossils. If a layer containing trilobite fossils was found above a layer containing grapholite fossils, what might scientists assume?

Scientists can use index fossils to date rock layers found in widely separated areas. An index fossil found in rock layers in different areas of the world indicates that the rock layers were probably formed during the same period.

Geologists also use index fossils to locate oil and natural gas deposits. These deposits are formed from plant and animal remains that have been changed by chemical processes over millions of years. Certain types of index fossils are found in the layers of sedimentary rock above oil and natural gas deposits.

Fossil Clues to the Past

Fossils also furnish scientists with important clues to the changes in climate and environment that occurred in the past. For example, fossils of alligatorlike reptiles have been found in Canada, which today

Science Notebook

Clues to Climate

It is hard to imagine that the bitter cold climate of Antarctica was once almost subtropical and that the ice-covered landscape supported thick forests. However, studies of fossils found in ocean-floor sediments off the coasts of Antarctica indicate just that.

Scientists with the Ocean Drilling Project have brought up samples of sediments that ranged in age from 60 million years old to the present. These samples reveal much about the complex history of climatic changes in Antarctica.

Fossil spores and pollen over 39 million years old indicated that beech tree forests once grew in Antarctica. Samples of sediment 37 to 60 million years old contained soil that is normally found in hot, humid climates. Thus scientists learned that prior to 37 million years ago the climate in the Antarctic was almost subtropical.

Fossils of freshwater organisms indicated that they were carried from Antarctic lakes to the ocean floor by rivers as recently as 20 million years ago. The fossils are evidence that at that time the climate was warm enough for unfrozen lakes to exist.

Scientists also found fossils of marine organisms that can live only in sunny coastal waters. Those found off the east coast were over 15 million years old, those from the west coast over 4.8 million years old.

What do these fossils indicate about the ice shelves that now cover the ocean off the east and west coasts of Antarctica?

has a relatively cold climate. Since alligators live only in very warm climates, these fossils indicate that Canada once had a tropical climate. Tropical plant fossils also have been discovered in Antarctica just 400 miles from the South Pole. Fossils of marine animals and plants have also been found in areas far from any ocean. These fossils indicate that such areas were once covered by an ocean.

Section 17.3 Review

1. Why are tar beds good sources of animal fossil remains?
2. List four types of fossils that can be used to provide indirect evidence of organisms.
3. How do geologists use fossils to date sedimentary rock layers?
4. Compare the process of mummification with the process of petrification.

Reviewing Chapter 17

Key Concepts

- According to the law of uniformitarianism, the forces that are changing the earth's surface today are the same forces that changed the earth's surface in the past. See page 275.
- Scientists use the law of superposition to determine the relative ages of rock layers. See page 276.
- A nonconformity, an angular unconformity, and a disconformity represent interruptions in the sequence of rock layers. See page 277.
- Scientists use the law of crosscutting relationships to determine the relative ages of rock layers. See page 278.
- Because they change over time, the rate of erosion and the rate of deposition are not always reli-

able bases for determining absolute age of rocks. See page 279.
- Varves are layers of sediments that form in glacial lakes. See page 280.
- Radioactive elements decay at measurable average rates that can be used to determine absolute age. See page 280.
- Entire organisms may be preserved through freezing or mummification, in amber, or in tar beds. See page 287.
- Fossilized traces of organisms include imprints, molds and casts, coprolites, and gastroliths. See page 288.
- Index fossils are found in rock layers of only one geologic age. See page 290.

Key Terms

absolute age (279)	evolution (286)	mold (288)	unconformity (277)
amber (287)	gastrolith (289)	mummification (287)	uniformitarianism
angular unconformity	half-life (282)	nonconformity (277)	(275)
(277)	index fossil (290)	paleontologist (286)	varve (280)
bedding plane (276)	law of crosscutting	paleontology (286)	
cast (289)	relationships (278)	petrification (287)	
coprolite (289)	law of superposition	relative age (276)	
disconformity (277)	(276)		

Review

On your paper, write the letter of the term that best completes each of the following statements.
1. The concept that the present is the key to the past is part of the theory of (a) unconformity (b) superposition (c) crosscutting relationships (d) uniformitarianism.
2. A sedimentary rock layer is older than the layers above it and younger than the layers below it, according to the law of (a) unconformity (b) superposition (c) crosscutting relationships (d) uniformitarianism.
3. A gap in the sequence of rock layers is (a) a

bedding plane (b) a varve (c) an unconformity (d) a uniformity.
4. An unconformity that results when new sediments are deposited on eroded horizontal layers is (a) an angular unconformity (b) a disconformity (c) a crosscut unconformity (d) a nonconformity.
5. A fault or intrusion is younger than the rock it cuts through according to the law of (a) unconformity (b) superposition (c) crosscutting relationships (d) uniformitarianism.
6. The age of a rock in years is known as

(a) index age (b) relative age (c) half-life age (d) absolute age.

7. Varves are formed by layers of (a) limestone mixed with coarse sediments (b) coarse sediments followed by fine sediments (c) shale followed by sandstone (d) sandstone followed by shale.

8. An atom with a mass number of 234 and an atomic number of 90 has (a) 90 neutrons and 144 protons (b) 234 neutrons and 90 protons (c) 144 neutrons and 90 protons (d) 117 neutrons and 117 protons.

9. The process whereby a radioactive nucleus emits two protons and two neutrons is called (a) petrification (b) superposition (c) radioactive decay (d) fossilization.

10. A radioactive element decays (a) at a measurable average rate (b) at a rate dependent

on temperature (c) only under great pressure (d) only at high temperatures.

11. The process whereby the remains of an organism are preserved by drying is (a) petrification (b) mummification (c) erosion (d) superposition.

12. Some remains of insects have been found preserved in (a) tar (b) amber (c) gastroliths (d) magma.

13. Molds filled with sediments sometimes produce (a) casts (b) gastroliths (c) coprolites (d) imprints.

14. Fossils that are found in many parts of the earth and that were formed by organisms that lived during a brief period of geologic time are known as (a) coprolites (b) molds (c) gastroliths (d) index fossils.

Application

On your paper, write answers to the following questions.

1. Hutton developed the theory of uniformitarianism by observing geologic changes on his farm. What changes might he have observed?

2. Give an explanation of the logic behind the law of superposition.

3. How might a scientist determine the original positions of the sedimentary layers in an angular unconformity?

4. Of two intrusions, one cuts through all the rock layers. The other is cut off and lies below several layers of sedimentary rock. Which intrusion is younger? Why?

5. A scientist is attempting to calculate the absolute age of sedimentary rock layers by using the average rate of deposition. However, between the third and fourth layers is a disconformity. What difficulties might this cause in determining the absolute ages of the layers below the disconformity?

6. What information about past climatic changes might scientists gather by studying varves?

7. 16 g of U-238 turn into 0.5 g of U-238 and 15.5 g of daughter products in how many years?

8. How are the processes of mummification and freezing similar?

9. Why might a scientist make an artificial cast?

10. Suppose that a certain type of fossil with unusual features is found in many areas of the earth. It represents a brief span of geologic time but occurs only in small numbers. Would the fossil make a good index fossil? Explain your answer.

Extension

1. Examine an exposed area of sedimentary rock in your community. Sketch your findings. Write a report based on your conclusions about the history of the rock bed.

2. Research two early methods for determining the age of the earth on the basis of the salinity of the oceans and the temperature of the earth. Report your findings to your class.

3. Find information about how moon rocks have been dated and what information was gained by studying them. Write a report about your findings.

4. Find out more about trilobite fossils. When and where did trilobites live, and in what type of rock are trilobite fossils found? Write an article based on your findings.

Chapter 18
A View of the Earth's Past

The rock record holds a fascinating story of the evolution of life on the earth. Fossils, such as those shown in the photo on this page, reveal a rich diversity of plant and animal life. They may range from evidence of a single-celled marine organism nearly 3.4 billion years old to the remains of the early ancestors of modern human beings. This chapter discusses the divisions of geologic time and the evidence of the evolution of living things throughout the earth's history.

Chapter Outline

This fossil starfish lived about 400 million years ago.

18.1 The Geologic Time Scale

Section Objectives

- **Summarize the development of the geologic column.**
- **List the major units of geologic time.**

The surface of the earth is constantly changing. Mountains form and erode; oceans rise and recede. As conditions on the earth's surface change, various organisms flourish and then die out. Evidence of these changes is recorded in the rock layers of the earth's crust. To describe the sequence and length of these changes, scientists have developed a geologic time scale. The geologic time scale is an outline of the development of the earth and of life on the earth.

The Geologic Column

Using the law of superposition and the study of index fossils, nineteenth-century scientists determined the relative ages of rock layers in areas throughout the world. No single area on earth contained a record of all geologic time. Therefore, scientists combined their observations from around the world to create a standard arrangement of rock layers. This ordered arrangement of rock layers, which is based upon the ages of the rocks, is called the **geologic column.** The geologic column represents a time line of the earth's history, with the oldest rocks at the bottom and the most recent rocks at the top.

Rock layers in the geologic column are distinguished from each other primarily by the kinds of fossils they contain and by the rock type. Fossils in the upper, more recent layers resemble modern plants and animals. Most of the fossils in the lower, older layers are of plants and animals that are very different from those living today. In fact, many of the fossils found in older layers are from species that have become extinct.

At the time the geologic column was developed, scientists based their estimates of the ages of rock layers on such things as average rates of sediment deposition. The development of radioactive dating methods, however, has enabled scientists to determine more accurately the absolute ages of rock layers in the geologic column.

Scientists can now use the geologic column to determine the age of a rock layer even if the rock contains no radioactive minerals. The rock layer being studied is compared with a layer in the geologic column that has the same fossil content, relative position, and, sometimes, physical characteristics. If the two layers match, they were very likely formed at about the same time. Why is relative position important in determining the ages of rock layers?

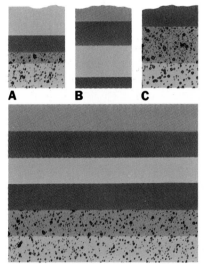

Geologic column

Figure 18–1. By combining observations of rock layers in areas A, B, and C, scientists can construct a geologic column.

Divisions of Geologic Time

Just as the calendar year is divided into months, weeks, and days, geologic time is divided into units. The geologic history of the earth is punctuated by major changes in the earth's surface or climate and by the extinction of various species. Geologists use these events as the basis for dividing the geologic time scale into smaller units.

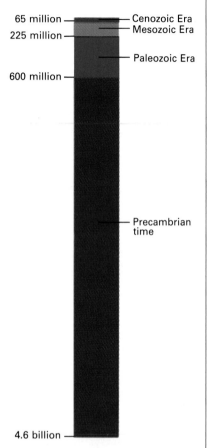

65 million — Cenozoic Era
— Mesozoic Era
225 million —

— Paleozoic Era

600 million —

— Precambrian
time

4.6 billion —

Figure 18–2. The present era, the Cenozoic, is the shortest era of geologic time, beginning 65 million years ago. Note its length in comparison to the other eras.

Rocks grouped within each unit contain similar fossil records. In fact, a unit of geologic time is often characterized by fossils of a dominant life-form.

Eras

The largest unit of geologic time is an **era.** There are four geologic eras. The earliest is the Precambrian Era. This era is more commonly referred to as **Precambrian time** because it is much longer than the other geologic eras. Also, very few fossils exist in Precambrian rocks, making it difficult to divide the 4 billion years of Precambrian time into smaller units. Precambrian rocks make up the lowest layers of the geologic column.

At the beginning of Precambrian time, the earth's crust was just beginning to solidify. Very early Precambrian rocks contain no evidence of life. However, a few fossils of bacteria and algae, thought to be the earth's first life-forms, have been found in rocks 3.5 billion years old. Rocks of late Precambrian time contain fossils of primitive worms, sponges, jellyfish, and corals. These fossils and other types of evidence have led most scientists to theorize that life began in the ocean.

The **Paleozoic Era** followed Precambrian time. The term *Paleozoic* comes from the Greek words meaning "ancient life." The Paleozoic Era lasted 375 million years. Rocks from this era contain fossils of a wide variety of primitive plants and animals.

The **Mesozoic Era,** a span of about 160 million years, followed the Paleozoic Era. *Mesozoic* means "middle life." Mesozoic rocks hold fossils of more-complex organisms such as reptiles and birds.

The present geologic era is the **Cenozoic Era.** *Cenozoic* means "recent life." Fossils of mammals are common in rocks of this era. The sequence of fossils in the geologic column shows that primitive forms of life were followed by more-complex forms.

Periods and Epochs

Some eras have been divided into time units called **periods.** Each period is characterized by specific fossils. A period is usually named for the location in which the rocks containing the identifying fossils were first found. For example, the Devonian Period of the Paleozoic Era was named for Devonshire, England. The term *Devonian* is applied to rock layers in the geologic column that developed at about the same time as the corresponding layers in Devonshire.

Rocks of the Cenozoic Era are the most recent and often the least deformed in the geologic column. They may contain an extremely detailed fossil record. This abundance of information has allowed scientists to divide the two periods of the Cenozoic Era into smaller time units called **epochs.**

Section 18.1 Review

1. How was the geologic column developed?
2. What are the divisions of geologic time?
3. Why is the geologic column useful to earth scientists?

18.2 Geologic History

Section Objectives

- **Identify the characteristics of Precambrian rock.**

- **Explain what scientists have learned from the geologic record about life during the Paleozoic era.**

- **Explain what scientists have learned from the geologic record about life during the Mesozoic era.**

- **Explain what scientists have learned from the geologic record about life during the Cenozoic era.**

History is a written record of past events. Just as the history of different civilizations is recorded in books, the geologic history of the earth is recorded in rock layers. The type of rock that makes up each layer and the fossils that are found in each layer reveal much about the conditions that existed when the layer formed. For example, limestone usually forms in bodies of water. The presence of a limestone layer in a region indicates that the area was once covered by water. Cross-bedded sandstone, on the other hand, often forms from sand dunes. Thus, the presence of this sandstone may indicate that a region was once a desert.

Fossils also indicate the kinds of organisms that lived during that geologic time. By examining rock layers and fossils, geologists and paleontologists have found a large amount of evidence that supports the **theory of evolution.** Additional evidence that supports the theory of evolution comes from the examination of living organisms by biologists. The theory of evolution states that organisms change over time and that new kinds of organisms are derived from ancestral types. The theory of evolution by natural selection was proposed by Charles Darwin, an English naturalist, in 1859. Natural selection is often called the *survival of the fit*.

One of the results of evolution by natural selection is that organisms are adapted to their environment. Because organisms are adapted to their environment, a change in the environment affects the organisms. Only those that are adapted to the change will survive. Organisms that cannot survive—in other words, those that are unfit to live and reproduce in the new environment—will die out.

Figure 18–3. Many animals that existed in the past are now extinct. Shown here are the fossilized bones of an extinct South American plant eater.

Figure 18–4. Stromatolites, fossil algae, are the most common Precambrian fossils.

Among the numerous types of environmental changes that affect the survival of organisms are major geologic and climatic changes. One example of a geologic change is a dramatic decrease in the amount of the earth's surface that is covered by water. At one time vast areas of the earth were covered by warm, shallow seas; these waters have since receded. An example of a major climatic change is a decrease in atmospheric temperature. Such a major decrease in temperature occurred at the time of the ice ages that have affected the earth periodically.

Based on evidence gathered from rocks and fossils, geologists try to determine how these changes affected the survival of various organisms. Scientists also attempt to learn what evolutionary changes may have taken place within organisms to enable them to survive environmental change. The rocks and fossils also reveal which organisms died out, perhaps because they were not adapted to change.

Precambrian Time

Precambrian time began with the formation of the earth nearly 4.6 billion years ago and ended about 600 million years ago. It makes up nearly 85 percent of geologic time. The Precambrian rock record is difficult to interpret. Most Precambrian rocks have been so severely deformed and altered by crustal activity that the original order of rock layers is rarely identifiable.

Large areas of exposed Precambrian rocks, called *shields,* are found on every continent. For example, a Precambrian shield covers much of eastern Canada. Precambrian shields are the result of several hundred million years of volcanic activity, mountain building, sediment formation, and metamorphism. Precambrian rocks were deformed and metamorphosed, causing the rocks to melt. Minerals collected near the surface of the earth as the rocks cooled. Nearly half of the deposits of valuable minerals in the world have been found in Precambrian shields. Among these minerals are nickel, iron, gold, and uranium.

Fossils are rare in Precambrian rocks, probably because Precambrian life-forms were soft-bodied. Such organisms lacked the bones, shells, or other hard parts that commonly form fossils. Also, Precambrian rocks are extremely old. Some date back nearly 3.8 billion years. Extensive crustal movements such as folding and faulting, volcanic activity, and erosion over this long period of time probably destroyed most Precambrian fossils.

Of the few Precambrian fossils that do exist, the most common are stromatolites, reeflike deposits produced by bacteria or algae. Stromatolites still form today in some shallow waters. The presence of stromatolite fossils in Precambrian rocks indicates that shallow seas may have covered much of the earth during some periods of Precambrian time. In North America no fossil remains of animals have been found in Precambrian rocks. However, some imprints of marine worms, jellyfish, and one-celled organisms have been found in the late Precambrian rocks of Australia.

The Paleozoic Era

The Paleozoic Era began about 600 million years ago and ended about 225 million years ago. Scientists theorize that at the beginning of the Paleozoic Era, the earth's landmasses were scattered in the world ocean. By the end of the Paleozoic Era, these landmasses had collided to form the supercontinent Pangaea. As you read in Chapter 4, this tectonic activity created new mountain ranges and lifted large areas of land above sea level.

Unlike Precambrian rocks, Paleozoic rocks contain an abundant fossil record. These fossils indicate that there was a dramatic increase in the number of plant and animal species on the earth at the beginning of the Paleozoic Era. As a result of this rich fossil record, geologists have been able to gather more facts about this era. Because of the wealth of information, they have divided the Paleozoic Era into seven periods.

The Cambrian Period

The Cambrian Period is the earliest period of the Paleozoic Era. From the fossils in Cambrian rocks, scientists have inferred that a variety of advanced forms of marine life first appeared during this period. These marine organisms quickly replaced the primitive Precambrian organisms as the dominant form of life. The explosion of Cambrian life may have been due in part to the warm, shallow seas that appear to have covered most of the continents during this period. Marine **invertebrates,** animals without backbones, thrived in these warm waters. The most common of the Cambrian invertebrates were the trilobites, hard-shelled animals that lived on the ocean floor. As you read in Chapter 17, scientists use trilobites as important index fossils for identifying layers of Cambrian rock throughout the world.

The second most common animals of the Cambrian Period were the brachiopods, a group of shelled animals. There is evidence that at least 15 different classes of brachiopods existed during this period. A few kinds of brachiopods still exist today, though they are

Figure 18–5. During the Paleozoic Era, various species of brachiopods, including this fossil brachiopod from the early Paleozoic Era, flourished in the warm, shallow seas.

Table 18—1. Geologic History of the Earth

Precambrian Era

Began: 4,600 million years ago
Duration: 4,000 million years
Characteristic organisms:

Jellyfish

Paleozoic Era

Cambrian Period

Began: 600 million years ago
Duration: 100 million years
Characteristic organisms:

Trilobite

Brachiopod

Ordovician Period

Began: 500 million years ago
Duration: 70 million years
Characteristic organisms:

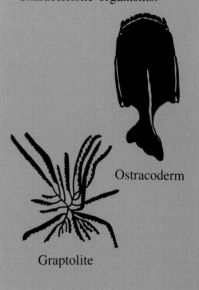

Ostracoderm

Graptolite

Paleozoic Era

Permian Period

Began: 280 million years ago
Duration: 55 million years
Characteristic organisms:

Lizardlike reptile

Pelycosaurs

Mesozoic Era

Triassic Period

Began: 225 million years ago
Duration: 30 million years
Characteristic organisms:

Ammonite

Icthyosaurus

Jurassic Period

Began: 195 million years ago
Duration: 60 million years
Characteristic organisms:

Pterosaurs

Stegosaurus

Paleozoic Era

Silurian Period

Began: 430 million years ago
Duration: 35 million years
Characteristic organisms:

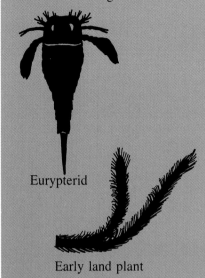

Eurypterid

Early land plant

Devonian Period

Began: 395 million years ago
Duration: 50 million years
Characteristic organisms:

Lungfish

Ichthyostega

Carboniferous Period

Began: 345 million years ago
Duration: 65 million years
Characteristic organisms:

Dragonfly

Swamp plants

Mesozoic Era

Cretaceous Period

Began: 135 million years ago
Duration: 70 million years
Characteristic organisms:

Angiosperm

Tyrannosaurus rex

Cenozoic Era

Tertiary Period

Began: 65 million years ago
Duration: 62.5 million years
Characteristic organisms:

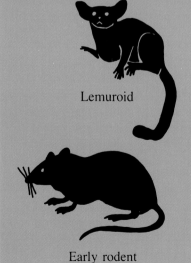

Lemuroid

Early rodent

Quaternary

Began: 2.5 million years ago
Duration: ?? million years
Characteristic organisms:

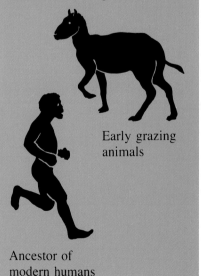

Early grazing
animals

Ancestor of
modern humans

rare. Other common Cambrian invertebrates include worms; jellyfish; snails; sponges; and cephalopods, which are ancestors of the modern squid and octopus. No evidence of land-dwelling plants or animals has been found among Cambrian fossils.

The Ordovician Period

During the Ordovician Period, the number of brachiopod species increased, and trilobites began to die out. Snails, clams, and other mollusks became the dominant life-forms. Large numbers of coral began to appear, although coral reefs did not occur until later. Colonies of tiny invertebrate animals called *graptolites* also flourished in the oceans during the Ordovician Period. Graptolite fossils, as discussed in Chapter 17, are useful index fossils for this period.

An important development of the Ordovician Period was the appearance of the ostracoderm, a primitive fish. Fossils of ostracoderms are the oldest fossils of **vertebrates,** animals with backbones, yet discovered. Ostracoderms were eellike; they had no jaws or teeth. Unlike modern eels, however, their bodies were covered with bony plates. During the Ordovician Period, as during the Cambrian and Silurian periods, there was no plant life as it is known today.

Horizons

Hypothesis: Search for a Deadly Star

For the past several years, a major subject of study at the Leuschner Observatory of the University of California has been the possible existence of a star called *Nemesis.* Named after the Greek goddess of doom, Nemesis is hypothesized to be a small, dark companion star of the sun.

In 1983 data suggesting the possible existence of Nemesis were gathered by University of Chicago paleontologists. J. John Sepkoski, Jr. and David Raup developed a series of computer-generated spindle diagrams, such as the ones shown here. The diagrams illustrate the number of families within animal classes known, by direct or indirect fossil evidence, to have been present during various time periods. The width of each spindle indicates the number of families within each class present during the corresponding eras.

A study of these diagrams and other data reveals mass extinctions of organisms on earth occur about every 26 million years.

Some scientists searched for a link between these extinctions and some cosmic event. They have found evidence suggesting that the extinctions occurred at about the same time that massive comet showers struck the earth. Scientists hypothesized

The Silurian Period

Marine life, both invertebrate and vertebrate, continued to thrive and evolve during the Silurian Period. Echinoderms, relatives of the modern starfish, and corals became more numerous. Animals such as spiders and millipedes evolved on land. Scorpionlike sea creatures, called *eurypterids,* were abundant during this period. Fossils of giant eurypterids nearly 2.7 m in length have been found in western New York. At the end of this period, plants began to grow on land.

The Devonian Period

The Devonian Period is often called the *Age of Fishes* because rocks from this period contain fossils of many kinds of bony fishes. One type of fish, called a *lungfish,* had the ability to breathe air. Other air-breathing fish called *rhipidistians* had strong fins which enabled them to crawl out of the water onto the land for short periods of time. The first true amphibian, *Ichthyostega,* probably evolved from these fish. *Ichthyostega,* which resembled a huge salamander, was probably the ancestor of modern amphibians such as frogs and toads. During the same period, land plants, such as giant horsetails, ferns, and cone-bearing plants, began to develop.

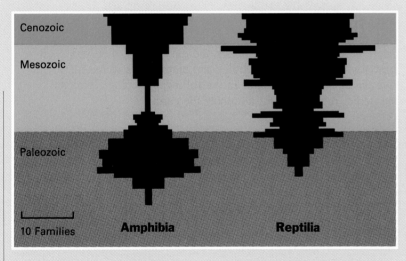

that these comet showers sent up great clouds of dust. According to this hypothesis, the dust became so dense that sufficient sunlight could not reach the earth. As a result, the planet became darker and colder, and much of the plant and animal life died out. This hypothesis is one possible explanation for the disappearance of the dinosaurs at the end of the Cretaceous period 65 million years ago.

Scientists have speculated about what could have caused these deadly comet showers. They are exploring the idea that every 26 million years or so, the hypothetical companion star Nemesis follows an orbit that takes it near the Oort cloud. The Oort cloud is solar system debris and comets orbiting the sun. The passage of Nemesis near this cloud would break up the debris, sending comet showers toward earth.

Astronomers use many instruments, including the 2.1 m telescope at Kitt Peak National Observatory (left) to observe the sky.

The next comet shower is not expected for at least another 14 million years. If scientists locate a star with an orbit that will carry it through the Oort cloud at that time, they will have found Nemesis.

How might scientists have discovered the 26-million-year cycle of mass extinctions?

The Carboniferous Period

The Carboniferous Period often is divided into the Mississippian and Pennsylvanian periods. During the Carboniferous Period, the climate was generally warm, and the humidity was extremely high all over the world. Forests and swamps covered much of the land. Scientists think that the coal deposits of Pennsylvania, Ohio, and West Virginia are the fossilized remains of these forests and swamps. During this period, the rock in which oil deposits are found was also formed. *Carboniferous* means "carbon bearing." A detailed discussion of coal formation appears in Chapter 11.

Amphibians and fish continued to flourish during the Carboniferous Period. Crinoids, relatives of modern starfish, were common in the oceans. Insects, such as giant cockroaches and dragonflies, were common on land. Toward the end of the Pennsylvanian Period, reptiles, the first vertebrates fully adapted to life on land, appeared. These early reptiles resembled large lizards.

The Permian Period

The Permian Period marks the end of the Paleozoic Era and the extinction of numerous Paleozoic life-forms. Nearly all the continents of the world had joined to form the supercontinent Pangaea. Collision of tectonic plates forced the Appalachian Mountains to rise so high that moist air could not rise over the mountain tops. On the northwest side of the mountains, areas of deserts and dry savanna developed. The shallow inland seas that had covered much of the earth evaporated. As the seas disappeared, many species of marine invertebrates, including trilobites and eurypterids, died out. However, fossils indicate that reptiles and amphibians were able to survive the changes in climate. In the following geologic era, the Mesozoic, reptiles would dominate the earth.

The Mesozoic Era

The Mesozoic Era began about 225 million years ago and ended about 65 million years ago. Geologic evidence shows that during the Mesozoic Era the surface of the earth changed dramatically. The supercontinent Pangaea began to break up into smaller continents. These new continents drifted and collided, uplifting mountain ranges such as the Sierra Nevada Range in California and the Andes Mountains in South America. The movements of the continents are discussed further in Chapter 19. Shallow seas and marshes covered much of the land. The climate became generally warm and humid.

Conditions during the Mesozoic Era favored the survival of reptiles. Dinosaurs, lizards, turtles, crocodiles, and snakes flourished during the three periods of the Mesozoic Era. As a result, this era is also known as the *Age of Reptiles*.

The Triassic Period

Dinosaurs first appeared during the Triassic Period. The word *dinosaur* comes from the Greek words meaning "terrible lizard." Dinosaurs varied greatly in size. Some species of dinosaurs were no

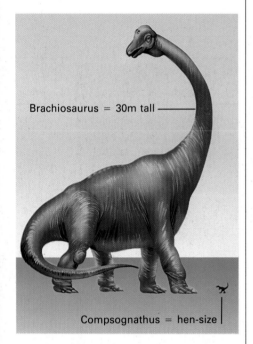

Brachiosaurus = 30m tall

Compsognathus = hen-size

Figure 18–6. *Brachiosaurus,* the largest known dinosaur, stood 30 m tall. One of the smallest dinosaurs, *Compsognathus,* stood less than 1 m tall.

larger than squirrels. Others weighed as much as ten elephants. Most of the dinosaurs of the Triassic Period were about 4 m to 5 m in length, and they could move quickly. These dinosaurs roamed through lush forests of cone-bearing trees and cycads, plants that resemble modern palm trees.

Reptiles called *icthyosaurs* and *plesiosaurs* inhabited the Triassic oceans. New forms of marine invertebrates also developed. The most common was the ammonite, a type of shellfish similar to the modern nautilus. Ammonite fossils are found throughout the world and serve as Mesozoic index fossils.

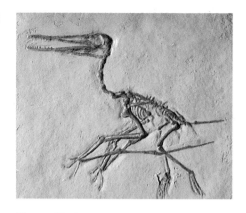

Figure 18–7. This fossil of a *ptero-dactyl* is an example of the flying reptiles common during the Jurassic Period.

The Jurassic Period

Dinosaurs became the dominant life-form of the Jurassic Period. Fossil records indicate that two major groups of dinosaurs evolved. These dinosaurs are distinguished by their skeletal structures. One group, called *saurischians,* included both herbivores, or plant eaters, and carnivores, or meat eaters. One of the largest saurischians was the herbivore *Apatosaurus.* Apatosaurs weighed up to 50 tons and reached a length of 25 m.

The other major group of Jurassic dinosaurs, called *ornithischi-ans,* were herbivores. One of the best known of the ornithischians was *Stegosaurus,* which had a single row of upright bony plates along its backbone. Stegosaurs were about 9 m long and stood about 3 m high at the hips.

Flying reptiles called *pterosaurs* or *pterodactyls* were common during the Jurassic Period. Like modern bats, pterosaurs flew on skin-covered wings. The fossils of the first known mammals, which were tiny rodentlike creatures, are found in rocks of this period. Also found in these rocks are fossils of the first true bird with feathers, *Archeopteryx.*

Figure 18–8. *Apatosaurs* and *Archeopteryx* were organisms that lived during the Jurassic Period.

The Cretaceous Period

Dinosaurs continued to dominate the earth during the Cretaceous Period. Among the most common was the carnivore *Tyrannosaurus rex.* Tyrannosaurs stood nearly 6 m tall. Their huge jaws held razor-sharp teeth 15 cm long.

Also among the common Cretaceous dinosaurs were ankylosaurs, whose bodies were covered with bony armorlike plates. Other Cretaceous dinosaurs were horned dinosaurs called *ceratopsians;* and *ornithopods,* which included duck-billed dinosaurs. Ankylosaurs, ceratopsians, and ornithopods were all herbivores.

Plant life seems to have evolved greatly by the Cretaceous Period. The first flowering plants, or angiosperms, appeared. The most common of these were trees, such as magnolias and willows. Later other trees such as maples, oaks, and walnuts became abundant. Angiosperms were so successful that they are still the dominant type of land plant.

The end of the Cretaceous Period was marked by the mass extinction of many species, including some marine reptiles and all dinosaurs. No dinosaur fossils have been found in rocks formed after the Cretaceous Period. As you read in Chapter 1, some scientists accept the meteorite-impact hypothesis as the explanation of this extinction. Other possible explanations include drastic changes in climate owing to the movement of continents and increased volcanic activity.

The Cenozoic Era

The Cenozoic Era began about 65 million years ago and includes the present period. By the beginning of the Cenozoic Era, the continents looked much as they do today, but they were located closer together. In general, the Cenozoic Era was a time of increased tectonic activity. As crustal plates collided, huge mountain ranges, such as the Alps in Europe and the Himalayas in Asia, were formed.

Figure 18–9. This skeleton is of a *Hipisodus,* a slim deer-like mammal about the size of a house cat. *Hipisodus,* a member of the hypertragulidae family, lived during the Oligocene Epoch and has no present day relatives.

During the Cenozoic Era, dramatic changes in climate appear to have occurred. At times huge glaciers covered nearly one third of the earth's land area. As temperatures fell during the ice ages, various species became extinct and other species appeared.

Fossils of mammals first appear in rocks of the late Mesozoic Era. However, mammals became the dominant life-form and underwent tremendous evolutionary change during the Cenozoic Era. Therefore, the Cenozoic Era is often called the *Age of Mammals.* How do you think mammals survived during the ice ages?

The Cenozoic Era has been divided into two periods—the Tertiary and the Quaternary. The Tertiary Period includes the time before the last major ice age. The Quaternary Period began with the last ice age and includes the present. The Tertiary Period and the Quaternary Period have been further divided into seven epochs. The Paleocene, Eocene, Oligocene, Miocene, and Pliocene epochs make up the Tertiary Period. The Pleistocene and Holocene epochs are in the Quaternary Period.

The Paleocene and Eocene Epochs

The fossil record indicates that during the Paleocene Epoch many new mammals evolved. There were small rodents, about the size of modern squirrels or mice, and a few small carnivores. Fossils of the first primates, called *lemuroids,* also are found in rocks of the Paleocene Epoch. Lemuroids were small, tree-dwelling mammals with thick fur and long tails.

Other mammals, including *Hyracotherium*, which is the earliest known ancestor of the horse, developed during the Eocene Epoch. Fossil records indicate that the first flying mammals and birds and the first whales and porpoises appeared during this epoch. Although the dinosaurs had died out, smaller reptiles continued to flourish. Worldwide temperatures dropped 4°C during the Eocene Epoch.

The Oligocene and Miocene Epochs

Geologic evidence indicates that during the Oligocene Epoch the climate continued to become cooler and drier. It became so dry that the Mediterranean Sea dried up, leaving only small, scattered salty lakes. Salt and other *evaporites* accumulated to a depth of 2,000 m on the floor of the sea. This change in climate favored the growth of grasses and cone-bearing and hardwood trees. Many of the earlier mammals became extinct. However, larger species of deer, pigs, horses, camels, cats, and dogs flourished.

The Miocene Epoch is often called the *Golden Age of Mammals.* The climate remained cool and dry. Great herds of primitive horses and camels roamed the plains, feeding on the abundant grasses. Fossil remains of members of the deer, rhinoceros, and pig families are commonly found in rocks of this epoch. *Baluchitherium*, a rhinoceroslike animal, lived during the Miocene Epoch. *Baluchitherium* is the largest known land mammal ever to have existed. It was nearly twice as large as a modern elephant. Miocene rocks also contain fossils of racoons, wolves, foxes, and the now extinct saber-toothed cat.

Figure 18–10. The tarsier, a relative of the lemuroids, is the sole modern survivor of a group of primates common during the Cenozoic Era.

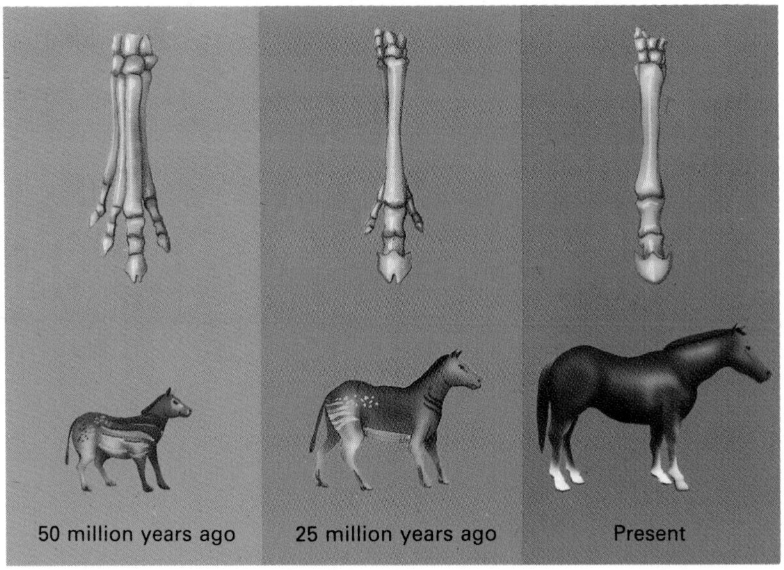

Figure 18–11. The ancient horse, *Hypocatherium*, first appeared early in the Cenozoic Era. The abundant fossil record allows scientists to reconstruct the evolutionary changes of the horse. Note the changes in size and in number of toes from 50 million years ago to the present.

50 million years ago   25 million years ago   Present

The Pliocene Epoch

During the Pliocene Epoch, hunting animals—members of the bear, dog, and cat families—evolved. These carnivores hunted the herds of grazing animals that inhabited the grassy plains. Fossils of the first modern horses are also found in rocks of this epoch. Study Figure 18–11, which traces the evolution of horses from the Eocene Epoch to the present time.

Before the end of the Pliocene Epoch, great climatic changes occurred, and polar ice began to spread. With so much water locked in ice, the sea level fell, and a land bridge probably appeared between Asia and North America. The uplifting of the earth's crust between North and South America formed another land bridge. Various species migrated between the continents across the land bridges.

The Pleistocene and Holocene Epochs

During the Pleistocene Epoch, four periods of glaciation occurred over much of Eurasia and North America. Some animals had special characteristics, such as the thick fur that covered woolly mammoths and rhinoceroses, that allowed them to endure the cold. Other species survived by moving to warmer regions. Less-adaptable species, such as giant ground sloths and dire wolves, became extinct. Fossils of the early ancestors of modern humans are found in rocks of the Pleistocene Epoch. These fossil remains date to nearly 2 million years ago. Scientists have evidence that indicates these early humans were hunters. Their successful hunting may have led to the extinction of many animals during the Pleistocene Epoch.

The Holocene Epoch, which includes the present, began about 11,000 years ago as the last ice age ended. As the glaciers melted, the sea level rose an estimated 45 m, and the coastlines took on their present shapes. The Great Lakes were also formed as the glaciers retreated. During the early Holocene Epoch, modern humans, or

Investigation

Geologic Time Scale

Geologic time spans some 4.6 billion years. Using a scale that represents time as distance, you can compare the length of eras and periods.

Science Process Skills Focus: constructing models, naming and labeling, ordering and sequencing

Materials

geologic time scale on pages 300–301; adding-machine tape, 5 m long; ruler; meter stick; 6 colored pencils

Procedure

1. Copy Table 18–I onto a piece of paper.
2. Complete the table, using the scale 1 cm equals 10 million years.
3. Lay the adding-machine tape flat on a hard surface. Use the meter stick and a pencil to mark off the beginning and end of Precambrian time according to the time scale you calculated. Do the same for the other eras. Label the eras and color each era a different color.
4. Study the geologic time scale on pages 300–301. Use the scale 1 cm equals 10 million years and calculate the scale length for each period listed. Label the periods on your scale.

Table 18–I.

Era	Length of time (years)	Scale length
Precambrian	4,000,000,000	
Paleozoic	375,000,000	
Mesozoic	160,000,000	
Cenozoic	65,000,000 (to present)	

5. Write in the major kinds of organisms that lived during each unit of time.

Analysis and Conclusions

1. Does the scale you made measure relative time or absolute time or both?
2. When did humans first appear? What is the scale length from that period to the present?
3. Add the lengths of the Paleozoic, Mesozoic, and Cenozoic eras. What percentage of the total geologic time scale do these eras combined represent? What percentage of the total time scale does Precambrian time represent?

Homo sapiens, developed agriculture and began to make and use tools of bronze and iron.

Compared with the entire geologic time scale, human history is extremely brief. If you think of the entire history of the earth as one year, the first organisms would have appeared in early May. Early humans would have appeared on December 31 at 7 P.M.; and modern humans, not until 11:55 P.M. that night.

Section 18.2 Review

1. Why are fossils rare in Precambrian rocks?
2. How did the formation of Pangaea affect Paleozoic life-forms?
3. How did the ice ages affect animal life during the Cenozoic Era?
4. Compare the effects of the Permian extinction with those of the Cretaceous extinction.

Reviewing Chapter 18

Key Concepts

- The geologic column is based on observations of the relative ages of rock layers throughout the world. See page 295.
- Geologic time has been divided into eras, periods, and epochs. See page 295.
- Precambrian rocks contain valuable minerals but few fossils. See page 298.
- The rock record of the Mesozoic Era reveals an environment that favored the development of reptiles. See page 304.
- The rock record of the Cenozoic Era includes the present period and reveals the rise of mammals as the predominant life-form. See page 306.

- **The rock record reveals the evolution of marine invertebrates and vertebrates during the Paleozoic Era. See page 300.**

Key Terms

Cenozoic Era (296)	invertebrate (299)	Precambrian time (296)	vertebrate (302)
epoch (296)	Mesozoic Era (296)		
era (296)	Paleozoic Era (296)	theory of evolution (297)	
geologic column (295)	period (296)		

Review

On your paper, write the letter of the term that best completes each of the following statements.

1. The geologic time scale is a (a) scale for weighing rocks (b) timer used by geologists (c) rock record of the earth's past (d) collection of the same kind of rocks.
2. Scientists have been able to determine the absolute ages of most rock layers in the geologic column by using (a) the law of superposition (b) radioactive dating (c) rates of deposition (d) rates of erosion.
3. An event that geologists would use in dividing the geologic time scale into smaller units is (a) a volcanic eruption (b) the arrival on earth of a huge meteorite (c) a major flood on a continent (d) the arrival of an ice age.

4. To determine the age of a specific rock layer, scientists might correlate it with a layer in the geologic column that has the same relative position, physical characteristics, and (a) fossil content (b) weight (c) temperature (d) density.
5. The earliest unit of geologic time is (a) Precambrian (b) Mesozoic (c) Paleozoic (d) Cenozoic.
6. *Paleozoic* means (a) "ancient life" (b) "middle life" (c) "recent life" (d) "primitive life."
7. Geologic periods may be divided into (a) eras (b) epochs (c) days (d) months.
8. *Cenozoic* means (a) "ancient life" (b) "middle life" (c) "recent life"

(d) "primitive life."

9. Precambrian time ended about (a) 4.6 billion years ago (b) 600 million years ago (c) 65 million years ago (d) 25 thousand years ago.

10. The most common fossils found in Precambrian rocks are (a) graptolites (b) trilobites (c) ostracoderms (d) stromatolites.

11. An important index fossil for the Cambrian Period is the (a) graptolite (b) trilobite (c) ostracoderm (d) stromatolite.

12. The first vertebrates appeared during (a) Precambrian time (b) the Paleozoic Era (c) the Mesozoic Era (d) the Cenozoic Era.

13. The *Age of Fishes* is the name commonly given to the (a) Cambrian Period (b) Ordovician Period (c) Silurian Period (d) Devonian Period.

14. The *Age of Reptiles* is the name commonly given to (a) Precambrian time (b) the Paleozoic Era (c) the Mesozoic Era (d) the Cenozoic Era.

15. Dinosaurs became the dominant life-form during the (a) Permian Period (b) Silurian Period (c) Jurassic Period (d) Tertiary Period.

16. The first flowering plants made their appearance during the (a) Cretaceous Period (b) Triassic Period (c) Carboniferous Period (d) Ordovician Period.

17. The *Age of Mammals* is the name commonly given to (a) Precambrian time (b) the Paleozoic Era (c) the Mesozoic Era (d) the Cenozoic Era.

18. Humans probably first appeared during the (a) Paleocene Epoch (b) Oligocene Epoch (c) Miocene Epoch (d) Pleistocene Epoch.

Application

On your paper, write answers to the following questions.

1. Explain how the law of superposition might have aided scientists in the development of the geologic column?

2. Why would it be difficult to divide Precambrian time into periods?

3. At one time paleontologists classified fossils by similarities in the structure of the hard parts of the organisms. Later the scientists found that this was an unreliable method of classifying fossils. On what basis might they have made this decision?

4. There are rich deposits of iron ore in northern Minnesota and Michigan. During which geologic era were these deposits probably formed? Explain your answer.

5. What information in the geologic record might lead scientists to infer that shallow seas covered much of the earth during the Paleozoic Era?

6. In 1966 two amateur fossil hunters found a fossil on the shore at Cliffwood, New Jersey. Scientists determined that the fossil was an insect intermediate between primitive ants and wasps. What types of information might they look for in this fossil discovery?

7. If you found a rock layer that contained dinosaur fossils, what assumptions could you make about the age of the rock layer?

8. Explain the basis scientists may have used for dividing the Cenozoic Era into the Tertiary and Quaternary periods.

Extension

1. Find out the locations of the Precambrian shields throughout the world. Shade in those areas on a map of the world.

2. Compare the evolution of life during the Paleozoic Era with that during the Mesozoic Era. How did environmental changes affect the development of organisms?

3. Much information about Mesozoic life-forms comes from the study of fossils found at Dinosaur National Monument in Utah and Colorado. Write a short report describing the monument and some of the fossils found there.

4. Research the discoveries made by British anthropologists Dr. Louis S. B. Leakey and Mary Leakey in Olduvai Gorge in Tanzania, Africa. Write a report based on your findings.

Chapter 19

The History of the Continents

For billions of years, the surface of the earth has been tectonically active. Lithospheric plates have drifted slowly around the planet, colliding with each other to form supercontinents and then splitting apart into smaller landmasses. The existing continents resulted from this ongoing process.

This chapter discusses the global movement of the continents. It traces the evolution of North America and shows how the rock layers of the Grand Canyon reveal the history of a region.

Chapter Outline

19.1 Movements of the Continents
Formation of Pangaea
Breakup of Pangaea

19.2 Growth of a Continent: North America
Beginnings of a Continent
Shaping the Continent
The Modern Continent

19.3 Formation of the Grand Canyon
History in Rocks
History in Fossils

Grand Canyon National Park as seen from Yaki Point.

19.1 Movements of the Continents

Section Objectives
- Identify the landmasses that made up Pangaea.
- Describe the breakup of the supercontinent Pangaea.

Early one morning in May, 1980, Mount St. Helens, a volcanic peak in the Cascade Range in southern Washington, erupted violently. The eruption blew away one side of the mountain. The nearby forest was almost instantly changed into a lifeless, barren landscape.

Sudden and dramatic events such as volcanic eruptions, earthquakes, and floods are reminders that the earth's surface is changing constantly. Changes such as these take place in minutes, hours, or days. However, most of the processes that shape the earth's crust occur very slowly. For example, changes in the locations of the continents and in the size and shape of the oceans have taken place over many millions of years. In fact, the earth's crust shows evidence of geological changes that have taken place over a period of about 4 billion years.

The rock record indicates that in the earth's long history there has never been a permanent or even typical arrangement of the continents and oceans. The continents and oceans as they exist today are the result of plate tectonics, or the movement of crustal plates. The theory of plate tectonics states that the lithosphere is made up of many plates that float on the asthenosphere. These plates drift slowly around the planet—usually no faster than a centimeter each year. What are some consequences of this plate movement?

Formation of Pangaea

Geologists find it difficult to trace the movements of the continents that occurred very early in the earth's history. However, by combining evidence from many scientific fields, a general picture can be constructed. There is little doubt that continents and oceans existed during the earliest periods of the earth's history. All of the continents existing today contain large areas of deformed Precambrian rocks

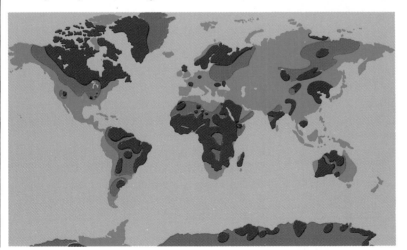

Figure 19–1. The major Precambrian cratons are shown in red. The exposed Precambrian rocks, or shields, are shown in gray.

called **cratons.** Precambrian rocks that have been exposed at the earth's surface are called *shields.* Cratons may represent ancient continents around which the modern continents have formed. The rocks of the cratons supply geologists with evidence of mountain building, volcanic activity, and sediment formation. However, the outlines of the continents and oceans as they existed during the Precambrian Era still are not known.

During the Paleozoic Era, there were probably two or three supercontinents that formed, broke up, and re-formed as lithospheric plates drifted slowly around the earth. The supercontinent Pangaea, thought by Alfred Wegener to be the continental mass from which the modern continents split off, formed about 225 million years ago. The collisions of the various landmasses that resulted in the formation of Pangaea also caused much mountain building. The Appalachian Mountains of eastern North America and the Ural Mountains,

Investigation

Plate Tectonics Theory

Although questions still remain about the movement of lithospheric plates, most scientists think that plate movement is caused by convection currents in the asthenosphere. You can demonstrate this theory by making a model of the asthenosphere and lithospheric plates.

Science Process Skills Focus: constructing models, observing, hypothesizing

Materials
water; shallow rectangular pan, about 30 cm × 40 cm; stove; dark-colored food coloring; 8 pieces of shirt cardboard, about 1 cm² each

Procedure
1. Fill the pan with water to 3 cm from the top and place on the stove.
2. Heat the water for 30 seconds over very low heat. Add a few drops of food coloring to the center of the pan. Observe and record what happens to the food coloring. Turn off the heat.
3. Label the cardboard pieces *1* through *8.* Carefully place pieces 1 through 4 as close together as possible in the center of the pan. Place a remaining cardboard piece in each corner of

the pan.
4. Turn on the heat. Sketch the pattern of movement for each cardboard piece. Turn off the heat. **Caution: Use a pot holder when handling heated objects.**

Analysis and Conclusions
1. In the plate tectonics model, what do each of the following represent: the water in the pan, the cardboard pieces, the stove?
2. In Step 2 what happened to the food coloring in the water? Explain your answer.
3. Describe what happened to the cardboard pieces as the water was heated in Step 4.
4. In what ways is the model you made an inaccurate representation of the movement of lithospheric plates?

which are the boundary between Europe and Asia, were formed during these Paleozoic Era collisions.

Pangaea, initially located in the polar region of the Southern Hemisphere, covered about 40 percent of the earth's surface. The remaining 60 percent of the surface was covered by a single large ocean called *Panthalassa*. A triangular body of water called the *Tethys Sea* cut into the eastern edge of Pangaea. North of the Tethys lay a portion of Pangaea called *Laurasia*. This area included the land that would eventually form North America and Eurasia. To the south of the Tethys lay the other part of Pangaea, which was called *Gondwana*. This land area eventually split to form South America, Africa, India, Australia, and Antarctica.

Because of its location in the southern polar region, the environment of Pangaea was harsh. Gondwana was often covered by an ice cap. Even the Sahara, one of the hottest places on the earth today, was once covered by a thick sheet of ice. As Pangaea began to drift northward toward the equator, however, much of the ice melted. This not only raised the level of the oceans but also triggered global warming. As the climate grew warmer, the number of species of land plants and animals increased.

Breakup of Pangaea

Two hundred million years ago, Pangaea probably began to break into two landmasses. Just north of the equator a great east-west rift split the supercontinent, nearly separating Laurasia and Gondwana.

Figure 19–2. By 225 million years ago, the earth's landmasses had joined together in the supercontinent called *Pangaea* (left). About 200 million years ago, Pangaea began to break up. (right).

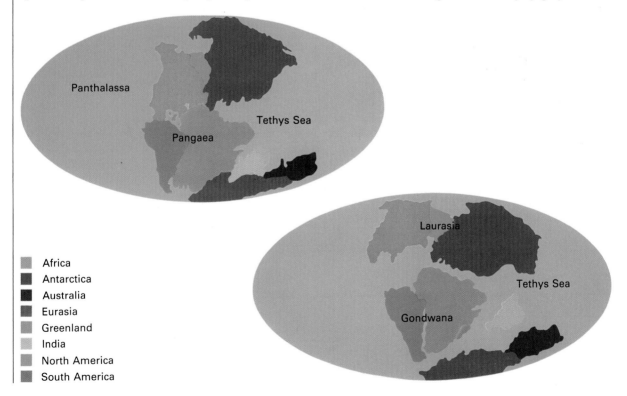

Africa
Antarctica
Australia
Eurasia
Greenland
India
North America
South America

They remained connected only where the southern tip of Spain touched the northwestern coast of Africa. East of this connection was the Tethys Sea; west of it the rift opened into Panthalassa. Ocean water poured into the rift, creating a narrow sea.

Geologic evidence indicates that as the rift opened, Laurasia began to rotate slowly in a clockwise direction. As this huge landmass moved, the sea located to its west began to widen. This body of water would become part of the Atlantic Ocean. The movement of Laurasia also resulted in the closing off of the Tethys Sea in the east. The Tethys Sea would become the Mediterranean Sea. The continued movement of Laurasia caused the east-west rift to diverge in a northward direction, separating North America from Europe. Ocean water filled the northward rift, forming the North Atlantic Ocean.

As Laurasia began to break apart, Gondwana also began to split. A rift opened in Gondwana splitting the landmass in two. The landmass that would become the continents of South America and Africa was located on one side of the rift. The landmass that consisted of India, Australia, and Antarctica was on the other side. India eventually broke away from Antarctica and began moving northward. Madagascar separated from Africa. About 100 million years ago, during the Cretaceous Period, a rift opened between Africa and South America, forming the South Atlantic Ocean. Over the last 65 million years Australia has separated from Antarctica, and India, continuing its northward movement, has collided with Asia.

Slowly the continents moved into their present positions. As the continents drifted around the earth, they collided with smaller pieces

Figure 19–3. About 65 million years ago, the South Atlantic Ocean began to form as South America split from Africa (left). Today the continents are still in motion (right).

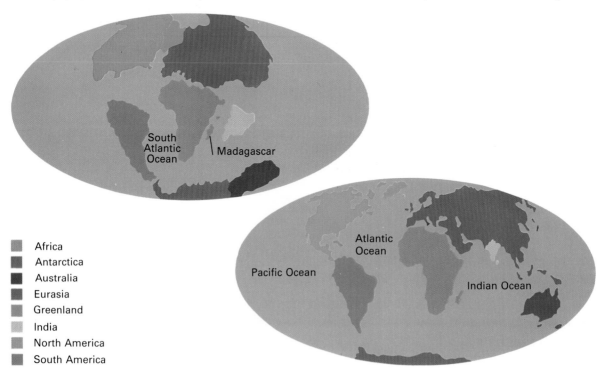

- Africa
- Antarctica
- Australia
- Eurasia
- Greenland
- India
- North America
- South America

Science Notebook

Continents of the Future

The movement of the crustal plates that carry the continents continues even today. For example, from actual measurements of plate movements, scientists now estimate that Hawaii and Japan are moving apart by as much as 12.7 cm per year.

If the plates continue to move as they are now doing, in about 150 million years the face of the earth will be very different. Africa will collide with Europe, closing the Mediterranean Sea and probably forming a high mountain range. A new ocean will be formed as the eastern side of Africa separates from the rest of the continent and moves eastward. As the North American and South American plates move westward and Europe and Africa move eastward, the Atlantic Ocean will become slightly larger. Antarctica will

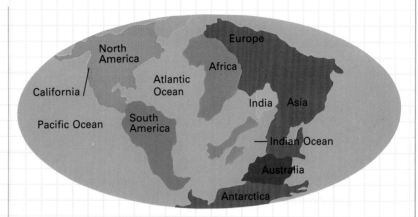

move north, eventually colliding with Australia.

In North America, the Baja Peninsula and the portion of southern California west of the San Andreas fault will have moved to where Alaska is today. If this plate movement occurs as predicted, Los Angeles will one day be located north of where San Francisco is today.

Scientists predict that 250 million years from now all the continents may come together again to form a supercontinent.

If the new supercontinent is located near the equator, how might the global climate and the level of water in the oceans be affected?

of continental crust and with volcanic islands on the ocean floor. These collisions welded new crust onto the continents and uplifted the land. Mountain ranges such as the Rockies, the Andes, the Alps, and the Himalayas were created. Continental drift also caused new oceans to open up and others to close.

Section 19.1 Review

1. What two major land areas made up the supercontinent Pangaea?
2. What processes that help to shape the earth's surface are evident in cratons?
3. Describe the formation of the North Atlantic Ocean and the South Atlantic Ocean.
4. Explain how the theory of plate tectonics relates to the formation and breakup of Pangaea.

Section Objectives

- **Summarize the changes in the North American continent that occurred during both the Precambrian time and the Paleozoic Era.**

- **Summarize the changes in the North American continent that occurred during the Mesozoic and Cenozoic eras.**

19.2 Growth of a Continent: North America

The ancient landmass around which the present North American continent developed probably first took shape during late Precambrian time. Since Precambrian rocks are the oldest in the geologic column, they are usually covered by younger rock layers. In the Grand Canyon, for example, Precambrian rocks are exposed only at the very bottom of the canyon. This is where the Colorado River has cut through the sedimentary layers to a depth of nearly 1 km.

In eastern Canada and parts of the northeastern United States, however, there is a large area where very old Precambrian rocks are exposed. This region is called the **Canadian Shield.** The Canadian Shield is the exposed portion of the craton around which the modern continent of North America has been built up. For 600 million years the Canadian Shield has been the only part of North America that has remained mostly intact.

To the south and west of the Canadian Shield is a central, stable region where the Precambrian rocks are buried under layers of sediments. The sediments are about 2.5 km thick in some places. They were deposited during the Paleozoic, Mesozoic, and Cenozoic eras when the land was covered by shallow seas. Surrounding the stable Canadian Shield and the platform around it are wide, mobile belts of mountains. These belts exist where the crust has been folded or faulted by plate movements and by the invasion of magma from below. Figure 19–4 shows a cross section of the continent, showing the position of the stable central region and the mountains characteristic of the mobile belts on either side.

Like similar rocks found all over the world, the Precambrian rocks of North America yield fossil evidence of only the simplest forms of life. Rocks of the Canadian Shield contain traces of microscopic bacteria. The fossils found in these rocks are thought to be at least 2 billion years old.

Figure 19–4. This cross section of the North American continent shows a central stable region bordered by two mountain belts.

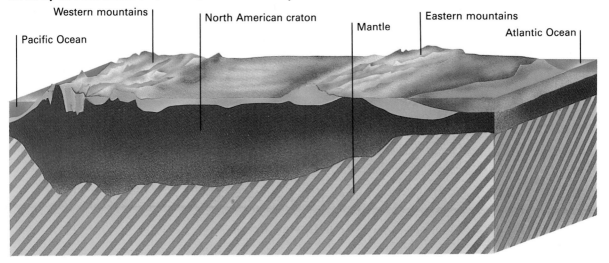

Western mountains
Pacific Ocean
North American craton
Mantle
Eastern mountains
Atlantic Ocean

Beginnings of a Continent

The rock structure of an existing continent can reveal a general outline of its geologic history. The rock structure of North America indicates that during the Paleozoic Era, much of what is now North America was part of a larger continent. This continent was located near the equator. A shallow sea covered much of what is now the United States.

Huge amounts of sediments were carried into the shallow water east and west of the present continent. The weight of the sediments caused the ocean floor to sink. More sediments were deposited and continued to accumulate on the sinking, shallow ocean floor. Eventually the sediments reached a thickness of over 12 km. Huge, Paleozoic, sediment-filled troughs called *geosynclines* developed. They now make up much of the older rock structures found along the eastern and western edges of North America. Thinner layers of Paleozoic sediments were also laid down over the stable region that is now the central United States.

As the Paleozoic Era ended 225 million years ago and the continents came together to form Pangaea, most of the warm, shallow seas disappeared. The marine invertebrates that had been so abundant for more than 400 million years vanished almost entirely. Their disappearance is thought to be related to the joining of the continents into Pangaea. The large inland sea that covered much of what was to become North America dried up. Left behind were the huge deposits of salt now found buried deep in Permian layers from Kansas to New Mexico. It is likely that the earth's climate also changed. How might the change in climate have contributed to the disappearance of marine life?

Shaping the Continent

The burst of tectonic activity accompanying the breakup of Pangaea played a large role in shaping North America. The basic form of the continent was determined by the rifts that split Laurasia from Gondwana and North America from Europe. The North America that separated from Laurasia during the Mesozoic Era did not have the same shape as the Paleozoic continent that became part of Pangaea. Much of what is now the Atlantic coastal area was once attached to the African plate. Some of the ancient European plate became welded onto parts of present-day New England and Canada. In addition, new pieces of continental crust, called *terranes,* were added. The terranes were carried along on oceanic crust and were welded to North America as the oceanic crust subducted beneath the continental crust.

Early in the Mesozoic Era, a large mountain range, the ancient Appalachians, existed along what is now the eastern United States. The Appalachians were created during the collision that formed Pangaea. On the western side of the continent, a sea covered the geosynclinal trough that had been created during Paleozoic times. The

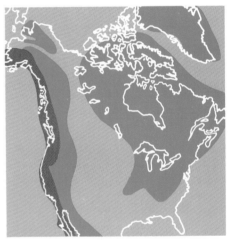

Figure 19–5. The top illustration shows that during the Paleozoic Era northeastern Canada was land (green area) and central North America was a wide sea (blue area). Sediment-filled troughs called *geosynclines* (brown areas) formed along the coasts. The bottom illustration shows that during the Cretaceous Period a shallow sea covered much of central North America.

central portion of the continent was periodically covered by a shallow sea. By the late Mesozoic Era, the eastern mountains had been worn down but were still evident. The Rocky Mountains began to rise east of the western geosyncline. This western uplift divided the continent into eastern and western land areas separated by an inland sea. The general landscape of the United States, characterized by eastern and western highlands separated by a region of low elevation, began to take shape.

The Modern Continent

By the beginning of the Cenozoic Era, about 65 million years ago, North America had begun to take on the shape it has today. The inland sea that covered much of the central region during Mesozoic times had drained away. Only the present coastal plain along the Atlantic and Gulf coasts remained submerged. The tectonic activity that took place during the Cenozoic Era created most of the major landforms of modern North America. Violent movements along plate boundaries in the western part of the continent caused the uplift of the Sierra Nevada range in California. These mountains, which first appeared during the Jurassic Period, were eroded until they were almost flat, then lifted again during the Cenozoic Era. East of the Sierra Nevada, a series of tilted fault-blocks created a large area of low mountains and wide valleys that are now called the *Basin and Range Province*. Uplift of a broad area around the Colorado River produced the Colorado Plateau. The Grand Canyon was created as the river cut into the rising land. Farther to the east, the Rocky Mountains continued to be pushed up. On the level platform to the east, sediments eroded from the Rockies were deposited on the platform, forming the Great Plains. Near the east coast of the continent, the ancient Appalachian Mountains continued to be worn down by weathering and erosion.

Volcanic activity has continued throughout the Cenozoic Era. Large areas of Washington, Oregon, Idaho, and northern California are covered with volcanic mountains and lava plateaus created by volcanic eruptions.

Figure 19–6. This east-west cross section of the United States, at about the latitude of Washington, D.C., shows the basic rock structures and the eras in which they were formed.

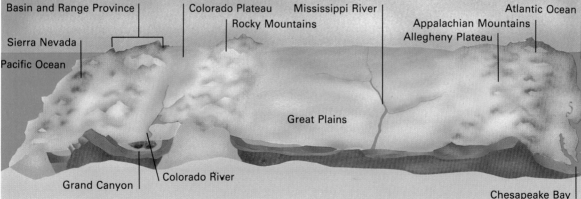

Basin and Range Province | Colorado Plateau   Mississippi River | Atlantic Ocean

Rocky Mountains | Appalachian Mountains
Allegheny Plateau

Sierra Nevada

Pacific Ocean

Great Plains

Grand Canyon | Colorado River

Chesapeake Bay

■ Mesozoic
■ Paleozoic
■ Cenozoic
■ Mostly Precambrian

IMPACT

Land Bridge to the Americas

Archeological evidence indicates that the first human inhabitants arrived in the Americas only about 20,000 years ago. Since humans inhabited other continents much earlier, their relatively recent appearance here is puzzling.

Many scientists now think that the first Americans may have migrated across a land bridge that connected North America and Asia during the Pleistocene Epoch.

Today the continents are separated by the Bering Strait. The Bering Strait is a narrow, shallow stretch of water that lies between Siberia and Alaska. At their closest points, the continents are just 80 km apart.

About 20,000 years ago, glaciers covered much of North America. In some places the ice sheets were 1.6 km to 3.2 km thick and held huge quantities of frozen water. As a result, the level of the oceans was nearly 84 m lower than it is today. Since the Bering Strait is currently only 42 m deep, the ocean floor between the continents would have emerged as the sea level dropped. The land bridge would then have been 1,000 m to 6,000 m wide.

The first Americans were probably hunters who followed herds of bison and mammoth across the land bridge into North America. By 6000 B.C. these nomads had reached the southern tip of South America.

The land bridge probably lasted several thousand years, until the glaciers retreated and the level of the oceans rose once again.

How did the arrival of humans probably affect the animal species already present in the Americas?

The ice ages that occurred during the last million years of the Cenozoic Era also helped shape the landscape of much of North America. When compared to the total time span of the earth's history, only a moment has passed since the glaciers last retreated.

Section 19.2 Review

1. How were the eastern and western geosynclines of North America formed?
2. Describe the continent of North America as it appeared at the end of the Mesozoic Era.
3. What important event during the last million years of the Cenozoic Era shaped much of the North American landscape?
4. Compare fossils found in Precambrian rocks with those found in rocks of later eras.

19.3 Formation of the Grand Canyon

Sedimentary rock layers are exposed in many areas. An exposed rock layer of any type is called an **outcrop.** An outcrop of sedimentary rock may be a small area only a few centimeters wide or a large cliff along the face of a valley. Some outcrops are created artificially. For example, where highways or railroads are cut through a hill, previously hidden rock layers are often exposed. Other outcrops occur naturally. Perhaps the best-known natural outcrop of sedimentary rock is the Grand Canyon in Arizona. There, the Colorado River has cut a huge gorge that is 6.4 km to 29 km wide and more than 1.6 km deep in some places. The exposed sedimentary layers in the Grand Canyon provide geologists an opportunity to study millions of years of geological history of this region of North America.

The Grand Canyon must have begun as a small valley cut by the Colorado River. Slowly, the land was lifted to form a high plateau. This uplifted region, now called the *Colorado Plateau,* occupies a large area in the southwestern United States. As the Colorado Plateau was uplifted, the river cut downward through the existing rock layers. Weathering processes acted on the slopes to help widen the gorge. The river probably has taken at least 10 million years to erode its way through the slowly rising plateau.

History in Rocks

Geologists studying the colorful scenery of the Grand Canyon can see in the exposed rocks a record of a part of the earth's history. Following the law of superposition, they assume that the rock layers in the Grand Canyon were deposited at different times, with the youngest deposits on top. There is no evidence that the great sheets of sedimentary rocks, whose edges are exposed in the walls of the canyon, have been overturned. Thus the upper layers can be judged to be younger than any layers found below. The absolute age of the layers must be determined from other evidence.

The bottommost layer of rock in the canyon is the **Vishnu schist.** The **Tapeats sandstone,** the **Bright Angel shale,** and the **Muav limestone** were deposited during the Cambrian Period. The next two geologic periods, the Ordovician and Silurian, are not represented by any rock layers in the canyon. The Devonian Period is only occasionally represented. Geologists think the Ordovician, Silurian, and Devonian periods were times of uplift and great erosion in the canyon. They also think that deposits from these periods were simply worn away.

The Mississippian Period is represented in the **Redwall limestone.** Geologists think redwall limestone was deposited when the Grand Canyon was submerged under a shallow sea. There are no deposits from the Pennsylvanian Period. The Permian Period, the

Figure 19–7. Geologists can study the exposed rocks of the Grand Canyon, such as these at Mather Point, to learn about the geologic history of the region.

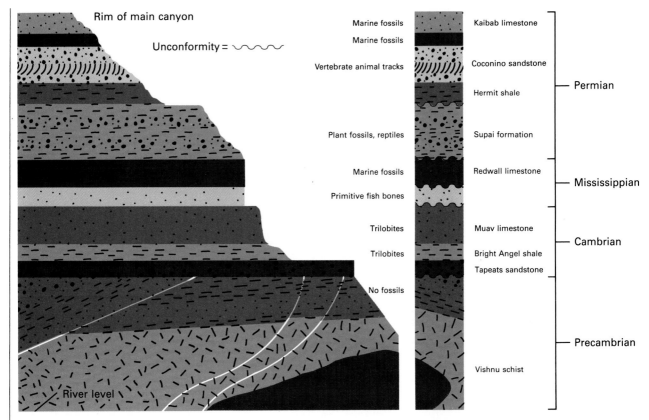

Figure 19–8. This cross section and geologic column of the Grand Canyon shows the positions of the various rock layers deposited throughout the geologic eras and the associated fossils.

last period of the Paleozoic Era, is represented by several deposits. These include the **Supai formation,** which consists of red sandstone and shale; the **Hermit shale,** the **Coconino sandstone,** and the **Kaibab limestone.**

The composition of these rock layers reveals more about the history of the canyon. For example, the light-colored layers you may have seen in many photographs of the Grand Canyon are composed of sandstone. Some of these sandstone layers show cross-bedding, which is characteristic of sand dunes. Therefore, there appears to be little doubt that these layers were laid down when the region was either an arid desert or a dune-covered beach. The layers of limestone and shale must have been deposited when the region was covered by a shallow sea. What do these facts tell you about the history of the Colorado Plateau?

History in Fossils

Through radioactive dating, geologists have established the absolute age of the fossils found in many of the rock layers of the Grand Canyon. The fossils are evidence of the kind of life that existed in this region during various periods. Limestone and shale, for example, often contain fossils of marine organisms such as clams, fishes, and corals. The Bright Angel shale contains many trilobites, marine invertebrates, and brachiopods.

The fossils show that a distinctive group of plants and animals lived during each period represented by the rock layers. Paleontologists can trace the changes in many of the organisms by examining their remains found in successive rock layers. Analysis of the fossils found in the older, lower rock layers through the younger, upper layers suggest a sequence of events. These fossils indicate that early, simple life-forms gradually either disappeared or evolved into more-complex forms. The position, composition, and fossil content of the rocks of the Grand Canyon give a detailed and lengthy history of that part of the earth's surface.

Career Focus: *Fossils*

"You don't have to go far away to find fossils. I've seen some fine specimens in the rock walls of downtown Chicago department stores."

Paleontologist

Mary Carman, who manages the paleontology collection at the Field Museum of Natural History in Chicago, seems to have "inherited" her interest in fossils.

"My family always loved fossils," she said. "Whenever we got the chance, we'd make field trips to look for Indian artifacts and fossils of all kinds."

Following in the footsteps of her great grandfather, an amateur geologist, and her grandmother, a chemistry teacher, Carman completed bachelor's and master's degree programs in geology at the University of Iowa. In her current position, she is responsible for all loans, acquisitions, and purchases of fossil specimens for the museum. She also coordinates displays and exhibits, leads field trips, and delivers lectures on a variety of topics associated with geologic history. In addition, she is conducting research on a type of micro-

scopic sea worm that existed some 200 to 600 million years ago. Fossils of these sea worms are valuable aids in locating oil deposits.

According to Carman, pictured left, one of the rewards of her job is examining newly discovered fossils, such as the one brought in to the museum recently by a Milwaukee collector.

"He had been fossil hunting in Iowa and found a fossil fish—a dipnoan—which is estimated to be some 375 million years old.

Section 19.3 Review

1. List the four bottommost layers of rock in the Grand Canyon. During what periods were they deposited?
2. What can the composition and fossil content of each layer of rock in the Grand Canyon reveal about the time when the layer was deposited?
3. If an outcrop shows alternating layers of sandstone and limestone, what can you assume about the geologic history of that region?

We were excited not only because it was in excellent condition, but because it was the first one of its kind to be found in North America. Although he wasn't sure what the fossil was, fortunately he knew it belonged in a museum."

Stratigrapher

Stratigraphers study the earth's rock layers to learn more about geologic history. Specifically, these scientists examine rock layers to determine their age, origin, history, and fossil and mineral content.

An important tool used by stratigraphers is the seismic survey, in which sound waves are used to detect different levels of subsurface rock. This information aids in locating rock layers where petroleum or natural gas might be found. The lower left photo shows stratigraphers at work.

Entry-level positions in stratigraphy require a bachelor's degree in geology, geophysics, or petroleum geology.

Paleontological Helper

Paleontological helpers keep records of rock and fossil specimens that have been collected in the field. They clean, wash, and prepare the samples, using special tools and cleaning solutions.

The paleontological helpers sort the specimens, compile background data on each item, and label the specimens. They then mount the specimens for study or experimental use. Most paleontological helpers work in research institutions or in large museums.

An entry-level position requires a bachelor's degree in geology or in a related field such as biology.

For Further Information

For more information on careers in geologic history, write the Paleontological Society, Department of Geology and Minerology, Ohio State University, 125 South Oval Mall, Columbus, OH 43210.

Chapter 19 Review

Key Concepts

- The supercontinent Pangaea formed during the late Paleozoic Era as lithospheric plates collided. Pangaea was made up of two major landmasses. See page 313.

- The breakup of Pangaea began during the Mesozoic Era. See page 315.

- The modern continent of North America is built up around a Precambrian craton. See page 318.

- By the beginning of the Cenozoic Era 65 million years ago, the North American continent had begun to take on the shape it has today. See page 320.

- The exposed rock layers of the Grand Canyon reveal the geologic history of the region over millions of years. See page 322.

- Scientists have used radioactive dating to establish the absolute ages of fossils found in the rock layers of the Grand Canyon. See page 323.

Key Terms

Bright Angel shale (322)	craton (314)	Redwall limestone (322)	Vishnu schist (322)
Canadian Shield (318)	Hermit shale (323)	Supai formation (323)	
Coconino sandstone (323)	Kaibab limestone (323)	Tapeats sandstone (322)	
	Muav limestone (322)		
	outcrop (322)		

Review

On your paper, write the letter of the term that best completes each of the following statements.

1. Areas of Precambrian rocks that may represent ancient continents are called (a) cratons (b) fossils (c) rifts (d) geosynclines.

2. Precambrian rocks that are exposed at the earth's surface are called (a) bedrock (b) vents (c) shields (d) geosynclines.

3. During the Paleozoic Era the continents were joined together in a huge landmass called (a) Panthalassa (b) Laurasia (c) Pangaea (d) Gondwana.

4. North America and Eurasia were once part of a landmass called (a) Panthalassa (b) Laurasia (c) Tethys (d) Gondwana.

5. South America, Africa, India, Australia, and Antarctica were once part of a landmass called (a) Panthalassa (b) Laurasia (c) Tethys (d) Gondwana.

6. Pangaea began to break up about (a) 600 million years ago (b) 400 million years ago (c) 200 million years ago (d) 100 million years ago.

7. The South Atlantic Ocean formed from a rift that opened about (a) 600 million years ago (b) 400 million years ago (c) 200 million years ago (d) 100 million years ago.

8. The area of exposed Precambrian rocks found in North America is called the (a) New England Shield (b) Canadian Shield (c) American Shield (d) Appalachian Shield.

9. During the Paleozoic Era, much of what is now the United States was covered by (a) a shallow sea (b) a desert (c) an ice cap (d) grasslands.

10. During the Mesozoic Era, new crust was added to the east and west coasts of North America in the form of (a) shields (b) terranes (c) fossils (d) geosynclines.

11. The Rocky Mountains began to form during (a) Precambrian time (b) the Paleozoic Era (c) the Mesozoic Era (d) the Cenozoic Era.

12. North America had taken on its present shape by the beginning of (a) Precambrian time (b) the Paleozoic Era (c) the Mesozoic Era (d) the Cenozoic Era.
13. The Colorado Plateau was uplifted during (a) Precambrian time (b) the Paleozoic Era (c) the Mesozoic Era (d) the Cenozoic Era.
14. The absence of certain layers in the rock record of the Grand Canyon may be the result of (a) flooding (b) cross-bedding (c) erosion (d) deposition.
15. The cross-bedded sandstone layers in the Grand Canyon indicate that the area may once have been covered by a (a) swamp (b) desert (c) glacier (d) forest.
16. The layers of limestone and shale in the Grand Canyon indicate that the area was once covered by a (a) shallow sea (b) desert (c) glacier (d) forest.
17. Fossils of marine organisms are often found in limestone and (a) sandstone (b) granite (c) lava (d) shale.
18. As a source of information on the history of the earth's surface, rocks are studied for their (a) position, color, and hardness (b) position, composition, and fossil content (c) composition, hardness, and color (d) fossil content, color, and hardness.

Application

On your paper, write answers to the following questions.

1. One hundred and fifty million years from now, the continents will all have drifted to new locations. How might these changes affect life on the earth?
2. Assume scientists know the rate at which North America and Europe are drifting farther apart on their respective continental plates. How might they determine when North America separated from Europe during the breakup of Pangaea?
3. If scientists discovered huge salt deposits in central Canada, what might they assume about the geologic history of that region?
4. If you found fossils embedded in glacial debris in Kansas, during which era might you assume the fossils were deposited? Why?
5. Geologists have found no deposits from the Pennsylvanian Period in the Grand Canyon. What might they assume about that period of the history of the canyon?
6. Suppose that during a certain period of time, the Grand Canyon had been covered by a sea. What kind of fossils would you expect to find in the sedimentary rock layers deposited during that period?

Extension

1. Write a report indicating how the breakup of Pangaea might have affected the distribution of animal species. Share your report with the class.
2. Shields provide the nucleus around which most continents have formed. Do research to find the name of each known shield and the continent on which it is located. Also find how the term *continental nucleus* applies to shields.
3. Trace a map of the continent of North America. Use different colors to shade in the terranes that have become part of the continent.
4. Describe in as much detail as possible what North America would look like if the Rocky Mountains had not been uplifted.
5. The Grand Canyon in Arizona and Zion and Bryce canyons in Utah are all located on the Colorado Plateau. However, the rocks of the Grand Canyon are the oldest. The rocks of Zion Canyon are next oldest, and those of Bryce Canyon are the youngest. Do research in the library to find out what geologic periods are represented by the rocks in each canyon. Draw a time line showing the periods during which the rocks in each canyon were formed.
6. Paleontologists have found traces of *Homo sapiens* in the Grand Canyon dating to the Pleistocene Epoch. Do research and write a report about the earliest human inhabitants of the Grand Canyon.

UNIT

6
Oceans

Monterey, California (background); Coral reef, Red Sea (inset)

Introducing Unit Six

When I was young, I had the opportunity to sail up the coast of Maine on a yacht. Between my general interest in science and the romantic aspects of sailing the sea, that one-week trip was all it took to draw me to the fields of oceanography and marine geology.

Now, for the first time, we have the remote sensing tools necessary to see and map the details of things I could only wonder about back then—submarine canyons that look like the Badlands of South Dakota and channels that resemble the Mississippi River. These and other new perspectives have helped us realize how dynamic our earth really is.

The unit you are about to study discusses some of the important strides already made in understanding the geology of our oceans—an underwater world of discovery covering three quarters of the earth's surface.

Bonnie A. McGregor

Bonnie McGregor
Marine Geologist
U.S. Geological Survey

20 The Ocean Basins

21 Ocean Water

22 Movements of the Ocean

Chapter 20
The Ocean Basins

Throughout much of recorded history, people regarded the ocean as a place of mystery and danger. They thought that fantastic monsters preyed on sailors and that the sea dropped off into space.

Exploration of the oceans has replaced these fears with scientific curiosity and understanding. The oceans, which cover nearly three quarters of the earth's surface, are no longer unfamiliar territory. In this chapter, you will study the features of the earth's that lie beneath the ocean.

Chapter Outline

This crinoid feather is found near the British West Indies.

20.1 The Water Planet

Section Objectives

- **Name the major divisions of the world ocean.**
- **Describe the goals of oceanography.**

Nearly three quarters of the earth's surface, about 362 million km², lies submerged beneath a body of salt water often called the world ocean. No other known planet has a covering of liquid water. Earth alone can be called the *water planet*.

The world ocean contains more than 97 percent of all the water on earth. Although the ocean is the most prominent feature of the earth's surface, it constitutes only a small part of the earth's total mass and volume. The mass of the earth's ocean is only about 1/4,000 of the mass of the earth as a whole. The volume of the solid earth is about 800 times greater than the volume of water in the world ocean.

Divisions of the Ocean

The world ocean is divided into three principal oceans: the Atlantic, Pacific, and Indian oceans. For convenience, geographers also refer to the Antarctic and Arctic oceans. However, the Antarctic Ocean is actually part of the Pacific Ocean, while the Arctic Ocean is an extension of the Atlantic Ocean. The term *sea* is applied to small areas of the ocean that are partially surrounded by land.

The division of the world ocean into specific oceans and seas is useful because each of these parts has special characteristics. For

Figure 20–1. The world ocean is divided into specific oceans and seas.

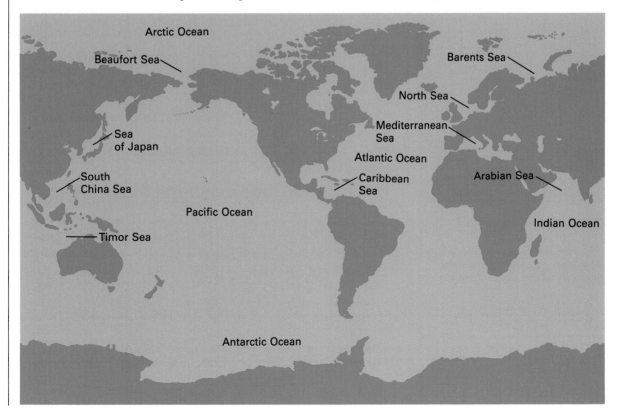

example, in the Arctic and Antarctic oceans, the water near the surface is less salty than that in the other oceans.

The Pacific Ocean occupies the largest area within the world ocean. The Pacific contains more than one half of the ocean water on earth. With an average depth of 3.9 km, the Pacific is also the deepest ocean. Next largest in size is the Atlantic Ocean. The Mediterranean, Caribbean, and Baltic seas, as well as the Gulf of Mexico, are all considered parts of the Atlantic Ocean. The Atlantic Ocean is the shallowest of the oceans, with an average depth of 3.3 km. The Indian Ocean is the third largest ocean and has an average depth of 3.8 km.

Exploration of the Ocean

The branch of earth science known as *oceanography* is the study of the physical characteristics, chemical composition, and life forms of the ocean. Oceanography got its start in the 1850's. An American naval officer, Mathew F. Maury, used records from navy ships to derive information about ocean currents, winds, depths, and weather conditions. These observations, which he then published, marked the beginning of the scientific study of the ocean.

In the late nineteenth century, the voyages of the British navy ship H.M.S. *Challenger* laid the foundation for the modern science of oceanography. From 1873 to 1876, a team of scientists aboard the *Challenger* crossed the Atlantic, Antarctic, and South Pacific oceans. The team measured water temperatures at great depths and collected samples of ocean water, sediments, and forms of marine life never before seen.

Today, many ships are equipped to perform oceanographic research. For example, the research ship *Glomar Challenger* carries equipment that can drill into rock located as deep as 4 km beneath the ocean surface. The samples drilled by the *Glomar Challenger* have provided scientists with valuable data about the **ocean basin.** The ocean basin is made up of the continental crust and the oceanic crust that lie beneath the ocean waters.

Submersibles

Underwater research vessels, called **submersibles,** also enable oceanographers to study the ocean depths. Some submersibles are piloted by people. The more advanced submersibles, however, are actually submarine robots similar to the *Jason Jr.,* shown in Figure 20–2. They can perform tasks ranging from photographing the ocean depths to mining the ocean floor.

Several types of submersibles are used for underwater research and exploration. One type of submersible is a **bathysphere.** It is a spherical diving vessel that was used in early ocean exploration. The bathysphere is carried to an area of the ocean where it is lowered slowly into the water on steel cables. The bathysphere, which carries passengers, remains connected to the research ship for communications and life support. Therefore, its movement and the tasks it can perform while under water are limited.

Figure 20–2. Robot submersibles like *Jason Jr.* send back detailed pictures of the ocean floor.

Figure 20–3. The bathyscaph *Alvin* has been used by oceanographers in deep-ocean research.

Another type of submersible, called a **bathyscaph,** is a self-propelled, free-moving submarine equipped for deep-ocean research. One well-known bathyscaph is *Alvin,* a two-passenger craft that has helped oceanographers make some exciting discoveries about the ocean. The *Alvin,* shown in Figure 20–3, has made more than 1,600 dives. During one of these dives, startled oceanographers found communities of unusual marine life. These organisms were living at depths and temperatures where it was thought no life could exist. Giant clams, blind white crabs, and long worms that live in tubes were just three of the strange life forms discovered. They were found near the volcanic vents along mid-ocean ridges. These animals adapted to life in a hostile environment.

Robot submersibles also are making important discoveries about the ocean. These robot craft enable oceanographers to study the ocean at great depths and for long periods of time. In 1985 oceanographer Robert Ballard used a seeing-eye robot vessel called *Argo* to locate the remains of the luxury liner *Titanic.* In 1912, just five days into its maiden voyage, the *Titanic* had sunk in deep water off the coast of Newfoundland after striking an iceberg. The *Jason Jr.,* highly maneuverable and equipped with very sensitive cameras, toured the inside of the sunken ship. *Jason Jr.* sent back pictures of barnacle-encrusted crystal chandeliers and delicate china cups with the *Titanic* crest still visible. Someday a similar submersible may tour the interior of an undersea volcano or explore the depths of a submarine canyon. Why are robot craft more practical for deep-ocean research than are craft designed to carry people?

Sonar

Most oceanographic research ships and submersibles are equipped with **sonar** to aid in mapping of the ocean basin, as shown in Figure 20–4. *Sonar* is an acronym for *sound navigation and ranging.* A sonar system consists of a transmitter and a receiver. The transmitter sends out a continuous series of sound waves from a ship to the ocean floor. The sound waves, traveling at the speed of about 1,500 m/sec, bounce off the ocean floor and are reflected back up to the

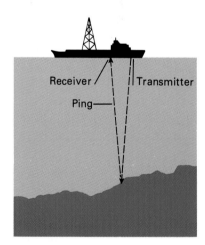

Figure 20–4. Active sonar sends out a pulse of sound. The pulse, called a *ping* because of the way it sounds, is reflected back when it strikes an object.

Investigation

Sonar

You can demonstrate the principle of sonar by calculating the rate at which pulses travel along a spring to determine distance.

Science Process Skills Focus: observing, interpreting data, predicting outcomes

Materials
toy spring coil or flexible spring, heavy string, meter stick, stopwatch, tape

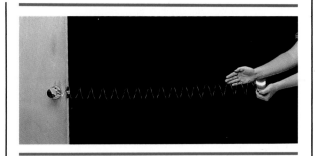

Procedure
1. Tie one end of the spring securely to a door knob, using heavy string. Pull the free end of the spring taut and parallel to the floor. Keep the tension of the spring constant throughout the investigation.
2. Use tape to mark the floor directly under the hand holding the spring. Measure and record the distance from your hand to the doorknob.
3. Note the time. Hold the spring taut and hit the spring sharply with the side of your free hand. Strike the spring as close to your hand holding the spring as possible.
4. Check the time again to see how long it takes for the pulse to travel to the doorknob and back to your hand.
5. Repeat Steps 2 through 4 three times. Each time hold the spring 60 cm closer to the doorknob than in the previous trial.
6. Calculate the rate of travel for each trial. To calculate, multiply the length of the spring between your hand and the doorknob by 2. Then divide by the number of seconds it took for the pulse to travel to the doorknob and back.

Analysis and Conclusions
1. Did the rate of travel of the pulse change during the course of the investigation?
2. If a pulse took 3 seconds to travel to the doorknob and back to your hand, what is the distance from the doorknob to your hand?
3. How is the apparatus you used like sonar? How is it different? Explain your answer.

receiver. Scientists measure the amount of time it takes for the sound waves to complete this round trip. In this way they can determine the depth of any area of the ocean basin. With this information scientists can make a profile of the ocean basin. One type of sound wave can penetrate even the sediment layers and shallow rock structures of the ocean basin. Its returning signal indicates the thickness of the sediments and the structure of the underlying rocks.

Section 20.1 Review
1. What are the major divisions of the world ocean?
2. What is oceanography?
3. Most submarines use sonar as an aid to navigation. How would sonar enable an underwater vessel to steer its way through the ocean depths?

20.2 Features of the Ocean Basins

Section Objectives

- **Describe the main features of the continental margins.**
- **Describe the main features of the ocean floor.**

Exploration of the ocean depths at many locations has revealed the shape and makeup of the ocean basin. The ocean basin can be divided into three major types of areas, as illustrated in Figure 20–5. The **continental margins** are those portions of the ocean basin made up of continental crust. The *ocean floor* is the portion made up of oceanic crust. Mid-ocean ridges are a worldwide system of interconnected underwater mountains.

Continental Margins

The line that divides the continental crust from the oceanic crust is not always obvious. Shorelines are not the true boundaries between the ocean floor and the continents. The real boundary lies some distance offshore beneath the ocean itself.

Continental Shelf

Every continent is bounded in most places by a zone of shallow water, where the ocean covers the edges of the continents. This part of the continent that is covered by the ocean is called the **continental shelf.** The shelf usually slopes gradually away from the shoreline, dropping only about 1.2 m every 100 m. The water above the gently sloping continental shelves is shallow. The average depth of the water covering the continental shelves is about 60 m. Although it is under water, a continental shelf is part of the continental margin rather than part of the ocean floor.

The width of the continental shelf varies. On the west coast of South America, the continental shelf is only a few kilometers wide. On the west coast of Florida, however, the shelf extends 760 km into the Gulf of Mexico. The widest continental shelves are those that extend out about 1,280 km from Siberia and Alaska into the Bering Sea. The continental shelf along the east coast of the United States has an average width of 170 km.

Figure 20–5. The ocean basin includes the continental margins, the ocean floor, and the mid-ocean ridge. Also notice the trench at the base of the continental margin.

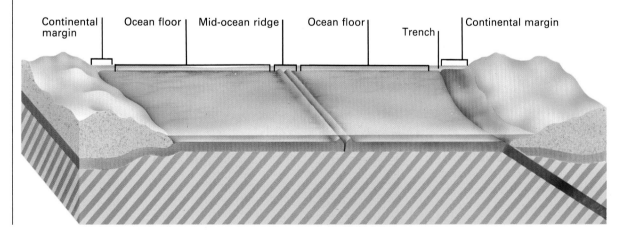

Continental margin | Ocean floor | Mid-ocean ridge | Ocean floor | Trench | Continental margin

The continental shelves were formed by changes in sea level. During the glacial periods, when glaciers held great amounts of water, the sea level fell. As the glaciers melted and the water was added to the ocean, the sea level rose.

When sea level was lower than it is today, parts of the continental shelf were exposed to weathering and erosion. Consequently, its surface is flat. In general, the continental shelf is a smoother version of the land surface at the shoreline.

Continental Slope

At the edge of a continental shelf is a steeper slope called a **continental slope.** The boundary between the continental crust and the oceanic crust is found at the base of the continental slope. Along the continental slope, the ocean depth increases to several thousand meters within a distance of a few kilometers.

IMPACT

Fishing on the Continental Shelves

Continental shelves provide some of the richest fishing grounds in the world. One of these is the Grand Banks, a stretch of shallow water 15 m to 300 m deep, lying southeast of the Canadian province of Newfoundland.

Fish and shellfish are plentiful in the waters of this broad continental shelf because of the ample food supply. The Gulf Stream and the Labrador Current meet at the Grand Banks. These two currents agitate the water, causing microscopic sea organisms to float freely. These organisms provide food for large numbers of small fish, which, in turn, provide food for larger fish.

The Grand Banks has long been a popular fishing area. As early as 1497, explorer John Cabot noted its excellent fishing. Today fleets from the United States, Canada, Japan, and Europe catch billions of kilograms of fish in the Grand Banks each year.

Fishing is crucial to the economy of the Atlantic provinces of Canada. Many people fish as a part-time occupation to help feed their families. Some farmers fertilize their land with calcium-rich crushed lobster shells. Grand Banks fishing has also helped develop industry in these provinces. Both Nova Scotia and Newfoundland have a number of seafood-processing plants.

Many nations located on seacoasts claim that their sovereign territories extend into the ocean—anywhere from 3 km to 300 km.

What is one possible reason for this practice?

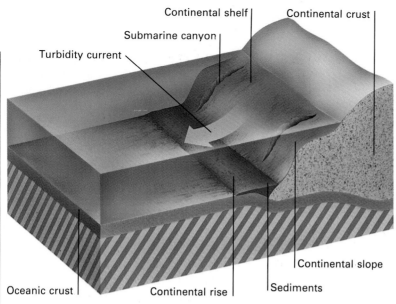

Continental shelf | Continental crust
Submarine canyon
Turbidity current
Continental slope
Sediments
Oceanic crust | Continental rise

Figure 20–6. A continental margin is made up of distinct areas and features. Note the turbidity current on the continental slope. Turbidity currents carry tons of sediment from the continental shelves to the ocean floor.

The continental shelf and continental slope may be cut by deep V-shaped valleys. These deep valleys in the continental slope are called **submarine canyons.** The origin of these deep canyons is not completely understood. However, they may have been caused by **turbidity currents.** These are currents that are very dense and carry large amounts of sediment down the continental slopes. Turbidity currents form when landslides of various materials run down a slope. Most likely, these landslides are triggered by earthquakes or by the force of gravity. The speed and composition of the landslides make them powerful agents of erosion—powerful enough, some scientists hypothesize, to cut canyons in the continental slopes.

Continental Rise and Trenches

Along the base of a continental slope may be sediments that have moved down the slope. This raised wedge of sediments at the base of the slope forms a feature called a **continental rise.**

In some places the continental slope plunges downward into a deep valley in the ocean floor. This deep valley is called a **trench.** A trench along a continental margin results when oceanic crust subducts beneath continental crust. Continental margins with trenches at their edges generally have narrow continental shelves, a very steep continental slope, and no continental rise. The depth of a trench is usually greater than 6,000 m. Most trenches are found in the Pacific Ocean. In addition to areas where oceanic crust is subducting beneath continental crust, where else would you expect to find trenches?

Ocean Floor

Like the earth's surface, the ocean floor, too, has its own distinctive features. These include broad, flat plains, submerged volcanic mountains, and gigantic volcanic mountain ranges. However, on the ocean floor the mountains are higher and the plains flatter than are those on any of the continents.

Abyssal Plains

The part of the ocean floor that begins at the edge of a continental margin is the **abyssal** (uh-BISS-uhl) **plain.** Abyssal plains, which cover about half of the deep-ocean floor, are extremely level areas. In fact, they are the flattest regions on earth. In some places, the elevation changes less than 3 m over more than 1,300 km.

Sonar reveals that the abyssal plains are made up of thick layers of sediments deposited on the ocean floor. Currents probably carry the sediments from the continental margins. The trenches found along the edges of some continents trap the sediments being carried

Career Focus: Oceans

"I've had a number of unusual experiences as an underwater photographer. Swimming next to 4,100-kg whales was just one of them."

Underwater Photographer

Chris Newbert's travels as an underwater photographer have taken him to such exotic places as the Coral Sea and the Palauan reefs in the South Pacific. A sport diver, Newbert has gone on to explore and photograph the great oceans of the world.

A self-taught photographer as well, Newbert, pictured left, has won numerous awards for his spectacular photos of marine life. To obtain these photos, Newbert has developed innovative techniques, including one that he calls "free-drift diving."

"I go far out to sea at night, dive down 30 to 60 meters, and just drift, photographing animals that migrate vertically," he said. When finished taking photos, he relies principally on a generator-powered light on his boat to guide him back to the surface.

In addition to having his photos published in books and magazines, Newbert has participated in photography exhibits throughout the United States. In 1984 he completed work as the photographer and author of his first book of underwater photography, *Within a Rainbowed Sea.* Newbert also conducts diving tours in many places around the world.

Not surprisingly, Newbert often encounters underwater companions. But he takes that in stride.

"Sharks, for example, will come after you from time to time. For the most part, though, I don't get too frightened."

down the continental slopes. Thus, abyssal plains are most widespread in oceans, such as the Atlantic, where there are no trenches along the continental margins. The Pacific Ocean, which has many trenches, has a less-extensive abyssal plain.

Seamounts

Scattered along the ocean floor are isolated volcanic mountains called **seamounts.** Most of these are at least 1,000 m high. Hundreds of seamounts dot the Pacific Ocean floor; they are much less common in the other oceans. Seamounts often rise above the surface

Oceanographer

The field of oceanography includes four main areas of study. Marine geologists study undersea rock formations and the characteristics of the ocean floor. Marine chemists investigate the chemical composition of seawater and of the materials that make up the ocean floor. Physical oceanographers chart the movements of waves, currents, and tides. Marine biologists study the animal and plant life of the ocean.

Oceanographers doing research often spend months at a time at sea collecting data. They may use specialized equipment such as submersibles and underwater cameras.

A bachelor's degree in oceanography is a minimum requirement for entry into this field.

Marine Surveyor

Marine surveyors gather data that can be used in mapping the ocean floor and shoreline. They also calculate water depths at specific locations.

Marine surveyors must be able to use a variety of specialized equipment. Among this equipment are such things as wire drags and sonic depth sounders to determine underwater elevations. The surveyors' findings are important for navigation and for the construction of dams, reservoirs, docks, and bridges.

Marine surveyors, shown in the photograph lower left, do much of their field work at sea. The data

are then analyzed and charted in the surveyors' office.

Most marine surveyors begin their career preparation with a bachelor's degree in engineering.

For Further Information

For more information on careers in marine science, write the National Oceanic and Atmospheric Administration, U.S. Department of Commerce, 11400 Rockville Pike, Rockville, MD 20852.

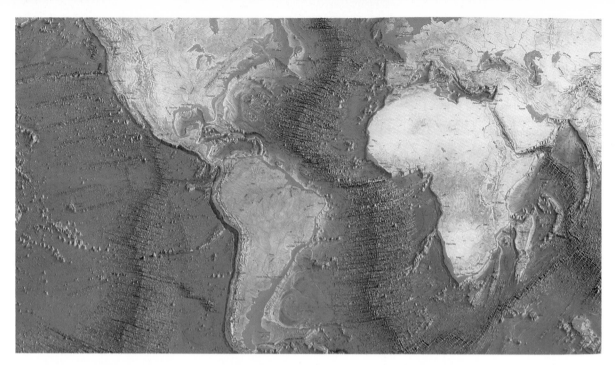

Figure 20–7. The Mid-Atlantic Ridge is clearly visible in this topographic map.

of the ocean, forming islands. The tops of some of these volcanic islands are flattened by waves. As the action of plate tectonics carries seamounts away from a mid-ocean ridge, the ocean crust sinks. As it sinks, it pulls the seamount beneath the ocean surface. These flat-topped, submerged seamounts are called **guyots** (GEE-oze).

Mid-Ocean Ridges

The most prominent features of the ocean floor are the *mid-ocean ridges,* shown in the satellite photo in Figure 20–7. They are a continuous series of underwater mountain ranges that run along the floors of all the oceans. According to the theory of plate tectonics, magma reaches the surface and builds up along a mid-ocean ridge, causing the ocean floor to be elevated. Mid-ocean ridges have a narrow depression, or *rift,* running along the center. This rift is evidence that the plates have moved away on each side of the ridge.

Each side of a mid-ocean ridge is covered with jagged peaks and valleys that form when magma pushes up through the oceanic plate on both sides of the rift. The new crust breaks into a series of tilted blocks separated by faults, called *transform faults,* that run perpendicular to the ridge. Older tilted blocks, which have moved farther from the ridge, are separated from the newly formed blocks by faults that run parallel to the rift.

Section 20.2 Review

1. Name the major features of a continental margin.
2. Why does the Pacific Ocean have less-extensive abyssal plains than does the Atlantic?
3. If sea level were to fall significantly, what would happen to the continental shelves?

20.3 Ocean-Basin Sediments

Section Objectives

- **Describe the formation of ocean-floor sediments.**

- **Explain how ocean-floor sediments are classified according to physical composition.**

Ocean-basin sediments differ greatly from one part of the ocean to another. The continental shelves and slopes are covered mainly with rock fragments. Most of these fragments have been carried into the ocean by rivers or have been washed away from the shoreline by wave erosion. The sediments are fairly well sorted according to size. Coarse gravel and sand are usually found close to the shore because these heavier rock fragments are not easily moved offshore by waves and currents. Much finer particles that are carried as suspended particles in ocean water are usually deposited on the ocean floor at a greater distance from shore.

The bottom sediments in the deep ocean, beyond the continental slopes, are different from those found in shallower water. Samples of the sediments on the deep-ocean floor are obtained by taking **core samples** or by scooping up sediments. Core samples are taken by driving long tubes into sediment layers. A core sample taken from the ocean floor is shown in Figure 20–8. What might scientists infer from a core sample containing a layer of sediment made up primarily of volcanic ash and dust?

Sources of Ocean-Floor Sediments

The study of sediment samples shows that most of the sediments on the deep-ocean floor are made up of materials that settle from ocean water. These materials may come from organic or inorganic sources.

Inorganic Sediments

Some of the ocean-floor sediments consist of rock particles that have been carried from the land to the oceans by rivers. When a river flows into the ocean, the river deposits its stream load as sediments. Most of these sediments are deposited along the shore and on the continental shelf. However, great quantities of these sediments occasionally slide down continental slopes to the ocean floor. A single landslide may move billions of kilograms of material down a slope. The tremendous force of the landslide creates powerful turbidity currents that spread the sediments over the ocean floor.

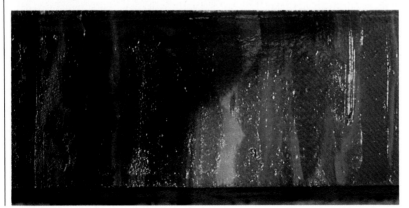

Figure 20–8. Notice the layers of sediment in this core sample drilled from the floor of the Red Sea. Core samples reveal important clues about the geologic history of the earth.

Some ocean-floor sediments consist of fine particles of rock, including volcanic dust, that have been blown great distances out to sea by the wind. These particles land on the surface of the water and eventually settle to the bottom. Volcanism on the ocean floor also produces layers of lava.

Icebergs provide sediments by carrying material from the land into the ocean. An iceberg is a large block of ice that has broken off from a glacier and floated out into the ocean. As a glacier moves slowly across the land, it picks up great quantities of rock. The rock becomes embedded in the glacier and moves with it. When the iceberg breaks off the glacier and melts, the rock material sinks to the ocean floor

Even meteorites contribute to ocean-floor sediments. Much of a meteorite burns up as it enters the earth's atmosphere. What remains falls to the earth's surface. Because so much of the earth's surface is ocean, most meteorite fragments fall into the ocean and become part of the sediments on the ocean floor.

Organic Sediments

The ocean floor is also covered with organic sediments. These are the remains of plants and animals. In many places on the ocean floor, almost all the sediments are organic.

The two most common substances in organic sediments are silica (SiO_2) and calcium carbonate ($CaCO_3$). Silica comes primarily from microscopic organisms, such as diatoms and radiolaria. Calcium carbonate comes mostly from the skeletons of tiny organisms called *foraminifera*. Animals, such as corals and clams, can also add calcium carbonate to ocean-floor sediments.

Chemical Deposits

Chemical deposits make up part of the ocean-floor sediments. Many chemical reactions take place in the ocean. During some of these reactions, solid materials are formed. These materials settle to the

Figure 20–9. The remains of diatoms (left) and radiolaria (right), both magnified greatly, are important components of organic sediments on the ocean floor.

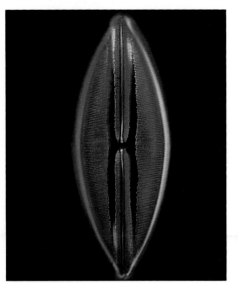
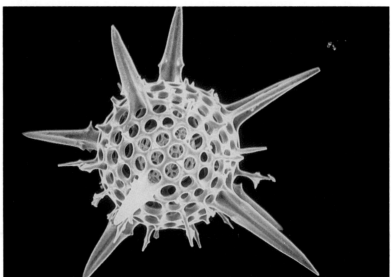

Figure 20–10. Undersea mining operations allow industry to tap ocean-floor mineral resources. Nodules, such as these mined from the East Pacific Rise are rich in minerals important to industry.

bottom. For example, potato-shaped lumps of minerals, or **nodules,** cover certain regions of the ocean floor. Nodules are composed mainly of oxides of manganese, nickel, and iron. They seem to have been formed by chemical changes affecting substances dissolved in ocean water. Other minerals found in ocean-floor sediments were formed directly in the ocean water. However, these minerals are not abundant, and how they were formed is not completely known.

Physical Classification of Sediments

Two general types of sediments are found on the ocean floor. **Muds** are very fine particles of rock that have settled to the ocean floor. One common type of mud is **red clay.** Red clay is at least 40 percent clay particles, mixed with silt, sand, and organic material. Actually, red clay can also be yellow-brown, blue, green, or gray.

About 40 percent of the ocean floor is covered with a soft, organic sediment called **ooze.** Most ooze looks and feels like sand. However, at least 30 percent of the ooze is made up of organic materials, such as microscopic particles of shells and the remains of diatoms, radiolaria, and foraminifera.

There are two types of ooze. One type, **calcareous ooze,** is mostly calcium carbonate. At depths below 4,500 m, calcium carbonate dissolves in the ocean water. Therefore, calcareous ooze is never found below that depth. The second type of ooze, **siliceous ooze,** is mostly silicon dioxide, which comes from radiolaria and diatoms. Most siliceous ooze is found around the continent of Antarctica because of the abundance of these two marine organisms in that location.

Section 20.3 Review

1. How do icebergs contribute to ocean-basin sediments?
2. Describe the composition of ooze.
3. There have been several recent attempts to develop an economical method of mining nodules from the ocean floor. Why is there such interest in ocean-floor nodules?

Reviewing Chapter 20

Key Concepts

- The world ocean can be divided into individual oceans and seas. See page 331.
- Oceanography is the study of the oceans and seas. See page 332.
- Features of the ocean floor include abyssal plains, seamounts, and mid-ocean ridges. See page 335.
- Ocean-floor sediments form from inorganic and organic materials as well as chemical deposits. See page 341.
- Based on physical characteristics, ocean-floor sediments are classified as mud or ooze. See page 343.

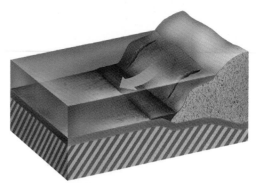

- **The continental margins include the continental shelf, the continental slope, and the continental rise. See page 337.**

Key Terms

abyssal plain (338)	continental rise (337)	nodule (343)	sonar (333)
bathyscaph (333)	continental shelf (335)	ocean basin (332)	submarine canyon (337)
bathysphere (332)	continental slope (336)	ooze (343)	
calcareous ooze (343)	core sample (341)	red clay (343)	submersible (332)
continental margin (335)	guyot (340)	seamount (339)	trench (337)
	mud (343)	siliceous ooze (343)	turbidity current (337)

Review

On your paper, write the letter of the term that best completes each of the following statements.

1. The largest of the oceans of the earth is the (a) Atlantic (b) Indian (c) Pacific (d) Antarctic.
2. The continental and oceanic crust that lies beneath the ocean waters makes up the (a) ocean basin (b) continental margin (c) continental slope (d) abyssal plain.
3. A self-propelled, free-moving submarine that is equipped for ocean research is a (a) turbidity (b) bathyscaph (c) bathysphere (d) guyot.
4. A system that is used for determining the depth of the ocean basins is (a) guyot (b) bathysphere (c) radiolaria (d) sonar.
5. Those portions of the ocean basin made up of continental crust are called (a) continental margins (b) abyssal plains (c) mid-ocean ridges (d) trenches.
6. The continental shelves were formed by (a) abyssal plains (b) turbidity currents (c) changes in sea level (d) the accumulation of sediments.
7. Deep valleys that sometimes occur in the continental slope are called (a) trenches (b) submarine canyons (c) abyssal plains (d) continental rises.
8. An accumulation of sediments at the base of a continental slope is called a (a) trench

(b) turbidity current (c) continental margin (d) continental rise.
9. Extensive abyssal plains rarely form in those parts of the ocean basin that have (a) trenches (b) continental margins (c) guyots (d) submarine canyons.
10. Volcanic mountains scattered along the ocean floor are (a) diatoms (b) seamounts (c) mid-ocean ridges (d) foraminifera.
11. Submerged flat-topped seamounts are (a) guyots (b) diatoms (c) submersibles (d) bathyspheres.
12. Large quantities of the inorganic sediment that makes up the ocean floor come from (a) landslides (b) earthquakes (c) diatoms (d) nodules.
13. Much of the silica on the ocean floor comes from (a) nodules (b) radiolaria and diatoms (c) guyots (d) foraminifera.
14. Potato-shaped lumps of minerals on the ocean floor are called (a) guyots (b) foraminifera (c) nodules (d) diatoms.
15. Very fine particles of rock that have settled to the ocean floor are called (a) muds (b) guyots (c) seamounts (d) nodules.

Application

On your paper, write answers to the following questions.

1. Suppose you are studying photographs of a newly discovered planet that has several large bodies of water. The three largest bodies of water—A, B, and C—are connected and occupy 70 percent of the surface of the planet. Two other bodies of water—D and E—are not quite so large and are partly surrounded by land. A sixth body of water—F—is completely surrounded by land. Which of these bodies of water resemble the earth's oceans? Which resemble seas? What type of body of water on earth does F most closely resemble?
2. The exploration of the ocean depths has been compared with the exploration of outer space. What similarities exist between these two environments and the attempts by people to explore them?
3. Suppose you were searching off the east coast of the United States for the wreckage of an old Spanish galleon. Explain how sonar could aid your search.
4. How might the continental shelves be affected if another ice age were to occur?
5. What might be the eventual fate of seamounts as they are carried along on the spreading oceanic crust?
6. A certain fish is known to exist only in one particular river in the central United States. Explain how the fossilized remains of this fish might become part of the sediments on the ocean floor.
7. Lava erupts from a volcano in the Cascade mountain range along the west coast of the United States. One component of the lava is minerals from nodules on the floor of the Pacific Ocean. Explain how these minerals could have become part of the lava.
8. Explain how it is possible that some red clays on the ocean floor have been found to contain material from outer space.

Extension

1. Prepare a brief report on the different types of submersibles. Your report should explain the special features of each type of submersible as well as how each has contributed to oceanographers' knowledge of the oceans.
2. Using modeling clay, prepare a model of a continental margin. Be sure that your model includes these features: a continental shelf, a continental slope, and a continental rise. Add a submarine canyon and a trench in appropriate places.
3. Create an imaginary walking tour of the ocean basins. Your tour should begin at the edge of a continent—perhaps at a beach on the east coast of Florida. Explain exactly what tourists should look for along the continental margin and the ocean floor on their way to the western coast of Africa.

Chapter 21
Ocean Water

Much of the earth's surface water exists in a liquid state. If the earth were much colder or hotter, the water would become a solid or a gas. The unique properties of liquid water make the oceans one of the most distinctive features of the earth. No other planet is known to have liquid water on its surface.

Chapter Outline

21.1 Properties of Ocean Water
Composition of Ocean Water
Salinity of Ocean Water
Temperature of Ocean Water
Density of Ocean Water
Color of Ocean Water

21.2 Life in the Oceans
Ocean Chemistry and Marine Life
Sunlight and Marine Life
Ocean-Life Zones

21.3 Ocean Resources
Fresh Water from the Ocean
Minerals from the Ocean
Food from the Ocean
Ocean-Water Pollution

A variety of ocean life flourishes in the Red Sea.

21.1 Properties of Ocean Water

Section Objectives
• **Describe the chemical properties of ocean water.**
• **Describe the physical properties of ocean water.**

Pure liquid water is tasteless, odorless, and colorless. Pure water is the basic substance in which solids and gases are dissolved, forming the solution called *ocean water,* or *sea water.* Besides the dissolved substances, small particles of matter and tiny organisms may be suspended in ocean water. Ocean water is a complex mixture that can sustain a variety of plant and animal life.

Ocean water has both chemical and physical properties. The chemical properties are those characteristics that determine the composition and enable water to dissolve other substances. The physical properties are those characteristics, such as temperature, density, and color, that can be used to describe ocean water. Scientists study all these properties of ocean water in trying to understand the complex interactions among the oceans, the atmosphere, and the land.

Composition of Ocean Water

Each year the earth's rivers carry about 400 billion kg of dissolved minerals into the ocean. Most of these minerals are salts. As water evaporates from the ocean, the dissolved minerals and salts remain behind. Thus, minerals and salts are not returned to the land in the water that falls as rain and snow during the water cycle.

Gases also enter ocean water dissolved in rivers and streams or directly from the atmosphere. Some gases in ocean water come from volcanoes on land and beneath the ocean. Gases from the earth's early atmosphere may also have dissolved in the ocean.

The relative amounts of the dissolved substances in ocean water have remained almost constant over millions of years. The processes that add new substances to the oceans seem to balance the processes that remove dissolved materials. For example, the deposition of sediments on the ocean floor removes billions of kilograms of material from the ocean each year. Marine plants and animals also use up many of these substances during their life processes.

Elements in Ocean Water

Ocean water is 96.5 percent pure water, or H_2O. Minerals and salts dissolved in ocean water contain about 75 of the known chemical elements, mainly in ionic form. The most abundant elements are chlorine (Cl^-) and sodium (Na^+). Other common elements are magnesium (Mg^{+2}), calcium (Ca^{-2}), and potassium (K^+). Table 21–1 shows the 10 major elements in ocean water. There are also trace elements, those elements that exist in very small amounts, in ocean water. Among the trace elements are gold, zinc, and phosphorus.

Dissolved Gases

Three principal gases in the atmosphere are nitrogen (N_2), oxygen (O_2), and carbon dioxide (CO_2). These three gases are also the principal gases dissolved in ocean water. Of the three, carbon dioxide

Table 21–1. Elements in Ocean Water	
Element	**Percent of Ocean water by mass**
Chlorine (Cl)	1.94
Sodium (Na)	1.08
Magnesium (Mg)	0.13
Sulfur (S)	0.09
Calcium (Ca)	0.04
Potassium (K)	0.04
Bromine (Br)	0.006
Carbon (C)	0.003
Strontium (Sr)	0.001
Boron (B)	0.0005

dissolves most easily in ocean water. Other atmospheric gases, which cannot dissolve as easily, are present in the ocean in smaller amounts. Usually the dissolved gases are in molecular form, just as they are in the atmosphere. More carbon dioxide and oxygen are dissolved in the ocean than in the atmosphere.

Temperature affects the amount of gas that dissolves in water. Unlike solids, which dissolve better in warm water, gases dissolve better in cold water. In colder regions water at the surface of the ocean will usually dissolve larger amounts of gases than will water in warm tropical regions.

Dissolved gases can leave the ocean and return to the atmosphere. If, for example, the water temperature rises, less gas will remain dissolved, and the excess gas will be released into the atmosphere. Thus, the ocean and the atmosphere are continuously exchanging gases as conditions change.

Salinity of Ocean Water

Dissolved salts make up about 3.5 percent of the mass of ocean water. Sodium chloride (NaCl), or common table salt, makes up about 78 percent of the dissolved salts. The remaining portion is composed mainly of six other salts, as shown in Figure 21–1. The amount of dissolved salts present in a sample of ocean water is described as its **salinity.** Salinity is the number of grams of dissolved salt in 1 kg of ocean water. For example, suppose that 1 kg of ocean water were evaporated and 35 g of salts remained. The salinity of this sample would be about 35 parts salt per 1,000 parts ocean water. This is written as *salinity = 35 0/00*. Ocean water with a salinity of 35 0/00 has almost 3.5 percent of its total mass made up of dissolved salts. What percentage of dissolved salts would be present in ocean water with a salinity of 40 0/00?

The process of evaporation, in which water changes from a liquid into a gas, increases the salinity of ocean water. Only the water

Figure 21–1. This chart shows the percentages of salts in ocean water.

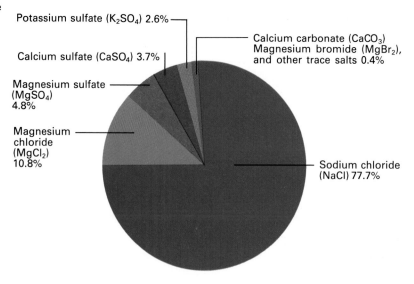

Potassium sulfate (K$_2$SO$_4$) 2.6%

Calcium sulfate (CaSO$_4$) 3.7%

Magnesium sulfate (MgSO$_4$) 4.8%

Magnesium chloride (MgCl$_2$) 10.8%

Calcium carbonate (CaCO$_3$) Magnesium bromide (MgBr$_2$), and other trace salts 0.4%

Sodium chloride (NaCl) 77.7%

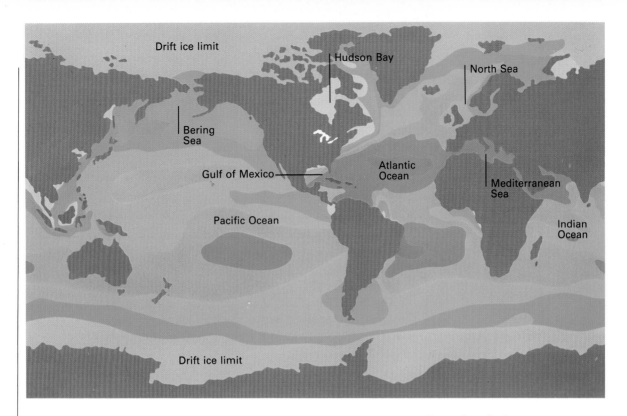

molecules are removed during evaporation. Dissolved salts and other solids remain in the ocean. When the rate of evaporation is high, the relative amount of dissolved solids in surface water increases. For this reason tropical waters have a higher concentration of dissolved solids at the surface than do polar waters. Salinity also varies with depth. Surface water has a higher salinity than does water at greater depths.

Over most of the surface of the ocean, salinity ranges between 33 0/00 and 36 0/00, with an average value of 34.7 0/00 for the world ocean. However, salinity at particular locations can vary greatly, as shown in Figure 21–2. The salinity of the Mediterranean Sea, for example, is about 40 0/00. The climate around the Mediterranean is hot and dry. The high temperature of the sea leads to high levels of evaporation. In contrast, the salinity of the Baltic Sea is only about 30 0/00 because of low temperatures and water draining from many rivers and runoff from glaciers. Because fresh water with low salt content is constantly being added to ocean water, the proportion of water to dissolved salts stays high and thus salinity remains low. Salinity also tends to be low in surface water near the equator because of heavy rainfall. In the Arctic Ocean, salinity is low because of the large amount of fresh water entering the ocean from numerous rivers and from melting ice.

Although the salinity of ocean water may differ from one location to another, the relative amounts of the dissolved solids in ocean water do not change significantly. This state of balance is brought about by the continuous mixing of all parts of the ocean. Any dissolved substance deposited in the ocean is eventually spread uniformly throughout the water.

Proportion of salt per 1000 parts of sea water

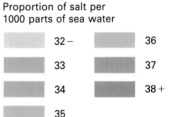

32 –		36	
33		37	
34		38 +	
35			

Figure 21–2. The average surface salinity of the world ocean varies from one location to another. Note areas of high and low salinity, for example the high salinity of the Mediterranean Sea and the low salinity of Hudson Bay.

Temperature of Ocean Water

Ocean water has the ability to absorb the long, invisible infrared wavelengths of sunlight. The absorbtion of these infrared rays heats the water. Thus infrared rays play an important role in determining the temperature of the ocean.

Infrared rays are completely absorbed within the upper regions of the ocean water. This means that the sun can directly heat only the upper region of the ocean. In the deeper regions of the ocean, the temperature of the water is usually about 2°C, which is just 4°C above the freezing point of salt water.

Surface Temperature

The movement of water in the ocean thoroughly mixes the warmed surface water. This mixing distributes heat downward to a depth of 100 m to 300 m. Thus, the temperature of this zone of surface water is relatively constant, decreasing only slightly with depth. However, the temperature of surface water does differ from one location to another.

The total amount of solar energy falling upon the surface of the ocean is much greater at the equator than at the poles. In tropical waters near the equator, average surface temperatures around 30°C are not unusual. Surface temperatures in polar oceans, however, usually drop below –2°C. Because ocean water freezes at about –2°C, vast areas of ice exist in both the Arctic and the Antarctic oceans. A floating layer of ice that completely covers the ocean surface is called **pack ice.** Usually the layer of ice is not more than 5 m thick, because the bulk of the ice insulates the water below and prevents it from freezing. The Arctic Ocean is covered by pack ice during most of the year.

In the middle latitudes, the surface temperature varies with the seasons. In some areas the surface temperature may vary by as much as 10°C to 20°C between summer and winter.

The Thermocline

Because the sun cannot directly heat ocean water below its uppermost regions, the temperature of the water drops sharply as the depth increases. In most places in the ocean, this sudden temperature drop begins not far below the surface. This zone of rapid temperature change is called a **thermocline.** The drop in temperature continues to a depth of 1,200 m.

The thermocline exists because the water near the surface becomes less dense as it is warmed by heat from the sun. This warm water cannot mix easily with the cold, dense water below. Thus, a thermocline marks the distinct separation between the warm surface water and the colder deep water. Below the thermocline the temperature of the water continues to drop, but very slowly. Changing temperature or shifting currents may alter the depth of the thermocline or cause it to disappear completely. Nevertheless, a thermocline is usually present beneath much of the ocean surface. Thermoclines that can definitely be measured are also observed in many lakes.

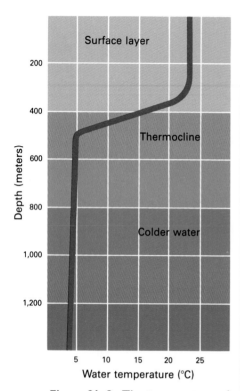

Figure 21–3. The temperature of ocean water decreases with increasing depth. Just below the surface is a thermocline, an area where the water temperature drops sharply.

Density of Ocean Water

Density is the mass of a substance per unit volume. For example, 1 cubic centimeter of pure water has a mass of 1 g. Its density is 1.00 g/cm^3. Two factors affect the density of ocean water: the amount of dissolved solids in the water and the temperature of the water. Dissolved solids—salts and minerals—add mass to the water in direct proportion to their amounts. The large amount of dissolved solids in ocean water makes it denser than pure fresh water. Ocean water has a density between 1.026 and 1.028 g/cm^3.

Ocean water also becomes more dense as it becomes colder and less dense as it becomes warmer. Temperature affects the density of ocean water more than the amount of dissolved solids in the water affect it. Thus, the densest ocean water is found in cold temperatures of the polar regions.

Investigation

Density

Temperature and salinity are two variables that influence the density of water. You can test these two variables using a simple straw float.

Science Process Skills Focus: hypothesizing, recording data, comparing and contrasting

Materials
glass jar, at least 1-L capacity; distilled water; ruler; plastic soda straw; modeling clay; scissors; table salt; freezer; thermometer

Procedure
1. Fill the jar with about 1 L of distilled water at room temperature.
2. Mark the length of the straw at 1-cm intervals. Fill 3 cm of one end of the straw with the modeling clay.
3. Float the straw upright in the jar. If the straw does not float upright, cut off the open end at 1-cm intervals until it floats upright.
4. As the straw floats in the jar, use the 1-cm markings to estimate the length of the straw that remains below the water surface. Record your observations.
5. Put the jar in the freezer until its temperature drops to 5–8°C. Remove the jar from the

freezer and repeat Step 4. Record your observations.
6. Wait until the water returns to room temperature. Add about 1/8 cup of salt to the water. Stir the water until the salt is completely dissolved. Repeat Step 4. Record your observations.

Analysis and Conclusions
1. In which trial was the water the most dense? The least dense? Explain your answer.
2. Based on your observations, how might you make the distilled water in Step 5 less dense?
3. Based on your observations, where would you expect the water in the ocean to be the least dense? The most dense?

Figure 21–4. In clear ocean water, blue light can penetrate up to 100 m before being absorbed. Until it is absorbed, blue light is reflected and makes the upper layers of ocean water appear blue. What can you infer about the depth of these vase and tube sponges?

Color of Ocean Water

The color of ocean water is determined by the way in which it absorbs or reflects sunlight. Because the ocean covers such a vast area of the earth's surface, most of the sunlight that reaches the earth falls upon it. Much of this sunlight penetrates the surface of the ocean and is absorbed by the water.

Most of the various wavelengths, or colors, of visible light are absorbed by water. Only the blue wavelengths tend to be reflected. If you look almost straight down into clear water, the water appears blue because blue is the last color to be absorbed.

The blue color of the ocean may not always be apparent. When light rays strike water at an angle close to horizontal, they are reflected rather than absorbed. This reflection causes the water to sparkle rather than to appear as a solid color. Also, many times the natural blue color of the ocean is clouded by small particles and organisms suspended in the water, causing the water to appear green or brown. No light of any wavelength can penetrate ocean water at depths below a few hundred meters. Therefore, only the upper regions appear to have color; the remainder is in total darkness.

Section 21.1 Review

1. Why does the water of the Arctic Ocean have relatively low salinity?
2. What is a thermocline?
3. How does temperature affect the density of ocean water?
4. Why would surface water in the North Sea be more likely to contain a higher percentage of dissolved gases than would surface water in the Caribbean Sea?

21.2 Life in the Oceans

Fossil evidence indicates that the first living organisms on the earth probably originated in the oceans more than 3 billion years ago. From that time to the present, marine organisms have changed, as have the physical and chemical properties of the ocean. Marine organisms depend on two factors for their survival: the nourishing substances available in ocean water and sunlight. Variations in the amount of either of these factors will affect the ability of marine organisms to survive and flourish.

Ocean Chemistry and Marine Life

Animals and plants living in the oceans help maintain the chemical balance of ocean water. These organisms remove from the water all the substances that they require for carrying out their life processes. At the same time, they return a tremendous variety of by-products to the water.

All life in the ocean is regulated by the life processes of plants. Plants absorb large amounts of substances containing carbon, hydrogen, oxygen, and sulfur. These elements are so highly concentrated in ocean water that the biological processes of marine organisms do not have any significant effect on their abundance. The elements nitrogen, phosphorus, and silicon are critical to the growth of plants. These substances are less abundant than oxygen and carbon dioxide. Therefore, heavy plant growth can reduce the concentration of these essential elements to nearly zero in areas of the ocean. When this occurs, the growth of plants in the depleted water will be greatly slowed or even stopped. How does this process compare with the growth of vegetation on land?

Marine plants and animals absorb from the ocean water the substances they need for life. These substances are returned to the water

Section Objectives

- **Explain how marine life alters the chemistry of ocean water.**
- **Explain why plankton can be called the foundation of life in the ocean.**
- **Describe the major zones of life in the ocean.**

Figure 21–5. Like many marine organisms, this *Argyropelecus* gets all it needs to maintain life, including food and oxygen, from the ocean water.

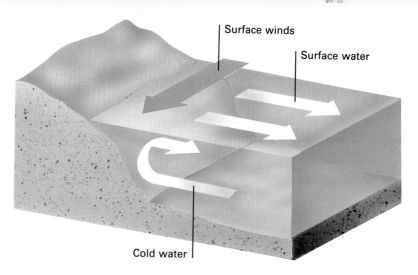

Surface winds

Surface water

Cold water

Figure 21–6. The illustration shows upwelling due to the movement of surface water. Up-welling occurs near a coastline.

when an organism dies and decays. Bacteria in the water digest these organic remains and again release the trapped substances. The decay processes occur at all depths. However, gravity slowly pulls the organic remains from the surface regions, where biological activity is greatest, to the ocean bottom.

In general, all of the elements necessary to marine life are consumed near the surface but released at great depths. Thus, the deeper water becomes a storage area for the vital materials needed to support life. Deep-dwelling animals use the material settling from the upper regions as a source of food. These important substances must be returned to the surface, however, before most organisms in the ocean can use them.

Several processes can cause deep ocean water to move upward, carrying with it the substances necessary for the survival of marine life. When wind blows steadily parallel to the shore along a coastline, surface water is moved farther out in the ocean. Deep water then moves upward to replace the surface water. This process, shown in Figure 21–6, is called **upwelling.** Upwelling commonly occurs off the western coasts of South America, California, and northern Florida. Areas of upwelling are rich fishing grounds.

In shallow water, wave action on the shore may be powerful enough to cause deep water to mix with surface water. Tides can also mix deep water with surface water. The distribution of life in the ocean depends, to a large extent, on the way life-supporting substances return to the surface from the ocean depths.

Sunlight and Marine Life

Almost all marine plants require sunlight as well as certain dissolved substances in the water. This means that plant growth in the ocean is restricted to the upper 80 m of water where light penetrates. Below about 80 m, there is never enough light to sustain plants. Within the zone where sufficient light exists, most regions of the ocean contain large amounts of free-floating, microscopic plants and animals called **plankton.** There are two main types of plankton: plants known as **phytoplankton** and animals called **zooplankton.** Phytoplankton remove dissolved materials from the water and use the energy from sunlight to carry on photosynthesis. The phytoplankton then serve as the food source for the zooplankton.

Both forms of plankton are eaten by larger ocean animals called **nekton.** Nekton are forms of ocean life that swim, such as fishes, dolphins, and squid. Because nekton can swim, they are able to search for food and avoid some predators.

In ocean food chains, plankton are consumed primarily by small fishes and squid. These, in turn, become food for adult fishes and other large marine animals. Some large animals, such as baleen whales, feed directly on plankton. Thus, phytoplankton are the first link in the complex food chains that support life in the ocean.

Organisms that live on the ocean floor are called **benthos.** Benthos include plants that grow in shallow waters and animals such as oysters, starfish, and crabs. Some types of benthos—sea anemones, for example—actually attach themselves to the ocean floor. Some types of tube worms are found only alongside volcanic vents in the ocean depths.

Ocean-Life Zones

The ocean can be divided into two general environments—bottom and water. The bottom environment is classified into five major zones and the water environment is classified into two major zones. The amount of sunlight, the temperature, and the water pressure determine the distribution of marine life within these seven zones.

The Intertidal and Sublittoral Zones

The first major bottom zone, the **intertidal zone**, lies between the low-tide and high-tide lines. Because of shifting tides and breaking waves, this zone is a relatively unstable environment for marine life. The waves scatter organisms, and the changing tides may leave the zone dry. Still, marine life flourishes here. Crabs, clams, mussels, sea anemones, and seaweed populate this zone.

The largest number of bottom-zone-dwelling organisms live on the shallow **sublittoral zone**. This continuously submerged zone, shown in Figure 21.8, is dominated by sea stars, brittle stars, and sea lilies.

The Bathyal, Abyssal, and Hadal Zones

The **bathyal zone** begins at the end of the continental shelf and

Figure 21–7. A type of zooplankton, known as *Nauplius* is shown left. Dinoflagellate, shown right, is a type of phytoplankton that lives in the upper zone of the ocean where light is available for photosynthesis.

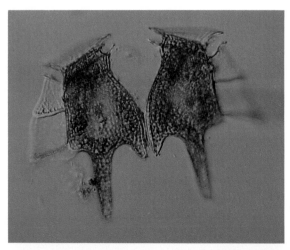

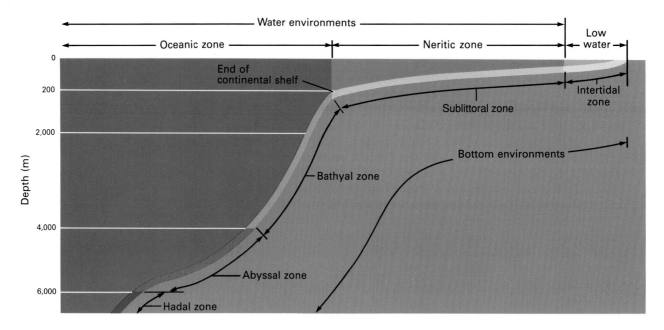

Figure 21–8. This diagram shows the seven major ocean-life zones. Note the relationship of the sublittoral zone to the neritic zone.

extends to a depth of 4,000 m. Because little or no sunlight reaches this zone, plant life is scarce. Among the animals that live in the bathyal zone are squid, octopus, and large whales.

The **abyssal zone**, which has no sunlight, begins where the bathyal zone ends and extends to a depth of 6,000 m. Organisms that live in the abyssal zone include anglerfish, devilfish, and sea cucumbers.

The term **hadal zone** is used for bottom environment areas deeper than 6,000 m. Life is sparse here and is mostly dependent upon food that falls from higher levels.

The Neritic and Oceanic Zones

The first water zone is the **neritic zone**. This zone rests above the sublittoral zone. The neritic zone has abundant sunlight, a fairly constant temperature, and relatively low pressure. These conditions make it ideal for marine life. Plankton and nekton fill its waters, which are the source of much of the fish and seafood that people eat.

The **oceanic zone** extends seaward beyond the continental shelf. Most organisms in this zone are the prey of some other organism.

Section 21.2 Review

1. In addition to the action of waves and tides, what other process causes deep ocean water to move upward?
2. Describe the two main types of plankton.
3. Describe the intertidal, neritic, bathyal, and abyssal zones of ocean life and name some organisms found in each.
4. Suppose a heavy cloud of volcanic ash were to reduce significantly the amount of sunlight reaching a specific part of the ocean. How might the reduction in the amount of sunlight received affect the marine life in that part of the ocean?

21.3 Ocean Resources

The sea has always been a source of food and a means of transportation. Recently, however, people have begun to realize its importance as a source of vital resources needed to support civilization. Increased knowledge of the oceans, combined with the earth's expanding population, has caused new interest in the sea as a source of fresh water, minerals, and food.

Fresh Water from the Ocean

Throughout the world the need for fresh water is increasing rapidly. Developing countries need huge supplies of fresh water for industry and irrigation. Even some nations that formerly had abundant supplies of fresh water are facing shortages.

The increasing demand for water can be met in two ways. First, the fresh water now available can be conserved to avoid waste. Second, the amount of available fresh water can be increased. The water supply can be increased by finding a way to convert ocean water to fresh water at a reasonable cost.

One means of increasing the fresh water supply is through *desalination*. As you read in Chapter 13, desalination is the extraction of fresh water from salt water. The most common method of desalination now used is **distillation,** which involves heating ocean water to remove the salt. Heat causes the liquid water to evaporate, leaving dissolved salts behind. When the water vapor condenses, the result is pure fresh water. However, evaporating liquid water often requires a great deal of costly heat energy. Distilling large amounts of fresh water will not be practical unless a way to use a less expensive form of energy can be developed.

Another method for desalinating ocean water involves freezing it. When ocean water freezes, the first ice crystals that form are free of salt. The salt remains in pockets of liquid water in the ice. The ice can then be removed and melted to obtain fresh water. This process requires only about one-sixth the amount of energy needed for distillation.

Other methods for desalinating ocean water include the use of special membranes that allow water to pass through while blocking the dissolved salts. Also, some chemicals combine temporarily with either water or salt and can thus be used to desalinate ocean water.

Minerals from the Ocean

Mineral deposits are another important resource of the ocean. On the first oceanographic research voyage, made by H.M.S. *Challenger* in 1872, strange black lumps of minerals called *nodules* were discovered on the ocean floor. Investigation has shown that huge areas of the ocean floor, particularly in the Pacific Ocean, are covered with these nodules.

Nodules are a valuable source of manganese, which is used in making certain types of steel. Nodules also contain large amounts of iron and small amounts of copper, nickel, and cobalt. Some nodules also contain phosphates that are useful as fertilizers.

The way these nodules are formed is not completely understood by scientists. However, measurement of their rate of growth suggests that they form quickly enough to supply the present worldwide need for the metals they contain. Unfortunately, the recovery of nodules is extremely difficult at the present time because they are found mostly in deep water. If an economical method of mining can be developed, the mineral-rich nodules may largely replace the sources of some ores on land.

Some valuable minerals are now extracted easily from the oceans. For many centuries salt has been obtained by evaporation of ocean water. The ocean is the main source of magnesium metal and bromine. These two elements are refined from their dissolved salts. However, the concentrations of most of the other useful minerals dissolved in the oceans are too small for extraction to be practical. For example, each cubic kilometer of ocean water contains about 6 kg of gold. However, obtaining just a few cents worth of gold would

Horizons

Technology: Efficient Desalination

Scientists have been striving for many years to develop a more efficient, less expensive method of turning ocean water into fresh water. To achieve this goal, chemist Mic Pleass enlisted the help of a special source—the ocean itself.

Although water covers more than 70 percent of the earth's surface, only 3 percent of it is fresh water. The other 97 percent is ocean water, too salty for drinking or irrigating crops. To make the ocean water usable, scientists have developed various systems to remove the salt through a process called *desalination*. The photograph at the left shows part of the inside of an early desalination plant.

The desalination system Pleass developed had its start in the Caribbean in the 1970's. At that time, islanders were paying as much as $20 for 3,800 L of fresh water—the equivalent of a supply for a family of four in the United States for four days. Approximately a third of the cost was attributed to the price of the fossil fuels needed to power land-based desalination plants. So Pleass decided to look to ocean waves for an alternate, cheaper source of power.

Working with Doug Hicks, a graduate student in oceanography, Pleass developed *Delbuoy*, a desalination system that is ocean-based and pow-

require processing about 4 million L of ocean water. Consequently, ocean water is not a practical source of gold.

The most valuable mineral resource taken from the ocean is the petroleum found beneath its floor. Huge deposits of oil and natural gas exist along the continental margins in many parts of the world. As the worldwide need for energy has grown, these fossil-fuel resources have become very important. As a result of new drilling techniques, oil and gas can be extracted as far as 100 km offshore and as deep as 100 m. It is also possible to drill wells to depths of several thousand meters along the continental shelves and on the continental rise. Off-shore drilling does present some serious problems, however. Oil spills and leaks can pollute the water and kill marine organisms and sea birds.

Food from the Ocean

Of all the resources that the ocean is capable of supplying, the one in greatest demand is protein-rich food. At present a large part of the population of the world has a starchy diet. Such a diet can maintain life, but the lack of protein to build strong tissues decreases the

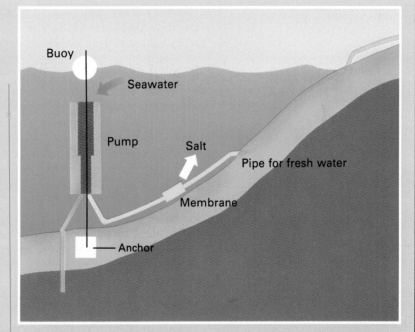

ered by waves. Simple in design, the Delbuoy system, pictured right, consists of a pump and a reverse-osmosis module. The pump is 3 m tall, anchored in 18 m of ocean water, and held upright by a buoy. The module contains a membrane through which water, but not salt, can pass. As the buoy rises and falls with the wave action, it operates the pump. The pump draws up ocean water, filters out the sand, and then sends the water through the membrane in the reverse-osmosis module. The salt-free water that passes through the membrane is pumped to storage tanks located on land. The salt is returned to the ocean.

The Delbuoy system is economical as well as efficient. The same 3,800 L of water that once cost $20 now costs only $5. With an improved Delbuoy system, the cost for this same amount of desalinated water could someday be as low as $1.

How might the Delbuoy system also reduce pollution of the environment?

Figure 21–9. Aquaculture establishments, such as this prawn farm in Hawaii, already provide a reliable, economical source of food.

ability of a person's body to fight disease. Perhaps half a billion people in the world suffer from some form of disease caused by a lack of protein in their diets. An important future source of high-protein food could be fish from the ocean.

In the future, the practice of **aquaculture,** the farming of the ocean, may become as important to food production as agriculture. Aquaculture involves developing and raising special breeds of marine animals and plants that yield large amounts of food. Aquaculture has already been used successfully to grow catfish and salmon on large aquatic farms. Similar methods might be used to breed other fish, shellfish, and plants in ocean farms. These farms would be closed-off areas of the ocean that provided a suitable environment for aquaculture.

Under the best conditions, an ocean farm could produce more valuable protein-rich food than a land farm of the same size. Fish and shellfish change their food supply into protein more efficiently than do most land animals. No marine animals, except marine mammals, use energy to keep an even body temperature. Because of the buoyancy of water, they also use less energy supporting their weight against the force of gravity. In agriculture only the top layers of soil can be used. In contrast, in ocean farms it may be possible to use the entire depth of the water to produce food. The ocean could even be fertilized by pumping the nutrient-rich bottom water to the surface. How would this type of artificial upwelling benefit an ocean farm?

Ocean-Water Pollution

The oceans have been used as a dumping ground for many kinds of wastes including garbage, treated sewage, and spent nuclear fuel. Until recently these wastes became very dilute or were destroyed as they spread throughout the ocean. The growth of the world population and the increase in industry, however, have changed the situation. The ability of the ocean to absorb wastes and renew itself cannot match the increasing wastes that are produced worldwide.

Figure 21–10. People have begun to recognize the importance of reducing ocean water pollution. This scientist is taking a water sample, which will be checked for impurities.

Coastal waters are in greatest danger. Pollution has destroyed the clam and oyster beds in some local areas. Sea birds have been found tangled in plastic products. Beaches are closed because of sewage and oil washed onto the sand.

Beside coastal waters and beaches, pollutants are now found in measurable amounts everywhere in the oceans. Traces of the insecticide DDT have been detected in the open ocean. Since lead was first added to gasoline 50 years ago, the amount of lead in the Pacific Ocean has increased tenfold. Both DDT and lead can cause problems in ocean food chains. As animals eat other animals, the amounts of DDT and lead build up in the bodies of the predators. In some areas of the world, the lead concentration in fish has made it inedible. Radioactivity from nuclear fallout also has been detected in the oceans. While none of these materials has yet reached harmful levels throughout the world, they can be harmful in local areas. The presence of certain pollutants indicates that the ocean cannot be used as a dumping ground forever.

Section 21.3 Review

1. Name and describe three methods for desalinating ocean water.
2. What ocean resource might eventually replace some minerals now mined on land?
3. What is aquaculture, and why is it important?
4. In what ways does ocean pollution threaten the quality of life on the earth?

Reviewing Chapter 21

Key Concepts

- Ocean water dissolves minerals, salts, and atmospheric gases. See page 347.
- The temperature, density, and color of ocean water vary around the earth. See page 350.
- Marine life helps maintain the chemical balance of ocean water. See page 353.
- Plant life regulates all ocean life. See page 353.
- There are seven zones of life in the ocean, each supports certain types of organisms. See page 355.
- The ocean is valuable as a source of fresh water, minerals, and food. See page 357.
- Water pollution threatens the ocean environment. See page 360.

Key Terms

abyssal zone (356)	hadal zone (356)	pack ice (350)	sublittoral zone (355)
aquaculture (360)	intertidal zone (355)	phytoplankton (354)	thermocline (350)
bathyal zone (355)	nekton (355)	plankton (354)	upwelling (354)
benthos (355)	neritic zone (356)	salinity (348)	zooplankton (354)
distillation (357)	oceanic zone (356)		

Review

On your paper, write the letter of the term that best completes each of the following statements.

1. The gas that dissolves most easily in ocean water is (a) carbon dioxide (b) argon (c) nitrogen (d) oxygen.
2. The amount of dissolved salts in a sample of ocean water is called its (a) salinity (b) nekton (c) plankton (d) density.
3. A floating layer of ice that completely covers an ocean surface is called (a) a glacier (b) a thermocline (c) pack ice (d) benthos.
4. In large bodies of water, a zone of rapid temperature change is called (a) a benthos (b) an abyssal zone (c) a bathyal zone (d) a thermocline.
5. When liquid water is warmed, its density (a) increases (b) decreases (c) remains the same (d) doubles.
6. Although most of the various wavelengths of visible light are absorbed by ocean water, the one wavelength that tends to be reflected is the color (a) violet (b) green (c) yellow (d) blue.
7. The process whereby surface water is blown farther out in the ocean and nutrient-rich deep water rises to take its place is called (a) desalination (b) distillation (c) upwelling (d) aquaculture.
8. Microscopic marine plants and animals are known as (a) plankton (b) benthos (c) nekton (d) tube worms.
9. Marine animals that can swim to search for food and avoid predators are called (a) phytoplankton (b) zooplankton (c) nekton (d) benthos.
10. Sea anemones are an example of (a) zooplankton (b) nekton (c) benthos (d) phytoplankton.
11. Perhaps the most unstable ocean-life zone is the (a) intertidal zone (b) abyssal zone (c) bathyal zone (d) neritic zone.
12. The ocean-life zone with abundant sunlight,

fairly constant temperature, and relatively low pressure is the (a) bathyal zone (b) neritic zone (c) abyssal zone (d) intertidal zone.
13. Some types of tube worms live around volcanic vents in the (a) bathyal zone (b) intertidal zone (c) abyssal zone (d) neritic zone.
14. The most common method of producing fresh water by desalinating ocean water is (a) distillation (b) evaporation (c) upwelling (d) aquaculture.
15. Lumps of minerals on the ocean floor are called (a) nekton (b) nodules (c) benthos (d) plankton.
16. Aquaculture is another name for (a) desalination (b) distillation (c) ocean farming (d) rapid temperature changes.
17. Scientists have been able to fertilize the ocean through (a) distillation (b) artificial upwelling (c) mining nodules (d) dissolving gases.
18. The increased use of gasoline has led to a tenfold increase in the Pacific Ocean of (a) plankton (b) upwelling (c) lead (d) salinity.

Application

On your paper, write answers to the following questions.
1. Suppose you are building a desalination system. In order to make your system as efficient as possible, you want to locate it in an area of the ocean where the salinity of the water is low. Explain why you would or would not choose each of the following locations: a site just off the coast of Israel; a site in Prudhoe Bay, off the coast of northern Alaska; a site off the coast of Sweden, on the Baltic Sea.
2. You have collected samples of ocean water from the three locations described in question 1. All three samples have been brought to the same temperature—neither hot nor cold. From which location would the water have the greatest density? Why?
3. Suppose that climatic and atmospheric conditions over one of the earth's oceans changed dramatically. The changes resulted in a complete absence of upwelling and wave action.
Explain what would happen to the marine life in this ocean and why.
4. How would a significant decrease in sunlight affect phytoplankton?
5. What impact would the decrease in sunlight discussed in question 4 have on other forms of marine life?
6. When oceanographers first explored the ocean floor along mid-ocean ridges, they discovered a variety of marine life, including sightless crabs. Explain why these crabs are not handicapped by their sightlessness.
7. You have decided to start an aquatic farm. In which of the zones of ocean life should you locate your farm? Why?
8. Some nations use certain areas of the oceans as disposal sites for used fuel rods from nuclear power plants. Explain what effect this might have on future marine life in those areas.

Extension

1. Prepare an illustrated report on phytoplankton and zooplankton. Describe several types of each and accompany the descriptions with drawings.
2. Look up information about intertidal zones. Write a detailed description of a typical intertidal marine community. Tell about the plants and animals that are part of this community and how they manage to survive in this difficult environment.
3. Research the new foods being produced through aquaculture. What is the nutritional value of these foods? Draw a world map and color in the countries that are developing aquaculture foods. Share the map and your research with your class.
4. Research pollution of North American coastal waters. Make a chart showing how a sample food chain is affected by lead pollution or trace the impact of another pollutant on one food chain.

Chapter 22

Movements of the Ocean

The power of the ocean is visible in its waves.

*T*he waters of the ocean are constantly in motion. Waves rise and fall in varied rhythms. Ocean currents— some near the surface and some deep below—flow like rivers. The level of the ocean moves up and down with the tides. This chapter discusses the causes and effects of these powerful movements of the ocean.

Chapter Outline

22.1 Ocean Currents

Section Objectives

• **Discuss how wind patterns affect surface currents.**

• **Explain how differences in the density of ocean water affect the flow of deep currents.**

The waters of the ocean move in streams called **currents.** Some of these currents are strong enough to affect the speed and direction of ships. As a result, people have studied ocean currents since the early days of sailing.

Learning what causes ocean currents has not been easy. These complex movements of water are often difficult to trace. Modern-day oceanographers identify ocean currents by studying the physical and chemical characteristics of the water. They try to follow the movement of the water to determine where the water has come from, where it is going, and what causes it to move.

The movement of water in the ocean is still not completely understood. However, as oceanographers slowly collect more-detailed data, they are better able to describe the general movements and probable causes of ocean currents. For example, oceanographers have determined that there are two types of currents. **Surface currents** move on or near the surface of the ocean and are driven by winds. **Deep currents** move beneath the surface of the ocean and are caused by differences in the density of the water.

Surface Currents

Ocean water, like any substance, can be set into motion only if it receives energy. Steady winds can provide this energy and move large masses of water near the surface of the ocean. Thus, wind is the driving force of surface currents. Almost all of the surface currents of the ocean result from global wind patterns. Because the transfer of energy from the wind to the water occurs at the upper layer of the water, that part of the ocean moves fastest. The velocity of the water decreases with depth so that at a depth of approximately 100 m the surface current is almost undetectable.

Currents and Wind Patterns

The global winds that most directly affect the flow of surface currents are called the **trade winds** and the **westerlies.** The trade winds are located just north and south of the equator. The westerlies are located in the middle and polar latitudes of the Northern Hemisphere and the Southern Hemisphere.

The trade winds affect the surface currents both north and south of the equator. In the Northern Hemisphere, the trade winds blow from the northeast. In the Southern Hemisphere, they blow from the southeast. As a result of the **Coriolis** (kore-ee-OE-luss) **effect,** trade winds near the equator blow toward the west. The Coriolis effect is the deflection of the earth's wind and ocean currents caused by the earth's rotation. As a result of the trade winds, huge circles of moving water are formed in the Northern and Southern hemispheres. In the Northern Hemisphere, the pattern of movement is clockwise. The direction of the movement in the Southern Hemisphere is counterclockwise.

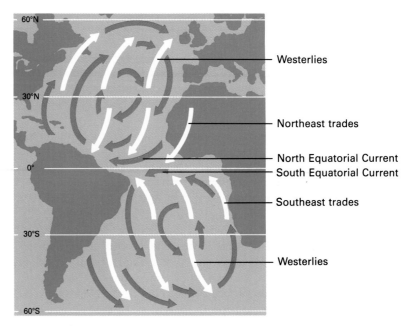

Figure 22–1. Global winds and the Coriolis effect together drive the surface currents of the oceans in great circular patterns. The white arrows represent global winds and the purple arrows represent ocean currents.

The westerlies affect the pattern of surface currents in the middle and polar latitudes, as shown in Figure 22–1. In the Northern Hemisphere, they blow from the southwest and push water at the ocean surface toward the east. In the Southern Hemisphere, they blow from the northwest. In which direction do the westerlies push the surface water of the ocean in the Southern Hemisphere?

Continental landmasses act as barriers to wind-driven surface currents. When a current flows toward a landmass, the current is deflected and divided, as shown in Figure 22–2.

Equatorial Currents
Equatorial currents are found in the Atlantic, Pacific, and Indian oceans. In each of these oceans, two warm-water currents, called the *North Equatorial Current* and the *South Equatorial Current,* move in a westward direction. Between these westward-flowing currents lies a weaker, eastward-flowing current, which is called the *Equatorial Countercurrent.*

Currents in the Northern Hemisphere
In the Atlantic Ocean, the North Equatorial Current pushes water against the east coast of North America and around the Gulf of Mexico. This swift, warm current is called the **Gulf Stream.** The Gulf Stream can move 100 million m^3 of water per second. By comparison, the Mississippi River moves 20,000 m^3 of water per second. The Gulf Stream moves from the Gulf of Mexico around the tip of Florida and north along the eastern coast of the United States. There it is strengthened by westerly winds and deflected to the east. In the North Atlantic, the Gulf Stream meets the cold-water Labrador Current, which flows out of Baffin Bay into the North Atlantic. When the warm air over the Gulf Stream blows over the cold water of the

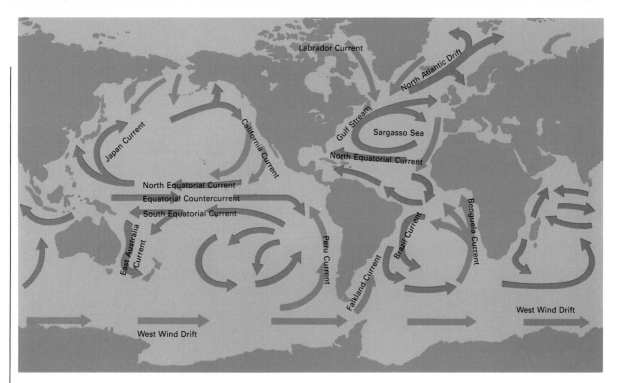

Figure 22–2. This map shows the major surface currents of the oceans of the world. Warm-water currents are shown in red; cold-water currents are shown in blue.

Labrador Current, dense fog often results. Near Greenland the Gulf Stream widens and decreases in speed until it becomes a vast, slow-moving warm current known as the *North Atlantic Drift*. A **drift** is a weak current. As the North Atlantic Drift approaches western Europe, it splits. One part becomes the Norway Current, which flows northward along the coast of Norway. The other part is deflected southward and becomes the Canary Current, which eventually rejoins the North Equatorial Current.

As you can see in Figure 22–2, the Gulf Stream, the North Atlantic Drift, and the North Equatorial Current completely circle the North Atlantic. At the center of this circle lies an area called the *Sargasso Sea,* a vast area of relatively calm, warm water. Light winds blow over the sea, and great quantities of brown seaweed called *sargassum* float on its surface. The pattern of currents around the Sargasso Sea seems to concentrate all kinds of floating debris. Even orange peels and plastic cups have been found there.

The pattern of currents in the North Pacific is similar to that in the North Atlantic. The warm Japan Current, the Pacific equivalent of the Gulf Stream, flows northward along the east coast of Asia. These waters then flow toward North America, eventually forming the cool-water California Current, which flows south along the California coast.

Currents in the Southern Hemisphere

In the Southern Hemisphere, surface currents also move in circular patterns. However, while the major circular currents in the northern oceans move clockwise, the currents in the southern oceans move counterclockwise.

In the most southerly regions of the Atlantic and Pacific oceans, constant westward winds produce the Antarctic Current, a powerful

Science Notebook

Shifting Currents

The flow of surface currents in the Indian Ocean is divided into two circular patterns, or gyres. One gyre circulates north of the equator, and the other, to the south of the equator. The southern gyre flows in a counterclockwise movement, as do all surface currents south of the equator. North of the equator, most currents flow in clockwise gyres.

However, the currents of the northern Indian Ocean change direction in response to the changing direction of strong seasonal winds called *monsoons*. The summer monsoon carries humid air and produces the heavy rains crucial to farmers in this part of the world. The drier winter monsoon produces little precipitation.

From April to October, a strong monsoon blowing from the southwest drives the surface waters of the northern Indian Ocean in a clockwise gyre. The waters flow eastward along the southern coast of Asia, forming the Monsoon Current. This current flows east along the coast of Indonesia, turns south, and joins the South Equatorial Current flowing westward.

From November to March, the southwest monsoon is replaced by a northeast wind. This monsoon changes the Monsoon Current flowing eastward along the Asian coast to the westward-flowing North Equatorial Current. At the equator the current turns toward the east, thus flowing parallel to, but in the opposite direction of, the South

Equatorial Current. The eastward flow of water is called the Equatorial Countercurrent.

How do you think early explorers used these currents to travel back and forth to the Far East?

cold-water current. No landmasses interfere with the movement of this current, and it completely circles Antarctica.

The warm surface currents in the Indian Ocean follow two patterns. In the southern part of the ocean, the currents follow a circular, counterclockwise pattern. In the northern part of the ocean, the currents are governed by winds called *monsoons* that change direction with the seasons.

Deep Currents

In addition to the wind-driven surface currents, the ocean has powerful currents that flow deep beneath its surface. These deep currents generally move more slowly than do the surface currents. Deep currents are produced as the water of the polar regions sinks and flows beneath warmer ocean water toward the equator.

The movement of polar waters is due to density differences. When water is cooled, it contracts, and the water molecules move closer together. This contraction makes the water denser, and, as a result, it sinks. When water is warmed, it expands and the water molecules move farther apart. Warm water is thus less dense and remains above the cold water.

The higher density of polar waters also results from an increase in the salinity of the water. Salinity may increase in polar regions where water is frozen in icebergs and pack ice. When polar water freezes, most of the salt remains in the unfrozen water. Thus dense and highly saline polar water sinks and forms a deep current that flows beneath less-dense ocean water. There is little mixing between the two currents. The deep-current water rises only when winds blow surface water aside, causing the deep water to rise toward the surface as an upwelling.

Antarctic Bottom Current

The temperature of the water near Antarctica is close to the freezing point of ocean water, –2°C. The salinity is also high—35 %. These two factors make the water off the coast of Antarctica the densest and coldest ocean water in the world. This dense, cold water sinks to the ocean bottom and slowly moves northward. It forms a deep-water current called the *Antarctic Bottom Current*. The Antarctic Bottom Current moves along the ocean bottom for thousands of kilometers, reaching into the northern oceans to a latitude of approximately 40°. Thus this dense, cold water forms the deepest ocean currents of the major oceans.

Deep Atlantic Currents

Oceanographers who study the deep currents in the Atlantic Ocean have discovered some general patterns. In a small region of the North Atlantic just south of Greenland, the water is exceptionally cold and has a high salinity. This cold, salty water sinks and forms a

Figure 22–3. The very dense and highly saline Antarctic water travels beneath less-dense ocean water, and forms the Antarctic Bottom Current.

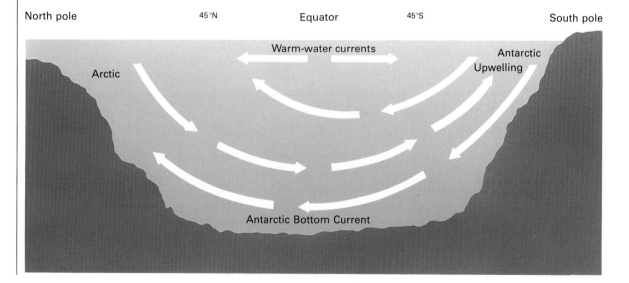

North pole   45°N   Equator   45°S   South pole

Warm-water currents

Arctic

Antarctic Upwelling

Antarctic Bottom Current

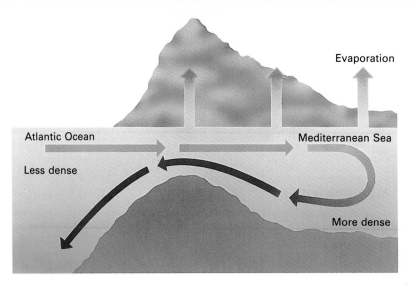

Figure 22–4. The highly saline dense water of the Mediterranean Sea forms a deep current as it flows over a high area at the mouth of the Mediterranean into the less dense Atlantic Ocean.

deep current that moves southward beneath the northward-flowing Gulf Stream. Most deep currents flow in a direction opposite that of the surface currents flowing above them. Near the equator this deep current divides. One part begins to flow northward again, while the rest of the current continues to flow southward to the Antarctic Ocean. The deep water from the Atlantic Ocean flows over the colder, denser Antarctic Bottom Current.

Deep Atlantic currents also exist near the Mediterranean Sea. Every summer an increase in evaporation and a decrease in rainfall make the water of the Mediterranean more saline and thus denser. This denser water sinks and flows out into the Atlantic Ocean, creating a deep current. In turn, water from the Atlantic, which is less saline and thus less dense, flows into the Mediterranean Sea.

Turbidity Currents

A *turbidity current* is a strong temporary current caused by an underwater landslide. Turbidity currents occur when large masses of sediment that have accumulated along a continental shelf or continental slope suddenly break loose and slide downward. The landslide mixes the nearby water with sediment. The sediment causes the water to become cloudy, or turbid, and denser than the surrounding water. The turbidity current gets under way as the dense water mass moves beneath the less dense, clear water. Turbidity currents flow with an agitated motion close to the sloping ocean floor, picking up more sediment as they flow.

Section 22.1 Review

1. What force drives most of the surface currents of the oceans?
2. Which winds affect the surface currents north and south of the equator?
3. Which winds affect the surface currents in the middle and polar latitudes?
4. What two factors affect the density of ocean water and cause deep currents?
5. Explain the role of density in turbidity currents.

22.2 Ocean Waves

Section Objectives

- **Describe the formation of waves and the factors that affect wave size.**

- **Discuss the interaction of the shore and the waves.**

If you visit the seashore, you can observe the most obvious movement of the ocean—its **waves.** A wave is the periodic up-and-down movement of water. Waves are a way of transferring energy. As you can see in Figure 22–5, a wave has two basic parts—a **crest** and a **trough.** The crest is the highest point of a wave. The trough is the lowest point between two crests. Scientists study these two features to determine other wave characteristics.

The **wave height** is the vertical distance between the crest and the trough of a wave. The *wavelength* is the horizontal distance between two consecutive crests or between two consecutive troughs. The **wave period** is the time it takes for one complete wavelength to pass a given point. The period of most waves ranges from two to ten seconds.

The speed at which a wave moves is calculated by dividing its wavelength by its period. This relationship is expressed by a simple formula:

$$\text{Wave speed} = \frac{\text{wavelength}}{\text{period}}$$

If a wave has a wavelength of 216 m and a period of 12 s, what is its speed?

Wave Energy

If there were no winds, the surface of the ocean would be almost as smooth as glass. Even the slightest breeze causes ripples in the ocean as a result of friction between the moving air and the water. The wind pushes directly against the side of a ripple. As the ripple receives more energy from the wind, it may grow into a wave. Although waves can be generated in different ways, the main source of wave energy is the wind. Therefore the longer the wind blows from the same direction, the more energy is transferred from the wind to the water. The more energy that is transferred, the larger the wave becomes. However, the wind seldom blows in one direction for very long.

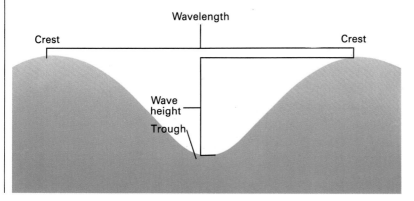

Figure 22–5. The vertical distance between the crest and the trough of a wave is the wave height. The distance from the crest of one wave to the crest of the following wave is the wavelength.

Usually the surface of the ocean is broken by many small waves moving in a number of directions in response to shifting winds. Larger waves, because of their large surface area, receive more energy from the wind than smaller waves do. Thus, larger waves tend to grow larger, and smaller waves quickly die out. One of a group of long, rolling waves that are all the same size is called a **swell.** Swells move in groups, one wave following the other. Swells that finally reach the shore may have been formed thousands of kilometers out in the ocean.

Water Movement in a Wave

Although the energy of a wave moves forward from water molecule to water molecule, the water itself moves very little. This surprising fact can be demonstrated by observing the movement of a cork floating on the water as a wave passes. The cork moves slightly forward as the wave approaches and then falls back an almost equal distance as the wave passes.

As a wave moves across the surface of the ocean, only the energy of the wave moves forward. The water does not move forward. The water particles within the wave move in a circular motion. During a single wave period, each water particle moves in one complete circle. At the end of the wave period, a circling water particle ends up almost exactly where it started.

As a wave passes a given point, the circle traced by a water particle on the ocean surface has a diameter that is equal to the height of the wave. Because waves receive their energy from wind pushing against the surface of the ocean, the energy received decreases as the depth of the water increases. As a result, water at various depths receives varying amounts of energy. Thus, the diameter of the circle that a particle moves in decreases with increasing depth, as shown in Figure 22–6. At a depth of about one half the wavelength, the circular motion is very slight. Below this point, there is almost no circular motion of water particles.

Figure 22–6. As a wave moves across the surface of the ocean, the water particles within the wave move in circular paths. The diameter of the circular path decreases with depth. Below a depth equal to one-half wavelength almost no circular motion occurs.

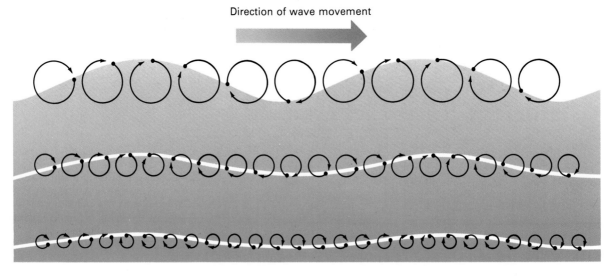

Direction of wave movement

Investigation

Waves

As a wave passes through water, the form of the wave moves across the surface. However, a wave does not carry water along with its motion. You can demonstrate this principle of wave movement with some common objects.

Science Process Skills Focus: demonstrating, observing, describing, recording data

Materials

sink or rectangular pan, at least 40 cm × 30 cm × 10 cm; water; ruler; cork stopper; adhesive tape; spoon

Procedure

1. Fill the pan with water to a depth of 7 cm. Float the cork near the center of the pan. On the side of the pan, mark the location of the cork with a small piece of tape as shown in the figure above.
2. Hold the spoon in the water and slowly and carefully move it up and down in the water.
3. Continue to make waves. Observe the movement of the cork. Sketch what you have observed.
4. Remove the cork from the pan. Again use the spoon to make waves. This time pay careful

attention to creating of a strong, steady series waves. Remove the spoon from the pan. Observe what happens when the waves reach the ends of the pan. Write down or sketch what you observed.

Analysis and Conclusion

1. Describe the motion of the cork when a wave passes.
2. How does the cork move relative to the mark on the side of the pan? Explain your answer.
3. When a wave breaks on the shore of a beach, it is clear that water is carried forward with the form of the wave. Based on your observations in Step 3, does this contradict your model?

Wave Size

Three factors determine the size of a wave. They are the speed of the wind, the length of time the wind blows, and the **fetch** of a wave. The fetch is the distance that the wind can blow across open water. Very large waves are produced by strong, steady winds blowing across a long fetch. Such conditions are most likely to occur during a storm. During a storm the steady high winds cause some waves to gather enough energy to reach great size. Strong, gusty winds produce choppy water with waves of various heights and lengths that may come from various directions. On calm days small, smooth waves move steadily across the surface.

The size of a wave will increase to only a certain height-to-length ratio before it collapses. The largest wave ever observed during a severe storm was 34 m high. Also, during storms with high wind speeds, the crest is blown off the wave, forming **whitecaps.**

Waves and the Shore

In shallow water near the shore, the bottom of the wave touches the ocean floor. A wave will touch the ocean bottom where the depth of the water is about one half the wavelength. Contact with the ocean floor creates friction, causing the wave to slow.

Breakers

The height of a wave changes as the wave approaches the shore. The water involved in the up-and-down motion of a wave extends to a depth of one-half wavelength. As the wave moves into shallow water, the bottom of the wave is slowed by friction. The top of the wave, however, continues to move at its original speed. The top of the wave gets farther and farther ahead of the bottom. Finally, the top of the wave topples over and forms a **breaker,** a foamy mass of water that washes onto the shore. The height of the wave when it topples over is one to two times the height of the original wave. Figure 22–7 illustrates the formation of breakers.

Shells left on the shore by the retreating wave mark the farthest point reached by the wave. Breaking waves scrape sediments off the bottom and move them along the shoreline of sandy beaches. The waves also erode the cliffs of rocky shorelines.

The size and the force of breakers are determined by the original wave height and the steepness of the ocean-basin floor close to shore. If the ocean basin is quite steep, the height of the wave increases rapidly, and the wave breaks with great force. Breakers of this type are common along the Pacific Coast. If the shore slopes gently, the wave rises slowly. The wave spills forward with a rolling motion that continues as the wave advances up the shore. This type of breaker is common along the Atlantic Coast.

Figure 22–7. Breakers begin to form where the depth of the water is about equal to one-half wavelength. Notice in the diagram (below) that wave height increases and wavelength decreases as waves approach the shore.

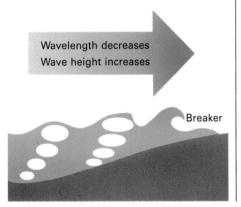

Wavelength decreases
Wave height increases

Breaker

Figure 22–8. Waves strike the shore head-on as a result of refraction. Notice the waves approaching the shore at an angle. These waves bend as they draw closer to the shore.

Refraction

Most waves approach the shore at an angle. When a wave reaches shallow water, however, it bends. This bending is called **refraction.** Refraction occurs when the part of the wave nearest the shore strikes the shallow bottom and slows down. The end of the wave that is still in deeper water moves forward at its regular speed. The wave gradually bends toward the beach and strikes the shore head-on. The original direction of the wave does not affect this bending. Note the changing direction of the waves in Figure 22–8.

Undertows and Rip Currents

Water carried onto a beach by breaking waves is pulled back into deeper water by an irregular current called an **undertow.** An undertow is seldom very strong, and it almost never creates a problem for swimmers.

Some people confuse the relatively weak undertow with the more dangerous but lesser-known **rip current.** Swift rip currents form when water returns to the ocean through breaks in underwater sandbars, which are parallel to the beach. A rip current is often strong enough to carry a swimmer out into deep water.

Rip currents generally occur in spurts that last only a few minutes at any given spot along the shore. Their presence can usually be detected by a gap in a line of breakers. A rip current can also be detected by the discolored water where sand has been stirred up by the current.

Longshore Currents

Sand bars are formed by another type of current called a *longshore current.* A longshore current forms as waves approach the beach at an angle. The longshore current is visible as a zigzag pattern moving parallel to the shore. Great quantities of sand are carried by the longshore current. If there is a bay or inlet along the shoreline where waves refract, sand will be deposited as the energy of the wave decreases. The shoreline becomes straighter as the longshore current deposits sandbars, spits, tombolos, or barrier islands.

Tsunamis

The most destructive waves in the ocean, *tsunamis*, are not powered by the wind. Tsunamis are seismic sea waves. Most are caused by earthquakes on the ocean floor, but some also can be caused by underwater landslides. Tsunamis are sometimes called *tidal waves*, which is misleading since tsunamis have no connection with tides.

Tsunamis are not very high, but they have a long wavelength. In deep water the wave height of a tsunami is usually less than a meter, but the wavelength may be as long as 240 km. A tsunami commonly has a wave period of about 15 min and an average speed of about 16 km/min. Because the wave height of a tsunami is so low in the open ocean, the tsunami cannot be felt by people aboard ships nor can it be detected from the air.

A tsunami has a tremendous amount of energy. The entire depth of the water is involved in the wave motion. All the destructive energy of this mass of water is delivered against the shore.

Near the shore the height of the tsunami greatly increases as its speed decreases. The wave may reach a height of 30 to 40 m as the water piles up. The arrival of a tsunami may be signaled by the sudden pulling back of the water along the shore. This occurs when the trough of the tsunami arrives before the crest. If the crest arrives first, a sudden, rapid rise in the water level occurs.

When the trough of a tsunami arrives first, the rapid withdrawal of water is followed by the arrival of the crest. The water then retreats with the arrival of the next trough. Many lives have been lost when people have misinterpreted such a retreat as the end of the tsunami. In Crescent City, California, following the Alaskan earthquake of 1964, the first wave of a tsunami crested 4 m above low tide. Three progressively smaller waves followed. People then returned to the shore, thinking the tsunami was over. However, a fifth wave struck in conjunction with the high tide, cresting at 6 m above low tide and killing 12 people.

The tsunami that was triggered by an earthquake in Chile in 1960 caused destruction on both sides of the Pacific Ocean. It destroyed villages along the coast of South America, crossed 17,000 km of ocean to Japan, and destroyed many villages there. Waves of this massive tsunami were still detected several days later.

Section 22.2 Review

1. Explain how wave height, wavelength, and wave period are determined from the crest and trough of a wave.
2. What factors determine the size of a wave?
3. Why do incoming waves bend toward the beach until they strike the shore head-on?
4. What is the cause of most tsunamis?
5. Explain why waves slow down in shallow water.

22.3 Tides

The daily changes in the level of the ocean surface are called **tides.** The force that causes the rise and fall of tides along coastlines was first identified in the late 1600's by Sir Isaac Newton. According to Newton's law of universal gravitation, the gravitational pull of the moon on the earth and its waters causes the rise and fall of the tides.

The gravitational pull of the moon is strongest on the side of the earth nearest the moon. As a result, the ocean on the side of the earth facing the moon bulges slightly, causing a high tide within the area of the bulge. This is called the *direct tide*.

At the same time, another slightly smaller tidal bulge is created on the opposite side of the earth. This is called the *opposite tide*. This bulge occurs because the ocean waters on the side of the earth opposite the moon are not as greatly affected by the gravitational pull of the moon as is the solid earth. Thus the ocean floor is pulled toward the center of the earth and away from the water. The net effect is a tidal bulge.

Low tides are formed halfway between the two high tides. Low tides form because ocean water flows away from the areas of low tide toward the areas of high tide.

Behavior of Tides

If the earth and the moon did not move, tides would always occur in the same locations. However, the earth rotates on its axis once every 24 hours. The moon moves through about 1/27 of its orbit around the earth in that same 24 hours. Thus, as the earth rotates, all areas of the ocean pass under the moon every 24 hr and 50 min.

The earth rotates from west to east. Therefore, the tidal bulges appear to move westward around the earth. Because there are two tidal bulges, most locations in the ocean have two high tides and two low tides daily. The difference between the levels of the high

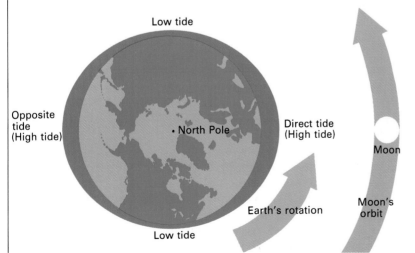

Figure 22–9. The direct tide and the opposite tide are high tides that form as the result of the gravitational pull of the moon. Most locations in the ocean have two high tides and two low tides daily.

tides and the low tides at a specific location is called the **tidal range.** The tidal range can vary widely from place to place. Because the moon rises about 50 min later each day, the cycle of high and low tides also occurs 50 min later.

The sun's gravitational pull also has an effect on the tides. However, the ability of the sun to produce tides is only one half that of the moon. Nevertheless, the sun's gravitational pull can strengthen or weaken the gravitational pull of the moon on the tides.

During the periods of the new moon and the full moon, the earth, the sun, and the moon are all aligned. The combined gravitational pull of the sun and the moon act on the earth, resulting in higher high tides and lower low tides. Therefore, the daily tidal range is largest at these times. During these periods, which occur twice a month, tides are called **spring tides.**

During the periods of the first and third quarters of the moon, the moon and the sun are at right angles with respect to the earth. For that reason the gravitational forces of the sun and moon do not act together but rather work against each other. As a result, the daily tidal range is small. Tides that occur during this time are called **neap tides.** How often do neap tides occur?

Horizons

Technology: Nova Scotia Tidal Power

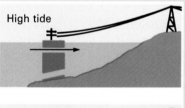

High tide

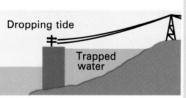

Dropping tide

Trapped water

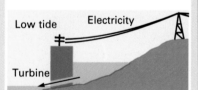

Low tide   Electricity

Turbine

The people of Nova Scotia, on the Atlantic Coast of Canada, are converting the energy of the remarkable tides in the Bay of Fundy into electricity.

Since the early 1900's, the tides in the Bay of Fundy had been considered a possible resource for generating electricity. However, a combination of low-priced fossil fuels and high-cost construction made tidal-power projects economically unsound. When fossil fuel prices rose dramatically in the late 1970's, utility companies began to look once again to the sea as a source of power.

One result of the renewed interest in tidal power can be found at Annapolis Royal,

Nova Scotia. This is the site of the first power plant in North America to harness tidal power. Completed in May 1984, the single-engine plant was designed and built by a local utility company, Nova Scotia Power. Notice the location and structure of this power plant in the photograph above right.

As the tidal bulges move around the earth, friction with the ocean floor slows the earth's rotation slightly. Scientists estimate that the average length of a day has increased by 10.8 min since the dinosaurs became extinct 65 million years ago.

Tidal Variations

Although the ocean is all one body of water, landmasses and irregularities in the ocean floor divide it into several basins. Tidal patterns are greatly influenced by the size, shape, depth, and location of the ocean basin in which they occur.

Along the Atlantic Coast of the United States, tides follow a semidiurnal, or twice daily, pattern. There are two high tides and two low tides each day, with a fairly regular tidal range. Along the shore of the Gulf of Mexico, however, tides follow a diurnal, or daily, pattern. There is only one high tide and one low tide each day. Along the Pacific Coast, the tides follow a mixed pattern, with an irregular tidal range. Pacific Coast tides often occur in the following sequence: a very high tide followed by a very low tide, then a medium high tide, followed by a slight retreat of water.

The operation of the power plant is diagramed left. When the tide rises, water is allowed to flow into a holding pond, where it is trapped. When the tide ebbs, the gates of the holding pond are opened, and the trapped water flows back into the ocean. As the water flows from the holding pond, it passes through the largest propeller-driven turbine in the world. This turbine is shown in the photograph below left. The water, traveling at the rate of about 4,390 m³/s, spins the propeller of the turbine. This spinning of the turbine activates a generator, which produces electricity. The electricity is then sent to nearby power stations for local distribution.

The hour-by-hour supply of electricity from this tidal-power plant varies with the fluctuating tidal cycle. However, over the course of a year, the plant generates about 50 million kilowatt-hours of electric power, enough to supply 100 average households. This power helps to reduce the need for power generated by more-expensive fossil and nuclear fuels.

Would more energy be produced by the power plant during spring tides or neap tides? Why?

Tidal patterns are also affected by **tidal oscillations** (AHSS-uh-LAY-shunz), slow, rocking motions of ocean water that occur as the tidal bulges move around the earth. Along straight coastlines and in the open ocean, the effects of tidal oscillations are not readily apparent. In some enclosed seas, such as the Baltic and the Mediterranean, tidal oscillations reduce the effects of the tidal bulges. As a result, these seas have a very small tidal range. In small basins and narrow bays, however, tidal oscillations may amplify the effects of the tidal bulges.

Tidal oscillations that greatly increase the tidal range are found in the narrow V-shaped Bay of Fundy, at the end of the Gulf of Maine in Nova Scotia. At times the tidal range in the Bay of Fundy exceeds 15 m. As the water oscillates and floods the bay at high tide, it raises the water level nearly 15 m. As the water moves back to the other end of the Gulf of Maine, the resulting low tide almost completely drains the bay. See Figure 22–10 for this effect. Nantucket Island, in contrast, has a relatively straight coastline and is thus barely affected by tidal oscillations. The tidal range off the coast of Nantucket Island is only a half meter.

Figure 22–10. The difference between low tide (top) and high tide (bottom) in the Bay of Fundy is nearly 15 m.

Figure 22–11. A tidal bore moves up the Amazon River at a speed of nearly 22 km/hr.

Tidal Currents

As the ocean water rises and falls with the tides, it flows toward and away from the coast. This movement of the water is called a **tidal current.** When the tidal current flows toward the coast, it is called a *flood current.* When the tidal current flows toward the ocean, it is called an *ebb current.* The areas along the coast that are affected by tidal currents are called *tidal flats.* Tidal flats may be narrow strips of beach or larger areas covering several square kilometers.

Tidal currents are not a powerful force in the open ocean. Tidal currents are strongest between two adjacent regions that have large differences in the height of the tides. In bays and along other narrow coastlines, tides may create rapid currents. Tidal currents off the coast of Brittany, in northwestern France, for example, accompany a high tide of 12 m. These tidal currents can attain a speed of 20 km/hr. Ship captains who attempt to navigate ships in such areas must take these powerful currents into account.

When a river enters the ocean through a long bay, the tide may rush into the river. The surge of water that rushes upstream, called a **tidal bore,** is shown in Figure 22–11. The water can rise as high as 5 m. In some cases, the tidal bore rushes upstream in the form of a large wave, which eventually loses energy. The tidal bore that pours into the Amazon River resembles a small, moving waterfall as it moves rapidly upstream. The tidal bores in the River Severn in England travel almost 20 km/hr and reach up to 33 km inland.

Section 22.3 Review

1. What causes tides?
2. What are spring tides? What are neap tides?
3. Where might tidal currents be of concern to ships that are approaching the land?
4. What factors influence the tidal patterns in a particular location?

Reviewing Chapter 22

Key Concepts

- Most of the surface currents of the ocean are the result of global wind patterns. These wind patterns are, in turn, affected by the presence of landmasses and the rotation of the earth. See page 365.

- Deep currents are produced as dense water near the North and South poles sinks and moves toward the equator beneath less-dense water. See page 368.

- The wind is the primary source of wave energy. See page 371.

- The gravitational effects of the moon and, to a lesser extent, the sun cause tides. See page 377.

Breaker

- **The size of a wave and the shape of the ocean-basin floor close to the shore determine the impact that a wave has on the shoreline. See page 374.**

- Tidal currents are not a powerful force in the open ocean but may create rapid currents in narrow bays. See page 381.

Key Terms

breaker (374)	Gulf Stream (366)	tidal bore (381)	undertow (375)
Coriolis effect (365)	neap tide (378)	tidal current (381)	wave (371)
crest (371)	refraction (375)	tidal oscillation (380)	wave height (371)
current (365)	rip current (375)	tidal range (378)	wave period (371)
deep current (365)	spring tide (378)	tide (377)	westerlies (365)
drift (367)	surface current (365)	trade winds (365)	whitecap (373)
fetch (373)	swell (372)	trough (371)	

Review

On your paper, write the letter of the term that best completes each of the following statements.

1. The waters in the ocean move in streams called (a) current (b) westerlies (c) waves (d) tides.

2. The effect of the earth's rotation on winds and ocean currents is called the (a) neap-tide effect (b) refraction effect (c) Coriolis effect (d) tsunami effect.

3. Two warm-water currents, the North and South Equatorial currents, flow (a) northward (b) westward (c) southward (d) eastward.

4. The warm, swift current that flows up the east coast of the United States and then into the North Atlantic is the (a) Japan Current (b) California Current (c) Labrador Current (d) Gulf Stream.

5. A weak current is known as a (a) swell (b) drift (c) tide (d) crest.

6. The vast area of relatively still, warm water located in the North Atlantic is referred to as the (a) Sargasso Sea (b) North Atlantic Drift (c) North Equatorial Current (d) Gulf Stream.

7. Deep currents are the result of (a) the Coriolis effect (b) changes in the density of ocean water (c) the trade winds (d) neap tides.

8. A strong, temporary current that is caused by an underwater landslide is called (a) a surface current (b) an equatorial current (c) a turbidity current (d) a Labrador current.

9. The periodic up-and-down movement of water is a (a) current (b) fetch (c) breaker (d) wave.
10. The highest point of a wave is its (a) trough (b) crest (c) period (d) length.
11. The time it takes for one complete wavelength to pass a given point is called the (a) wave speed (b) wave height (c) trough (d) wave period.
12. Groups of long, rolling, waves of the same size are called (a) swells (b) tsunamis (c) breakers (d) whitecaps.
13. The length of open water across which the wind blows to create a wave is the (a) trough (b) sargassum (c) fetch (d) wave period.
14. Waves that have their crests blown off by the wind form (a) breakers (b) whitecaps (c) crests (d) tsunamis.
15. When the faster-moving top part of a wave topples over the slower bottom part, the result is a foamy mass of water called a (a) whitecap (b) rip current (c) breaker (d) crest.
16. Sand bars may be formed by (a) undertows (b) rip currents (c) tsunamis (d) longshore currents.
17. Seismic sea waves are called (a) tsunamis (b) rip currents (c) undertows (d) turbidity currents.
18. The daily changes in the elevation of the ocean surface are called (a) waves (b) tides (c) swells (d) breakers.
19. The difference in the level between high tide and low tide is the (a) neap tide (b) spring tide (c) tidal range (d) tidal oscillations.
20. The movement of water toward and away from the coasts due to tidal forces is called a (a) tidal bore (b) tidal current (c) tidal range (d) tidal oscillation.

Application

On your paper, write answers to the following questions.

1. During winter in the northern Indian Ocean, winds called *monsoons* blow in a direction opposite to the direction that they blow during summer. What effect do these winds have on the surface currents?
2. What could you do to create a deep current in a pan of slightly salty water? Explain how you would go about this.
3. From the deck of a stationary ship on the open ocean you notice that at least 15 minutes pass between the rise of one wave and the next. Can you assume that the sea is very calm? Explain why or why not.
4. Suppose that a retaining wall is built along a shoreline. What will happen to waves as they pass over the retaining wall?
5. Imagine that you are fishing from a small boat anchored off the shore of the Gulf of Mexico. You are lulled to sleep by the gently rocking boat but wake up suddenly to find your boat on wet sand. What has happened?
6. Along the shoreline of a particular bay, high tide rises to about 20 m, and low tide recedes to about 10 m. The bay is wide, with a large opening into the ocean. Should a tidal-power plant be built across the bay? Explain your answer.

Extension

1. Read about tsunamis that have occurred over the past 100 years. Write a short report on your findings.
2. Obtain a timetable that gives the average times for daily high and low tides for a shore location near where you live. If the times are irregular, find out what causes the irregularities. Report your findings. If the times are regular, report on the factors that influence this regularity.
3. Find out how much electricity is provided by the La Rance, France, tidal power plant project. Find out what impact the project has had on the environment of the area. Report on your findings.

UNIT
7
Atmospheric Forces

Cumulus clouds (background); Kitt Peak National Observatory, Arizona (inset)

Introducing Unit Seven

Climate is not governed by political boundaries. *Regardless of where you live, weather events happening halfway around the world today are likely to affect you within a week.*

Fitting together the cause-and-effect patterns that influence the atmosphere of the entire globe is an important area of research and study today. From small matters of convenience—Shall I take along my umbrella?—to the largest concerns of human life and safety, weather is a significant, dramatic, and often violent factor in everyone's life.

In this unit, you will explore the global atmospheric forces that bring everyone in the world the same important phenomenon—their local weather.

Kenneth N. Bergman

Kenneth Bergman
Climatologist
Climate Analysis Center
National Weather Service

23 The Atmosphere

24 Water in the Atmosphere

25 Weather

26 Climate

385

Chapter 23

The Atmosphere

Have you ever thought of yourself as an organism that lives in a sea of gases, as a fish lives in water? People live within a narrow region where the earth's atmosphere meets the land. The conditions within this limited space have a large effect on people's lives. In this chapter you will learn how natural forces and processes act on this narrow region to influence conditions such as weather, air pressure, and wind.

Chapter Outline

The wind provides energy to move this windsurfer over the water.

23.1 Characteristics of the Atmosphere

Section Objectives

- **Discuss the composition of the earth's atmosphere.**
- **Explain how two types of barometers work.**
- **Describe the layers of the atmosphere.**
- **Identify the weather conditions that increase the effects of air pollution.**

The atmosphere is the mixture of gases and particles that surrounds the earth. The atmosphere influences almost every living thing. You breathe the gases of the atmosphere. The temperature of the atmosphere determines how you dress and what many of your daily activities will be.

As explained in Chapter 1, the study of the atmosphere is called *meteorology*. Meteorologists study all the characteristics of the atmosphere. They also study **weather** and **climate**. Weather is the general condition of the atmosphere at a particular time and place; it includes temperature, air movements, and moisture content. The general weather conditions over a long period of time, either yearly or seasonally, are the climate. Meteorology, however, is not limited to weather and climate changes. Meteorologists may specialize in certain areas such as agriculture, aviation, forestry, and health.

Composition of the Atmosphere

The atmosphere, or air, is a mixture of chemical elements and compounds. As Figure 23–1 shows, the most abundant elements in air are the gases nitrogen, oxygen, and argon. The most abundant compounds in air are the gases carbon dioxide and water vapor. However, the graph does not include water vapor because the amount of water vapor varies greatly under different conditions.

Water vapor is added to air by evaporation. Most water vapor comes from the oceans, but some also comes from lakes, ponds, streams, and the soil. Plants give off water vapor during transpiration, one of their life processes. At the same time that water vapor is being added to the atmosphere by evaporation, it is being removed by condensation and precipitation. The percentage of water vapor in air varies, depending on factors such as time of day, location, and season. Moist air may contain as much as 4 percent water vapor. Dry air has less than 1 percent water vapor. Because the amount of water vapor in air is variable, the composition of air is usually given for dry air.

Another important substance in the atmosphere is a form of oxygen called **ozone,** although it is present only in very small amounts. Oxygen (O_2) has two atoms per molecule, whereas ozone has three. What would the chemical formula for ozone be?

The ozone in the atmosphere is important because it protects the earth's inhabitants by absorbing harmful ultraviolet rays of the sun. Without ozone in the atmosphere, people would be severely burned by ultraviolet rays. Unfortunately, a number of human activities damage the ozone layer. Gases from aerosol spray cans, for example, break down ozone. Hydrocarbons from the burning of supersonic aircraft fuel also break down ozone.

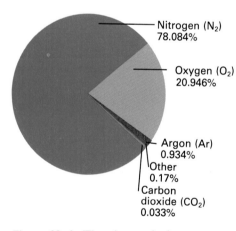

Nitrogen (N_2) 78.084%

Oxygen (O_2) 20.946%

Argon (Ar) 0.934%

Other 0.17%

Carbon dioxide (CO_2) 0.033%

Figure 23–1. The pie graph shows the composition of dry air by volume. Other gases included in the 0.17% are neon (Ne), helium (He), and xenon (Xe).

In addition to gases, the atmosphere usually contains various kinds of solid particles, which are called *atmospheric dust*. Atmospheric dust includes mineral particles lifted from soil by winds, ash from fires, volcanic dust, and microscopic organisms. It may also include particles from meteors that have vaporized. When tiny drops of ocean water are tossed into the air as sea spray, the drops evaporate. Left behind in the air are tiny crystals of salt, another component of atmospheric dust. Large dust particles remain in the atmosphere only briefly because their weight causes them to fall. However, many particles, which are so small they cannot be seen, remain suspended in the atmosphere for months or years.

All over the earth and up to an altitude of about 100 km, the composition of dry air is nearly the same. Although nitrogen and oxygen are always being added to, as well as removed from, the atmosphere, the relative amounts of these gases do not change significantly.

Oxygen in the Atmosphere

The amount of oxygen in the atmosphere is the result of natural processes that maintain the chemical balance of the atmosphere. Animals, bacteria, and other plants remove oxygen from the air as part of their life processes. Forest fires, the burning of fuels, and the weathering of rocks also use up oxygen. Living things, burning, and weathering could quickly use up most atmospheric oxygen if it were not for various processes that add oxygen to air. Land and ocean plants produce large quantities of oxygen. During photosynthesis plants use sunlight, water, and carbon dioxide to produce their food. Oxygen is released as a product of photosynthesis. The amount of oxygen produced by plants each year nearly equals that consumed by both animals and plants. Thus, the oxygen content of the air is in a state of balance. It does not change significantly over hundreds or even thousands of years.

Figure 23–2. This illustration shows the use and production of oxygen and carbon dioxide in the atmosphere. The oxygen-carbon dioxide cycle maintains a stable amount of oxygen in the atmosphere.

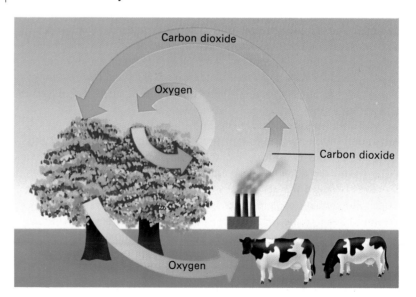

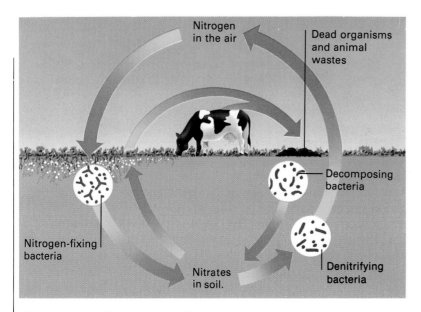

Figure 23–3. The nitrogen cycle maintains a stable amount of nitrogen in the air.

Labels in figure:
Nitrogen in the air
Dead organisms and animal wastes
Decomposing bacteria
Nitrogen-fixing bacteria
Nitrates in soil.
Denitrifying bacteria

Nitrogen in the Atmosphere

The amount of nitrogen in the atmosphere is maintained through the **nitrogen cycle.** During the cycle, nitrogen moves from the air to the soil to animals and back again to the air. As Figure 23–3 shows, nitrogen is removed from the air mainly by the action of nitrogen-fixing bacteria. These microscopic organisms live in the soil and on the roots of certain plants. The bacteria chemically change nitrogen from the air into nitrogen compounds, which are vital to the growth of all plants. When animals eat the plants, the nitrogen compounds enter their bodies. These compounds are then returned to the soil through animal excretions or by the decay of dead organisms. In the soil, the processes involved in decay release nitrogen and return it to the atmosphere. A similar nitrogen cycle takes place among water-dwelling plants and animals.

Atmospheric Pressure

Gravity pulls the gases of the atmosphere toward the earth's surface and holds them there. Due to the pull of gravity, 99 percent of the total mass of atmospheric gases is found within 32 km of the earth's surface. The remaining 1 percent extends upward for hundreds of kilometers but gets increasingly thinner at high altitudes. In other words, there is less air at these higher altitudes.

A 1-cm^2 column of air that reaches from sea level to the top of the atmosphere weighs 1.03 kg and has a force of 10.1 newtons. In other words, at sea level on every square centimeter of the earth's surface the atmosphere presses down with a force of 10.1 newtons, and it weighs 1.03 kg. The ratio of the weight of the air to the area of the surface on which it presses is called **atmospheric pressure.** Since there is less air at higher altitudes, there is less weight pressing down on surfaces at those altitudes. Thus, the atmospheric pressure is lower at higher altitudes.

You have probably experienced the effects of changes in air pressure. If you have traveled up a tall mountain or if you have risen very quickly in an elevator, you may have had a popping sensation in your ears. This sensation is due to the decreased air pressure on the outside of your eardrum. When the air pressure on both sides of the eardrum is equalized, the popping stops.

Mercurial Barometer

An instrument that measures atmospheric pressure is called a **barometer.** One type of barometer is called a *mercurial barometer.* Atmospheric pressure presses on the liquid mercury in a well at the base of the barometer. The pressure squeezes the mercury up to a certain height inside a tube. The height of the mercury inside the tube depends on the amount of atmospheric pressure. The greater the atmospheric pressure is, the higher the mercury rises.

Air pressure measured by a mercurial barometer is expressed by how high the mercury rises in the tube. A reading of 760 mm on the

Investigation

Barometric Pressure

A barometer measures changes in atmospheric pressure. You can construct a simple aneroid barometer with some common objects.

Science Process Skills Focus: constructing models, recording data, interpreting data

Materials

plastic wrap; coffee can with a diameter of about 10 cm, open at one end; rubber band; drinking straw, about 10 cm long; adhesive tape; cardboard, 10 cm wide and at least 8 cm taller than the can

Procedure

1. Refer to the figure in making your barometer.
2. Secure plastic wrap tightly over the open end of the can with a rubber band.
3. Tape one end of the straw onto the plastic wrap near the center, as shown in the figure.
4. Fold the cardboard so that it stands upright at least 3 cm above the top of the can.
5. Place the cardboard so that the free end of the straw just touches the front of the cardboard. Mark an *X* where the straw touches.
6. Draw three horizontal lines on the cardboard: level with the *X*, 2 cm above, and 2 cm below

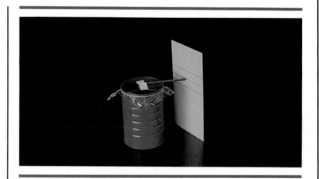

the *X*. Label the lines as shown.
7. Position the cardboard so that the straw touches the *X* again. Tape the base of the cardboard in place.
8. Observe the level of the straw at least once a day over a 5-day period. Record your results.

Analysis and Conclusions

1. What factors affect how your model works? Explain your answer.
2. What does an upward movement of the straw indicate? A downward movement?
3. Compare your results with the barometric pressure listed in your local newspaper. What kind of weather can you associate with high pressure? With low pressure?

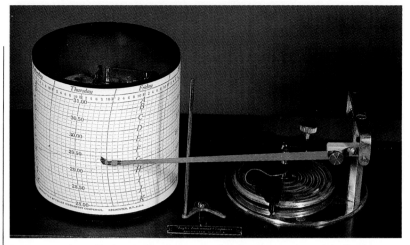

Figure 25–4. Aneroid barometers measure air pressure by the compression and expansion of a sealed metal box. When an aneroid barometer is constructed to keep a continuous record of air pressure, such as the one shown here, it is called a *barograph*.

barometer is called **standard atmospheric pressure** and indicates the atmospheric pressure measured at sea level. Standard atmospheric pressure, or the reading of 760 mm of mercury, is sometimes referred to as *one atmosphere*.

Official weather maps use another measurement of air pressure, called *millibars* (mb). One millibar is equal to about 0.001 of standard atmospheric pressure.

Aneroid Barometer

The type of barometer most commonly used today does not contain mercury or any other liquid. It is called the *aneroid barometer*. The word *aneroid* means "without liquid." Inside an aneroid barometer is a sealed metal container from which most of the air has been removed. When the atmospheric pressure increases, the sides of the container bend inward. When the pressure decreases, the sides bulge out again. These changes are indicated by a moving pointer on a scale. The movement of the pointer along the scale is controlled by the changing shape of the container. The scale is usually marked to show the pressure either in millimeters of mercury or in millibars. Aneroid barometers can be constructed to keep a continuous record of atmospheric pressure.

An aneroid barometer can also measure altitude above sea level. When used for this purpose, it is called an *altimeter*. The scale then registers elevation instead of pressure. At high altitudes the atmosphere is less dense and exerts less pressure. Thus a lowered pressure reading can be interpreted as an increased altitude reading. To accurately measure altitude, an altimeter must be corrected for local weather conditions.

Layers of the Atmosphere

As altitude increases, air pressure decreases rapidly but smoothly. Thus, there are no sharp pressure changes that separate the atmosphere into layers. The atmosphere does, however, show distinct differences in temperature with increasing altitude. The temperature differences mainly result from the way solar energy is absorbed as it moves downward through the atmosphere. Based upon temperature differences, scientists identify four layers of the atmosphere.

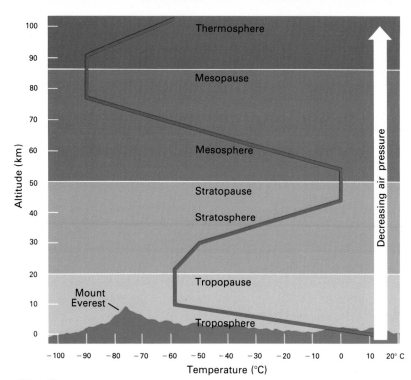

Figure 23–5. The red line indicates the temperature at each altitude of the atmosphere.

The Troposphere

The atmospheric layer closest to the earth's surface is called the **troposphere.** The word *troposphere* comes from a Greek root meaning "change." The term aptly describes this atmospheric layer in which almost all weather change occurs. Nearly all the water vapor and carbon dioxide in the atmosphere is found in the troposphere. Temperature within this layer decreases as altitude increases because there is increasing distance from the warming effect of sunlight absorbed by the earth's surface. The temperature within the troposphere decreases at the average rate of 6.5°C/km. But, at an average altitude of 10 km, the temperature stops decreasing. In this zone, called the **tropopause,** the temperature remains nearly constant. The tropopause is the upper boundary of the troposphere. The altitude of this boundary is not constant; it changes with latitude and with the season of the year. For example, at the equator the tropopause is found at an altitude of about 17 km. However, at the poles the altitude of the tropopause varies between 6 km and 8 km.

The Stratosphere

The layer of the atmosphere called the **stratosphere** extends upward from the tropopause to an altitude of 50 km. In the lower stratosphere, the temperature is about −60°C. In the upper stratosphere, the temperature begins to increase as the altitude increases. The air in the stratosphere gets warmer as a result of the direct absorption of solar energy by ozone. Almost all the ozone in the atmosphere is found within the stratosphere.

The temperature of the ozone layer rises steadily to an altitude of about 50 km, where the stratosphere reaches its highest temperature. This high-temperature zone, called the **stratopause,** marks the upper boundary of the stratosphere.

The Mesosphere

Above the stratopause and extending to an altitude of about 80 km is the atmospheric layer called the **mesosphere.** In this layer the temperature decreases as the altitude increases. In fact, the mesosphere is the coldest layer of the atmosphere, dropping to a temperature of nearly –100°C. The upper boundary of the mesosphere, called the **mesopause,** is marked by an increase in temperature.

The Thermosphere

In the atmospheric layer above the mesopause, called the **thermosphere,** the temperature increases steadily with altitude. In the thermosphere, nitrogen and oxygen atoms absorb solar energy. This process explains the high temperatures in the thermosphere. There are not enough data about temperature changes in the thermosphere to determine its upper boundary.

In the very thin air of the thermosphere, a thermometer cannot accurately measure the temperature. A thermometer measures the temperature of the particles, or the energy of the moving molecules, that strike it. Because the air in the thermosphere is so thin, the particles move quickly but are far apart. Therefore they do not strike the thermometer often enough to produce an accurate temperature reading. Special instruments are needed to measure temperature accurately in the thermosphere. These instruments have recorded temperatures of more than 2,000°C in the thermosphere.

The lower region of the thermosphere, at an altitude of 80 km to 550 km, is often called the **ionosphere.** In the ionosphere solar rays absorbed by atmospheric gases cause the atoms of gas molecules to lose electrons and to produce ions and free electrons. The ionosphere gets its name from these ions. The ions and the free electrons are concentrated into four layers. The layers of free electrons can reflect radio waves back to the earth, as shown in Figure 23–6.

Above the ionosphere is the region where the earth's atmosphere blends into the almost-complete vacuum of interplanetary space. This zone of indefinite altitude, called the **exosphere,** extends for thousands of kilometers above the earth.

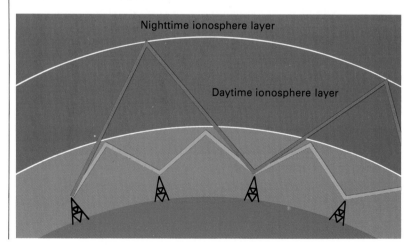

Nighttime ionosphere layer

Daytime ionosphere layer

Figure 23–6. Radio waves can be transmitted around the world by reflecting them off of the ionosphere. At night the radio waves can travel farther because the lowest ion layer disappears and the waves are reflected off a higher ion layer.

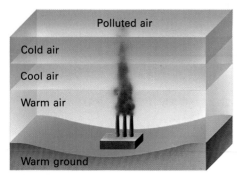

Normal

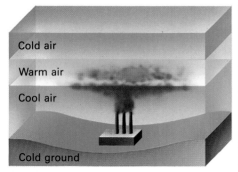

Inversion

Figure 23–7. Normal conditions are shown top, a temperature inversion is shown bottom. During a temperature inversion, polluted cool air becomes trapped beneath a warm-air layer.

Air Pollution

Any substance in the atmosphere that is harmful to people, animals, plants, or property is an **air pollutant.** Many substances commonly found in air, such as sulfur dioxide, carbon monoxide, lead, and hydrocarbons, are known to be harmful when breathed by humans.

On several occasions severe air pollution has caused a large number of deaths. Less dramatic but perhaps more serious are the effects on the health of people who breathe polluted air during their entire lives. Studies have shown that constant exposure to pollution decreases people's ability to resist many illnesses.

The main source of air pollution is the burning of fossil fuels. Before automobiles became common, most air pollution resulted from the burning of coal by industry and in homes. Today most air pollution comes from the burning of coal and petroleum fuels. When these fuels burn, the sulfur that is released forms harmful sulfur dioxide gas. The operation of automobile engines produces several harmful substances such as certain hydrocarbons, nitrogen oxides, carbon monoxide, and lead.

Acid rain is another harmful side effect of the burning of fossil fuels. Gases emitted by the burning of fossil fuels form acids when combined with water in the air. These acids fall to the earth as acid rain. Over time acid rain poisons fish, ruins soil, and kills crops and trees.

Air pollution can become an even more serious problem as a result of certain weather conditions. A common cause of periodic air pollution is the layering of warm air on top of cool air. Warm air, which is less dense, can rise over cool polluted air and can trap the cool air beneath it. Meteorologists refer to this condition as a **temperature inversion.** In some areas, topography may make air pollution even worse by keeping the polluted inversion layer from dispersing. Los Angeles, for example, is the site of frequent temperature inversions. Cool Pacific air becomes polluted, is covered by a warm-air layer, and is prevented from moving by mountains that border the city. Under conditions in which air cannot circulate up and away from an area, trapped automobile exhausts produce *smog.* Smog is a type of pollution composed of smoke and fog.

Air pollution can be controlled only by preventing pollutants from being released into the atmosphere. Federal and local laws aimed at reducing the amount of pollutants produced by automobiles and industry have been passed.

Section 23.1 Review

1. What are the most abundant elements in dry air?
2. What does a barometer measure?
3. Where in the atmosphere do most weather changes occur?
4. Which industrial city—one on the Great Plains or one near the Rocky Mountains—would have less-frequent incidents of temperature inversion? Why?

23.2 Solar Energy and the Atmosphere

Section Objectives

- **Explain how radiant energy reaches the earth.**
- **Describe how visible light and infrared energy warm the earth.**
- **Summarize the processes of radiation, conduction, and convection.**

The earth's atmosphere is heated by the transfer of energy from the sun in several ways. Some of the heat in the atmosphere comes directly from the rays of the sun as they pass through the atmosphere. Some heat enters the atmosphere indirectly as the earth's surface absorbs solar energy and then gives it off as heat.

Radiation

All of the energy that the earth receives from the sun travels through the space between the earth and the sun as **radiation.** Radiation includes all forms of energy that travel through space as waves. Light is the form of radiation that can be seen with the eyes. However, there are many other forms of radiation that cannot be seen, such as radio waves.

Radiation moves through space in the form of waves at a very high speed—300,000 km/s. The distance from one wave crest to the next, as shown in Figure 23–8, is called the *wavelength* of a wave. The various types of radiation differ in the length of their waves. Visible light, for example, consists of waves with various wavelengths that you see as different colors. Wavelengths shorter than those of visible light include ultraviolet rays, X rays, and gamma rays. Longer wavelengths include **infrared** (IN-fruh-RED) **waves** and radio waves. The waves that make up all forms of radiation are called *electromagnetic waves.* The complete range of wavelengths makes up the **electromagnetic spectrum.**

Figure 23–8. This illustration shows the electromagnetic spectrum from short gamma-ray waves to long radio waves.

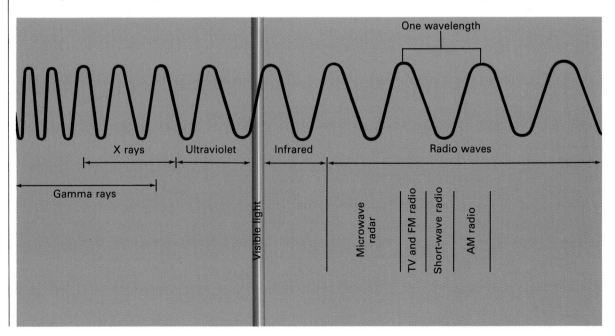

One wavelength

X rays | Ultraviolet | Infrared | Radio waves

Gamma rays

Visible light

Microwave radar

TV and FM radio

Short-wave radio

AM radio

Almost all the energy reaching the earth from the sun is in the form of electromagnetic waves. A small amount of solar energy is carried to the earth by atomic particles emitted by the sun. Before the sun's radiation reaches the solid part of the earth, it passes through the earth's atmosphere. The atmosphere affects this radiation in several ways. First the molecules of nitrogen and oxygen in the air absorb the very short wavelengths of X rays, gamma rays, and ultraviolet rays. This absorption occurs in the mesosphere and higher layers of the atmosphere. As solar energy is absorbed, molecules and atoms of nitrogen and oxygen lose electrons and become positively charged ions. In the stratosphere ultraviolet rays act upon oxygen molecules to form ozone.

Thus, almost all of the shorter wavelengths are absorbed in the upper atmosphere. The small amount of ultraviolet radiation that does reach the earth's surface causes sunburn of skin exposed to

Science Notebook

Visible Light and the Atmosphere

The atmosphere has a number of effects on the light that you see. The wavelengths of visible light that are most readily scattered by the fine dust in the air are the shorter ones. You see these wavelengths as blue and violet. Thus the sky looks blue when the air is clear. Larger particles, such as water droplets in clouds or fog, scatter most of the wavelengths of visible light. This effect makes the sky look white—a combination of all colors. At high altitudes, where the atmosphere is less dense, there are few particles to cause scattering. As a result, the sky looks black.

A rainbow is another example of the effects of the atmosphere on visible light. A rainbow is caused by the separation of sunlight into a range of colors by raindrops. Refraction, or bending, and reflection of light rays in the raindrops separate each ray of white light into the entire spectrum of colors. You can see these colors only when you face the water droplets with the sun behind you.

A mirage is an optical illusion created by light rays. Light rays are refracted as they strike the boundary between a layer of hot air and a layer of cool air. You may have seen a mirage on a summer day when water seemed to appear on the dry pavement of a highway. This mirage occurs when a layer of cool air reflects the sky onto a layer of hot air close to the ground. When the light rays enter the hot air, they are bent upward, causing the sky to appear as a pool of water on the pavement. This is called an *inferior mirage*.

A distant mountain that seems to be suspended in the sky is called a *superior mirage*. This type of mirage forms when light rays pass from hot air into cool air.

Are the light rays in a superior mirage bent upward or downward?

direct sunlight for too long a time. Most of the rays that reach the lower atmosphere have longer wavelengths—those of visible, infrared, and radio waves. Most incoming infrared radiation is absorbed by carbon dioxide, water vapor, and other complex molecules in the troposphere. Only small amounts of visible light and radio waves are absorbed as they pass through the atmosphere.

Scattering

Clouds, dust, and gas molecules in the atmosphere affect the path of radiation from the sun, causing scattering. Scattering means that water droplets and dust suspended in the atmosphere reflect and bend the rays. This bending sends the rays out in all directions without changing their wavelengths. In clear, cloudless air, scattering is also caused by the reflection of light off gas molecules. Scattering sends some of the radiation back into space. The remaining radiation continues downward toward the earth's surface. As a result of scattering, sunlight reaching the earth's surface comes from all directions. Scattering is what makes the sky appear blue and makes the sun appear red at sunrise and sunset. Short-wavelength rays, such as blue light, are more easily scattered than long-wavelength rays are. The sun appears red when it is low in the sky because more blue light rays are scattered by the atmosphere. Thus more of the longer-wavelength red light rays reach the earth's surface, giving the sun its red color.

Reflection

Of the total amount of solar energy reaching the earth's atmosphere, about 20 percent is absorbed by the atmosphere. About 30 percent is scattered back into space or is reflected from clouds or the earth's surface. The remaining 50 percent is absorbed by the surface.

When solar energy reaches the earth's surface, the surface either absorbs the energy or reflects the energy. The kind of surface on which the radiation falls determines to a large extent whether absorption or reflection occurs. Table 23–1 shows the amount of incoming solar radiation reflected by various surfaces.

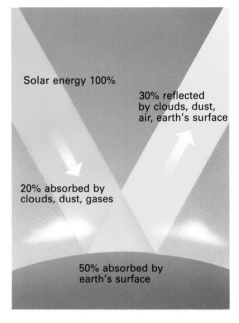

Solar energy 100%

30% reflected by clouds, dust, air, earth's surface

20% absorbed by clouds, dust, gases

50% absorbed by earth's surface

Figure 23–9. Most solar energy is absorbed by the earth's surface and by atmospheric particles and gases. The remainder is reflected back into space.

Table 23–1 Reflection and Absorption

| Surface | Percentage of solar radiation: | |
	Reflected	Absorbed
Soils	5-10	95-90
Desert	20-45	80-55
Grass	16-26	84-74
Forest	5-20	95-80
Snow	40-95	60- 5
Water (high sun angle)	3-10	97-90
Water (low sun angle)	10-80	90-20

Because 30 percent of the total solar energy that reaches the earth's atmosphere is either reflected or scattered, the earth is said to have an average reflectivity of 0.3. The fraction of solar radiation reflected by a particular surface is called its **albedo.** The earth has an albedo of 0.3. The albedo of the moon is 0.07. What percent of the total solar energy reaching the moon is reflected?

Absorption and Infrared Energy

The solar radiation that is not reflected by the earth's surface is absorbed by the earth. Part of the absorbed radiation is composed of the infrared rays that have penetrated the atmosphere. When you feel the warmth of the sun, you are feeling infrared rays. The rocks, soil, water, and other earth materials are heated when they absorb infrared rays and visible light. The heated materials then produce their own infrared rays from the heat energy. These infrared rays have much longer wavelengths than the infrared rays that reach the earth's surface directly from the sun. The infrared rays with shorter wavelengths pass through the gases of the atmosphere. The infrared rays produced by the warmed materials of the earth's surface are mostly absorbed by water vapor and carbon dioxide in the atmosphere.

The Greenhouse Effect

The absorption of long-wavelength infrared rays from the earth's surface traps heat energy from the sun and prevents it from escaping back into space. As a result, the lower atmosphere becomes warmed. The warmed lower atmosphere causes the earth's surface to have a much higher temperature than it would if there were no atmosphere. This process by which the atmosphere absorbs radiation has been compared to a greenhouse. People once believed that the glass of a greenhouse allowed the short wavelengths of infrared rays from the sun to pass through. They also believed that the glass prevented the longer infrared rays emitted by the warmed surfaces within the

Figure 23–10. The visible and infrared rays of incoming sunlight pass through the water vapor and carbon dioxide of the atmosphere. Most of the longer infrared rays sent out by the warmed surfaces on the earth are trapped by these same substances.

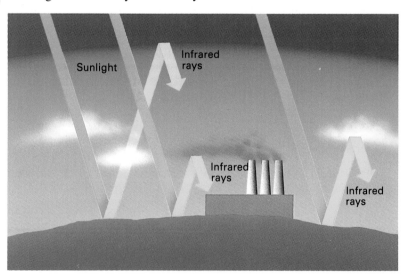

greenhouse from escaping to the outside. However, experiments have shown that a greenhouse is heated mostly because air warmed by the infrared rays is prevented from escaping. Thus, a greenhouse does not become heated in the same way as does the atmosphere. Nevertheless, the process by which the atmosphere traps infrared rays over the earth's surface is usually called the **greenhouse effect.** By this process, the energy from the sun warms the air after having first been absorbed at the earth's surface. The atmosphere is heated mostly in the lowest layer, the troposphere.

The average temperatures over all the earth's surface do not change much from year to year. Normally the amount of solar energy trapped is about equal to the amount that escapes into space. However, human activities change this balance and may cause the average temperature of the atmosphere to increase. For example, the burning of fossil fuels releases carbon dioxide into the air. Carbon dioxide absorbs infrared rays very effectively. Measurements have shown that the amount of carbon dioxide in the atmosphere has been increasing in recent years and seems likely to continue to increase in the future. Increases in the amount of carbon dioxide will intensify the greenhouse effect and may cause the atmosphere to become warmer. Such general warming might cause climate changes in many parts of the world. Scientists are doing research on the possible effects of a global warming.

Variations in Temperatures

Radiation from the sun does not heat the earth evenly in all places at all times. How warm the atmosphere becomes in any region on the earth's surface depends upon several factors. Location is one factor. Average temperatures are higher near the equator than near the poles. The direct rays of the sun striking near the equator are more effective in heating an area than the slanting rays striking the polar regions. Slanting rays spread their energy over a larger area than do direct rays.

Elevation is another factor in regional temperature variations. High elevations, such as mountaintops, become warmer than do nearby valleys during the day but cool more quickly at night. The thinner air at high altitudes contains less water vapor and carbon dioxide to trap the heat. Desert temperatures usually show large changes between day and night because there is little water vapor present to hold the heat.

The temperature of water changes less than the temperature of land does when solar energy is absorbed or when heat is given off. Water thus has a moderating effect. Under similar conditions, regions close to large bodies of water generally have more moderate temperatures. In other words, they will be cooler during daytime and warmer at nighttime than inland regions with the same general weather conditions. Also the location of an area in relation to the wind patterns makes a difference. A region that receives winds off the ocean waters has a different temperature from a similar region in which the winds blow from the land toward the ocean.

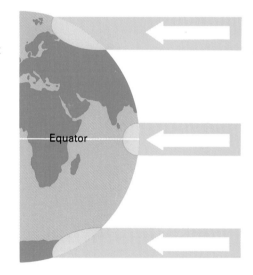

Figure 23–11. Temperatures are higher at the equator because solar energy is concentrated in a small area. Farther north and south, the same amount of solar energy is spread out over a larger area and temperatures are lower.

There is also a delay in the heating of the atmosphere both seasonally and daily. The sun is highest and its rays most direct in the Northern Hemisphere from March 21 to September 21. You might, as a result, expect this period to be the warmest time of the year. However, it is usually late June before enough heat has been absorbed and reradiated from the ground to warm the atmosphere. Only then does the warm, summer weather begin. For similar reasons, the warmest hours of the day are usually at midafternoon although the sun is strongest between 12:00 noon and 1:00 PM.

Conduction and Convection

While most heating of the atmosphere comes from radiation, a small amount results from **conduction** and *convection*. In conduction, the molecules in a substance vibrate as they become heated. The vibrating molecules cause other molecules to begin vibrating. The motion makes the substance warm. Solid substances, in which the molecules are close together, make good conductors because heat can be transferred quickly from molecule to molecule. Air is warmed when it comes in contact with anything hotter than the air itself. However, the molecules of air are far apart. As a result, air is a poor conductor of heat. Some heating of the lower part of the atmosphere takes place through conduction when air comes into contact with the warmed surface of the earth.

Convection, as you learned in Chapter 4, involves the movement of gases or liquids when they are heated unevenly. The movement of air due to convection takes place when some of the air is heated by radiation or conduction. As that air is heated by radiation or conduction, it becomes less dense and tends to rise. Nearby cooler air, which is more dense, tends to sink. As it sinks, the cooler air pushes the warm air up. The cold air is, in turn, warmed and then it rises also. This continuous cycle of cold air sinking and warm air rising helps warm the earth's surface relatively evenly. Because heated air is less dense than cool air is, it exerts less pressure on the earth than the same volume of cooler air does. Consequently, the atmospheric pressure is generally lower beneath a body of warm air than it is under cool air. As dense, cool air moves into a low-pressure region, the less dense, warmer air is pushed upward. The general movement of cool air is always toward regions of lower pressure. These pressure differences, which are the result of the unequal heating that causes convection, create winds.

Section 23.2 Review

1. What type of radiation causes sunburn?
2. Why is the atmospheric pressure lower beneath a body of warm air than beneath a body of cold air?
3. You decide not to be outside during the warmest hours of a warm summer day. When will the warmest hours probably be? How do you know?

23.3 Winds

Because the earth receives more solar energy at the equator than at the poles, there is a belt of low air pressure at the equator. The heated air in the region of the equator is constantly rising. At the poles the colder air is heavier and tends to sink. This sinking of cold air creates regions of high atmospheric pressure.

Pressure differences in the atmosphere at the equator and at the poles create a general movement of air worldwide. Notice in Figure 23–12 that air moves from high-pressure belts toward low-pressure belts. Therefore, air near the earth's surface generally flows from the poles toward the equator. At higher latitudes the rising warm air cools, and there is a general return flow of air from the equator toward the poles.

Global Winds

Everything that moves over the surface of the earth is affected by the rotation of the earth on its axis. The rotation causes surface winds in the Northern Hemisphere to turn to the right and those in the Southern Hemisphere to turn to the left. This motion is called the *Coriolis effect,* after the nineteenth-century French mathematician who first described it. A ball thrown on the earth's surface will curve only slightly due to the Coriolis effect. The ball remains in the air much too short a time to be affected. The ocean currents and winds of the world, however, are strongly affected. Winds that would otherwise blow directly from a high-pressure area toward a lower-pressure area, are turned by the Coriolis effect.

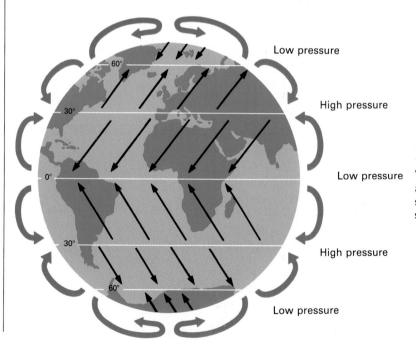

Low pressure

High pressure

Low pressure

High pressure

Low pressure

Figure 23–12. The arrows around the perimeter of the globe show the general movement of air. Winds blow from high-pressure areas to low- pressure areas. The slanting arrows across the globe show the Coriolis effect.

Chapter 23   The Atmosphere   **401**

A low-pressure belt exists at the equator because the heated air there tends to rise. As the warm air rises and begins to move toward the poles, the Coriolis effect deflects the winds to the east. Around 30° latitude some of the air moves downward toward the earth's surface, forming a belt of high pressure. At the earth's surface, the descending air turns and flows both northward and southward. The air that does not descend at 30° latitude continues toward the poles, where it then descends. The air flowing from the equator completes three looping patterns of flow called **wind cells.** The Northern and Southern hemispheres each have three wind cells.

Trade Winds

The wind cell moving toward the equator between 30° and 0° latitude creates surface winds. These winds are called *trade winds*. Like all winds, they are named according to the direction from which they flow. In the Northern Hemisphere, the trade winds flow from the northeast and are therefore called the *northeast trades*. The northeast trades are deflected to the west by the Coriolis effect. In the Southern Hemisphere the trade winds are called the *southeast trades* and are also turned to the west.

Horizons

Technology: Energy from the Wind

Through modern technology, one of the oldest means of producing mechanical power, the windmill, or wind turbine, shown left, is making a comeback as an energy supplier.

The advantages of using wind as a power source are that it consumes no fossil or nuclear fuel and there are no waste products. Also, the wind itself costs nothing and is available in limitless supply. Among the disadvantages of wind power are the unpredictable speed and direction of the wind.

However, the fossil fuel crisis of the 1970's renewed interest in solving the problems associated with producing power from the wind. Since then a number of advances in wind-generated power have been made. These include the reintroduction of the Darrieus turbine.

Invented in 1920 by a French engineer, the Darrieus,

The trade wind systems of the Northern and Southern hemispheres meet at the equator in a narrow zone called the **doldrums.** The air in this zone moves mainly upward, and near the earth's surface the winds are weak and undependable. What problems might sailing ships have in the doldrums?

A belt of high pressure in the vicinity of 30° latitude is created by the descending air. This high pressure belt is called the **horse latitudes,** or *subtropical highs*. The surface winds here are weak and changeable. The name *horse latitudes* was given to this region in the days when sailing ships carried horses from Europe to the New World. Horses were often thrown overboard to save food and water for the sailors when the ships were unable to move from this zone because of the lack of wind.

Prevailing Westerlies

In the subtropical highs some of the descending air that moves toward the poles is also deflected. This deflection creates another wind cell in both the Northern Hemisphere and the Southern Hemisphere. These winds are known as the *prevailing westerlies*. In the Northern Hemisphere, the westerlies are southwest winds. In the Southern

shown lower left, looks like an upside-down eggbeater. The Darrieus operates on a vertical axis rather than a horizontal axis like standard propeller wind turbines. Therefore, it is mechanically simpler than earlier wind generators. It is also easier to maintain because all of its mechanical and control parts are located on the ground.

There have been other advances in wind-generator technology besides the Darrieus turbine. These advances include the use of both lighter and less-costly materials and the development of faster airfoil blades. Instead of being flat like earlier blades, an airfoil is shaped like a wing, as shown in the picture. With this shape an airfoil can use the lift force of the wind more effectively.

Notice the picture of a typical windmill farm above. Between 1982 and 1986, about 13,000 wind generators were installed in three areas along the crest of the Diablo Mountains in California. One million barrels of oil would have been needed to supply the energy produced by these generators during that time.

What problems would there be in running the Darrieus turbine in areas around the equator and at 30° latitude?

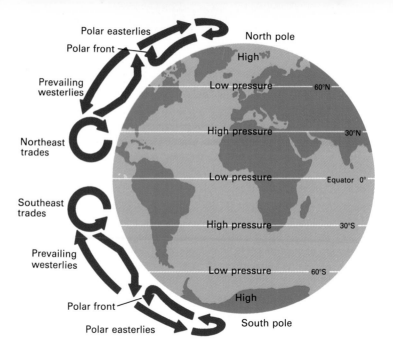

Figure 23–13. Both the Northern Hemisphere and the Southern Hemisphere have three wind cells. Wind cells are the result of pressure differences in the atmosphere at the equator and at the poles.

Hemisphere, they are northwest winds. The prevailing westerlies are located in a belt between 40° and 60° latitude. This middle latitude wind cell is not as strong as the one formed near the equator. The westerlies are not as constant as are the trade winds.

Polar Easterlies

A third wind cell exists near each of the earth's poles, as shown in Figure 23–13. North and south of the belt of westerlies, around 65° latitude, is a belt of low pressure. These **subpolar lows** result when warm air is lifted by cold polar air moving toward the equator. Over the polar regions themselves, descending cold air creates areas of high pressure. The general surface movement of the cold polar air masses is toward the equator. Surface winds created by these cells are turned to the west by the Coriolis effect, becoming the **polar easterlies.** The continuously low temperatures in the polar regions result in only small pressure differences. Consequently, polar easterlies are usually weak winds.

Wind and Pressure Shifts

As the sun's vertical rays shift northward and southward during the year, the positions of the pressure belts and wind belts also shift. Although the sun's rays move 23.5° in latitude, the average shift for the pressure and wind belts is about 6° of latitude. However, even this small change means that some areas are in different wind belts during the year. In southern California, for example, westerly winds prevail in the winter, but trade winds are dominant in the summer.

Jet Streams

In the lower regions of the atmosphere over both the Northern Hemisphere and the Southern Hemisphere are bands of high-speed winds. These winds are the **jet streams.** Because the temperature of polar air and middle latitude air differs so greatly, the pressure also differs greatly. The cold polar air is much denser than the warmer air of the middle latitudes. The resulting pressure differences produce the polar jet streams. These bands of winds, found at an altitude of 10 km to 15 km, are about 100 km wide and 2 km to 3 km deep. The polar jet streams may reach speeds of almost 500 km/hr. These winds do not blow steadily but instead change speed and position. In the subtropical regions, very warm equatorial air meets the cooler air of the middle latitudes, creating the subtropical jet streams. Unlike the polar jet streams, the subtropical jet streams do not change much in speed or position. The polar jet streams and the subtropical jet streams flow from west to east.

Local Winds

At any particular place, the movements of air are influenced by local conditions. A local feature that produces temperature differences often causes a local wind. Local winds are not usually part of the global wind patterns. Gentle winds that extend over distances of less than 100 km are called *breezes*.

Land and Sea Breezes

Equal surface areas of land and water may receive the same total amount of energy from the sun. However, the land surface reaches a higher temperature than the water does. During daylight hours, therefore, a sharp temperature difference develops between a body of water and the land along its shore. This temperature difference is apparent in the air above the land and water. The warmer air above the land rises, and the cool air from above the water moves in to replace it. A cool wind moving from water to land, called a **sea breeze,** generally begins in the late afternoon. Overnight the land cools more than the water does, and the sea breeze is replaced by a **land breeze.** A land breeze flows from the cooler land to the warmer water.

Mountain and Valley Breezes

In the daytime, mountains heat more quickly than surrounding valleys do. The slopes of mountains easily absorb the sun's energy. The valleys, because they are generally covered with forests and other vegetation, absorb energy more slowly. In the daylight hours, a gentle **valley breeze** blows up the slopes. The valley breeze is caused by cooler air from the valleys moving up to replace the warmer mountain air. At night the mountains cool more quickly than the valleys do. Then, cooler air descends on the mountain slopes, creating a **mountain breeze.**

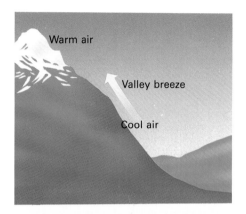

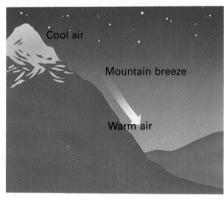

Figure 23–14. In the daytime, air near mountaintops heats more rapidly than air in valleys heats. The warm air rises and draws cool air from the valley up the slope as a valley breeze. At night, air at the mountaintop cools more quickly than valley air. The cool air moves down the mountain slope as a mountain breeze.

Section 23.3 Review

1. What are the results of the Coriolis effect?
2. What surface winds make up the middle latitude circulation cells?
3. On a camping trip on the Oregon coast, you decide to hike to the ocean, but you are not sure of the direction. The time is 4:00 PM. How might the breeze help you find the ocean? Why?

Reviewing Chapter 23

Key Concepts

- The earth's atmosphere is a mixture of gaseous elements and compounds and suspended solid and liquid particles. See page 387.

- Atmospheric pressure can be measured by a barometer. See page 390.

- The atmosphere can be divided into a number of layers. The layers vary in temperature and composition. See page 391.

- Certain substances, when released into the air, can be harmful to people, animals, plants, and property. See page 394.

- Most of the energy that reaches the earth from the sun comes in the form of electromagnetic waves. See page 396.

- Visible light and infrared rays from the sun penetrate the earth's atmosphere and heat materials such as rocks, soil, and water on the surface. See page 398.

- Heat is transferred within the atmosphere by three processes—radiation, conduction, and convection. See page 400.

- Air pressure differences and the earth's rotation influence the various global wind patterns. See page 401.

- A surface feature, such as a body of water, a mountain, or a valley, can influence local wind patterns. See page 405.

Key Terms

air pollutant (394)	greenhouse effect (399)	ozone (387)	temperature inversion (394)
albedo (398)	horse latitude (403)	polar easterlies (404)	thermosphere (393)
atmospheric pressure (389)	infrared waves (395)	prevailing westerlies (403)	tropopause (392)
barometer (390)	ionosphere (393)	radiation (395)	troposphere (392)
climate (387)	jet streams (404)	sea breeze (405)	valley breeze (405)
conduction (400)	land breeze (405)	standard atmospheric pressure (391)	weather (387)
doldrum (403)	mesopause (393)	stratopause (392)	wind cell (402)
electromagnetic spectrum (395)	mesosphere (393)	stratosphere (392)	
exosphere (393)	mountain breeze (405)	subpolar low (404)	
	nitrogen cycle (389)		

Review

On your paper, write the letter of the term that best completes each of the following statements.

1. During one part of the nitrogen cycle, nitrogen is removed from the air mainly by nitrogen-fixing (a) bacteria (b) waves (c) minerals (d) crystals.

2. Atmospheric pressure measured at sea level is (a) 99 percent (b) 1.03 kg (c) 32 km (d) 1.03 kg/cm^2.

3. A barometer measures (a) atmospheric pressure (b) wind speed (c) ozone concentration (d) wavelengths.

4. Almost all of the water and carbon dioxide in the atmosphere is in the (a) exosphere (b) ionosphere (c) troposphere (d) stratopause.

5. Radio stations can increase the area they reach by bouncing radio waves off the (a) stratosphere (b) mesosphere (c) ionosphere (d) troposphere.

6. Around Los Angeles, frequent temperature inversions result from cool, polluted air being trapped by (a) acid rain (b) a layer of colder air (c) mountains (d) the ocean.
7. Almost all the energy reaching the earth from the sun is in the form of (a) atomic particles (b) electromagnetic waves (c) ultraviolet rays (d) gamma rays.
8. Raindrops may separate sunlight into a range of colors, thereby causing (a) a mirage (b) an inferior mirage (c) acid rain (d) a rainbow.
9. The process in which the atmosphere traps warming infrared rays over the earth's surface is called the (a) greenhouse effect (b) Coriolis effect (c) doldrums (d) wind cell.
10. Heat can be transferred within the atmosphere in three ways—radiation, conduction, and (a) scattering (b) temperature inversion (c) weathering (d) convection.
11. A looping pattern of wind flow is known as (a) the Coriolis effect (b) a wind cell (c) a trade wind (d) a prevailing westerly.
12. A gentle wind covering less than 100 km is called a (a) jet stream (b) subpolar low (c) breeze (d) trade wind.

Application

On your paper, write answers to the following questions.
1. Explain how houseplants can increase the amount of oxygen in your home.
2. During a jet flight over the North Pole and toward a region in the middle latitudes, the pilot adjusts the altimeter. Why is this adjustment necessary?
3. Some aerosol sprays are banned in the United States. Which of the four layers of the atmosphere does this ban help protect? Explain your answer.
4. After several weeks of heavy rainfall in a certain area of the country, many of the fish in a local lake die. In addition, soils are found to be highly acidic and nearby trees are losing their leaves. What kind of pollution may have caused these problems? What is the source of this pollution?
5. In a drive across the desert with your family, you see a distant sand dune that appears to be floating on air. How can you explain what you see?
6. You hear a talk about the earth's weather. The speaker says, "Infrared rays coming from the earth's surface heat the atmosphere much like a greenhouse is heated." Explain why that statement is incorrect.
7. You are listening to your local weather report. The forecaster explains that a mass of cool air is moving into a low-pressure region over your area. Can this change be explained mainly by conduction or convection? Why?
8. In what ways would a knowledge of the global wind belts have helped a sixteenth-century explorer sailing between Spain and the northern part of South America?
9. What effect do you think the jet streams might have on airplane travel?
10. If there is a breeze blowing from the ocean to the land on the coast of Maine, about what time of day is it? How do you know?

Extension

1. Contact your local Environmental Protection Agency to find out what is being done in your area to control air pollution. Report your findings to the class.
2. Do research to find out more about rainbows. Draw diagrams showing how rainbows form and why a rainbow forms an arc. Present both diagrams to the class.
3. Do research on the Beaufort wind scale. Assign a Beaufort number to the morning and evening winds in your area each day for a week. Plot the results on a graph with the time and day on one axis and the Beaufort number on the other axis. Share your graph with the class and lead a discussion about any trends you notice.

Chapter 24
Water in the Atmosphere

Water vapor is one of the most important substances in the earth's atmosphere. Although the amount of water vapor in the atmosphere is quite low, water vapor has major influences on weather conditions and climate. Suppose that all the water vapor in the atmosphere were to condense and fall to the earth as rain. The resulting worldwide layer of water would not be as thick as this book.

In this chapter, you will learn how water enters and leaves the atmosphere.

Chapter Outline

24.1 Atmospheric Moisture
Heat Energy and Water
Humidity
Dew Point

24.2 Clouds and Fog
Cloud Formation
Types of Clouds
Fog

24.3 Precipitation
Forms of Precipitation
Causes of Precipitation
Cloud Seeding
Measuring Precipitation

The top clouds are cirrus; the lower clouds are cumulus clouds.

24.1 Atmospheric Moisture

Section Objectives

• **Explain how water vapor enters the air.**

• **Explain the meaning of humidity and describe how it is measured.**

• **Describe what happens when the temperature of air decreases at or below the dew point.**

Water exists in the atmosphere mainly in its gaseous form called *water vapor.* Water vapor is a tasteless, odorless, and invisible gas. However, you can detect the presence of water vapor by observing the spout of a tea kettle in which water is boiling. Notice the mist rising from the spout of the kettle. If you look carefully, you will see that the mist starts forming an inch or more above the spout. The space between the spout and the mist is filled with water vapor. The mist itself is condensed water vapor and is a liquid, not a gas.

Water in the atmosphere exists in two forms other than water vapor. Water can be a solid such as ice, or it can be a liquid in the form of water droplets.

Heat Energy and Water

Each water molecule of ice is strongly attracted to others around it. All the molecules of ice are held almost stationary in a definite crystalline arrangement. If you add enough heat to ice, the ice will change from a solid to a liquid to a gas. In other words, water changes from ice to liquid water to water vapor. When heat energy is added to ice, the molecules within the ice begin to move more rapidly, causing the ice to melt. In liquid water the molecules are able to move around each other, but they remain close together. When liquid water is heated, the movement of molecules speeds up, causing the molecules to collide with one another. Such collisions can cause the molecules to move rapidly enough to evaporate from the surface of the liquid. These evaporated molecules become the gas, water vapor.

Ice can change directly into water vapor without first becoming a liquid. Water vapor can also change directly into ice. The process by which a solid can change directly to a vapor or a vapor change directly back to a solid is called **sublimation.** When the air is dry and the temperature is below freezing, ice and snow may sublimate into water vapor.

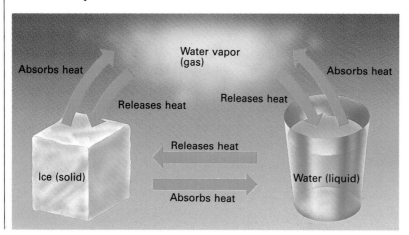

Figure 24–1. Water exists in three forms. Notice that water either absorbs heat or releases heat in changing from one form to another.

Although some water vapor enters the atmosphere through sublimation, most of the water vapor in the atmosphere comes from evaporation. Water absorbs heat energy from the sun. This heat causes molecules of water to evaporate and pass into the air. Most of this evaporation takes place in the regions around the equator, where the largest amounts of solar energy are received. The ocean is the principal source of atmospheric moisture. Billions of kilograms of water evaporate each day from the surface of the ocean. Evaporation from lakes, ponds, streams, and soil add water vapor to the atmosphere. Plants also give off water vapor, and a small amount of moisture comes from volcanoes and from the burning of fuels.

When water evaporates and enters the atmosphere, its molecules move much more rapidly than they did when they were in the liquid form of water. The motion is actually energy stored in the molecules and is called **latent heat.** *Latent* means "hidden." Latent heat will be released when water condenses and returns to the liquid form. Latent heat also is involved when water freezes or thaws.

Humidity

The amount of water vapor in the atmosphere is referred to as its **humidity.** As water molecules evaporate into the air, the humidity of the air increases. When the air holds all the water vapor it can at a given temperature, the air is **saturated.** The amount of water vapor that a volume of air can hold increases as the temperature rises. As the temperature goes down, the capacity of a volume of air to hold water vapor decreases. Figure 24–2 shows this relationship between air temperature and capacity to hold water vapor.

Relative Humidity

A common way to express the amount of water vapor in the atmosphere is by **relative humidity.** Relative humidity is a ratio. It compares the mass of water vapor in the air with the mass of water vapor that the air can hold at its saturation point. For example, suppose that 1 m³ of air at a temperature of 20°C contains 13.9 g of water vapor. Meteorologists know that at the same temperature 1 m³ of air can contain 17.1 g of water vapor. They can thus calculate the relative humidity with this equation.

$$\frac{\text{(present) } 13.9 \text{ g/m}^3}{\text{(saturated) } 17.1 \text{ g/m}^3} \times 100 = \text{(relative humidity) } 81\%$$

At a certain temperature, relative humidity changes as moisture enters or leaves the air. It also changes if the amount of moisture in the air remains the same but the temperature of the air changes. If the moisture remains the same, the relative humidity decreases as the temperature rises and increases as the temperature falls.

Measuring Relative Humidity

A **psychrometer** is an instrument used to measure relative humidity. A psychrometer, shown in Figure 24–3, consists of two identical

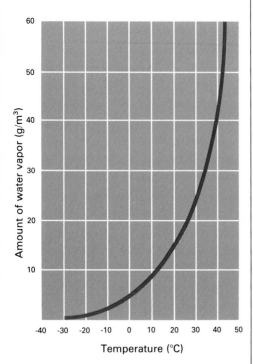

Figure 24–2. This graph shows the amount of water vapor that a given volume of air can hold at various air temperatures.

thermometers. The bulb of one thermometer is covered with a damp wick while the bulb of the other thermometer remains dry. The person measuring relative humidity holds the psychrometer by a handle and whirls it. This whirling circulates the air around both thermometers. The water in the wick of the wet-bulb thermometer starts to evaporate. Evaporation requires heat, so heat is withdrawn from the thermometer. Consequently, the temperature reading of the wet-bulb thermometer is lower than that of the dry-bulb thermometer. When the air is dry, the rate of evaporation in the wick of the wet-bulb thermometer increases. The drier the air is, the more rapidly water evaporates from the bulb of the wet-bulb thermometer and the more quickly the wet bulb cools. The rate of evaporation will decrease as the air becomes more saturated.

Approximate relative humidity can be determined by using a table such as Table 24–1. Locate the column that represents the temperature difference of the thermometer readings and the row of the dry-bulb temperature. The value found where the column and row intersect is the relative humidity. If the difference between the thermometer readings is 5°C and the dry-bulb temperature is 12°C, what is the relative humidity?

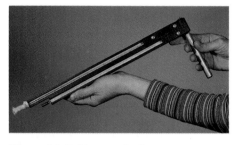

Figure 24–3. Meteorologists use a psychrometer and a relative humidity table, such as Table 24–I, to determine relative humidity. Relative humidity is always expressed as a percentage.

Table 24–1: Relative Humidity (in percentage)

Dry-bulb reading (°C)	Difference between wet-bulb reading and dry-bulb reading (°C)									
	1	2	3	4	5	6	7	8	9	10
0	81	64	46	29	13					
2	84	68	52	37	22	7				
4	85	71	57	43	29	16				
6	86	73	60	48	35	24	11			
8	87	75	63	51	40	29	19	8		
10	88	77	66	55	44	34	24	15	6	
12	89	78	68	58	48	39	29	21	12	
14	90	79	70	60	51	42	34	26	18	10
16	90	81	71	63	54	46	38	30	23	15
18	91	82	73	65	57	49	41	34	27	20
20	91	83	74	66	59	51	44	38	31	24
22	92	83	76	68	61	54	47	41	34	28
24	92	84	77	69	62	56	49	44	37	31
26	92	85	78	71	64	58	51	47	40	34
28	93	85	78	72	65	59	53	48	42	37
30	93	86	79	73	67	61	55	50	44	39

Another instrument used to measure humidity is based on the fact that human hair stretches when the moisture in air increases. A piece of human hair will stretch about 2.5 percent when the relative humidity increases from 0 to 100 percent. The **hair hygrometer** is an instrument that records the changing length of a bundle of hairs when humidity changes. A variation in length causes a pointer to move on a scale or causes a pen to move on a graph, showing changes in relative humidity.

To measure humidity at high altitudes, an electrical instrument called a *radiosonde* is carried up into the atmosphere by a balloon. A radiosonde is triggered by passing an electrical current through a moisture-attracting chemical substance. The amount of moisture changes the electrical conductivity of the chemical substance. The change can then be measured and expressed as the relative humidity of the surrounding air.

Investigation

Relative Humidity

Relative humidity is a ratio of the amount of water vapor in the air to the amount of water vapor the air can hold at that temperature. Using a relative humidity table, you can determine relative humidity.

Science Process Skills Focus: observing, recording data, predicting outcomes

Materials
thermometer; cheesecloth or light cotton material, about 5 × 5 cm; rubber band; water; small cardboard fan; Relative Humidity Table on page 411

Procedure
1. Lay the thermometer on a table for a few minutes, until it adjusts to room temperature. Observe and record the dry-bulb temperature.
2. Secure the cheesecloth around the bulb of the thermometer with a rubber band. Moisten the cheesecloth with room-temperature water.
3. Hold the thermometer firmly at the top. Using your free hand, rapidly fan the bulb of the thermometer with the cardboard for 1 minute. Observe and record the wet-bulb temperature.
4. Repeat Steps 1 through 3 two more times. For each trial, subtract the temperature on the wet-

bulb thermometer from the temperature on the dry-bulb thermometer. Use the table on page 411 to determine the relative humidity for each trial. Calculate the average relative humidity for the three trials. Record your results.

Analysis and Conclusions
1. How do you account for the difference between the readings of the dry-bulb and wet-bulb thermometers?
2. What is expressed by a relative humidity figure of 60 percent?
3. What might cause someone to feel hot and uncomfortable on a warm, humid day? Explain your answer.

Specific Humidity

Meteorologists use **specific humidity** to express the actual amount of moisture in air. Specific humidity is the number of grams of water vapor in 1 kg of air. For example, the very moist air in tropical regions might have a specific humidity of about 18 g/kg. The cold, dry air in polar regions, on the other hand, often has a specific humidity of less than 1 g/kg. Because specific humidity is measured only in units of mass, it is not affected by changes in temperature or pressure. Only a change in the amount of water vapor in the air changes the specific humidity.

Dew Point

You have learned that warm air can hold more water vapor than cold air can. The temperature to which air must be cooled to reach saturation is called the **dew point.** At any temperature lower than the dew point, water vapor begins to condense to a liquid or to change into a solid by sublimation. The dew point temperature of air depends upon the amount of water vapor already present in the air. When the air is nearly saturated with a relative humidity of nearly 100 percent, only a small temperature drop is needed for air to reach its dew point.

Air may cool to its dew point by *conduction* when the air contacts a cold surface. During the night, grass, leaves, and other objects near the ground lose heat. Their surface temperatures often drop to the dew point of the surrounding air. Air, which normally remains warmer than do surfaces near the ground, cools to the dew point when it contacts objects such as the cooler grass. The resulting form of condensation, which is shown in Figure 24–4, is called **dew.** You can see dew as tiny water droplets on many surfaces that have cooled. Dew is more likely to form on clear nights when there is little wind to disperse it.

If the dew point falls below the freezing temperature of water, water vapor will sublimate into solid ice crystals, or **frost,** shown in Figure 24–4. Since frost forms from water vapor without first becoming liquid, it is not frozen dew. Frozen dew, which is relatively uncommon, forms as clear beads of ice. Frost often indicates that the temperature is low enough to damage growing plants.

Figure 24–4. The top picture shows dew, which forms as the cold ground cools the night air. The bottom picture shows frost, which forms when the dew point drops below freezing.

Section 24.1 Review

1. What is the principal source of water vapor in the atmosphere of the earth?
2. What term describes the temperature at which air reaches its saturation point?
3. How does a change in temperature affect relative humidity and specific humidity?
4. At a given temperature, 1 m³ of air can hold 10 g of water vapor. What is the relative humidity at that temperature if 1 m³ of air is holding 9 g of water?

24.2 Clouds and Fog

Clouds and fog are visible masses of water particles or ice particles suspended in the atmosphere. People commonly think of clouds as being high in the sky and fog as being close to the ground. However, clouds are not limited to high altitudes. Some clouds develop close to the surface of the earth. Both clouds and fog originate from water vapor in the air.

Cloud Formation

Clouds result from the condensation of water vapor throughout a large volume of air. For water vapor to condense, a solid surface must be available on which condensation can take place. The open air, of course, contains no large solid surfaces. However, it does contain millions of suspended particles of ice, salt, dust, and other solid matter. These suspended particles, called **condensation nuclei,** provide the necessary surfaces for cloud-forming condensation. Because the condensation nuclei are so small, less than 0.001 mm in diameter, they remain suspended in the atmosphere for a long time. Water molecules become attached to the particles, and as the molecules collect, water droplets are formed.

Another condition necessary for cloud formation is that the air be saturated. When air temperature reaches the dew point, the air is saturated, and water vapor begins to condense. Because the amount of water vapor that air can hold decreases as temperature goes down, cooling leads to condensation. Several processes may bring about the cooling necessary for clouds to develop.

Convective Cooling

Because the higher layers of air compress the lower layers, the lower layers are more dense and have a higher pressure. When air rises into a region with a lower atmospheric pressure, its molecules move farther apart. This work of expanding the distance between molecules of air uses potential energy that is stored in compressed air. As this energy is used, the temperature of the air falls. The lowering of the temperature of a mass of air due to its rising and expanding is called **convective cooling,** as shown in Figure 24–5. Just as expansion causes air to cool, compression causes air to become hotter.

Adiabatic Temperature Changes   Changes in temperature that result solely from the expansion or compression of air are called **adiabatic** (AD-ee-uh-BAT-ik) temperature changes. The adiabatic temperature changes associated with cloud formation take place at predictable rates. The temperature of dry air decreases about 1°C for every 100 m the air rises. However, air usually contains moisture. Moisture influences the rate of adiabatic change. Condensing water vapor in rising air gives off heat, which is absorbed by the air. This condensation slows the rate at which the air cools. The cooling rate

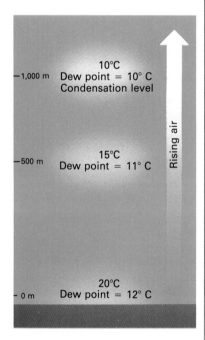

Figure 24–5. Notice that the rising dry air cools at the rate of 1° C for each 100 m. The dew point drops 0.2° C for each 100 m. In this illustration, temperature and dew point are the same at 1,000 m. Above that altitude, condensation begins and clouds form.

10°C
Dew point = 10° C
Condensation level
—1,000 m

15°C
Dew point = 11° C
—500 m

20°C
Dew point = 12° C
— 0 m

Rising air

of air varies with the amount of moisture it contains. The average rate of cooling for moist air is about 0.7°C per 100 m. Sinking air, whether moist or dry, heats at the rate of 1°C for every 100 m of descent. For air in which water is changing its form, the rate of adiabatic temperature change is about 0.6°C.

Condensation Level Most clouds form through convective cooling. Figure 24–5 shows the typical sequence of cloud formation by convective cooling. On a sunny day, one area of the earth's surface may absorb more solar energy than do the areas around it. The air above this area of extra warmth becomes heated through conduction. The warmed, expanding air rises by convection and undergoes adiabatic cooling. When the air reaches a level where its temperature is lower than its dew point, the moisture in the air condenses to form a cloud. This level at which condensation forms, called the *condensation level*, is marked by the bottom of the clouds. Further condensation allows clouds to expand above the condensation level.

Forceful Lifting

Often air does not rise spontaneously; rather, an event occurs that forces the air to rise. The forced upward movement of air commonly results in the cooling of air and in cloud formation. Air can be forced upward when a moving mass of air meets sloping terrain, such as a mountain range. The rising air expands and cools, and clouds form. As shown in Figure 24–6, entire mountaintops can be covered with clouds formed in this way. The large areas of clouds evident during storms are also formed by forceful lifting. These clouds are formed when a mass of warm air is pushed above a denser mass of cooler air. This process will be discussed further in Chapter 25.

Figure 24–6. Clouds can form as air is pushed up along a mountain slope and cooled below its dew point.

Figure 24–7. All clouds are a form of one of the three major types. These types, shown from top to bottom, are stratus clouds, cumulus clouds, and in the top half of the bottom photo, cirrus clouds. The cumulus cloud shown in the middle photo is a form known as stratocumulus.

Temperature Changes

Clouds will also form when one body of moist air mixes with another body of moist air with a different temperature. The mixing of the two bodies of air into one will cause the air's temperature to change. This temperature change may cool the combined air below its dew point. The result is condensation and cloud formation.

Advective Cooling

Another cooling process associated with cloud formation is **advective cooling.** Advective cooling is cooling produced when wind carries warm, moist air across a cold ocean or region of land. The cold water or land absorbs heat from the air, and the air cools. If the air is cooled below its dew point, condensation and fog or cloud formation takes place. Clouds formed by advective cooling are often very low in the atmosphere.

Types of Clouds

There are three major types of clouds. All other clouds are combinations or variations of these three types. The most convenient method of classifying clouds is by their altitude and shape. The three types are **stratus clouds, cumulus clouds,** and **cirrus clouds.**

Stratus Clouds

The lowest clouds in the sky are stratus clouds. *Stratus* means "sheetlike" or "layered." The base of one of these clouds may be as low as 1.8 km from the earth's surface. Stratus clouds form where a layer of warm, moist air lies above a layer of cool air. When the overlying warm air cools below its dew point, wide clouds appear. Look at the stratus clouds shown in Figure 24–7. These clouds cover large areas of sky, often blocking out the sun. The amount of rain that falls from stratus clouds is usually very little.

Two variations of stratus clouds, shown in Figure 24–8, are known as *nimbostratus* and *altostratus*. The terms *nimbo* and *nimbus* mean "rain." *Nimbostratus* clouds, unlike other stratus clouds, do bring heavy rains or snow. The term *alto* means "high." You already know the meaning of the word *stratus*. How would you describe the clouds known as *altostratus?*

Cumulus Clouds

Above the stratus clouds, between 2.4 km and 13.5 km from the earth's surface, are cumulus clouds. Cumulus means "piled" or "heaped." Look at the cumulus clouds in Figure 24–7 and in Figure 24–8. Cumulus clouds are thick with high, billowy tops. Cumulus clouds form when large bodies of air rise and cool. As the cooling air reaches its dew point, the clouds form. The flat base characteristic of some cumulus clouds shows where condensation began. The height of a cumulus cloud depends on the speed of the upward movement and the amount of moisture in the air. The height of the cloud indicates how high the cooling air rose. The lowest cumulus clouds are known as *stratocumulus*.

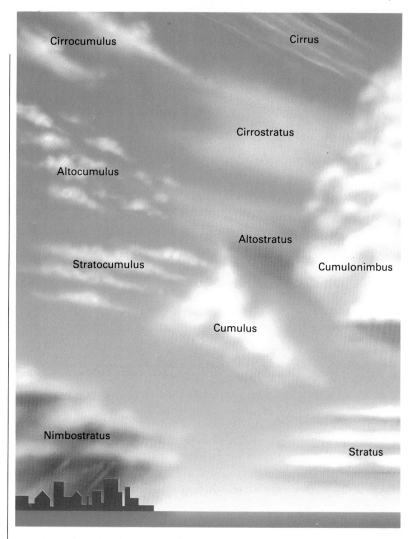

Figure 24–8. Meteorologists classify cloud types by altitude and shape.

Cumulus clouds usually form in fair weather. On hot, humid days, cumulus clouds reach their greatest heights. The large, thick clouds thus produced are *cumulonimbus clouds,* or thunderclouds, and are often accompanied by rain, lightning, and thunder.

Cirrus Clouds

The highest clouds in the sky are wispy, feathery cirrus clouds, shown in Figure 24–7. *Cirro* and *cirrus* mean "curly." Cirrus clouds form between 7 km and 13 km above the earth's surface. The low temperatures at such high altitudes form clouds that are made up of ice crystals. Sunlight can easily pass through these thin clouds.

Locate the extremely high, billowy clouds, called *cirrocumulus clouds,* in Figure 24–8. Cirrocumulus clouds are composed entirely of ice crystals. These clouds form quite rarely. Cirrocumulus clouds often appear just before a snowfall or a rainfall. Long, thin clouds called *cirrostratus clouds* lie at about 11.5 km from the surface of the earth. A halo seems to appear around the sun or moon when either is viewed through a cirrostratus cloud. This halo effect is caused by the bending of light rays as they go through the ice crystals. Like cirrocumulus clouds, cirrostratus clouds are rare.

Figure 24–9. Radiation fog, or ground fog—such as this fog in the Napa Valley in California—forms at night. The cold ground cools the air below its dew point, creating thick bodies of low-lying fog.

Fog

Fog, like clouds, is the result of the condensation of water vapor in the air. The chief difference between the fog and clouds is that fog forms very near the surface of the earth when air close to the ground is cooled. For example, you may be familiar with one type of fog that results from the nightly cooling of the earth. The layer of air in contact with the earth becomes chilled below its dew point, and its water vapor condenses into droplets. This type of fog is called **radiation fog,** because it results from the loss of heat by radiation. This type of fog is also called *ground fog*. Radiation fog usually forms on calm, clear nights. It is thickest in valleys and low places because the dense, cold air in which it forms sinks to the lower elevations. Radiation fog is often quite thick around cities, where there are greater numbers of smoke and dust particles to act as condensation nuclei.

Another type of fog, called **advection fog,** forms when warm, moist air moves across a cold surface. Advection fog is common along coasts, where warm, moist air from above the water moves in over a cooler land surface. Also, dense fogs may form over the ocean when warm, moist air is carried over cold ocean currents.

Two other types of fog often form inland. An **upslope fog** is formed by the lifting and adiabatic cooling of air as it rises along land slopes. Upslope fog is really a kind of cloud formation at ground level. A type of fog known as **steam fog** usually forms over inland rivers and lakes. Steam fog is a shallow layer of fog formed when cool air moves over a warm body of water.

Section 24.2 Review

1. What is adiabatic cooling?
2. What is the chief difference between fog and clouds?
3. On a hot and humid summer day, you notice enormous, billowy clouds with broad, flat bases. What can you predict about the weather? Explain how you know.

24.3 Precipitation

Any moisture that falls from the air to the earth's surface is *precipitation*. Precipitation may be liquid or solid. Types of liquid precipitation include drizzle and rain. Types of solid precipitation include snow, sleet, and hail.

Forms of Precipitation

Drizzle is a form of liquid precipitation consisting of drops smaller than 0.5 mm in diameter. The drops are close together and, because they are so small, they fall very slowly. Drizzle results in only a small amount of total precipitation. Rain consists of drops between 0.5 mm and 5 mm in diameter. Raindrops larger than 5 mm break up into smaller ones as they fall.

The most common form of solid precipitation is snow, which consists of ice particles. These particles may fall as small pellets, as individual crystals, or as crystals that combine into snowflakes. Most snow takes the form of snowflakes. Snowflakes may be as small as several millimeters in diameter or as large as several centimeters. Snowflakes tend to be small at low temperatures and become larger as the temperature nears 0°C. This size difference occurs because colder air is usually less moist. Consequently, less moisture is available for ice-crystal growth.

Extremely cold temperatures near the ground sometimes produce ice pellets called **sleet,** which form when rain falls through a layer of freezing air. Occasionally rain does not freeze until it strikes a surface near the ground. In this event it forms a thick layer of sheet ice, or glaze ice, as shown in Figure 24–10. The weight of ice often is great enough to break tree limbs and power lines. Conditions that produce glaze ice occur in what is called an ice storm.

Figure 24–10. Glaze ice forms as rain freezes on surfaces near the ground, such as on this plant.

Hail is solid precipitation in the form of lumps of ice. These lumps can be nearly spherical or irregularly shaped. Hail usually forms in cumulonimbus clouds. Convective currents within the clouds carry raindrops to high levels, where they freeze. As the frozen raindrops fall, they accumulate additional layers of liquid water on their surface. The hail may be carried up again into freezing air. Also, it may fall back through another layer of near-freezing moist air. Either way, the coating of water freezes, and another layer of ice forms. If the process is repeated a number of times, the hailstones may accumulate many layers of ice. Hailstones often consist of alternate layers of clear and cloudy ice. The clear ice forms as a hailstone passes through a layer of very moist air. The cloudy layer forms when water droplets with air bubbles between them freeze to the surface. Some hailstones become quite large. The impact of large hailstones can damage crops and property.

IMPACT

The Destructive Power of Hail

Hailstones start out as tiny crystals of ice falling from clouds. Winds carry the crystals back up into cooler atmospheric regions. There, layer after layer of supercooled water is added to the surface of the crystals. This water freezes, making the crystals larger. Hailstones have grown to the size of grapefruit.

The largest hailstone ever recorded had a circumference of over 44 cm and a mass of 758 g. Because of its size, this hailstone hit the ground at an estimated speed of 160 km/hr.

One hailstorm lasting only a few minutes can cause tremendous damage. For example, hail has been responsible for at least one airplane crash. It has torn holes in the roofs of houses and automobiles and in the cabins of aircraft. Hail has killed livestock and wildlife and stripped plants of their leaves.

However, the most consistent damage caused by hail is damage to crops. In 1978 hail destroyed all the crops in 40 out of 53 villages in one region of Syria. This single storm left 60,000 people on the brink of starvation.

Each year in the United States, hail destroys more than $225 million worth of wheat, corn, soybeans, and other crops. Hailstorms can have a devastating effect on individual farmers. Unlike drought, hail does not cause gradual damage. Instead, a family's entire crop, representing the work of an entire year, and their investment can be wiped out within minutes.

If you break a hailstone in half, you will see a layered texture. Explain why.

Causes of Precipitation

A cloud produces precipitation when its droplets or ice crystals become large enough to fall as rain or snow. Most cloud droplets have a diameter of about 20 micrometers. A micrometer equals 0.000001 m. A cloud droplet is thus smaller than the period at the end of this sentence. Droplets of this size easily remain suspended in the air. Even slight air movements prevent droplets from falling. Before a cloud droplet falls as precipitation, it must increase to about 100 times its normal diameter.

Coalescence

One way cloud droplets reach the precipitation stage is in a process called **coalescence** (KOE-uh-LESS-uns). Coalescence involves differences in size between cloud droplets. Generally, the original size of a cloud droplet depends on its condensation nucleus. Large nuclei tend to form large cloud droplets. Large droplets do not remain suspended in the cloud as long as small ones do. Instead, large droplets drift downward and, in doing so, collide and combine with smaller droplets. Each large droplet continues to coalesce until it contains at least a million times as much water as it did originally. By this time, the mass of the droplets is great enough to fall, and the coalesced droplets fall as raindrops.

Supercooling

Precipitation may also form in clouds that contain water vapor, ice crystals, and water droplets that have gone through **supercooling.** Supercooled water droplets have a temperature of less than 0°C. The temperature of water droplets in these cold clouds may fall as low as −10°C. Yet even at this low temperature the water droplets do not freeze. They cannot freeze because there are too few **freezing nuclei**

Figure 24–11. During coalescence, as cloud droplets fall, they collide and combine with small droplets. The resulting larger droplets fall as rain.

Figure 24–12. The temperature at which ice crystals form determines their shape.

on which ice can form at these temperatures. Freezing nuclei are kinds of condensation nuclei with crystalline structure similar to that of ice. Because only a few ice crystals will form, most of the water in the cloud remains as supercooled water droplets. Water molecules evaporate from the supercooled water droplets and condense on the ice crystals. The ice crystals increase in size until they gain enough mass to fall as rain or snow.

Meteorologists believe that almost all the rain and snow in the middle and high latitudes of the earth results from ice crystals in supercooled clouds. In the tropical regions, rain is commonly formed by the process of coalescence.

Cloud Seeding

Knowing how ice crystals form in clouds, meteorologists are able to take the first steps toward producing rain when it is needed. The method they use in attempting to cause or increase precipitation is called **cloud seeding.** In cloud seeding, freezing nuclei are added to supercooled clouds. In one method of cloud seeding, silver-iodide crystals, which resemble ice crystals, are used as freezing nuclei. Silver-iodide vapor is released into clouds from either burners on the ground or flares dropped from aircraft. In another method of cloud seeding, powdered dry ice is dropped from aircraft. The dry-ice particles cool the cloud droplets and cause ice crystals to form.

In some experiments with cloud seeding, seeded clouds produced more precipitation than unseeded clouds did. In other experiments, there was no significant difference in the amount of precipitation produced by seeded and unseeded clouds. In some cases, in fact, cloud seeding appeared to cause less precipitation. Meteorologists have thus concluded that cloud seeding may increase

Figure 24–13. Special equipment attached to the underside of cloud-seeding planes, (inset) releases freezing nuclei into clouds. Meteorologists hope that cloud seeding may help form rain clouds, such as the one shown, and bring rainfall to drought-stricken areas.

precipitation under some conditions but decrease it under others. Research is under way to identify the conditions that produce increased precipitation. This information may help scientists improve their techniques for seeding clouds. Cloud seeding may eventually become a way to overcome the many problems associated with droughts. Successful cloud seeding may also help control severe storms. Meteorologists may be able to release precipitation from clouds before a storm can become too large.

Measuring Precipitation

A **rain gauge** is an instrument for measuring the amount of rainfall. One type of rain gauge consists of a wide-mouthed funnel placed over a cylindrical container. Rainwater passes through the funnel into the container below. In many rain gauges the mouth of the funnel is much larger than the container under it. Why do you think a rain gauge is constructed in this manner?

Another type of rain gauge consists of a funnel and a small divided bucket. Rain caught in the funnel fills one side of the bucket. The bucket then tips, dumping the water and allowing the other half to fill up. Each time one side of the bucket fills with 0.25 mm of rainwater, it tips and sets off an electrical device that records the amount. The rainwater dumped from the bucket is collected and weighed to check the accuracy of the record. A third type of gauge catches the water in a large bucket that is weighed continuously. The weight is recorded directly on a graph as centimeters of rain.

Rain gauges measure only the precipitation that falls in one spot. It is difficult to establish the total amount of rain that falls over a large area. Even over short distances, precipitation may vary by sizable amounts. This is especially true in mountainous regions.

Snow is measured by both the depth of accumulation and the water content. The depth of snow is measured with a measuring stick. The water content is determined simply by melting the snow and measuring the amount of water that results. The amount of water contained in any given volume of snow depends upon the type of snow. For example, as much as 20 cm of dry snow may be needed to produce 1 cm of liquid water. But only 6 cm of wet snow may be needed to produce 1 cm of liquid water. On the average, 10 cm of snow will produce about 1 cm of liquid water.

Figure 24–14. A rain gauge is made up of a funnel and a cylindrical container.

Section 24.3 Review

1. Which form of liquid precipitation is made up of drops that are smaller than 0.5 mm in diameter?
2. In what parts of the world are raindrops commonly formed by the process of coalescence?
3. The usual purpose of cloud seeding is to create precipitation. How then might it help prevent a major storm?
4. If the bucket of a tipping-bucket rain gauge has emptied ten times, how much rain has fallen? Explain how you know.

Reviewing Chapter 24

Key Concepts

- Most of the moisture in the air comes from evaporation from the oceans. See page 410.

- Humidity is the amount of water vapor in the air. See page 410.

- When air temperature reaches dew point, water vapor condenses to a liquid or it changes to a solid by sublimation. See page 413.

- Clouds form when a body of air rises and cools below its dew point. See page 414.

- The three major types of clouds are stratus, cumulus, and cirrus. See page 416.

- Radiation, advection, upslope, and steam fogs form in different ways. See page 418.

- Forms of precipitation include drizzle, rain, snow, sleet, and hail. See page 419.

- Precipitation may be caused by coalescence or by supercooling. See page 421.

- Cloud seeding is one way meteorologists attempt to produce rain artificially. See page 422.

- A rain gauge measures liquid precipitation. Snow is measured by its depth and water content. See page 423.

Key Terms

adiabatic (414)
advective cooling (416)
advection fog (418)
cirrus cloud (416)
cloud seeding (422)
coalescence (421)
condensation nuclei (414)

convective cooling (414)
cumulus cloud (416)
dew (413)
dew point (413)
freezing nuclei (421)
frost (413)
glaze ice (419)
hail (420)

hair hygrometer (412)
humidity (410)
latent heat (410)
psychrometer (410)
radiation fog (418)
rain gauge (423)
relative humidity (410)
saturated (410)
sleet (419)

specific humidity (413)
steam fog (418)
stratus cloud (416)
sublimation (409)
supercooling (421)
upslope fog (418)

Review

On your paper, write the letter of the term that best completes each of the following statements.

1. The process by which ice changes directly into water vapor is referred to as (a) advection (b) conduction (c) condensation (d) sublimation.

2. Most of the evaporation from the oceans takes place around (a) the equator (b) the poles (c) 60° latitude (d) 30° latitude.

3. Relative humidity is always expressed (a) in g/m³ (b) as a percentage (c) in g/kg (d) in degrees Celsius.

4. To express the actual amount of moisture in the air, meteorologists use (a) latent heat (b) relative humidity (c) specific humidity (d) dew point.

5. When air temperature drops, its capacity for holding water is (a) slightly higher (b) much higher (c) about the same (d) lower.

6. The tiny water droplets that result when air is cooled by contact with a cold surface are called (a) dew (b) frost (c) humidity (d) steam fog.

7. Changes in temperature that result solely from the expansion or compression of air are called (a) adiabatic (b) supercooling (c) latent cooling (d) advection.

8. Clouds form when the water vapor in air con-

denses as (a) the air is heated (b) the air is cooled (c) snow falls (d) the air is superheated.

9. Low, sheetlike clouds are called (a) cirrus clouds (b) stratus clouds (c) cumulus clouds (d) cirrocumulus clouds.

10. The term *nimbo* or *nimbus* added to the name of any form of cloud means (a) high (b) billowy (c) rain (d) layered.

11. The fog that results from the nightly cooling of the earth is called (a) steam fog (b) upslope fog (c) advection fog (d) radiation fog.

12. The fog produced when warm, moist air moves over a cold surface is called (a) steam fog (b) upslope fog (c) advection fog (d) radiation fog.

13. Rain that freezes when striking the ground produces (a) sleet (b) glaze ice (c) hail (d) frost.

14. Clouds in which the water droplets remain liquid below 0°C are said to be (a) saturated (b) supersaturated (c) superheated (d) supercooled.

15. In one method of cloud seeding, silver-iodide crystals are used as (a) freezing nuclei (b) dry ice (c) cloud droplets (d) a superheater.

16. A wide-mouthed funnel and a cylindrical container are used in making an instrument called a (a) hygrometer (b) rain gauge (c) psychrometer (d) barometer.

Application

On your paper, write answers to the following questions.

1. Where would the air contain most moisture—over Panama, at latitude 10°N, or over Antarctica? Explain your answer.

2. What is the relative humidity when the air holds three fourths of the moisture it can hold when saturated? Explain your answer.

3. One body of air has a relative humidity of 97 percent. Another has a relative humidity of 44 percent. At the same temperature, which body of air is closer to its dew point? Explain your answer.

4. Why would polluted air more likely form clouds than clean air would?

5. You are in school and you hear thunder outside. Describe the clouds that you would probably see if you looked out the window.

6. You are camping in a valley on a calm, clear night. You awake and notice a thick fog. What type of fog is it, and how did it form?

7. You see lumps of ice the size of golf balls falling from the sky. Are you seeing sleet, hail, or freezing rain? How do you know?

8. In the tropical regions, raindrops are commonly formed by coalescence. Little precipitation there forms by supercooling. Why might this be true?

9. During a severe drought in a farming region, a meteorologist recommends cloud seeding to produce rain. Would cloud seeding be a sure way to end the drought? Explain your answer.

10. One day in January, 6 cm of very wet snow falls on your area. If all this snow melted quickly, how deep would the water be? Explain how you know.

Extension

1. Look up the meanings and origins of the following terms used to describe clouds: *stratus, cumulus, cirrus.* Present your findings to the class.

2. At a convenient time of day, perhaps just after school or in the early evening, observe and sketch the cloud formations from a particular location. Do this at the same time and place every day for a week. Identify the cloud formations and share your drawings with other members of your class.

3. Do research to find out about a region that has little precipitation. Find out why the precipitation is so low. Find out how the dry climate influences plant and animal life and how people survive in the region. Report your findings to the class.

Chapter 25

Weather

If you study pictures of the earth taken from a satellite, you would see evidence that the atmosphere is far from quiet and peaceful. In this chapter you will learn how storms, lightning, and other atmospheric phenomena are produced and how meteorologists predict the weather.

Chapter Outline

Lightning illuminates the night sky.

25.1 Air Masses

Differences in air pressure at different places on the earth create wind patterns. The region around the equator receives much more solar energy than do the regions at the poles. Because warm air rises and cold air sinks, the heated equatorial air rises. As the air rises, it creates a low-pressure belt. Conversely, cold air near the poles sinks, creating high-pressure belts.

Air moves from areas of high pressure to areas of low pressure. Therefore, there is a worldwide movement of surface air from the poles toward the equator. At higher altitudes the cooler air returns from the equator toward the poles. Temperature and pressure differences on the earth's surface alter this general wind pattern, however. These conditions create three *wind cells* in the Northern Hemisphere and three in the Southern Hemisphere. The earth's rotation also influences the wind pattern by causing the deflection of winds called the *Coriolis effect,* as described in Chapter 23. This deflection influences the direction of the prevailing winds.

In areas where air pressure differences are small, air can remain relatively stationary. If the air remains stationary or moves slowly over a uniform surface, it takes on the characteristic temperature and *humidity* of that region. Such a large body of air— one with uniform temperature and moisture content—is called an **air mass.** An air mass is sometimes thousands of kilometers in diameter. A region in which air masses form must have a fairly uniform temperature and moisture content, such as the arctic regions or the ocean. Air masses that form over the arctic regions of North America are very cold and dry. Air masses that form over the tropical oceans are warm and moist.

Section Objectives

- **Explain how an air mass forms.**
- **List and describe the types of air masses that usually affect the weather of North America.**

Figure 25–1. The movement of an air mass often is indicated by clouds.

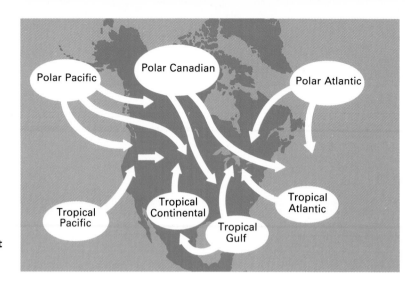

Figure 25–2. This map indicates the general direction of movement of the seven air masses that influence the weather in North America.

Types of Air Masses

Air masses are classified according to their source regions. The source regions also determine the temperature and the humidity of the air mass. The source regions for cold air masses are polar areas; these air masses are labeled *P*. The source regions for warm air masses are tropical areas; these air masses are labeled *T*. Air masses formed over the ocean are called *maritime* and are labeled *m*. Air masses formed over land are called *continental* and are labeled *c*. Maritime air masses tend to be moist; and continental air masses, dry. The combination of tropical or polar air and continental or maritime air results in four main types of air masses. These air masses are **maritime polar** (mP), **maritime tropical** (mT), **continental polar** (cP), and **continental tropical.** What letters designate a continental tropical air mass?

An air mass may remain over its source region for days or weeks. Eventually it will move into other regions because of the overall wind pattern.

North American Air Masses

Air masses that strongly affect the weather of North America come from seven regions. These regions and the general directions of the air movements are shown in Figure 25–2. An air mass usually brings with it the weather of its source region, but it may change as it moves away. For example, cold, dry air may become warmer and moister as it moves from land over the warm ocean. As the lower layers of air are warmed, the air rises; this warmed air can create heavy clouds and *precipitation*.

Polar Air Masses

Three polar air masses influence weather in North America. These air masses are called *polar Canadian, polar Pacific,* and *polar Atlantic*.

Polar Canadian air masses are classified as cP air masses. They form in northern Canada over land that is covered by ice and snow.

The cP air masses generally move southeastward across Canada and into the northern United States. Occasionally these cool air masses reach as far south as the Gulf Coast of the United States. In the summer they usually bring cool, dry weather. In the winter they bring very cold weather to the northern United States.

Polar Pacific air masses, classified as mP air masses, originate over the northern Pacific Ocean and southwestern Alaska. These air masses are very moist but not extremely cold. The Pacific Coast is the area most affected by them. In the winter, polar Pacific air masses bring rain and snow to the Pacific Coast. In the summer, they bring cool, often foggy weather. These air masses lose much of their moisture as they move eastward across the coastal mountain ranges, the Sierra Nevada and the Rocky mountains. Thus, they bring cool and dry weather to the mid United States.

Polar Atlantic air masses, classified as mP air masses, form over the northern Atlantic Ocean. They move in a generally eastward direction toward Europe. However, they may pass over the northeastern portion of North America, bringing cold, cloudy weather to New England. In the winter, they usually bring cold, cloudy weather and light precipitation. In the summer, these air masses often produce cool weather with low clouds and fog.

Tropical Air Masses

Four tropical air masses influence the weather in North America. These air masses are called *tropical continental, tropical gulf, tropical Atlantic,* and *tropical Pacific.*

Tropical continental (cT) air masses flow over North America only in the summer. They form over Mexico and the southwestern United States and generally move northeastward, usually bringing clear, dry, and very hot weather.

Tropical gulf and tropical Atlantic air masses are both classified as mT air masses. These air masses form over the warm waters of the Gulf of Mexico and the South Atlantic Ocean, respectively. They move northward across the eastern United States, bringing mild, often cloudy, weather in the winter. In the summer, they bring hot, humid weather and many thunderstorms.

Tropical Pacific air masses are also mT air masses. They form over the warm areas of the Pacific Ocean, but rarely reach the Pacific Coast. However, in the winter they may bring very heavy precipitation to the coast and the southwestern deserts.

Section 25.1 Review

1. Define *air mass*.
2. What is the name of the air mass that forms over the warm water of the Gulf of Mexico? What letters designate the source region of this air mass?
3. Suppose that snow is falling on the Pacific Coast area. What type of air mass is probably responsible for this weather? What letters designate the source region of this air mass?

- **Compare the characteristic weather conditions associated with cold fronts and with warm fronts.**
- **Describe how a wave cyclone forms.**
- **Describe the stages in the development of hurricanes, thunderstorms, and tornadoes.**

25.2 Fronts

When two air masses meet, temperature differences usually keep the air masses separate. The air of a cool air mass is dense and does not mix with the less dense air of a warm air mass. Thus, a definite boundary, called a **front,** usually forms between air masses. A typical front is about 100 km long, but some fronts may be several thousand kilometers long. Changes in the weather usually take place along the various types of fronts.

Types of Fronts

In order for a front to form, one air mass must collide with another air mass. The kind of front that forms depends upon how the air masses are moving. When a cold air mass overtakes a warm air mass, a **cold front** is formed, as shown in Figure 25–4. The moving cold air lifts the warm air. If the warm air is moist, clouds will form. Large cumulus and cumulonimbus clouds typically form along a fast-moving cold front. Storms created along a cold front are usually short-lived and violent. A long line of heavy thunderstorms, called a **squall line,** may occur and advance just ahead of a fast-moving cold front. A slow-moving cold front lifts the warm air ahead of it more slowly than does a fast-moving front. For that reason, a slow-moving cold front produces less-concentrated cloudiness and precipitation.

A warm air mass that overtakes a cooler air mass produces a **warm front.** The less dense warm air rises over the cooler air. The slope of a warm front is gradual, as you can see in Figure 25–4. Because of this gentle slope, clouds may extend far ahead of the surface location, or base, of the front. A distinct pattern of clouds precedes the approaching base of a warm front. At the beginning of the pattern are cirrus clouds. Behind the cirrus clouds are cirrostratus

Figure 25–3. This characteristic pattern of clouds precedes the base of a warm front.

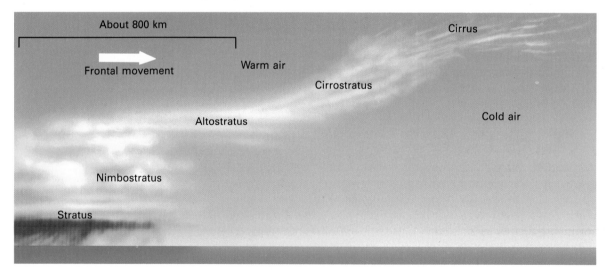

clouds, followed by altostratus, then nimbostratus, and, finally, stratus clouds at the base of the front. A warm front generally produces heavy precipitation over a large area. A warm front may produce violent storms if the body of warm air advancing over the cooler air is very moist.

Sometimes when two air masses meet, neither is displaced. The two air masses move parallel to the front between them. The front formed between the air masses is called a **stationary front,** because it does not move. The weather around a stationary front is similar to that produced by a warm front.

An **occluded front,** as shown in Figure 25–4, forms when a fast-moving cold front overtakes a warm front, lifting the warm air completely off the ground. The advancing cold front then comes into contact with cool air that develops beneath the lifted warm air. The warm front is completely cut off, or occluded, from the ground by colder air. The warm air is held parallel to the ground in the upper levels of the atmosphere. Figure 25–4 shows the symbols commonly used for the various types of fronts.

Polar Fronts and Wave Cyclones

Over each of the earth's polar regions is a cap of cold air. The boundary at which the cold polar air meets the warmer air of the middle latitudes is called a **polar front.** A polar front circles the earth at about 40° to 60° latitude in each hemisphere. The polar fronts move closer to the equator in the winter and back toward the poles in the summer. In the winter, the average position of the polar front in North America is slightly south of Florida. In the summer, it is north of the Great Lakes.

Waves often develop along the polar fronts. A wave is a bend formed in a cold front or a stationary front. These waves are similar to the waves that moving air produces when passing over a body of water. However, they are much larger, often hundreds of kilometers in length. High-speed *jet-stream* winds, which you read about in Chapter 23, help develop a wave. As a wave develops in the polar front, meteorologists can detect shifts in the position and motion of the jet stream.

The waves along the boundary of a polar front are the beginnings of low-pressure storm centers called **wave cyclones.** A wave cyclone consists of a very large body of air—up to 2,500 km in diameter. Its winds blow in circular paths around the low-pressure region at the center. Wave cyclones form not only along polar fronts but also along other cold or stationary fronts, and they influence weather in the middle latitudes particularly.

Stages of a Wave Cyclone

In the first stage of a typical wave cyclone, there is a stationary front between a warm air mass and a cold air mass. At this stage the winds usually move parallel to the front. However, the winds on one side of the front blow in the opposite direction from the winds on

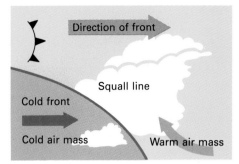

Cold front

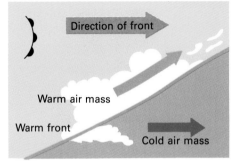

Warm front

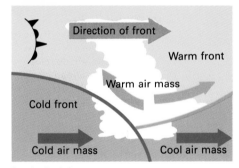

Occluded front

Figure 25–4. Cold and warm air masses collide to form a cold front (top), a warm front (center), and an occluded front (bottom). Note the symbols meteorologists use to designate these fronts.

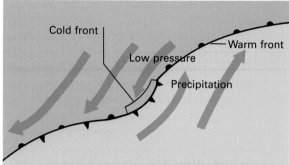

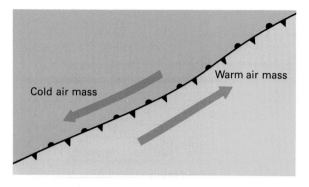

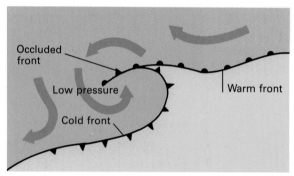

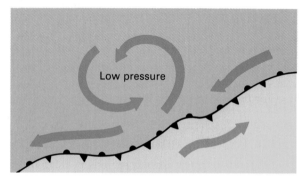

Figure 25–5. Wave cyclones occur along a cold or a stationary front (top, left). A wave cyclone begins when a bulge of cold air develops and advances slightly ahead of the rest of the front (top, right). As the fast-moving cold front overtakes the slow-moving warm front, an occlusion forms (bottom, left) and the storm reaches its highest intensity. Eventually the system uses all its energy and the wave cyclone dies (bottom, right).

the other side. A wave develops along the stationary front as cold air that is moving toward the equator pushes into the warm air, forming a cold front. At the same time, warm air that is moving poleward pushes into the cold air, forming a warm front. The result of this air movement is a slow-moving warm front and a fast-moving cold front turning around a central point. Clouds and precipitation spread along both the warm and cold fronts.

A low-pressure region develops at the central point where the two fronts come together, as you can see in Figure 25–5. The less dense warm air is lifted as it meets the cold air along the warm front. It is also pushed up by the advancing cold air behind the cold front. As a result of these lifting actions, an occluded front develops. The occluded front usually brings the storm to its highest intensity, with winds spinning around the central low-pressure region. Eventually the wave cyclone uses up the moisture and energy from the warm air and dies down.

A wave cyclone usually takes one to three days to develop. During this time the air masses are in motion, and the disturbance moves with them. A specific sequence of cloud formation is associated with the development of a wave cyclone. In North America wave cyclones generally move in an easterly direction at speeds of about 32 to 64 km/hr, and they spin counterclockwise. North American wave cyclones generally follow one of two paths. One path begins in the North Pacific Ocean and moves northward toward Alaska; it then continues southward along the western coast of North America. The other path originates in eastern North America and crosses the North Atlantic Ocean to Europe.

Anticyclones

The air of an **anticyclone** moves outward from a center of high pressure, unlike a wave cyclone, which moves inward toward a low-pressure center. Because of the Coriolis effect, the circulation of air around an anticyclone is clockwise in the Northern Hemisphere. In general, whereas wave cyclones bring cloudy, stormy weather, anticyclones bring fair weather, since their sinking air does not promote cloud formation.

Hurricanes

Some storms behave similarly to wave cyclones. However, the enormous power of the storms is concentrated over a small area, and they are usually much more violent and destructive than are wave cyclones. A severe tropical storm, with winds as strong as 240 km/hr, is called a **hurricane.** Hurricanes develop over warm, tropical oceans near the equator. Like the wave cyclones of the middle latitudes, hurricane winds spiral in toward the storm center. Seldom more than 700 km in diameter, hurricanes are much smaller than wave cyclones but much more powerful. These whirling storms are the most violent storms that occur on the earth. The greatest number of hurricanes—an average of 20 per year—occurs in the North Pacific Ocean.

A hurricane begins when very warm, moist air over the ocean rises rapidly. When moisture in the rising warm air condenses, a large amount of energy in the form of latent heat is released. This heat increases the force of the rising air. Moist tropical air continues to be drawn into the column of rising air, releasing more heat and sustaining the process. The entire storm system spins as a result of the Coriolis effect. An average hurricane has an energy content equal to the total amount of electricity used in the United States for six months.

A fully developed hurricane consists of a series of thick cloud bands spiraling upward into the center of the storm, as shown in Figure 25–6. Precipitation is very heavy. Winds increase in velocity

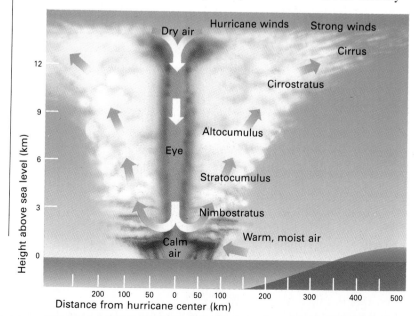

Figure 25–6. Thick clouds swirl upward around the center of a fully developed hurricane.

toward the center, or *eye,* of the storm, reaching speeds of 160 km/hr. The eye itself, however, is a region of calm, clear air. The winds of a hurricane usually last 9 to 12 days. During this time the hurricane moves with the prevailing winds. Hurricanes that form over the Pacific Ocean are called **typhoons.**

Thunderstorms

A storm accompanied by thunder, lightning, and strong winds is called a **thunderstorm.** Thunderstorms often occur when a small section of air in a warm, moist mT air mass is heated and rises. High surface temperature on the ground influences the rising of this warm, moist air. For that reason, thunderstorms occur most commonly in the late afternoon or early evening. Mountains or cold air in an occluded front also may cause the air to rise.

IMPACT

Weather Watch

In August 1969 Hurricane Camille slammed into the Gulf Coast of the United States. Winds up to 200 km/hr hammered at coastal towns in Mississippi and Louisiana, ripping out power lines, flattening houses, and leaving 258 people dead.

A decade later equally fierce Hurricane Frederic touched down in the same region. Yet this time the death toll was much lower: only a few lives were lost. Thanks, in part, to improved weather forecasting, nearly half a million people had been evacuated safely.

Each year severe storms batter many parts of the United States, causing loss of life and an estimated $20 to $30 billion in damage. Hurricanes, tornadoes, blizzards, and floods all take their toll.

The impact of these storms is evident in the wreckage they leave behind. Yet these disasters have also had another, less obvious, impact.

In response to the destructive potential of storms such as these, people have looked for more accurate ways to predict the weather. This search has led to the development of the science of meteorology and the growth of weather forecasting.

Over the past 20 years technological advances have vastly improved the accuracy of weather forecasts. New systems of radar, atmospheric sensors, weather satellites, and computers now make it possible for meteorologists to track killer storms and issue warnings.

What areas of the earth should meteorologists monitor to detect developing hurricanes? Explain why.

A thunderstorm develops in three distinct stages. The first stage of a thunderstorm is called the *cumulus stage*. At this stage warm, moist air rises until the water vapor within it condenses and forms a cumulus cloud approximately 7.5 km high. In the next stage, called the *mature stage,* the rising warm, moist air swells higher and higher. The cumulus cloud grows until it becomes a cumulonimbus cloud. The cloud reaches a height of approximately 12.2 km, where the top spreads out in an anvil shape. Heavy showers of ice crystals, water drops, and, occasionally, hailstones fall from the cloud. This precipitation produces a generalized cooling effect. While some of the air continues to rise, much of the air is dragged downward by the falling ice and raindrops, causing a strong downdraft. During the *final stage,* the strong downdraft stops air currents from rising. The thunderstorm dies out as the supply of water vapor in the air decreases. Why does the decrease in available water vapor cause the storm to subside?

During a thunderstorm clouds discharge electricity in the form of lightning. The released electricity heats the air, causing it to expand rapidly. The rapid expansion and collapse of the air produces the loud noise known as *thunder*. For lightning to occur, the clouds must have areas with distinct electrical charges. The upper part of the cloud usually carries a positive electrical charge, while the lower part carries both positive and negative charges. Lightning occurs as a huge spark that travels between the two parts of the cloud when the difference in their electrical charges becomes great. Scientists do not fully understand why clouds in thunderstorms have these differently charged regions. One theory states that large raindrops or ice particles carrying negative charges sink to the bottom of the cloud. Lighter, positively charged particles, such as cloud droplets and ice crystals, collect in the upper parts of the clouds. However, this theory has not yet been supported by scientific evidence.

Each year lightning from thunderstorms kills many people. A few precautions can lessen the risk of being struck by lightning. The first rule is to avoid any high location in open land. Lightning follows the shortest path between the cloud and the ground. Therefore, it is most likely to strike the highest object in a particular location, such as a tree, an isolated building, or a standing person. All metal objects and bodies of water should be avoided during a thunderstorm. However, the interiors of buildings, especially those with metal frames, and the interiors of automobiles are relatively safe. Metal structures are safe because the lightning is conducted to the ground along the metal exterior.

Tornadoes

The smallest, most violent, and shortest-lived severe storm is a **tornado.** A tornado is a whirling, funnel-shaped cyclone. A tornado forms when a thunderstorm meets high-altitude, horizontal winds. These winds cause the rising air in the thunderstorm to rotate. One of the storm clouds may develop a narrow, funnel-shaped, rapidly

Figure 25–7. Tornadoes, such as this one, which occurred in Waukesha county, Wisconsin, can be recognized by their funnel-shaped, rapidly spinning extension.

spinning extension. The extension reaches downward and may or may not actually touch the ground. If the tip of the funnel does touch the ground, it generally moves in a wandering path faster than a person can run. Frequently the funnel rises and touches down again a short distance away. The tornado generally covers a path not more than 100 m wide. Usually everything in that path is destroyed.

The destructive power of a tornado is due to the speed of the winds whirling within the funnel. Although the speed of these winds has never been measured, their great destructive power indicates that they may reach speeds up to 480 km/hr. Most of the injuries caused by tornadoes occur when people are trapped in collapsing buildings or are struck by objects flung by the wind. If you are outside during a tornado, you should immediately find a low place in the ground, lie down, and cover your head.

Tornadoes occur in many locations. However, they are most common in the midwestern region of the United States during the late spring or early summer. Tornadoes that occur over the ocean are called **waterspouts.** Waterspouts are usually smaller and less powerful than are tornadoes occurring over land.

Section 25.2 Review

1. What kind of front forms when a cold air mass overtakes a warm air mass?
2. What are the two paths commonly traveled by wave cyclones in North America?
3. If you are canoeing on a lake when a thunderstorm breaks out, are you in danger of being struck by lightning? Explain your answer.

25.3 Weather Instruments

Section Objectives

• Describe the types of instruments used to measure air temperature and wind speed.

• Describe the instruments used to measure upper-atmospheric weather conditions.

Weather observations are based, in part, on measurements of *atmospheric pressure,* humidity, and precipitation. In Chapter 23 you read about *barometers,* which are used to measure atmospheric pressure. In Chapter 24 you read about *psychrometers* and *hair hygrometers,* which are used to measure *relative humidity,* and *rain gauges,* which are used to measure precipitation. Weather observations are also based on measurements of temperature and winds, which are made by special instruments.

Measuring Air Temperature

An instrument used to measure temperature is called a *thermometer.* Three types of thermometers are used by meteorologists. A common type of thermometer uses a liquid—usually mercury or alcohol—sealed in a glass tube. A rise in temperature causes the liquid to expand and move up the tube. A drop in temperature causes the liquid to contract and move down the tube. A scale marked on the glass tube indicates the temperature. Both the Celsius (C) and Fahrenheit (F) scales are commonly used in the United States.

A second type of thermometer is called a **bimetal thermometer.** This thermometer consists of a bar made up of two strips, each of a different metal. The metals, such as brass and iron, expand different amounts when heated. The bar will curve when heated and straighten again when cooled. An instrument called a **thermograph** measures temperature changes by recording the movement of the bar. In a thermograph a pen is attached to the bar. The pen point is placed against a chart on a rotating drum. As the drum rotates, the bending and straightening of the bar, and thus the temperature changes, are recorded on the chart.

A third type of thermometer is an **electrical thermometer.** As the temperature rises, the electric current flowing through certain materials increases. The electric current flowing through these materials is translated into temperature readings. This type of thermometer is especially useful when an observer cannot be present.

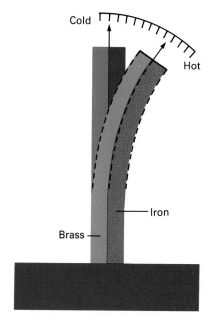

Figure 25–8. A bimetal thermometer curves when heated and straightens when cooled.

Measuring Wind Speed and Direction

An instrument called an **anemometer** (AN-uh-MOM-uh-ter) measures wind speed. A typical anemometer consists of small cups attached by spokes to a shaft that rotates freely. The wind pushes against the cups, causing them to rotate. This rotation triggers an electrical signal that registers the wind speed—in m/sec, mph, or knots—on a dial. Some anemometers record the wind speed on a graph.

The direction of the wind is determined by a **wind vane.** Often a wind vane is shaped like an arrow with a large tail. The wind vane turns freely on a pole as the tail catches the wind. Thus, the arrow always points into the wind. A wind is described according to the direction from which it comes, so that *westerly* describes a wind that blows from the west. Wind direction may be described precisely as one of 32 directions, or points, on the compass. Wind direction also may be recorded in degrees, beginning with 0° at north and moving clockwise around to 360°. East is designated as 90°, and south, as 180°. How would a west wind be indicated in this system?

Measuring Upper-Atmospheric Conditions

Thermometers, anemometers, and wind vanes measure conditions of the atmosphere near the earth's surface. However, these conditions are only a part of the complete weather picture. An instrument commonly used by meteorologists to investigate weather conditions in the upper atmosphere is the *radiosonde,* also discussed in Chapter

Investigation

Wind chill

Wind chill is a term used by meteorologists to describe the cooling power of moving air. You can demonstrate the relationship between temperature and wind speed with a simple model.

Science Process Skills Focus: demonstrating, identifying cause-and-effect relationships

Materials

shallow pan, about 30 cm across; water; thermometer; electric fan; clock or watch with second hand

Procedure

1. Place the pan on a level table and fill to a depth of about 1 cm with water at room temperature.
2. Lay the thermometer in the center of the pan with the bulb submerged. Be sure you can read the temperature without touching the thermometer.
3. Do not disturb the pan for 5 minutes, then read and record the water temperature.
4. Position an electric fan a few centimeters from the edge of and facing the pan. **Caution: Do not get the fan or cord wet.**
5. Turn on the fan to a low speed and observe

and record the water temperature every minute until the temperature remains constant.
6. If the fan has different speed settings, repeat Step 6 with the fan at a higher speed.

Analysis and Conclusions

1. How does the moving air affect the temperature of the water?
2. The moving air is the same temperature as the still air in the room. What causes the water temperature to change?
3. On a cool, windy day, what would be the best way to dress to stay comfortable outdoors? Explain your answer.

24. Carried aloft by a helium-filled balloon, a radiosonde measures relative humidity, air pressure, and air temperature. These measurements are sent out by a radio signal. A special radio on earth receives and records the information. The path of the balloon is also tracked to determine the direction and speed of high-altitude winds. When the balloon reaches an extremely high altitude, it bursts and the radiosonde is parachuted back to earth.

Another instrument for determining weather conditions in the upper atmosphere is **radar.** Radar is an electronic device that sends out, or transmits, a pulse of radio waves in the form of a beam. Objects that cross the beam reflect it back to the transmitter, which receives the returning beams. Radar can detect objects that are too small or too far away for the human eye to see. For example, particles of water in the atmosphere reflect radar pulses. Thus, precipitation and other weather disturbances are visible on a radar screen, or radarscope. Radar can also indicate the precise location and extent of a storm and can help meteorologists track the development of a storm system. However, radar cannot detect cloud droplets and dust particles. Another device—the laser—is used to detect the presence of these particles and to measure the heights of clouds. The laser emits a beam of light that meteorologists use like a ruler to measure the height of the cloud.

Instruments carried by weather satellites also provide important information about the upper atmosphere. For example, cameras on satellites photograph clouds. You have probably seen such photographs on a television weather report. Satellite photographs provide weather information for regions where observations cannot be made on the ground. The direction and speed of the wind at the level of the clouds can also be measured by examining a continuous sequence of cloud photographs. Satellite photographs made by using infrared light reveal temperatures at the tops of clouds and at the surface of the land. Satellite instruments can also measure conditions such as the temperature of the ocean surface and the height of waves.

Meteorologists also use computers to understand the weather. Before computers were available, it was very difficult, and sometimes nearly impossible, to solve the mathematical equations describing the behavior of the atmosphere. In addition to solving many of these equations, computers can provide information that is useful in predicting weather changes. Computers can also store weather records for quick retrieval. In the future, powerful computers may greatly improve weather forecasts and provide a much better understanding of the atmosphere.

Figure 25–9. Today meteorologists use a variety of instruments to gather and analyze weather data.

Section 25.3 Review

1. How does a bimetal thermometer work?
2. What does a radiosonde measure?
3. If a wind vane is pointing toward the west, from what direction is the wind blowing? Explain your answer.

25.4 Forecasting the Weather

Predicting the weather has challenged people for thousands of years. People of early civilizations attributed control of weather conditions such as wind, rain, and thunder to gods. More than 4,000 years ago, people attempted to forecast the weather using the positions of the stars as the basis for their forecasts.

Modern weather forecasting began with the invention of more-sophisticated scientific instruments, such as the thermometer and the barometer. The invention of the telegraph in 1844 enabled meteorologists to share information about weather conditions quickly, and it led to the development of national weather services. In 1870 the United States formed a weather-forecasting agency as part of the Army Signal Corps. Twenty years later the agency was organized into the Weather Bureau. In 1970 it was renamed the National Weather Service. Because accurate weather forecasting is so important, nations around the world cooperate in gathering and exchanging weather data.

Every six hours weather observers at stations all over the world report weather conditions. These observers record the barometric pressure and whether it is rising or falling. They record the speed and direction of surface wind. They measure precipitation, record the temperature, and determine the humidity. They note the type, amount, and height of cloud cover. Observers also record visibility and general weather conditions. In addition, many larger observation stations send up radiosonde equipment to determine conditions in the upper atmosphere. Each station puts the information in an international code and sends it to a central collection center. Weather centers around the world exchange the weather information they have collected.

The World Meteorological Organization (WMO) sponsors a program—World Weather Watch—to promote the rapid exchange of weather information. The organization helps developing countries establish or improve their meteorological services. It also offers advice on the effect of weather on natural resources and on human activities such as farming and transportation. WMO was founded in 1873 and is now part of the United Nations.

Making a Weather Map

Meteorologists usually prepare weather maps—the basic tools of weather forecasting—at the centers where coded weather information is received. They revise these maps several times a day as new weather data are collected. Weather observations from the various stations reporting to the center are first translated into internationally recognized symbols. Clusters of symbols are plotted on the map around each reporting station, showing the conditions at that station. Such symbols around a station on a map are called a **station model.**

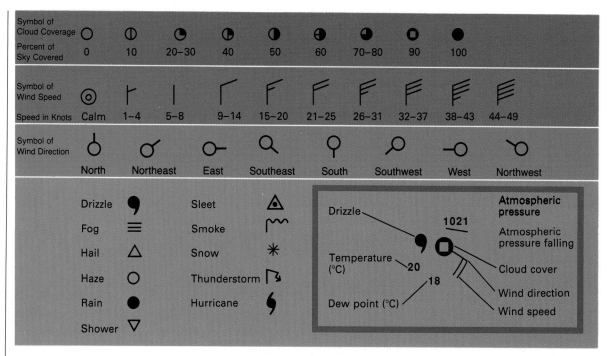

Symbol of Cloud Coverage	○	◑	◔	◔	◑	◕	●	▣	●
Percent of Sky Covered	0	10	20–30	40	50	60	70–80	90	100

Symbol of Wind Speed	◎	⊢																
Speed in Knots	Calm	1–4	5–8	9–14	15–20	21–25	26–31	32–37	38–43	44–49

Symbol of Wind Direction	◯	◯	◯	◯	◯	◯	◯	◯
	North	Northeast	East	Southeast	South	Southwest	West	Northwest

Drizzle	,	Sleet	◬
Fog	≡	Smoke	⌇
Hail	△	Snow	✳
Haze	○	Thunderstorm	↱
Rain	●	Hurricane	↰
Shower	▽		

Figure 25–10. Meteorologists use these symbols to indicate weather conditions , wind speed, cloud cover, and wind direction. A station model (inset) shows conditions around a weather station.

Common weather symbols describe cloud cover, wind speed, wind direction, and weather conditions such as type of precipitation and storm activity. These symbols and a station model incorporating them are shown in Figure 25–10. Notice that the symbols for cloud cover, wind speed, and wind direction are combined in one symbol.

Other information included in the station model is the air temperature and the *dew point*. As explained in Chapter 24, the dew point is the temperature to which the air must be cooled to become saturated with water vapor. When the air is nearly saturated, there is little difference between its temperature and the dew point. Therefore, a comparison of these figures is an indication of the relative humidity at that time.

The station model also indicates the atmospheric pressure in millibars. The position of a straight line under this figure—horizontal or angled up or down—shows whether the atmospheric pressure is steady or is rising or falling.

Lines are then drawn connecting points of equal atmospheric pressure. These lines are called **isobars.** The spacing and shape of the isobars help meteorologists interpret their observations about the speed and direction of the wind. Closely spaced isobars indicate a rapid change in pressure and high wind speeds. Widely spaced isobars generally indicate a slow change in pressure and low wind speeds. Isobars that form circles indicate where there are centers of high or low air pressure. Such centers are usually marked *H* for *high* or *L* for *low*. What configuration of isobars would represent a storm system?

The weather map is completed by marking in the location of fronts and areas of precipitation. Fronts are indicated by sharp

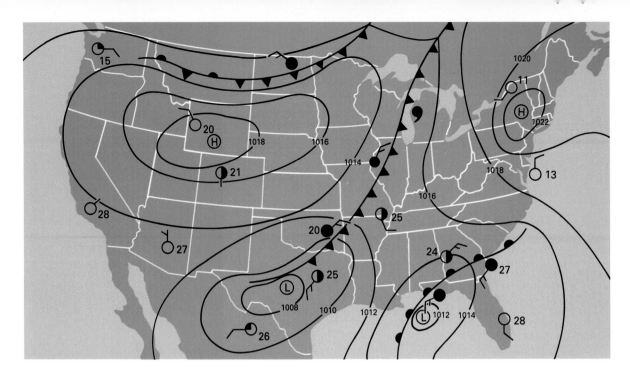

Figure 25–11. A typical weather map shows isobars, highs and lows, fronts, wind direction and speed, temperature, and precipitation.

changes in winds, temperature, or humidity. Areas between stations reporting precipitation are usually colored green. Use the weather map in Figure 25–11 to interpret the weather in your area the day the weather map was made.

Types of Forecasts

Meteorologists make two types of forecasts: daily and long-term. Daily forecasts predict weather conditions for 36 hours. Long-range forecasts usually cover periods ranging from 5 to 30 days. Extended forecasts usually cover periods longer than 30 days.

Daily and Long-Term Forecasts

To forecast the weather, meteorologists study the most recent weather map. Maps for the previous 24 hours are compared with the newest one. This comparison allows meteorologists to follow the progress of large weather systems. Satellite photographs and radar reports of precipitation supply additional information. The intensity and path of weather systems are plotted on updated maps. Meteorologists usually use computers to prepare maps indicating conditions at many levels of the atmosphere. Computers can also make maps showing the possible weather conditions for the next several days. However, meteorologists must carefully interpret these maps because computer predictions are based on generalized descriptions. Using all the information available, meteorologists then forecast the weather. Temperature, wind direction, wind speed, cloudiness, and precipitation can usually be forecast accurately. However, it is often difficult to predict precisely when precipitation will occur and the exact amount that will fall.

Accurate weather forecasts can be made for two to three days. Accuracy decreases with each additional day. Extended forecasts are

Science Notebook

Project GALE

Modern weather forecasting has a number of limitations. For example, severe cyclonic winter storms along the Atlantic Coast of the United States often take forecasters by surprise. These storms, such as the one shown, which hit Washington, DC, in 1983, immobilize cities and cause destruction.

In 1986 a group of 200 meteorologists began a project to learn more about these unpredictable Atlantic storms. The project, named *GALE* for **G**enesis of **A**tlantic **L**ows **E**xperiment, used what is called a *mesoscale* approach. That is, the researchers studied weather phenomena on a small scale, concentrating on the eastern coast of the United States. Normal weather observations, they believed, were made too infrequently and covered too wide an area.

The GALE team used radar, portable ground stations, weather ships, and four times the number of weather balloons launched daily by the United States Weather Service.

The GALE researchers found that a combination of local features is responsible for turning small storms into big ones. One of those features is the Appalachian Mountains, which trap cold air from the northeast. Another important feature is the Gulf Stream, which flows along the eastern coast of the United States. The contrast between the warm maritime air influenced by the Gulf Stream and the cool mountain air leads to the development of storms.

In what ways might the GALE approach and findings be helpful to other weather forecasters?

---

made by computer analysis of slowly changing large-scale movements of the air. These changes help meteorologists predict the general weather pattern. Meteorologists cannot forecast the long-term weather at a particular place, however. The distance between stations is too great to provide the necessary information.

Controlling the Weather

Scientists are currently investigating methods of controlling rain, hail, hurricanes, cyclones, and lightning. So far the most successful method has been the production of rain by *cloud seeding*. In this process freezing nuclei are added to supercooled clouds, causing rain to fall. Scientists in the Soviet Union have use cloud seeding successfully with potential hail clouds, thus causing rainfall rather than hail. Cloud seeding is discussed further in Chapter 24.

Hurricanes have also been seeded with freezing nuclei, and some meteorologists think that this process may reduce the intensity of a storm. Attempts to control middle-latitude cyclones in this way have not been successful, however. Meteorologists know so little about the structure of tornadoes that they have not yet tried to control these storms.

Attempts have also been made to control lightning. The release of large quantities of ions near the ground can modify the electrical properties of small cumulus clouds. However, it is not known whether this method would affect the electrical properties of large

Career Focus: *Weather*

"One of the most interesting things about meteorology is that it's not an exact science. For researchers, there's always a challenge."

Meteorologist

Not long ago, summaries of weather observations and information about forecast accuracy were only available 18 months after the fact—too late to be of much practical use. Today, thanks to the work of meteorologists like Mary Heffernan of the National Weather Service, much summary and forecast data are available within 24 hours.

As a specialist in systems development, Heffernan has used her undergraduate training in computer science to design the software needed for this more timely forecast-evaluation program.

"The system not only allows us to collect data more quickly, but helps us get feedback on forecast accuracy almost immediately, when it's most useful," she said.

This quick feedback allows weather forecasters to improve their forecasting methods.

Heffernan, pictured left, is now involved in developing another state-of-the-art support system for weather forecasters of the 1990s. Called *AWIPS-90,* for *Advanced Weather Interactive Processing System,* this new program will make it possible for computers to more quickly combine weather data from a variety of sources including radar and satellites. The National Weather Service has a model of the program in operation.

clouds. Seeding of potential lightning storms with silver-iodide nuclei has seemed to modify the occurrence of lightning, although no conclusive results have been obtained.

Section 25.4 Review

1. What is a station model?
2. Why are new and 24-hour-old weather maps compared?
3. Explain which region on a weather map—one with widely spaced isobars or one with closely spaced isobars—has stronger winds.

"Trying to understand what's happening in the atmosphere isn't always easy," said Heffernan, "but when you can see a need, define it, develop a solution, and finally see it implemented, the rewards are great."

Severe-Storm Forecaster

Severe-storm forecasters work for the National Weather Service in stations such as the one in Kansas and the one in Florida. Kansas forecasters predict severe storms and tornadoes. Florida forecasters predict hurricanes that might threaten the East and Gulf coasts of the United States.

Severe-storm forecasters, such as the one shown left, gather information from many sources, including satellites and radar. The collected data are then analyzed. If necessary, storm watches and warnings are issued.

This career requires a bachelor's degree in meteorology and experience in weather forecasting and analysis.

Weather Observer

Weather observers are also known as meteorological technicians. They observe, measure, and record weather conditions. Weather observers are also responsible for analyzing data, verifying the accuracy of the data, and then supplying the results to meteorologists or other weather information services.

Weather observers must be able to read instruments that measure temperature, wind velocity, air pressure, humidity, and rainfall.

Weather observers do not need a college degree, but a high school background in math and science is essential. Many people learn these skills in technical training programs.

For Further Information

For more information on careers in meteorology, write the U.S. Department of Commerce, National Oceanic and Atmospheric Administration, National Weather Service, Silver Spring, MD 20910.

Reviewing Chapter 25

Key Concepts

- An air mass is a large body of air with uniform temperature and humidity. See page 427.
- Cold and warm fronts are associated with characteristic weather conditions. See page 430.
- A wave cyclone is a storm with a low-pressure center. See page 431.
- Hurricanes, thunderstorms, and tornadoes are violent, destructive storms. See page 433.
- Thermometers are used to measure air temperature. Anemometers are used to measure wind speed. See page 437.
- Radiosonde, radar, satellite equipment, and computers are used to measure upper-atmospheric weather conditions. See page 438.
- Meteorologists prepare weather maps based on information from weather stations around the world. See page 440.

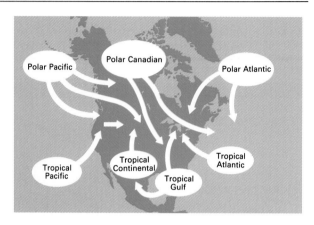

- **Polar air masses and tropical air masses influence the weather of North America. See page 428.**
- Meteorologists make daily and long-term forecasts of the weather. See page 442.

Key Terms

air mass (427)	continental tropical (428)	maritime tropical (428)	thunderstorm (434)
anemometer (437)		occluded front (431)	tornado (435)
anticyclone (433)	electrical thermometer (437)	polar front (431)	typhoon (434)
bimetal thermometer (437)		radar (439)	warm front (430)
	front (430)	squall line (430)	waterspout (436)
cold front (430)	hurricane (433)	station model (440)	wave cyclone (431)
continental polar (428)	isobar (441)	stationary front (431)	wind vane (437)
	maritime polar (428)	thermograph (437)	

Review

On your paper, write the letter of the term that best completes each of the following statements.

1. A region where air masses can form must be fairly (a) cold (b) warm (c) hilly (d) uniform.
2. In an air mass designated *cP*, the *c* stands for (a) continental (b) cold (c) coastal (d) cool.
3. Polar Canadian air masses generally move (a) southeasterly (b) northerly (c) northeasterly (d) westerly.

4. The air masses that sometimes bring heavy rains to the deserts of the southwestern United States are called (a) polar Canadian (b) polar Atlantic (c) tropical Pacific (d) tropical continental.
5. The type of front formed when two air masses move parallel to the front between them is called (a) stationary (b) occluded (c) polar (d) warm.

6. A front that is completely closed off from the ground by cold air is called (a) cold (b) occluded (c) polar (d) warm.
7. The winds of a wave cyclone blow in circular paths around a (a) front (b) low-pressure region (c) high-pressure region (d) jet stream.
8. The eye of a hurricane is a region of (a) hailstorms (b) great turbulence (c) calm, clear air (d) strong winds.
9. In the mature stage of a thunderstorm, a cumulus cloud grows until it becomes a (a) stratocumulus cloud (b) altocumulus cloud (c) cumulonimbus cloud (d) cirrocumulus cloud.
10. Tornadoes that occur over the ocean are called (a) waterspouts (b) typhoons (c) waves (d) hurricanes.
11. A wind with a direction designated as 90° is blowing from the (a) north (b) south (c) east (d) west.
12. An instrument commonly used to investigate upper-atmospheric weather conditions is the (a) anemometer (b) wind vane (c) radiosonde (d) thermograph.
13. The lines on a weather map connecting points of equal atmospheric pressure are called (a) isobars (b) highs (c) lows (d) fronts.
14. It is generally difficult to accurately predict (a) wind speed (b) amount of precipitation (c) wind direction (d) temperature.

Application

On your paper, write answers to the following questions.
1. If the air in your region is warm and dry, what type of air mass could be responsible? What letters designate this air mass?
2. Suppose you are traveling with your family through the desert in the southwestern United States and a heavy rain begins to fall. What type of air mass may have brought the rain?
3. Suppose the air is warm and moist. You hear on the weather report, however, that a fast-moving cold front will reach your region the next day. What kind of weather conditions can you expect?
4. People on Vancouver Island, off the west coast of Canada, hear reports of a wave cyclone in Alaska. Is it likely that the wave cyclone will reach their area? Explain why.
5. Suppose a hurricane is passing through a Caribbean island. Suddenly the rain and winds stop and the air becomes calm and clear. Is it safe to go outside? Explain your answer.
6. Is it safe to be on the street in an automobile during a tornado? Explain your answer.
7. In what direction would a wind of 315° make a wind vane point?
8. An air-traffic controller is monitoring nearby airplanes by radar. The controller warns an incoming pilot of a storm a few miles away. How did the radar screen help the controller detect the storm?
9. On a weather map you see a circle that is half darkened and a straight vertical line extending upward. What can you say about the weather in that area for that day?
10. In February 1983, 63.5 cm of snow fell on the Atlantic Coast of the United States in about 48 hr. Weather forecasters had not expected this storm. Why are such storms unpredictable?

Extension

1. Research one particularly destructive hurricane, thunderstorm, or tornado. Report your findings to the class. Use visual aids such as maps and charts in your report.
2. Cut the weather map from your local newspaper for seven days in a row. Trace the paths of fronts that passed over your area during that time. Compare the predicted cloud cover, preciptitation, and temperatures for each day with the conditions that actually occurred. Report your findings to the class.
3. Listen to the weather report for a week. Make a note each time the report mentions an approaching front and the weather it is expected to bring. Predict the weather for the next week and present your forecst to the class.

Chapter 26

Climate

On a December morning while residents of New York City are shoveling out after a snowstorm, residents of Los Angeles awake to another warm, sunny day. Daily weather conditions and seasonal weather patterns vary greatly around the world. For example, New York has a humid climate, warm summers, and cold winters, while Los Angeles has a dry climate and mild temperatures year-round.

In this chapter you will learn about the factors that influence weather conditions and that produce predictable global climatic patterns.

Chapter Outline

26.1 Factors That Affect Climate
Latitude
Heat Absorption and Release
Topography

26.2 Climate Zones
Tropical Climates
Polar Climates
Middle-Latitude Climates
Local Climates

These varied landscapes in New Zealand have different climates.

26.1 Factors That Affect Climate

The average weather conditions of a region, or the weather patterns that occur over many years, are referred to as *climate*. Usually scientists describe climate in terms of the average monthly and yearly temperatures and the average amount of precipitation. Average temperatures are calculated by adding two or more temperature readings and dividing by the number of readings. For example, the average daily temperature is calculated by averaging the high and low temperatures of the day. The monthly average is determined by averaging the daily averages. The yearly average temperature can be calculated by averaging the monthly averages.

Another way scientists describe temperature is by indicating the **temperature range.** Temperature range is the difference between the highest and lowest temperatures of a day or month. The yearly temperature range is the difference between the highest and lowest monthly averages.

Using only average temperatures to describe climate can be misleading. For example, both St. Louis, Missouri, and San Francisco, California, have an average yearly temperature of about 13°C. However, St. Louis has a climate with cold winters and hot summers. San Francisco has a generally mild climate all year with mild winters and cool summers.

Another major weather condition, precipitation, is described as the average precipitation a region receives in a year. However, average yearly precipitation alone does not accurately describe climate. For example, the average yearly precipitation for New York City and for Miami, Florida, is almost the same. In Miami, however, the rain falls mostly during the rainy season from May to October. In New York City, various forms of precipitation fall throughout the year. Clearly, the climates of these two cities differ.

An accurate description of climate must include several factors that influence both temperature and precipitation. These factors include latitude, heat absorption and release, and topography.

Latitude

A major influence on the climate of a region is its latitude, or distance from the equator. Latitude determines the amount of solar energy received by and the prevailing wind patterns of the region.

Solar Energy

The amount of solar energy that the earth receives depends on two factors. They are the angle at which the rays of the sun strike the earth and the number of hours of daylight. The angle at which the sun's rays strike a region is determined by its latitude and by the tilt of the earth's axis. At the equator the rays always strike the earth at almost a 90° angle.

Section Objectives

- **Explain how latitude determines the amount of solar energy received on earth.**
- **Describe how the different rates at which land and water are heated affect climate.**
- **Explain the effects of topography on climate.**

In equatorial regions both day and night are about 12 hours long throughout the year. The result is steady high temperatures year-round and a yearly temperature range of only 3°C or 4°C in most areas. There are no summers or winters—only dry or rainy seasons.

At higher latitudes the sun's rays strike the earth at an angle lower than 90°. The rays do not heat the earth as much because their energy is spread over a wider area. Thus, average yearly temperatures in these locations are lower than those at the equator. Also, the lengths of the days and the nights vary. For example, at 45° north or south of the equator, the hours of daylight vary from about 16 hours in the summer to about 8 hours in the winter. Therefore, the yearly temperature range is large—more than 30°C in some regions.

In the polar regions, the sun sets for only a few hours each day in the summer and rises for only a few hours each day in the winter. Thus, the annual temperature range is very large, but the daily temperature ranges are very small.

When the average daily temperatures in various parts of the world are plotted on a map, they form a series of zones, as shown in Figure 26–1. The boundaries of these zones roughly follow the parallels of latitude. The average daily temperatures are highest near the

Figure 26–1. The zones of average daily temperature (° Celsius) in January (top) and July (bottom) roughly follow the lines of latitude.

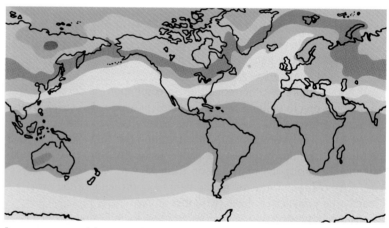

January normal temperature

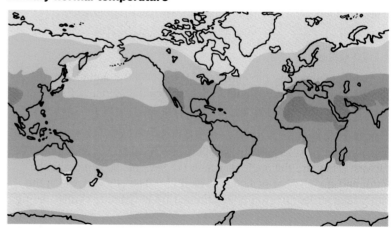

July normal temperature

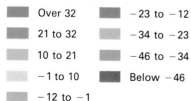

Degrees in Centigrade

Over 32	−23 to −12
21 to 32	−34 to −23
10 to 21	−46 to −34
−1 to 10	Below −46
−12 to −1	

IMPACT

El Niño

From late 1982 to mid-1983, unusually violent weather created havoc around the world. Hurricanes and storms battered the Pacific United States, Hawaii, and Tahiti. Floods ravaged the southeastern United States, Cuba, Ecuador, and northern Peru. Droughts spread through Central and South America, Australia, the Philippines, Indonesia, and India.

More than a thousand people died. Loss of property and crops soared to billions of dollars.

A seemingly unrelated event that occurred at the same time helped explain the unusual weather. Large numbers of plants and fish that lived off the coast of Peru died. The cause was the presence of a warm Pacific current known as *El Niño.*

El Niño appears every three to ten years. It begins with a weakening of the trade winds that usually push warm water toward the western Pacific Ocean. The weakened trade winds allow the warm water to flow eastward instead. This warm water heats the cool coastal waters of the west coast of South America to more than 27°C. Plants and fish die because they cannot live in the warmer water.

The effects of El Niño are not limited to the oceans. Ocean temperatures strongly affect atmospheric temperatures and, thus, overall weather patterns. Storms, droughts, and other unexpected weather can result. El Niño can affect up to 70 percent of the earth's weather.

What other weather phenomenon can disrupt the global pattern of air movements?

equator and decrease with distance from the equator. However, the irregular shape of the temperature zones indicates that the amount of solar energy is not the only factor affecting climate.

Wind Patterns

Latitude also determines global wind belts that affect a region and thus the general direction of the wind in any particular location. The latitude ranges of these global wind belts are described in Chapter 23. Winds affect many weather conditions, such as humidity, precipitation, temperature, and cloud cover. Hence, regions with different prevailing winds often have different climates. The global wind pattern is also influenced by storms and local weather.

Within the different global wind belts are various regions of low and high pressure. In the equatorial belt of average low pressure—the doldrums—the air rises continuously and thus loses moisture. As

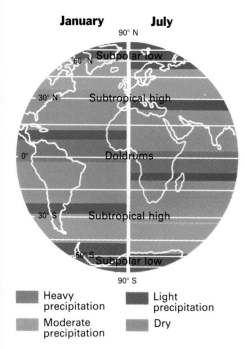

January   July

90° N

60° N   Subpolar low

30° N   Subtropical high

0°   Doldrums

30° S   Subtropical high

60° S   Subpolar low

90° S

■ Heavy precipitation   ■ Light precipitation
■ Moderate precipitation   ■ Dry

Figure 26–2. During winter in the Northern Hemisphere global wind and precipitation belts shift to the south.

a result, the regions within this latitude range of low pressure generally receive heavy precipitation. The amount of rainfall is most abundant in a belt around the equator and decreases steadily with increasing latitude. In the areas around 20° to 30° north and south latitude—the subtropical highs—the air is mostly dry and sinking, and little precipitation occurs.

Closer to the poles, beginning at around 45° to 55° latitude, is a belt of low average precipitation. In these regions—the subpolar lows—warm tropical air meets cold polar air, and *wave cyclones* frequently develop. At latitudes above 50° to 55°, average precipitation decreases in the cold, dry polar air masses.

With the changing seasons, the global wind pattern shifts in a north-south direction, as shown in Figure 26–2. As the wind and pressure belts shift, the belts of precipitation associated with them also shift.

Heat Absorption and Release

The way solar energy strikes the earth and is absorbed or reflected also influences the surface temperature. Land heats faster and to a higher temperature than water does. One reason for this difference is that the land surface is solid and basically unmoving, while the water surface is liquid and continuously changing. Waves, currents, and other movements continuously replace warm surface water with cooler water from the ocean depths. This action prevents the surface temperature of the water from increasing rapidly. The surface temperature of the land, on the other hand, can continue to increase as more solar energy is received.

Land and water also absorb and release heat at different rates. The **specific heat** of water is higher than that of land. Specific heat is the amount of heat needed to raise the temperature of 1 g of a substance 1°C. A given mass of water requires more heat than does the same mass of land to increase its temperature the same number of degrees. Even if not in motion, water warms more slowly than land does. Water also releases heat more slowly than land does.

The average temperatures of land and water at the same latitude vary also because of differences in the loss of heat through evaporation. Evaporation affects water surfaces much more than it does land surfaces.

Ocean Currents

The amount of heat absorbed or released by the air is influenced by the temperature of ocean currents with which the air comes in contact. If winds consistently blow toward the shore, currents have a stronger effect on air masses. For example, the combination of a warm Atlantic current and steady westerly winds gives northwestern Europe an unusually high average temperature for its latitude. On the other hand, the warm Gulf Stream has little effect on the east coast of the United States because westerly winds usually blow warm maritime air away from the coast.

Seasonal Winds

Heat differences between the land and the oceans sometimes cause winds to shift seasonally in certain regions. During the summer the land heats more quickly than the ocean does. If a low-pressure center develops over the land, the warm air rises and is replaced by cool air from the ocean. Thus the wind moves landward. During the winter the land loses heat more quickly than the ocean does, and the cool air flows away from the land. Thus the wind moves seaward. Such seasonal winds are called **monsoons.** They are strongest over the large landmasses near the equator. For example, monsoons in southern Asia result from the heating and cooling of the northern Indian peninsula. In the summer, winds carry moisture to the land from the ocean, bringing heavy rainfall. In the winter, winds bring dry weather, sometimes even drought. Monsoon conditions also occur in eastern Asia and in equatorial regions with long coastlines.

Investigation

Evaporation

Evaporation affects temperature. The converse is also true. You can demonstrate how temperature affects the rate of evaporation of water.

Science Process Skills Focus: observing, identifying cause-and-effect relationships

Materials

portable lamp with an incandescent bulb, preferably with flexible neck; 4 small shallow bowls; thermometer; clock or watch; ruler; water

Procedure

1. Copy Table 26–I onto a piece of paper.
2. Place the lamp at one end of a table or counter and turn it on. Place one bowl directly under the light, a second bowl 25 cm away from the lamp, a third 50 cm away, and a fourth 75 cm away.
3. Lay the thermometer across the top of the first bowl. Wait 3 minutes, then read and record the temperature. Repeat this step for the other three bowls.
4. Hold the ruler upright in the center of the first bowl. Add water to the bowl until it reaches the 1-cm mark. Repeat this step for the other three bowls.

Table 26–I:

Bowl	Temperature	Amount of water evaporated
1		
2		
3		
4		

5. Leave the bowls of water in place and keep the light burning for 24 hours. Then measure the amount of water in each bowl. Record your results on the table.

Analysis and Conclusions

1. At what distance from the lamp did the greatest amount of water evaporate? The least?
2. Explain the relationship between temperature and the rate of evaporation.
3. Explain why puddles of water dry out much more quickly in summer than they do in fall or winter.

Figure 26–3. Air passing over a mountain range loses its moisture as it rises and cools. The dry air is warmed 1°C per 100 m as it descends on the other side of the mountain.

Rising, cooling air

Descending, warming air

Topography

The topography, or shape of the land, also influences climate. The altitude, or height above sea level, produces distinct temperature changes. Average temperature decreases as altitude increases in the troposphere. For every 100 m increase in altitude, the average temperature decreases 1°C. Even along the equator, for example, the peaks of high mountains are cold enough to be covered with snow.

Mountains influence the temperature and moisture content of passing air masses. When a moving air mass enters a mountainous region, it rises and cools. As the air rises, it loses most of its moisture through precipitation. As the air descends on the other side of a mountain, it is warmed about 1°C every 100 m. Air flowing down mountain slopes, therefore, is usually warm and dry, as shown in Figure 26–3. One such wind is the **foehn** (FUHN)— a warm, dry wind that flows down the northern slopes of the Alps Mountains. Similar warm winds that flow down the eastern slopes of the Rocky Mountains are called **Chinooks.** A Chinook can raise the air temperature very rapidly in a short period of time. In 1900 a Chinook raised the temperature in a small town in Montana 17°C in three minutes. Would you expect to find more vegetation on the side of a mountain facing toward or away from the prevailing winds? Why?

Some winds that blow down mountain slopes are not warm. These winds are so cold that they remain cold even after heating. The **mistral,** which blows from the North down the Alps to the Mediterranean Sea, is a stormy, cold wind. It is sometimes strong enough to knock over chimneys. Another cold northern wind, the **bora,** blows from the mountains of Yugoslavia to the Adriatic Sea.

26.1 Section Review

1. List two factors that determine the amount of solar energy that an area receives. On what do these factors depend?
2. Does land or water heat more quickly? Why?
3. Compare monsoons and Chinooks.

26.2 Climate Zones

A geographic region that has a predictable temperature range and other predictable weather conditions is called a *climate zone*. The earth has three major climate zones. The warm zone immediately around the equator is the zone of **tropical climates.** Tropical climates have an average monthly temperature of at least 18°C, even during the coldest month of the year. Tropical climates are influenced by the continental and maritime tropical air masses, which develop close to the equator.

At the other extreme are the **polar climates.** In these regions the average monthly temperature is never higher than 10°C. Continental and maritime polar air masses originate in these areas.

Between the tropical and polar climate zones is the zone of temperate climates, or **middle-latitude climates.** The average monthly temperature of these climates is no warmer than 18°C in the coldest month and no cooler than 10°C in the warmest month. In middle-latitude climate zones, the weather changes often because both tropical and polar air masses move across these regions. The middle-latitude climate zones are also frequently exposed to cyclonic storms, with strong winds and heavy rains, that are produced along the polar front. The general boundaries of the three major climate zones are shown in Figure 26–4.

Tropical Climates

Each principal climate zone has specific average temperatures. However, there are several different types of climates within these zones because of differences in the amount of precipitation. For example, within the tropical climates are three types of climates: **rain-forest, tropical-desert,** and **savanna climates.**

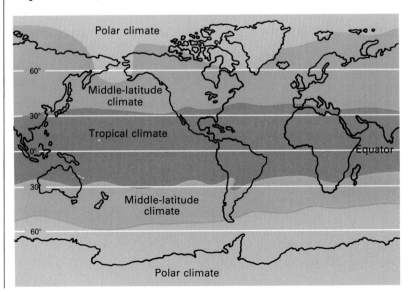

Figure 26–4. The general boundaries of the major climate zones are shown on this map.

Science Notebook

The Tropical Rain Forest: An Endangered Habitat

Tropical rain forests are one of the earth's most complex and varied natural habitats. Approximately one third of all the plant and animal species on earth live in rain forests. Scientists have identified 1,100 species of trees, shrubs, and grasses on less than 2.6 km^2 of land in the Colombian rain forest. In contrast, all of the British Isles, which are located in a middle-latitude climate, support only about 1,450 species.

The plants in rain forests are sources for many medicines and for products used in agriculture and industry. For example, the copaiba tree produces a liquid similar to diesel fuel. The wild plants of the rain forests also may become an important source of new crops if cultivated varieties lose their resistance to diseases and insects.

Complete destruction of the earth's rain forests, however, is occurring at an alarming rate. Scientists predict that at the present rate, all tropical rain forests will be destroyed by the end of the century. Forests are burned to provide land for agriculture and urban development. In addition to destroying the forests, the burning of the trees adds carbon dioxide to the atmosphere. This carbon dioxide contributes to the greenhouse effect, which may be causing a rise in global temperatures. Also common are slash-and-burn farming practices: Forests are cut and burned down, crops are planted, then the land is

abandoned when the minerals in the soil have been used up.

Approaches to reclaiming the rain forests include using chemical fertilizers and crop rotation to restore the fertility of the soil.

How does destruction of the rain forests affect the tropical climate?

Rain-Forest Climate

The warm, humid regions within 5° to 10° on either side of the equator are covered with dense vegetation called *rain forests*. For that reason the climates of these regions are known as *rain-forest climates*. The moist, rising easterly wind produces an annual rainfall that is usually greater than 250 cm. The yearly temperature range is very small—about 3°C. Central Africa, the Amazon River basin of South America, and parts of Central America and Southeast Asia have rain-forest climates.

Tropical-Desert Climate

Warm, dry weather conditions occur in regions about 2,570 km north and south of the equator. The northern boundary lies along the Tropic of Cancer at 23.5° north of the equator, and the southern boundary lies along the Tropic of Capricorn at 23.5° south of the equator. These areas have tropical-desert climates. Tropical-desert climates are influenced by the dry, sinking air masses of the subtropical highs. They include some of the earth's driest deserts. Annual rainfall of tropical-desert climates is less than 25 cm. The largest belt of tropical deserts extends across North Africa and southwestern Asia and includes the Sahara, Libyan, Arabian, and Thar deserts. Smaller tropical deserts in the Northern Hemisphere are located in the Sonoran region of northern Mexico and in the southwestern part of the United States. The Kalahari Desert in south central Africa and much of the interior of Australia also are tropical deserts.

Savanna Climate

A third type of tropical climate, the savanna climate, results when an area has both a rain-forest climate and a tropical-desert climate at different times. During different seasons the precipitation belts shift toward the poles, producing very wet summers and very dry winters in these regions.

The weather conditions favor the growth of the type of plants common to a *savanna*. As shown in Figure 26–5, a savanna consists mainly of open areas of coarse grasses that generally grow in clumps. Widely scattered on the grassland are drought-resistant trees and shrubs. Savanna climates are found in the areas that border the rain forests of South America and Africa. Parts of Southeast Asia and northern Australia also have savanna climates. In Southeast Asia the differences in the seasons are extreme because of the alternating monsoon rains and dry periods.

The Hawaiian Islands have both rain-forest and savanna climates. These islands are located in the belt of moist trade winds that blow almost continuously from the northeast toward the equator. The trade winds bring heavy precipitation to the windward sides of the islands. However, because of the mountains, many parts of the large islands are not affected by the winds. The areas that receive less rainfall generally have a savanna climate.

Figure 26–5. The vegetation in a savanna climate consists of drought-resistant grasses, shrubs, and trees, such as those in Australia.

Polar Climates

Climates that are influenced by polar air masses—polar climates—occur in the regions located between 55° north latitude and the Arctic Circle. There are two types of polar climates: the **subarctic climate** and the **tundra climate.**

Subarctic Climate

All the land across North America, Europe, and Asia that lies between 55° and 65° north latitude, including most of Alaska, has a subarctic climate. Dry continental polar air masses control this climate. In subarctic climates, the yearly precipitation is only 25 cm to 50 cm, which is just a little more than falls on tropical deserts. Winters are severe and summers are short. The subarctic climates have a yearly temperature range that is unusually large. In fact, the largest yearly temperature range on earth—61°C—was recorded in subarctic Yakutsk, Siberia. In places with subarctic climates, the vegetation consists mostly of sparse forests of pine, fir, spruce, and other cone-bearing trees.

Tundra Climate

The northern part of Alaska and other land areas at the latitude of the Arctic Circle have a tundra climate. This climate is named for the landscape common to the region, the *tundra*. A tundra has no trees; its ground is covered with mosses, lichens, and small flowering plants. It also has large expanses of rocky land with no vegetation. The yearly temperature range is not as great as that of subarctic climates. This is because areas with tundra climates are near the ocean, which holds heat during the winter. The warmest months—July and August—have average temperatures of only 4°C. Only about 25 cm of precipitation are received in a year, mostly as snow.

The areas of tundra climates that are located between 60° latitude and the poles are called **polar deserts.** Polar deserts have very dry, cold air and receive little precipitation. Parts of polar deserts are moist because they are covered with ice, but dry desertlike conditions are more common.

Figure 26–6. Although the Alaskan tundra has a cold average yearly temperature, many small plants bloom during the summer and fall.

Middle-Latitude Climates

North America stretches from Central America, in the tropical climate zone, to northern Canada and Alaska, in the polar climate zone. However, the United States and most of Canada lie in the middle latitudes.

The climates of the various parts of the United States differ greatly. Cyclonic storms bring most of the precipitation that falls on the United States and Canada. Because this precipitation falls unevenly across these two countries, they have several different types of middle-latitude climates.

Marine West-Coast Climate

Between 40° and 60° north latitude, the northwestern coastline of the United States has a **marine west-coast climate.** This land is in the belt of prevailing westerly winds of cool, moist maritime polar air. As these winds move east from the Pacific Ocean, the west coast receives precipitation. The northwestern coasts of Washington and Oregon and northern California, where mountains block the movement of moist air toward the east, receive a great deal of moisture. The average yearly precipitation is 50 cm to 75 cm. The average temperature is a relatively cool 20°C in summer and a relatively mild 7°C in winter. The yearly temperature range is 13°C. Most regions with marine west-coast climates are covered with dense forests of cone-bearing trees.

Mediterranean Climate

Regions along the southern Pacific coast of the United States have a climate like that of the northern coast of the Mediterranean Sea. This climate, thus, is called a **Mediterranean climate.** Regions with a Mediterranean climate are located between regions with a tropical-desert climate and those with a marine west-coast climate. They generally lie between 30° and 40° north latitude. These regions have dry summers and wet winters. In the summer, the dry air of the subtropical highs blows over the region. In the winter this pressure belt shifts southward, bringing cyclonic storms to the region. Almost all of the yearly rainfall—an average of 25 cm—falls during the mild winter months. The yearly temperature range is small—only 7°C—with average summer temperatures of 21°C and average winter temperatures of 14°C.

Middle-Latitude-Desert and Steppes Climates

Two types of middle-latitude climates are found between 35° and 50° north latitude in the interiors of Asia and North America. They are the **middle-latitude-desert** and the **middle-latitude-steppes climates.** Much of the land in the far western United States, other than the coasts and mountains, has a middle-latitude-desert climate. Little precipitation—less than 25 cm—falls annually in these deserts. Unlike tropical deserts, middle-latitude deserts have a winter season that may be quite cold and a summer season that ranges from warm to very hot. The vegetation of middle-latitude deserts consists of widely scattered drought-resistant shrubs and cacti.

Figure 26–7. The heavy rains of a marine west-coast climate help to maintain the growth of dense forests, such as this one in Prairie Creek Redwoods State Park in California.

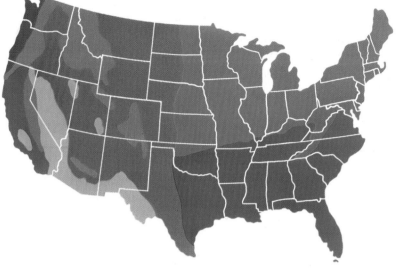

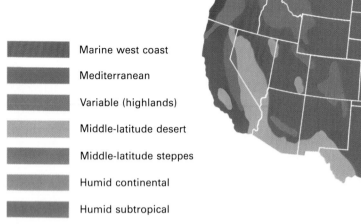

Marine west coast

Mediterranean

Variable (highlands)

Middle-latitude desert

Middle-latitude steppes

Humid continental

Humid subtropical

Figure 26–8. This map shows the distribution of the middle-latitude climates in the continental United States.

From the western United States, the dry air moves eastward. When it reaches the central part of the continent, it begins to pick up moisture from the tropical air moving northward.

At elevations of 1,200 m to 2,100 m, steppes gradually replace barren western deserts. Steppes receive 25 cm to 50 cm of rain a year, which supports a dense growth of grasses. The yearly temperature range is high—24°C. The average summer temperature is 23°C, and the average winter temperature, –1°C. As shown in Figure 26–8, the plains on the eastern side of the Rocky Mountains have a steppes climate. Would the climate on the western side of the Rocky Mountains be drier or wetter than that on the eastern side? Why?

Humid Continental Climate

The areas farther inland from the steppes and extending to the east coast in North America and Asia have a **humid continental climate.** The location of the North American humid continental climate is shown in Figure 26–8. Areas with this type of climate are subject to cold, dry polar air masses and warm, moist tropical air masses. Summers are usually warm and humid as tropical air masses move north. Winters are commonly very cold as polar air masses move south. When these air masses meet, weather conditions may change rapidly and violently. Seasonal changes are great, with yearly temperature ranges as high as 30°C. Average summer temperatures are as high as 25°C, and average winter temperatures, as low as –5°C. Yearly average precipitation, mostly from cyclonic storms and summer thunderstorms, is at least 75 cm. Dense forests of hardwood and softwood trees are found in a humid continental climate.

Humid Subtropical Climate

The southeastern coasts of continents located at around 30° north or south latitude have a **humid subtropical climate.** The southeastern coast of the United States, shown in Figure 26–8, is one of these areas. In the summer, moist, tropical air masses move north across this region. These air masses usually bring warm, humid weather. The tropical air often brings heavy rains also. In the winter, continental polar air masses moving from inland regions may bring brief but intense cold. For example, Charleston, South Carolina, has an average summer temperature of 27°C and an average winter temperature of 10°C. However, the temperature occasionally has plunged to –15°C in Charleston. The yearly temperature range in humid subtropical climates is a relatively small 17°C. Annual precipitation in a humid subtropical climate is between 75 cm and 165 cm. The land in a humid subtropical climate is usually covered with a dense growth of grasses and trees.

Local Climates

The climate in any particular place may be influenced by local conditions as well as by the major factors that have been discussed. The elevation of the land is the most important factor affecting local weather conditions.

Large lakes and forests also influence local climates. Lakes, like oceans, moderate temperatures. They can also cause an increase in precipitation on the shore farthest from the prevailing wind. For example, the eastern shore of Lake Michigan generally has more moderate temperatures, more cloudiness, and higher precipitation than the western shore does. Forests affect local climates by reducing the speed of the wind and by increasing the humidity.

Cities are climate zones. In a city the average temperature is 1°C to 2°C higher than that in surrounding rural areas. There are several reasons for this phenomenon. During the day vegetation absorbs solar energy and gives off water vapor by transpiration. Because cities contain far less vegetation than rural areas do, less transpiration occurs, and, therefore, more solar energy is available to heat the air. In addition, at night the air over cities is warmed by radiation from the materials in streets and buildings that have been heated during the day. Heavy traffic and some of the energy used for heating, lighting, and industry may also raise the air temperature in cities. More precipitation falls within cities and in areas crossed by winds that have blown over the cities than in rural areas. Dust, smoke, and other pollutants, carried into clouds by rising warm city air, form nuclei around which raindrops condense.

26.2 Section Review

1. Where do tropical-desert climates occur?
2. In which type of climate does the largest yearly temperature range occur?
3. List the five types of middle-latitude climate that occur in the continental United States and Canada.
4. A city in a middle-latitude-desert climate might have the weather conditions of a steppes climate. Why?

Reviewing Chapter 26

Key Concepts

- Latitude determines the angle at which the sun's rays strike the earth. See page 449.
- The different rates at which land and water are heated, the temperature of ocean currents, and seasonal winds all affect climate. See page 452.
- The temperature and moisture content of the air are influenced by altitude and the presence of mountains. See page 454.
- The three types of tropical climates are based on different amounts of precipitation. See page 455.
- Regions with subarctic and tundra climates receive little precipitation. See page 458.
- Five types of middle-latitude climates occur in the continental United States and Canada. See page 459.
- Industries, radiation from building materials, and lack of vegetation modify the climates of cities. See page 461.

Key Terms

bora (454)	Mediterranean climate (459)	monsoon (453)	temperature range (449)
Chinook (454)	middle-latitude climate (455)	polar climate (455)	tropical climate (455)
foehn (454)		polar desert (458)	tropical-desert climate (455)
humid continental climate (460)	middle-latitude-desert climate (459)	rain-forest climate (455)	tundra climate (458)
humid subtropical climate (460)	middle-latitude-steppes climate (459)	savanna climate (455)	
marine west-coast climate (459)	mistral (454)	specific heat (452)	
		subarctic climate (458)	

Review

On your paper write the letter of the term that best completes each of the following statements.

1. At the equator the sun's rays always strike the earth (a) at a low angle (b) at nearly a 90° angle (c) 18 hours each day (d) no more than 8 hours each day.
2. Nights are longest in the winter and shortest in the summer (a) at the equator (b) at high altitudes (c) in the middle of the ocean (d) at the poles.
3. Water cools (a) more slowly than land does (b) more quickly than land does (c) through the process of transpiration (d) because of waves and currents.
4. Ocean currents influence temperature by (a) eroding shorelines (b) heating or cooling the air (c) washing warm, dry sediments out to sea (d) dispersing the rays of the sun.
5. Winds that blow in opposite directions in different seasons because of the differential heating of the land and the oceans are called (a) Chinooks (b) mistrals (c) monsoons (d) cyclones.
6. When a moving air mass enters a mountainous region, it (a) stops moving (b) slows and sinks (c) rises and cools (d) reverses its direction.
7. Tropical deserts exhibit all of the following characteristics except (a) location along the Tropics of Cancer or Capricorn (b) dense plant growth (c) influence of the subtropical highs (d) extremely dry conditions.

8. A tropical climate is characterized by very wet summers and very dry winters is called (a) a Mediterranean climate (b) a savanna climate (c) a trade-wind climate (d) an equatorial climate.

9. Compared with subarctic climates, regions with tundra climates have (a) more trees (b) a higher altitude (c) more daylight (d) a smaller yearly temperature range.

10. In regions with Mediterranean climates, almost all the yearly precipitation falls (a) during monsoons (b) in the summer (c) in the winter (d) during hurricanes.

11. The weather conditions tend to fluctuate rapidly throughout the year in a (a) subarctic climate (b) middle-latitude-desert climate (c) Mediterranean climate (d) humid continental climate.

12. The various structures and activities in cities affect the local climate by (a) decreasing the average temperature (b) increasing both the average temperature and precipitation (c) increasing the average temperature and decreasing the precipitation (d) decreasing the precipitation.

Application

On your paper write answers to the following questions.

1. The Milankovitch theory states that a periodic change in the tilt of the earth's axis was a factor in the onset of the ice ages. Use what you know about the factors affecting climate to explain how this might have occurred.

2. Imagine you are going to build a vacation house near a coast with a warm offshore current. What must you investigate to determine how the current will affect the temperature on land? Explain your answer.

3. Large lakes tend to moderate the temperature of the land nearby. Explain why.

4. Why do monsoons occur near the equator?

5. Assume you like the mountains but do not like humid weather. Would you be more comfortable on the side of the mountains facing toward or away from the prevailing winds? Explain why.

6. If you enjoy a warm, dry climate all year but do not like the vegetation in deserts, in what climate or climates would you live? Assume that you would not mind moving one or more times a year.

7. Explain why the vegetation in areas with tundra climates is sparse even though these areas receive precipitation that is adequate to support plant life.

8. Explain why the same winds that influence the climate of the middle-latitude deserts in the southwestern United States bring abundant rainfall to the Midwest.

9. Why do weather conditions change rapidly in humid continental climates and remain relatively constant in the other middle-latitude climates?

10. Explain why the classification of climates often fails when you think in terms of only a specific location.

Extension

1. At the library locate old copies of the local newspaper. Record the daily high and low temperatures for a summer month and a winter month during the past year. Graph the information and calculate the yearly temperature range. Report your findings.

2. Use the information you have learned about climate zones to draw a map showing variation in vegetation around the world. Include areas with little or no vegetation as well as those with various types of vegetation. Share your map with the class.

3. Identify factors that might influence the local climate in your area. Compare the daily temperature and amount of precipitation in your area with those in a nearby area that is not influenced by those factors. For example, if you live in a city, collect data on the conditions in a surrounding suburb or rural area. Discuss your findings with the class.

8
Studying Space

Trifed Nebula in Sagitarius (background); Astronaut Bruce McCandless floating in space (inset)

Introducing Unit Eight

We are small creatures on a small planet in a small corner of the universe. So far, we are unique—at least within this solar system.

Exploring and understanding the circumstances on earth and in space that allow us to exist and continue existing are especially important today. In the last century of our planet's 4.5-billion-year history, technology has given us the opportunity to affect the dynamic systems of the earth. For the first time, we have the ability to affect our protective environment and to alter or maintain the atmosphere that sustains life on our planet.

In this unit, you will study the moon, sun, planets, and stars. Although they may seem remote, they are important to our lives on earth. None of us is untouched by their presence.

Kathryn D. Sullivan

Kathryn Sullivan
Astronaut
National Aeronautical and Space Administration

27 Stars and Galaxies

28 The Sun

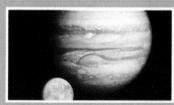

29 The Solar System

30 Moons and Rings

<h1>Chapter 27</h1>

Stars and Galaxies

*O*f all the stars in the sky, none appears as bright as the sun. Certainly, to people on the earth, the sun is the most important star: the earth could not exist without it. However, the sun has about the same size, temperature, and brightness as billions of other stars. The sun is an ordinary star with one important difference: It is orbited by an inhabited planet. In this chapter you will learn about the stars and galaxies of the universe.

Chapter Outline

This is the Dumbbell Nebula in Vulpecula.

27.1 Characteristics of Stars

Section Objectives

• **Describe how astronomers determine the composition and surface temperature of a star.**

• **Explain why stars appear to move to an observer on the earth.**

• **Name and describe the way astronomers measure the distance from the earth to the stars.**

• **Explain the difference between absolute magnitude and apparent magnitude.**

Life on the earth is dependent upon the sun, the star nearest to the earth. A **star** is a body of gases that gives off a tremendous amount of radiant energy in the form of light and heat. Except for its relationship to the earth, the sun is similar to billions of other stars.

From the earth most stars in the night sky appear to be tiny specks of white light. However, if you look closely at the stars, you will notice that they vary in color, as listed in Table 27–1. For example, the star Betelgeuse shines red, the star Rigel shines blue, and the star Arcturus shines orange-red.

Stars vary in size and mass as well as in color. Some stars are less than 20 km in diameter, far smaller than the earth. Other stars have a diameter 1,000 times that of the sun. The sun, a medium-sized star, has a diameter of about 1,392,000 km. Most stars that are visible in the night sky are medium-sized stars.

Most stars also have about the same mass as the sun, which is about 330,000 times more massive than the earth. Some small stars have only 1/50 of the sun's mass. Large stars may have up to 50 times the sun's mass. Stars also differ in composition, temperature, distance from the earth, and brightness.

Composition and Temperature

No spacecraft from the earth has directly studied any star other than the sun. Consequently, astronomers learn about stars primarily by analyzing the light the stars emit. Astronomers direct starlight through a *spectroscope,* a device for separating light into different colors or wavelengths. Starlight passing through a spectroscope produces a display of colors and lines called a *spectrum.* There are three types of spectra: emission, or bright-line; absorption, or dark line; and continuous.

Of particular use to scientists are the dark-line spectra of stars. Analysis of their dark-line spectra reveals certain characteristics,

Table 27–1 Classification of Stars

Color	Surface temperature (°C)	Example
Blue	Above 30,000	Rigel
Blue-white	15,000–30,000	Algol
White	8,000–11,000	Vega
White-yellow	About 7,500	Canopus
Yellow	About 5,500	Sun
Orange	About 4,000	Arcturus
Red	About 3,000	Betelgeuse

such as composition and temperature. Every chemical element has a characteristic spectrum. The colors and lines in the spectrum of a star indicate which elements make up the star. Through spectrum analysis scientists have learned that the stars are composed of many elements found on the earth. In most stars hydrogen is the most common element, and helium is the second most common. Elements such as iron, sodium, and calcium make up the remaining mass of the stars.

The surface temperature of a star is indicated by its color, as shown in Table 27–1. Most stars are in the range of 2,800°C to 24,000°C. Some are hotter and some are cooler. Usually a star shining with a predominantly blue light has an average surface temperature of 35,000°C. However, the surface temperatures of some blue stars are as high as 50,000°C. Red stars are the coolest, with average surface temperatures of 3,000°C. Yellow stars, such as the sun, have surface temperatures of about 5,500°C.

Motion

Two kinds of motion are associated with stars–actual motion and apparent motion. Because stars are so far away, their actual motion can be seen only with high-powered telescopes or other instruments. The apparent motion of stars, motion visible to the unaided eye or through low-powered telescopes, is due to movement of the earth. You can photograph the apparent motion of the stars from the earth. Aim a camera at the northern sky on a clear evening and leave the shutter open for an hour. The result will be a photograph similar to the one shown in Figure 27–1.

Figure 27–1. The stars appear as circular trails in this long-exposure photograph. The trails result from the rotation of the earth on its axis.

Figure 27–2. This photograph shows an exterior view of the Kitt Peak National Observatory in Arizona. Astronomers check the instrument at the focus of the 2.1-m telescope at the observatory (inset).

Almost all of the streaks on the photograph are recordings of the apparent motion of stars. The circular trails left by most stars suggest they are moving around a central star called Polaris, the North Star. Actually the circular pattern results from the rotation of the earth on its axis. Polaris is located almost directly above the North Pole and does not appear to move.

The earth's revolution around the sun causes the stars to appear to move in a second way. As the earth orbits the sun, stars located on the side of the sun away from the earth are obscured by the sun. Thus different stars are visible during different seasons. The visible stars appear to shift slightly every night. Over a period of time, all stars appear to move westward across the sky and then to disappear below the western horizon.

Some stars are always visible in the night sky. They lie so close to Polaris that their apparent nightly orbit does not carry them below the horizon. From the Northern Hemisphere, these stars can always be seen circling the North Star. The circling stars are called **circumpolar** stars. The stars of the Little Dipper are circumpolar for most observers in the Northern Hemisphere.

Most stars actually move in at least three ways. First, they rotate on an axis. Second, they revolve around another star. Third, they move away from or toward the earth.

From the spectrum of a star, astronomers can learn more about the motion of that star. As you read in Chapter 1, the spectrum of a star moving toward or away from the earth appears to shift. The apparent shift in the wavelength of light emitted by a light source moving toward or away from an observer is called the *Doppler effect*. The spectrum of a star moving toward the earth is shifted toward its blue end. This shift, called **blue shift,** occurs because the light waves from a star appear to be shorter as the star moves toward the earth.

A star moving away from the earth has a spectrum that is shifted toward the red end of the spectrum. This shift, called **red shift,** occurs because the wavelengths of light appear to be longer. Most distant galaxies, or large groups of stars, have red-shifted spectra, indicating that these galaxies are moving away from the earth.

Distance to the Stars

The distances between the stars and the earth are measured in **light-years.** A light-year is the distance that light travels in one year. Since the speed of light is 300,000 km/s, light travels about 9.5 trillion km in one year. Light from the sun takes about 8 minutes to reach the earth, so the sun is 8 light-minutes from earth. By contrast, the star system nearest the earth is the Alpha Centauri system. The closest star in this system is Proxima Centauri. It is 4.3 light-years away, nearly 300,000 times the distance from the earth to the sun. Sirius, the brightest star seen from the earth, is 9 light-years away. Polaris is about 680 light-years away.

Astronomers use many different methods to determine distances between the stars and the earth. The only direct method is called **parallax.** As the earth circles the sun, observers are able to study the stars from slightly different angles. During a six-month period, a nearby star will appear to shift slightly relative to more-distant stars. The closer a nearby star is, the greater will be the amount of shift. From the amount of shift, astronomers can calculate the distance to any star within 300 light-years. Because the angle of shift is very small and extremely difficult to measure, scientists usually use photography to measure the shift. The star is photographed at the beginning and end of a six-month period, and its position in relation to other stars is studied each time.

Figure 27–3. Observers on the earth see nearby stars against those in the distant background. The movement of the earth causes nearby stars to appear to move.

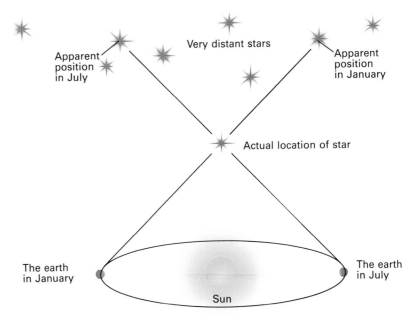

Investigation

Parallax

You can demonstrate the principle of parallax by viewing an object from several locations.

Science Process Skills Focus: observing, interpreting, inferring

Materials

5 paper plates, about 15 cm in diameter; colored marker; thread; scissors; adhesive tape; ladder

Procedure

1. Color the surface of one of the plates with the marker.
2. Cut five 1-m lengths of thread. Tape one end of each piece of thread to the edge of a paper plate.
3. Tape the free end of each piece of thread to the ceiling in various positions about 30 cm apart. Position the plates in a random pattern as shown in the figure.
4. Stand directly in front of and directly facing the colored plate at a distance of several meters. Close one eye and sketch the position of the colored plate relative to the other plates and the background.
5. Take several steps back and to the right of your original position. Repeat Step 4.

6. Repeat Step 5 two more times.

Analysis and Conclusions

1. Compare your drawings. Did the colored plate change position as you viewed it from different locations? Explain your answer.
2. What kind of results would you expect if you continued to repeat Step 4 at greater and greater distances? Explain your answer.
3. If you noted the positions of several stars with a powerful telescope, what would you expect to observe about their positions if you sighted the same stars several months later? Explain your answer.

You can demonstrate how parallax works by extending your arm in front of you and holding up your thumb. Close one eye and note the position of your thumb against the background. Then open that eye and close the other one. Now note the position of your thumb. Your thumb will seem to have moved across the background by about 3°. Because your eyes are a few centimeters apart, each eye sees your thumb from a different angle. Try repeating this exercise with your thumb close to your face. Your thumb will appear to jump from one side of your face to the other. The closer your thumb is to your face, the greater is its apparent motion. Similarly, the closer a star is to the earth, the greater is its apparent motion.

Astronomers estimate how bright a more distant star is by studying its spectrum. They compare this estimate of true brightness with the apparent brightness of the star. From these two measurements, astronomers can calculate the distance of a star from the earth.

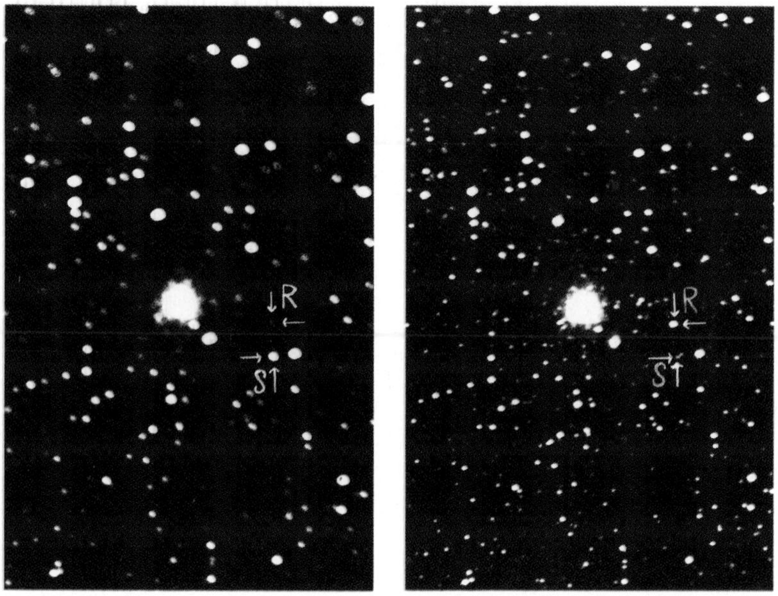

Figure 27–4. These two photographs show the variable stars R Scorpii and S Scorpii. In the photograph on the left, R Scorpii is faint and S Scorpii is bright. In the photograph on the right, R Scorpii is bright and S Scorpii is faint.

Some special stars serve as distance indicators. For example, a **Cepheid** (SEE-fee-ud) **variable** star brightens and fades in a regular pattern. The brightness of some Cepheids varies by as much as 600 to 700 percent. This change in brightness is caused by a rhythmic swelling and shrinking of the star. Most Cepheids have regular cycles ranging from 1 day to 100 days. Cepheids with longer cycles are brighter. Astronomers can measure the cycle of a Cepheid and can then estimate its true brightness. By comparing its apparent brightness and true brightness, astronomers can calculate the distance to the Cepheid variable. This, in turn, tells them the distance to the galaxy in which the Cepheid is located.

Stellar Magnitudes

Over 3 billion stars can be seen through a telescope. Of these, only about 5,000 are ever visible to unaided observers on the earth. The visibility of a star depends on its brightness and its distance from the earth. Astronomers use two scales to describe the brightness of a star. One scale is based on how bright the star appears from the earth. The other scale is based on how bright the star would be if all stars were the same distance from the earth.

Apparent Magnitude

The brightness of a star as it appears from the earth is called its **apparent magnitude.** Astronomers use special instruments that are attached to telescopes to measure the light from a star. The measurement is then assigned a number on the scale shown in Figure 27–5. The brightest stars have the lowest numbers; the dimmest stars, the highest numbers. The most powerful telescopes can detect stars with an apparent magnitude of about 23. These stars are about 100 million times fainter than the average star seen without a telescope.

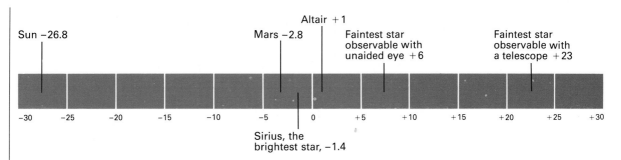

Figure 27–5. The sun, with an apparent magnitude of –26.8, is the brightest object in the sky. All other objects are dimmer, so their apparent magnitudes are higher.

The faintest star that can be seen by the unaided eye has an apparent magnitude of 6. A star of this brightness is called a sixth-magnitude star. A first-magnitude star is one of the brightest stars in the sky.

The apparent magnitudes of a few stars, some planets, the moon, and the sun are negative numbers because they are brighter than first-magnitude stars. For example, the brightest star in the night sky, Sirius, has an apparent magnitude of –1.4. The sun has an apparent magnitude of –26.8.

Absolute Magnitude

The apparent magnitude of a star depends on how much light the star emits and how far the star is from the earth. The true brightness, or **absolute magnitude,** of a star is how bright the star would appear if it were seen from a distance of 32.6 light-years.

Assuming the sun were 32.6 light-years away, it would be a fifth magnitude star. Therefore, the absolute magnitude of the sun is +5. Most stars have absolute magnitudes between –5 and +15. Thus, the sun is in the middle of the range of absolute magnitudes.

Each star has both an apparent magnitude and an absolute magnitude. The relationship between these two measures of brightness depends on the distance between the earth and the star. Stars that are less than 32.6 light-years from the earth appear brighter than they would if they were 32.6 light-years away. Consequently, these stars have apparent magnitudes that are lower than their absolute magnitudes. For example, the sun is only a fraction of one light-year from the earth. It has an apparent magnitude of –26.8 and an absolute magnitude of +5.

Stars that are more than 32.6 light-years from the earth appear dimmer than they would if they were only 32.6 light-years away. Consequently, these stars have apparent magnitudes that are higher than their absolute magnitudes. How far away is a star with an apparent magnitude of +7 and an absolute magnitude of +7?

Classification of Stars

Plotting the surface temperatures of stars against their absolute magnitudes reveals an interesting pattern. The graph illustrating this pattern is a Hertzsprung-Russell diagram, or **H-R diagram.** The graph is named for Ejnar Hertzsprung and Henry Russell, the astronomers

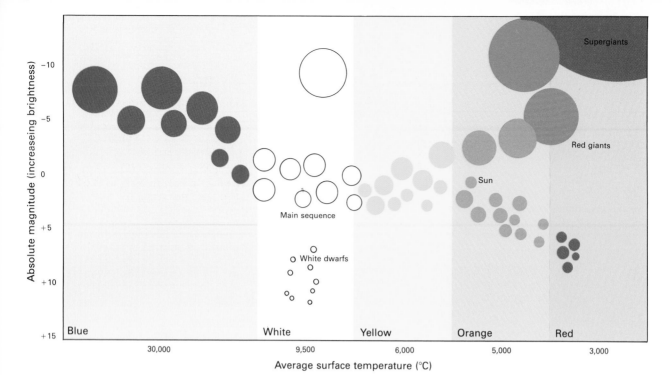

Figure 27–6. The Hertzsprung-Russell diagram shows the relationship between the surface temperatures and the absolute magnitudes of stars.

who discovered the pattern. As you can see in Figure 27–6, the brightness of most stars increases as their surface temperatures increase.

The majority of stars fall within a band running through the middle of the H-R diagram. The band extends from cool, dim, red stars at the lower right to hot, bright, blue stars at the upper left of the H-R diagram. Stars within this band are called **main-sequence stars.** The sun and almost every star visible in the night sky are main-sequence stars.

A group of cool, bright stars appears near the upper right corner of the H-R diagram. These stars are very large. A cool star can be very bright only if it has a large surface area to give off a large amount of light. These huge stars, such as Arcturus, are called **red giants.** Some are so large that they are known as **supergiants.** Antares, with a diameter of more than 2.7 million km, is an example of a supergiant.

The stars found in the lower left of the H-R diagram are hot but dim because of their small size. They are known as **white dwarfs.** A typical white dwarf star is about the same size as the earth.

Section 27.1 Review

1. What do astronomers analyze to determine the color and surface temperature of a star?
2. Why do stars appear to move westward across the sky?
3. How do astronomers measure the distance to stars that are less than 300 light-years from the earth?
4. Assume that a star has an apparent magnitude of +2 and an absolute magnitude of +4. What do you know about the distance of that star from the earth?

27.2 Stellar Evolution

Since a typical star exists for billions of years, astronomers will never be able to observe one star throughout its entire life. Instead, astronomers have developed theories about the evolution of stars by studying stars in different stages of development.

A star begins as a cloud of gas and dust, or **nebula** (NEB-yuh-luh; pl: nebulae), such as the one shown in Figure 27–7. A nebula is usually composed of about 65 percent hydrogen, 30 percent helium, and 5 percent heavier elements. The particles of material within a nebula have a very weak gravitational attraction for one another. When some force, such as the explosion of a nearby star, compresses some of the particles, the nebula begins to contract.

According to the law of universal gravitation, gravitational force increases as distance decreases. Therefore, as the density of the particles increases, their gravitational attraction for one another increases. As these particles come together, a sphere of matter builds up within the cloud.

Gravitational forces cause the nebula to continue to shrink. As the nebula becomes smaller, it begins to spin more rapidly. You may have seen the effect of a decreasing diameter on the speed of a spinning object, such as an ice skater. As a spinning skater pulls his or her arms in closer to the body, the rate of spin increases.

The shrinking, spinning nebula begins to flatten into a disk of matter with a central concentration called a **protostar**. The increase in temperature in the center of a protostar has two causes. One cause

Section Objectives

- **Describe how a protostar develops into a star.**
- **Explain how a main-sequence star generates energy.**
- **Describe the possible evolution of a star during and after the red-giant stage.**

Figure 27–7. The Crab Nebula resulted from the explosion of a star. Chinese astronomers first saw the explosion almost 1,000 years ago.

is collision. As the particles move toward the center of the nebula, they collide. Whenever solid objects collide, some of the energy of their motion is converted into heat energy. You can demonstrate this principle by rubbing your hands together. The motion of your hands against each other warms them.

Pressure also causes the temperature in the nebula to increase. As the nebula shrinks and the force of gravity pulls matter toward its center, the pressure in the core of the nebula increases. All materials become warmer when compressed. You can observe this principle by again watching an ice skater. The pressure of the skate blade on the ice heats the ice beneath the blade and causes it to melt.

The contracting and heating up of the nebula continues for several million years. As the collisions and pressure push the temperature in the core of the protostar to near 15,000,000°C, *nuclear fusion* begins. Nuclear fusion is the process through which small atomic nuclei combine into larger atomic nuclei, releasing energy. When fusion takes place, a protostar begins to generate energy and is considered a star.

A nebula may produce more than a single star. Often two or more stars created out of the same nebula revolve around each other,

Horizons

Technology: New Solar System?

Located about 469 trillion km from the earth, Beta Pictoris was originally considered just another distant star. Then, in 1984, it took center stage, right in the middle of what is possibly a new solar system.

Many scientists think that a solar system is formed when the gravitational force of a star captures interstellar particles. As these particles revolve around the star, they begin to come together as bodies of matter. These bodies of matter may eventually become planets.

According to astronomers Bradford Smith and Richard Terrile, this process appears to be going on around Beta Pictoris. Smith works for the University of Arizona. Terrile works for the NASA Jet Propulsion Laboratory. In 1984 Smith and Terrile became curious about Beta Pictoris after reviewing data from IRAS, the Infrared Astronomy Satellite. To satisfy their curiosity, the astronomers devised a system to take the first visible-wavelength, computer-enhanced pictures of Beta Pictoris, shown right. Smith and Terrile's system was built around three optical instruments: a telescope at Las Campanas Observatory in Chile, (left) a planetary camera, and a coronagraph. The coronagraph is an instrument that can detect dim interstellar bodies.

and the force of gravity keeps them close together. A nebula may also produce planets. Astronomers think that the earth and the other planets formed from the same nebula at about the same time as the sun. Therefore, planets may have formed around other stars at about the same time as these stars formed.

Main-Sequence Stars

The second and the longest stage in the life of a star is the main-sequence stage. During this stage energy is generated in the core of the star as hydrogen atoms fuse into helium atoms. Fusion releases enormous amounts of radiant energy. For example, when one gram of hydrogen is converted to helium, the energy released is enough to keep a 100-watt light bulb burning for 200 years. This energy moves outward in much the same way as the energy that rises upward through boiling water. The star does not expand, however, because the force of gravity pulls matter inward. The energy from fusion balances the force of gravity, making the star stable in size. A main-sequence star maintains a stable size as long as it has an ample supply of hydrogen to fuse into helium.

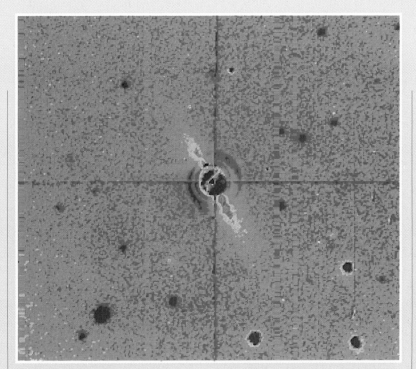

The resulting pictures revealed several features that are thought to be common to developing solar systems.

First, Beta Pictoris is surrounded by a halo of light that scientists think is actually a disk of particles 130 billion km wide. The distribution of the particles in the disk fits the scientific model for beginning solar systems. The disk appears less dense toward the center. This suggests that still-forming bodies are sweeping up carbon, ice, and silicon debris as they build into larger spheres that eventually may become planets.

What researchers have learned suggests to them that Beta Pictoris is the first of many developing solar systems yet to be discovered.

Assuming that a solar system is forming around Beta Pictoris, how does that process differ from the probable formation of the solar system around the sun?

Red Giants

A star enters its third stage when almost all of the hydrogen atoms within its core have fused into helium atoms. Without hydrogen as a source of fuel, the core of a star contracts under the force of its own gravity. This contraction increases the pressure in the core of the star. The higher pressure causes the helium atoms in the core to fuse into carbon atoms. Hydrogen fusion continues to take place in a shell surrounding the helium core.

The combined hydrogen fusion and helium fusion release energy, which causes the outer shell of the star to expand greatly. The shell of gases grows cooler as it expands. The star is no longer a main-sequence star. Instead, it has become a red giant or a supergiant. Red giants are 10 times bigger than the sun. Supergiants are at least 100 times bigger than the sun.

The stages in the life of a star cover an enormous period of time. Scientists estimate that over a period of 5 billion years, the sun, a main-sequence star, has converted only 5 percent of its original hydrogen.

Dwarf Stars

The end of helium fusion marks the end of the red-giant stage in the evolution of a medium-sized star. With energy no longer available from fusion, the star enters its final stages and begins to cool and dim in a series of contractions. Each time the star contracts, it loses some of its gases. The dying star may shed its whole outer atmosphere in an expanding shell of gases called a **planetary nebula.** Gravity causes the last of the matter in the star to collapse inward. What is left is a hot, dense core of matter—a white dwarf. White dwarfs shine for billions of years before they cool completely.

Figure 27–8. The cool, outer shell of expanding gas around a dying star is a planetary nebula. The ring nebula, Lyra is shown here.

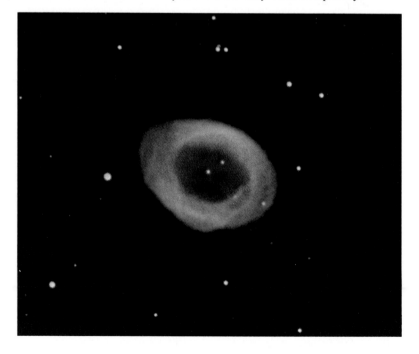

Black Dwarfs   As white dwarfs cool, they become fainter and fainter. When a white dwarf no longer emits energy, it becomes a **black dwarf.** Black dwarfs do not give off a measurable amount of heat or light energy. This is the final stage for many stars. Why are black dwarfs virtually invisible?

Novas   Some white dwarfs do not evolve quietly into black dwarfs. During the process of cooling, one or more large explosions may occur that release energy, gas, and dust into space. A white dwarf that has such an explosion is called a **nova.** The explosion may cause the star to become many thousands of times brighter. To an astronomer, a nova may appear up to one million times brighter than the sun. Then, sometimes within only a few days, the nova begins to fade back to its normal brightness. A white dwarf may become a nova several times before it becomes a black dwarf.

Astronomers think that novas probably occur in white dwarfs that revolve around a main-sequence star or a red giant. White dwarfs are denser than main-sequence stars or red giants. Therefore, the white dwarf has a greater surface gravity than its companion. As gases from the companion star accumulate on the white dwarf, the pressure builds until the white dwarf explodes as a nova.

Supernovas

Stars with masses 16 to 30 times that of the sun may produce explosions up to 100 times brighter than novas. In 1054 Chinese astronomers saw an explosion in the sky that was so bright they could see it during the day for three weeks. The amount of energy radiated during that time was equal to the energy produced by the sun over a period of 500 million years. What the Chinese astronomers saw was a **supernova,** a star that has a tremendous explosion and blows itself apart.

Supernovas occur in stars much larger than those that produce novas. After the red-giant stage, these larger stars collapse with a gravitational force much greater than that of smaller stars. The collapse produces such high pressures and temperatures that nuclear fusion begins again. This time carbon atoms in the core of the star fuse into heavier elements such as nitrogen. These heavier elements then fuse into iron.

Fusion continues until the core is almost entirely iron. At this point there is not enough energy for the iron to fuse into other elements. Nuclear fusion stops. The iron begins to absorb huge amounts of energy from gravitational attraction. Pressure builds up until the star explodes. During the explosion the energy released about equals the amount of energy radiated by an ordinary star over its lifetime.

Neutron Stars   After an explosion the core of a supernova may contract into a very small but incredibly dense ball of neutrons, called a **neutron star.** A spoonful of matter from a neutron star would weigh 100 million tons on the earth. A neutron star with more mass than the sun may have a diameter of only about 32 km. Neutron stars rotate very rapidly.

Figure 27–9. This supernova, photographed in April 1987, occurred in the Southern Hemisphere. It was the first supernova visible to the naked eye in nearly 4 centuries.

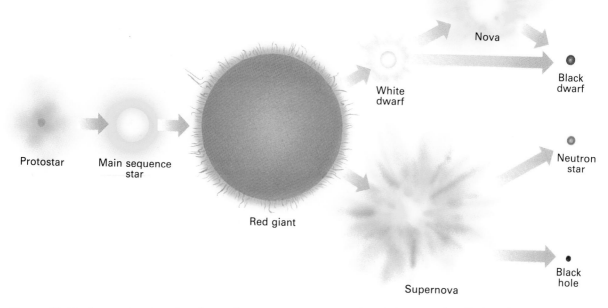

Protostar

Main sequence star

Red giant

White dwarf

Nova

Black dwarf

Neutron star

Supernova

Black hole

Figure 27–10. A star the size of the sun becomes a black dwarf at the end of its life cycle. A larger star may become a neutron star or a black hole.

Some neutron stars emit two beams of radiation that sweep across space like the light beams from a lighthouse. These neutron stars are called **pulsars.** Astronomers detect the radiation from pulsars as radio waves.

Black Holes Stars that are 30 to 50 times more massive than the sun may also become supernovas. However, because of size, they do not become neutron stars. Because they are even larger than the stars that become neutron stars, these stars contract with even greater force. The force of the contraction crushes even the dense core of the star, leaving what astronomers think is a hole in space, or a **black hole.** The gravity of a black hole is so great that not even light can escape from it. The velocity needed to escape a black hole is greater than the speed of light.

Since black holes do not give off light, locating them is difficult. However, astronomers theorize that a black hole can be observed by its effect on a companion star. Matter from the companion star is pulled into the black hole, disappearing forever from the universe. Just before the matter is pulled in, X rays are given off. Astronomers try to locate black holes by detecting these X rays. In the 1970's, astronomers identified what they think is a black hole within the constellation Cygnus. Today astronomers speculate that a black hole may be at the core of each galaxy.

Section 27.2 Review

1. List two reasons why the temperature of a protostar increases.
2. What is the process that generates energy in the core of a main-sequence star?
3. What form of fusion occurs in a red giant?
4. What causes a nova explosion?
5. Explain why only very large stars can form black holes.

27.3 Star Groups

When you look into the sky on a clear night, you see what appear to be individual stars. These visible stars are only some of the billions of stars that make up the universe. However, only one in four of these stars is actually a single star. Astronomers estimate that about one third of the stars are double stars and the rest are groups of three or more stars. In addition to these double-star and triple-star systems, there also are groups and clusters of stars.

Constellations

Using a star map and observing carefully, you can identify some of the star groups that form star patterns or regions. Although the stars that make up a pattern seem to be close together, they are not all the same distance from the earth. In fact, they may be very distant from one another.

If you look at the same area for several nights, the positions of the stars in relation to one other do not appear to change. Because of the tremendous distance from which the stars are viewed, they appear fixed in their patterns. However, if you observe each pattern at exactly the same time on successive nights, it will appear to shift a bit from night to night. For more than 3,000 years, people have observed and recorded these shifting but seemingly fixed patterns, which are called **constellations.**

Astronomers recognize 88 constellations. Some constellations are named for real or imaginary animals, such as Ursa Major, the great bear, and Draco, the dragon. Other constellations are named for ancient gods or legendary heroes, such as Hercules and Orion. Most

Section Objectives

- **Describe the characteristics that identify a constellation.**
- **Describe the three main types of galaxies.**
- **Explain the big bang theory.**

Figure 27–11. These charts show the constellations visible in the Northern Hemisphere. Different constellations are visible in the Southern Hemisphere.

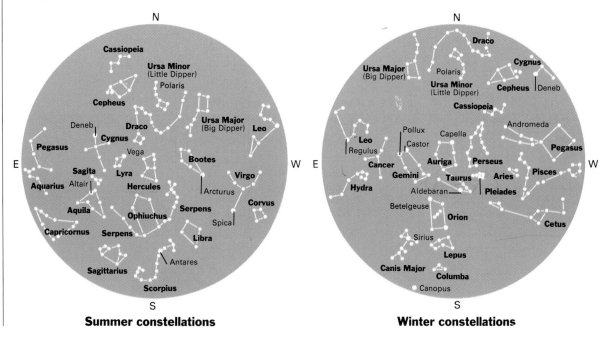

Summer constellations

Winter constellations

constellations, however, do not actually look like the figures for which they are named.

Since 1928 astronomers have used the 88 constellations to divide the entire sky into sectors. Just as you can use a state map to locate a particular town, you can use a map of the constellations to locate a particular star.

Astronomers label the stars within each constellation according to apparent magnitude. The brightest star in a constellation is labeled *alpha*, or α; the second brightest is called *beta*, or β; and so on. Thus, the brightest star in Scorpius is called *Alpha Scorpii* or α *Scorpii;* the second brightest, *Beta Scorpii.*

Many bright stars in the sky also have individual names that are not connected with the constellation in which they are found. For example, Alpha Scorpii, which has a red glow similar to that of the planet Mars, is also called Antares, meaning ''rival of Mars.''

Galaxies

The major components of the universe are **galaxies.** These large-scale groups of stars are bound together by gravitational attraction. A typical large galaxy is about 100,000 light-years in diameter and contains 100 billion stars.

Besides billions of stars, a galaxy also contains gas and dust clouds, or nebulae. Astronomers have discovered that some nebulae are bright. Through spectroscopic studies astronomers have learned that there are two kinds of bright nebulae: those that glow from the hot gases within and those that shine by reflecting the light of nearby stars. Other nebulae are dark, such as the Horsehead Nebula shown in Figure 27–12. Dark nebulae themselves are visible as dark areas amidst the stars. Dark nebulae absorb the light of more distant stars behind them.

Astronomers have discovered at least 1 billion galaxies in the known part of the universe. Two galaxies, the Large Magellanic Cloud and Small Magellanic Cloud, are the closest neighbors to the earth's galaxy, which is the **Milky Way Galaxy.** Even so the Large and Small Magellanic clouds are 150,000 light-years away. Within 3

Figure 27–12. The distinctive shape of this dark nebula, the Horsehead Nebula, is reflected in its name.

Figure 27–13. The three main types of galaxies are spiral (left), elliptical (center), and irregular (right).

million light-years of the Milky Way Galaxy are about 17 other galaxies. These galaxies and the Milky Way Galaxy collectively are called the **Local Group.** If each of these galaxies is of average size, how many stars are in the Local Group?

Types of Galaxies

In studying the vast number of galaxies, astronomers found that galaxies could be classified by shape into the three main types shown in Figure 27–13. One type, called a **spiral galaxy,** has a nucleus, or center, of bright stars and flattened arms that spiral around the nucleus. The spiral arms contain millions of stars. Some spiral galaxies have a bar of stars that runs through the center. These galaxies are called **barred spiral galaxies.**

Galaxies of the second type vary in shape from nearly spherical to flattened disks. These galaxies are called **elliptical galaxies.** They are very bright in the center and do not have spiral arms. Elliptical galaxies contain very little dust and gas and are generally older than the other types of galaxies.

The third type of galaxy, called an **irregular galaxy,** has no particular shape. Irregular galaxies tend to be smaller and fainter than other types of galaxies. Some astronomers think that the irregular shapes of these galaxies might have been caused by gigantic explosions at their centers. Thus, the stars forming an irregular galaxy are unevenly distributed in the galaxy.

The Milky Way

If you look into the night sky, you will see a cloudlike band of stars that stretches across the sky. Because of its milky appearance, this part of the sky is called the *Milky Way*. The Milky Way is part of a spiral arm of the Milky Way Galaxy. The Milky Way Galaxy is a spiral galaxy in which the sun is but one of billions of stars. Each star seems to have its own motion. Some stars seem to be moving toward the sun, while others seem to be moving away from it.

The Milky Way Galaxy has an average diameter of about 100,000 light-years. At its nucleus the galaxy is 2,000 light-years thick, with the sun located about 30,000 light-years from the center.

As do all spiral galaxies, the Milky Way Galaxy rotates. Viewed from above, the Milky Way Galaxy rotates counterclockwise. The sun, which is located in one of the spiral arms, revolves around the center of the galaxy at a speed of about 250 km/s. At this speed, it completes one rotation in about 200 million years.

Star Clusters   Besides single stars the Milky Way Galaxy contains star clusters. These groups of hundreds of stars may be **open clusters** or **globular clusters.** A globular cluster has a spherical shape, while an open cluster is more loosely shaped. Usually globular clusters contain more stars than do open clusters. Also, globular clusters are distributed around the central core of the galaxy.

Binary Stars   Most stars in the galaxy are **binary stars** or **multiple-star systems.** Binary stars are pairs of stars that revolve around each other. Multiple-star systems have more than two stars. In a multiple-star system, two stars may revolve rapidly around a common center of gravity. At a greater distance from the pair of stars, a third star revolves more slowly around the central two.

Formation of the Universe

The *big bang theory* is the most widely accepted theory explaining the formation of the universe. According to the big bang theory, all matter and energy of the universe was once packed into a dense, hot sphere. Scientists do not agree about the size of the sphere. About 17 billion years ago, a gigantic explosion, the so-called *big bang,* took place. Matter and energy were propelled outward in all directions, with some parts moving faster than others. As the matter and energy moved outward from the center, the force of gravity began to have an effect. Matter began to condense, forming the galaxies. The galaxies continued to move outward from the center. They are still moving outward today, as shown in Figure 27–14.

In 1960 astronomers discovered objects in the universe that are about 12 billion light-years from the earth. Light from these objects must travel for 12 billion years to reach the earth. Therefore, we see

Figure 27–14. Before the big bang, all matter and energy in the universe was concentrated in a sphere. When the big bang occurred, matter was sent outwards in all directions. Galaxies began to form. The universe continues to expand today.

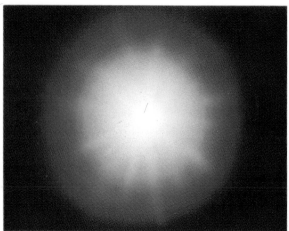

Science Notebook

Quasars

Quasars are among the most puzzling objects in the sky. Viewed through an optical telescope, a quasar appears as a small, faint star. The word quasar is a shortened term for *quasistellar radio sources.* The prefix *quasi* means "similar to," and the word *stellar* means "star."

Quasars are the most distant objects that have been observed from the earth. Quasars have large red shifts. The Doppler shift of one quasar indicates that it is moving away from the earth at a speed of more than 270 million m/s. That speed is 90 percent of the speed of light. The red shift of the quasar (large object) and the faint galaxy (small object) are similar in the computer-enhanced color photograph shown here.

Many quasars are hundreds of times brighter than the brightest galaxies. The absolute magnitude

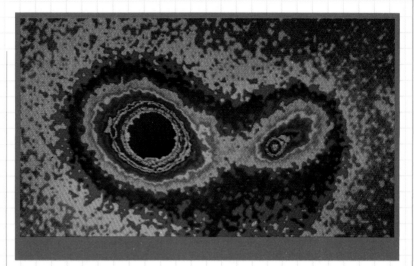

of the brightest quasar is –28. In addition, quasars appear to be pinpoints of light, unlike galaxies, which include billions of bright objects. Scientists do not understand how such relatively tiny objects can emit so much energy. One hypothesis is that a quasar has a giant black hole in its center, which pulls mass from the surrounding space. The energy

that is observed on earth is emitted as the matter falls into the black hole.

Astronomers have found one quasar that is about 15 billion light-years away.

If the big bang occurred 17 billion years ago, how long after the big bang did the light now reaching the earth leave that quasar?

these objects as they were 12 billion years ago. These starlike objects, called **quasars,** give off radio waves and X rays that can be detected from the earth. Quasars were among the first objects formed after the big bang.

Section 27.3 Review

1. Why do the patterns of the stars appear to shift slightly from night to night?
2. List the three basic types of galaxies.
3. What type of galaxy is the Milky Way Galaxy?
4. Describe a quasar.
5. Imagine that astronomers found a galaxy that was 20 billion light-years from the earth. How would they have to revise the big bang theory to account for such a discovery?

Reviewing Chapter 27

Key Concepts

- To determine the composition and surface temperature of a star, astronomers study the spectrum of the star. See page 467.

- The stars appear to move around a central point and to move westward across the sky. See page 468.

- To measure the distance to a star from the earth, astronomers use direct and indirect methods. See page 470.

- Apparent magnitude and absolute magnitude are two ways of characterizing the brightness of a star. See page 472.

- A protostar is considered a star when it begins to generate energy. See page 475.

- A main-sequence star generates energy through hydrogen fusion. See page 477.

- A red giant is a large, cool star with a core in which helium fusion is occurring. See page 478.

- A constellation is a star group that can be identified. See page 481.

- Astronomers have identified three main types of galaxies. See page 483.

- The big bang theory is the most widely accepted explanation of the formation of the universe. See page 484.

Key Terms

absolute magnitude (473)	circumpolar (469)	Milky Way Galaxy (482)	pulsar (480)
apparent magnitude (472)	constellation (481)	multiple-star system (484)	quasar (485)
barred spiral galaxy (483)	elliptical galaxy (483)	nebula (475)	red giant (474)
binary star (484)	galaxy (482)	neutron star (479)	red shift (470)
black dwarf (479)	globular cluster (484)	nova (479)	spiral galaxy (483)
black hole (480)	H-R diagram (473)	open cluster (484)	star (467)
blue shift (469)	irregular galaxy (483)	parallax (470)	supergiant (474)
Cepheid variable (472)	light-year (470)	planetary nebula (478)	supernova (479)
	Local Group (483)	protostar (475)	white dwarf (474)
	main-sequence star (474)		

Review

On your paper, write the letter of the term that best completes each of the following statements.

1. In the majority of stars, the most common element is (a) oxygen (b) helium (c) hydrogen (d) sodium.
2. The color of the hottest stars is (a) red (b) yellow (c) green (d) blue.
3. Stars appear to move in circular paths around Polaris because (a) the earth rotates on its axis (b) the earth orbits the sun (c) the stars revolve around Polaris (d) Polaris is the center of the Milky Way Galaxy.
4. Stars that are visible throughout the year are called (a) supernovas (b) circumpolar stars (c) pulsars (d) quasars.
5. The change in position of a nearby star compared with the position of a faraway star is called (a) parallax (b) red shift (c) blue shift (d) the Cepheid variable.
6. The brightest stars have apparent magnitudes

that are (a) over $+20$ (b) between $+10$ and $+19$ (c) between $+1$ and $+9$ (d) negative numbers.

7. The absolute magnitude of a star is (a) the relative brightness of the star (b) the true brightness of the star (c) the comparative brightness of the star (d) the apparent brightness of the star.

8. A protostar becomes a star when it begins to (a) develop a red shift (b) generate energy (c) shrink and spin (d) explode as a nova.

9. A main-sequence star generates energy by fusing (a) nitrogen into iron (b) helium into carbon (c) hydrogen into helium (d) nitrogen into carbon.

10. A dying star can shed some of its gases as a (a) planetary nebula (b) white dwarf (c) globular cluster (d) black dwarf.

11. After the red-giant stage, medium-sized stars become (a) supernovas (b) neutron stars (c) black holes (d) white dwarfs.

12. Black holes are difficult to locate because they (a) move very quickly (b) do not give off light (c) have very low gravity (d) are far away from any stars.

13. A pattern of stars is called a (a) galaxy (b) nebula (c) pulsar (d) constellation.

14. Stars appear in fixed locations in the sky because they (a) are so far from the earth (b) do not move (c) are all moving toward the earth (d) are all in the same galaxy.

15. The basic types of galaxies are (a) spiral, elliptical, and irregular (b) barred, elliptical, and open (c) spiral, quasar, and pulsar (d) open, binary, and globular.

16. Quasar formation is associated with (a) nuclear fusion (b) main-sequence stars (c) the explosion of a supernova (d) the big bang.

Application

On your paper, write answers to the following questions.

1. If the spectrum of a star indicates that the star shines with a red light, approximately what is the surface temperature of the star?

2. Why are the constellations that can be seen from the earth visible during different seasons of the year?

3. What do you know about the distance of a star from the earth if the star has a large parallax?

4. Explain why Polaris is considered to be a very bright star even though it is not a bright star in the earth's sky.

5. Why does heat build up more rapidly in a large protostar than in a small one?

6. Explain why an old main-sequence star will be composed of a higher percentage of helium than will a young main-sequence star.

7. Suppose that a scientist has discovered a red-dwarf star. Describe the likely size and surface temperature of such a star.

8. The planets are much closer to the earth than are the stars. How does this explain why the stars appear to be stationary and the planets appear to move across the sky?

9. If a galaxy is very old and contains little dust or gas, what type of galaxy is it likely to be?

10. Suppose that all galaxies and stars began to show blue shifts. What would this indicate about the size of the universe?

Extension

1. The modern scale of star magnitudes is based on a system established in 129 B.C. by Hipparchus of Nicea. Conduct library research to find out how the modern scale differs from the one introduced by Hipparchus.

2. In February 1987 astronomers saw a supernova explosion in the Large Magellanic Cloud. Because Ian Shelton was the first astronomer to notice it, the explosion is known as Shelton's Supernova. Write a report on Shelton and his historic discovery.

3. Make a storyboard showing the development of the universe according to the widely accepted big bang theory.

Chapter 28

The Sun

Imagine, if you can, the heat and light that would be produced by 300,000 candles. Then think of 300,000 candles squeezed into a single square inch of space. That is the amount of heat and light given off by every square inch of the sun's surface—a surface that is 34 times as great as that of the earth. Energy from the sun makes life on the earth possible. In this chapter you will learn about the structure of the sun and how it produces such huge amounts of energy.

Chapter Outline

28.1 Structure of the Sun
The Core
The Inner Zones
The Sun's Atmosphere

28.2 Solar Activity
Sunspots
Prominences and Solar
    Flares
Auroras

28.3 Formation of the Solar System
Formation of the Sun
Formation of the Planets
Formation of the Earth

The sun's light energy is blocked from the earth during an eclipse.

28.1 Structure of the Sun

Section Objectives

- **Explain how the sun converts matter into energy in its core.**
- **Compare the radiative and convective zones of the sun.**
- **Describe the three layers of the sun's atmosphere.**

Throughout most of human history, people thought that the sun's energy came from fire. People knew that burning a piece of coal or wood produced heat and light. They assumed that the sun, too, burned some type of fuel to produce its energy. Not until this century did scientists discover the source of the sun's energy.

The tremendous heat and light of the sun make it appear to be a dazzling, brilliant ball with no distinct features. The sun's brightness can damage your eyes. NEVER LOOK DIRECTLY AT THE SUN. Astronomers use special scientific instruments to study the sun. Spectroscopic analysis of the sun's rays shows that all the known elements are found in the sun. However, most exist in trace amounts only. Hydrogen makes up 90 percent of the sun's content, and hydrogen and helium together make up 98 percent. Studies indicate that the sun has several regions and distinct features.

The sun has three basic regions: the core, the inner zones, and the atmosphere. The inner zones and the atmosphere are each further divided into smaller layers. Each layer of the sun has specific characteristics. However, the boundaries between layers are not distinct. Each layer blends gradually into the next. If you could travel through the sun, you could not tell when you left one layer and entered the next.

The Core

At the center of the sun is the core. The core makes up 10 percent of the sun's diameter, which is 1,300,000 km. The temperature of the sun's core is about 15,000,000°C. No liquid or solid can exist at that high a temperature. The core, like the rest of the sun, is made up entirely of gas.

The sun's mass is 300,000 times greater than the earth's mass. Consequently, the force of gravity is much greater on the sun than on the earth. The sun's gravity is so strong that the center of the sun is 10 times denser than iron.

The enormous pressure and heat of the sun change the structure of atoms within the core. On the earth, atoms normally consist of a nucleus surrounded by one or more electrons. The nucleus, composed of protons and neutrons, remains unchanged even if the number of electrons changes. Within the core of the sun, however, the heat and pressure strip electrons away from the atomic nuclei. The exposed nuclei can then be changed by nuclear reactions. The most common nuclear reaction occurring inside the sun is nuclear fusion. As a result of nuclear fusion, atomic nuclei are combined to form larger atomic nuclei.

Hydrogen Fusion

The nuclei of hydrogen atoms are the primary elements in the fusion occurring in the sun. The sun contains a greater number of hydrogen atoms than it does of any other kind of atom. A hydrogen atom, the

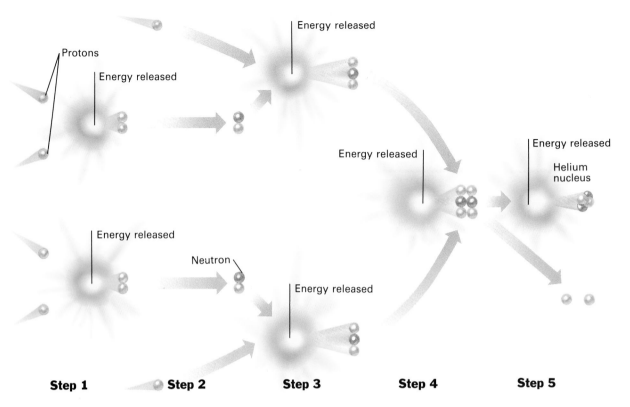

Step 1   **Step 2**   **Step 3**   **Step 4**   **Step 5**

Figure 28–1. In the core of the sun, the nuclei of hydrogen atoms fuse into helium. The process converts mass into energy.

simplest of all atoms, usually consists of one electron and a nucleus of one proton. In the sun's core, hydrogen atoms have lost their electrons. Thus, only the protons remain.

The process of nuclear fusion that produces most of the sun's energy consists of five steps, as shown in Figure 28–1. First, two hydrogen nuclei, or protons, collide and fuse. Second, one of these protons changes into a neutron. Third, another proton combines with the proton-neutron pair, producing a nucleus made up of two protons and one neutron. Fourth, two of these nuclei collide and fuse. Finally, the resulting cluster throws off two protons. The remaining combination of two protons and two neutrons is the nucleus of a helium atom. During each step of the reaction, energy is given off.

The fusion of hydrogen nuclei may take place in a slightly different series of steps, but the final product is always a helium nucleus. The helium nucleus has about 0.7 percent less mass than did the hydrogen nuclei that combined in forming it. The loss of mass occurs in the hydrogen-to-helium reaction when some of the mass of the hydrogen has been converted into energy.

Mass into Energy

In 1905 the famous physicist Albert Einstein proposed that a large amount of energy could be produced from a very small amount of matter. At the time he proposed this, the existence of nuclear fusion was unknown. In fact, scientists had not yet discovered the nucleus of the atom. Einstein's work was based entirely on his theory of relativity, a revolutionary view of the physics of matter, energy, and time. Einstein's proposal involved the equation $E = mc^2$. E represents the energy produced; m, the mass, or the amount of matter, changed; and c, the speed of light, which is 300,000 km/s.

Einstein's equation can be used to calculate the amount of energy produced from a given amount of matter. With this equation, astronomers are able to explain the vast quantities of energy that the sun produces, using only a very small amount of fuel. The sun changes more than 600 million tons of hydrogen into helium every second. Yet this is a small amount compared to the total mass of hydrogen in the sun. At this rate how much hydrogen is changed into helium each day?

Reactions other than hydrogen fusion also take place in the sun's core. For example, a complicated reaction among nuclei of carbon, nitrogen, and oxygen atoms also produces helium and energy.

Investigation

Solar Viewer

Solar telescopes enable astronomers to study the sun without looking at it directly. Using some common objects, you can make a solar viewer that functions somewhat like a solar telescope.

Science Process Skills Focus: observing, describing, comparing, measuring

Materials
shoebox with lid; ruler; scissors; adhesive tape; 2 index cards, 8 cm × 13 cm; safety pin; piece of aluminum foil, about 4 cm × 4 cm

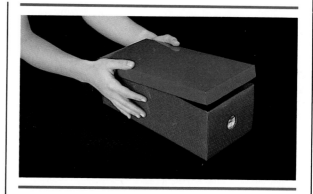

Procedure
1. Cut a hole with a diameter of about 3 cm in the center of one end of the shoebox. Tape an index card on the inside of the box on the side opposite the hole.
2. Use the safety pin to make a tiny hole in the center of the foil. Tape the foil over the hole in the shoebox. Put the lid on the shoebox.
3. Hold the box as shown in the figure, toward the sun. Lift the lid and observe the image of the sun projected on the index card. **Caution: Never look directly at the sun. Direct sunlight can damage your eyes.** Write down what you observe.
4. Hold the second index card inside the box at various distances from the pinhole. Write down what you observe at each distance.
5. Repeat Step 3 several times. Make the pinhole slightly larger each time. Write down what you observe for each diameter.

Analysis and Conclusion
1. Does the image of the sun change in size or brightness as the distance between the pinhole and index card is changed?
2. What happens to the image as you make the pinhole larger?
3. How is your solar viewer like a solar telescope? How is it different?

The Inner Zones

Before reaching the sun's atmosphere, the energy produced in the core moves through two inner zones. The zone surrounding the core is called the **radiative zone.** The temperature in this zone is about 2,500,000°C. In the radiative zone, energy moves from atom to atom in the form of waves, or radiation. These waves transfer energy across a space.

Around the radiative zone is the **convective zone** with a temperature of about 1,100,000°C. Energy produced in the core moves through this zone by *convection,* the transfer of energy by moving liquids or gases. Hot gases carry heat energy to the sun's surface. As the atoms of the hot gases move outward and expand, they collide with the atoms of other gases and lose heat. The cooling gases become denser than the other gases and sink to the bottom of the convective zone. There, the cooled gases are heated by the energy from the radiative zone and rise again. Thus heat is transferred to the sun's surface as the gases continuously rise and sink.

The Sun's Atmosphere

Surrounding the convective zone is the sun's atmosphere. Although the sun is made of gases, the term *atmosphere* refers to the uppermost region of solar gases. This region has three layers.

The innermost layer of the solar atmosphere—the **photosphere,** or light sphere—is made of gases bubbling up from the convective zone. The temperature in the photosphere is about 6,000°C. This layer has a grainy appearance, called **granulation,** which results as the gases rise from and sink to the convective zone. Much of the energy given off from the photosphere is in the form of visible light. The visible light is what is seen from the earth. Therefore, the photosphere is considered the surface of the sun.

Figure 28–2. False colors have been added to this photograph of the sun to show temperature differences.

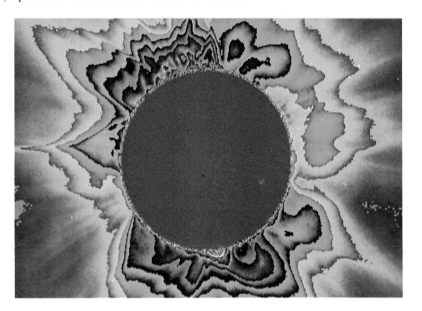

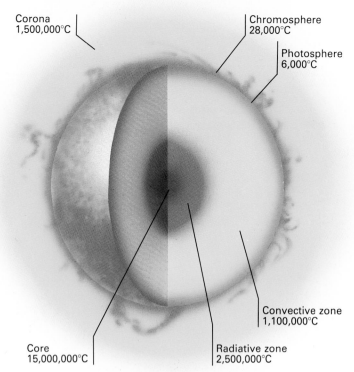

Corona
1,500,000°C

Chromosphere
28,000°C

Photosphere
6,000°C

Convective zone
1,100,000°C

Core
15,000,000°C

Radiative zone
2,500,000°C

Figure 28–3. The boundaries of the sun's regions and layers that surround the sun's core are not distinct. Each layer blends gradually into the next.

Above the photosphere lies the **chromosphere,** or color sphere, a thin layer of gases that seems to glow with a reddish light. Its temperature is about 28,000°C. The gases of the chromosphere move away from and toward the photosphere. Thus they transfer energy to the outermost layer. The upward movement sometimes forms narrow jets of hot gas that shoot outward from the chromosphere and then fade away within a few minutes. Some of these jets reach heights of 16,000 km.

The outermost layer of the sun's atmosphere—the **corona** (kuh-ROE-nuh), or crown—blends into space. This layer is a huge cloud of gas with a temperature of about 1,500,000°C. Although the corona is relatively thin, it prevents most of the atomic particles from the sun's surface from escaping into space. However, some electrically charged atomic particles, or ions, stream out into space through holes in the corona. These particles, called the **solar wind,** flow to distant parts of the universe.

The chromosphere and the corona are normally not seen from the earth because of the brightness of the photosphere. Occasionally, however, the moon moves between the earth and the sun and blocks out the light of the photosphere. The corona then becomes visible.

Section 28.1 Review

1. Describe the most common type of nuclear fusion in the sun.
2. What are the two end products of fusion in the sun?
3. Will the amount of hydrogen and helium in the sun increase or decrease over the next few million years?
4. What layer of the sun's atmosphere can normally be seen?
5. How does the transfer of energy in the radiative zone differ from the transfer of energy in the convective zone?

Section Objectives

- **Explain how sunspots are related to powerful magnetic fields on the sun.**

- **Compare prominences and solar flares.**

- **Describe how the solar wind can cause auroras on the earth.**

28.2 Solar Activity

The gases that make up the inner zones and the atmosphere of the sun are in constant motion. The energy produced in the sun's core and the force of gravity combine to cause the cyclical rising and sinking of gases. The gases also move because the sun rotates on its axis. Being a ball of hot gases rather than a solid sphere, the parts of the sun rotate at different speeds. Places closest to the sun's equator take only 25.3 earth days to make one rotation. Points near the poles take 33 earth days. For simplicity, astronomers calculate the average of the sun's rotation to be 27 earth days.

Sunspots

The combination of the up-and-down movement of gases within the sun's convective zone and the movement of the sun's rotation produces magnetic fields. These magnetic fields slow down activity in the convective zone. Slower convection means that less gas is transferring heat from the core of the sun to the photosphere. Therefore, regions of the photosphere near strong magnetic fields are up to 3,000°C cooler than surrounding areas.

The cooler areas of the sun appear darker than the areas surrounding them. Cool, dark areas of gas within the photosphere that are caused by powerful magnetic fields are called **sunspots.** Large sunspots can be more than 100,000 km in diameter, which is several times the size of the earth. Photographs of sunspots are shown in Figure 28–4.

Astronomers have carefully observed sunspot activity. They have found that sunspots initially appear in groups about midway between

Figure 28–4. Sunspots usually appear in pairs—one magnetic north and one magnetic south. The photograph on the left has been artificially colored to show this polarity. The photograph on the right is of the same sunspots in visible light.

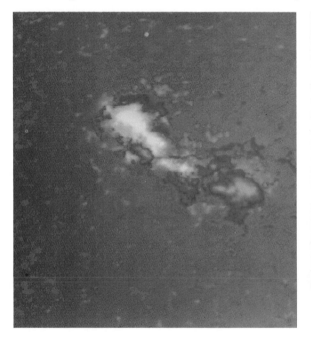
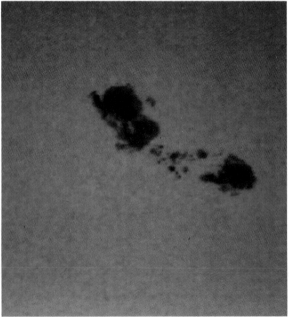

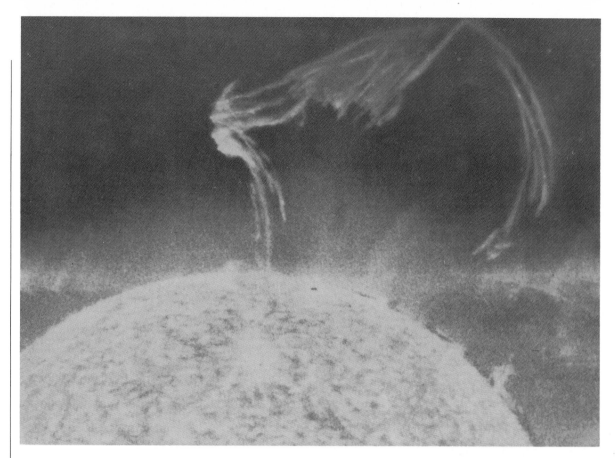

the sun's equator and poles. As sunspots slowly disappear, they also seem to move toward the sun's equator. The movement of sunspots was one of the first indications to astronomers that the sun rotates on its axis. The number of sunspots also varies according to an average 11-year cycle called the **sunspot cycle.**

A sunspot cycle begins when the number of sunspots is very low but is beginning to increase. Astronomers may not see any sunspots for several weeks. Then gradually they see more and more. The number of sunspots increases for a few years until it reaches a peak. At a peak, 100 or more sunspots may be visible. After the peak the number of sunspots begins to decrease until it reaches a minimum. Another 11-year cycle begins when the number of sunspots begins to increase again. If the number of sunspots was very low in 1977, when did the next low point in the cycle occur?

Figure 28–5. A solar prominence can arch half a million kilometers above the sun's surface.

Prominences and Solar Flares

The magnetic fields that cause sunspots also create other disturbances in the solar atmosphere. Great clouds of glowing gases, called **prominences,** form huge arches that reach high above the sun's surface, as shown in Figure 28–5. Some arches are as high as 90,000 km. Each prominence follows curved lines of magnetic force from one sunspot area to another. Most prominences last for several weeks or several months. A few can be seen for almost a year. Others die out within one hour.

The most violent of all solar disturbances is a **solar flare,** a sudden outward eruption of electrically charged atomic particles. Solar flares may extend upward several thousand kilometers within minutes. Few eruptions last more than an hour. Solar flares usually occur near sunspots. During a peak in the sunspot cycle, five to ten solar flares may be visible each day.

Many of the particles from a solar flare are flung out so forcefully that they escape into space. These particles increase the strength of the solar wind. As the gusts of solar-wind particles enter the atmosphere of the earth, they can generate a sudden disturbance in the earth's magnetic field. Such disturbances are called *magnetic storms*. Although several small magnetic storms may occur each month, the average number of severe storms is less than one per year. These powerful storms have been known to disrupt radio communications on the earth.

IMPACT

Solar Flares

One night in 1570, a man living in central Europe looked out on his town. He was horrified by the sight of "two great pillars . . . [with] fire running down the pillars." The frightened man was actually seeing an aurora.

Although auroras and other effects of solar flares no longer cause fear, they still influence people's lives. For example, solar flares send out intense ultraviolet rays that reach the earth in about 8 min. They can disrupt the ionosphere, the part of the atmosphere that reflects radio waves. Such disruptions can cause interference with radio transmission. In 1972 solar flares resulted in so much interference that ships on the St. Lawrence River could not contact each other.

Solar flares also disturb the earth's magnetosphere, producing magnetic storms. These storms, caused by bursts of atomic particles from the sun, can cause compass needles to fluctuate, making them difficult to read. Magnetic storms often produce strong power surges along electrical wires. Some surges are strong enough to burn out power stations and overload telephone lines.

Occasionally the energy from solar flares has been used by people. In 1859 some telegraph companies operated their telegraph equipment using power surges caused by magnetic storms.

If the energy of magnetic storms could be harnessed, what benefits might result?

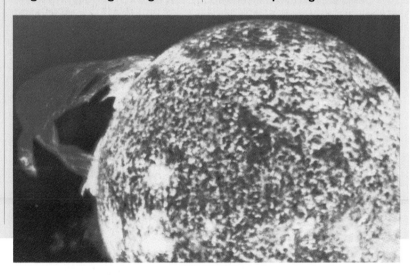

Figure 28–6. Auroras can fill the entire sky with a colorful curtain of light.

Auroras

On the earth the most spectacular effect of a magnetic storm is the appearance in the sky of bands of light called **auroras** (uh-RORE-uhz). Figure 28–6 shows an example of an aurora. When the electrically charged particles of the solar wind approach the earth, they are guided toward the earth's magnetic poles by the earth's *magnetosphere*. The magnetosphere is the space around the earth that contains a magnetic field. The electrically charged particles strike the gas molecules in the upper atmosphere, thereby producing green, red, blue, or violet sheets of light. Because of the effect of the magnetosphere, auroras are usually seen close to the magnetic poles. Depending upon which pole they are near, auroras are also called *northern lights* or *southern lights*.

Auroras usually occur between 100 km and 1,000 km above the earth's surface. They are most frequent in March, April, September, and October. In the northern United States, auroras are visible about five times a year. A display often lasts up to three hours.

Although auroras are most often visible in polar regions, they are sometimes visible near the equator. In 1921 the residents of some South Pacific islands were able to see the southern lights.

Section 28.2 Review

1. Why are sunspots cooler than surrounding areas?
2. How long is the sunspot cycle?
3. How are prominences different from solar flares?
4. What causes auroras?
5. If the strength of the sun's magnetic field decreased from its current level, why would the temperature of sunspots increase?

- **Explain the nebular theory of the origin of the solar system.**
- **Describe how the planets developed.**
- **Describe the formation of the land, the atmosphere, and the oceans of the earth.**

28.3 Formation of the Solar System

Scientists have long debated the origins of the **solar system.** The solar system includes the sun and the bodies revolving around the sun. In the 1600's and 1700's, many scientists thought that the sun formed first and threw off the materials that later formed the **planets.** Planets are the nine major bodies orbiting the sun. In 1796, however, the French mathematician Marquis Pierre Simon de Laplace advanced a new hypothesis. It stated that the sun and the planets condensed out of the same spinning *nebula,* or cloud of gas and dust. The hypothesis also stated that the entire solar system formed at approximately the same time. Laplace's hypothesis developed into what became known as the **nebular theory.**

Formation of the Sun

According to the current nebular theory, the big bang spread matter throughout the expanding universe. The cloud of gas and dust that eventually developed into the solar system is called the **solar nebula.** The solar nebula was as large as the current solar system.

About 4 billion to 5 billion years ago, shock waves from a nearby supernova or some other force caused the solar nebula to contract and become denser. A star—the sun—began to form in the center of the solar nebula. In Chapter 27 you read about the origin of a typical star. The sun is thought to have developed by this same process. Heat from collisions and pressure from the force of gravity caused the center of the solar nebula to become denser and hotter. When the temperature at the center of the nebula became great enough, hydrogen fusion began, and the sun formed. About 99 percent of the matter in the solar nebula became part of the sun.

Formation of the Planets

While the sun was forming in the center of the solar nebula, planets were forming in the outer regions, as shown in Figure 28–7. The small bodies of matter in the solar nebula are called **planetesimals** (PLAN-uh-TESS-uh-mulz). Some planetesimals joined together through collisions and through the force of gravity to form much larger bodies called **protoplanets.** The protoplanets acted like giant magnets, pulling in other planetesimals from the solar nebula.

Eventually the protoplanets condensed into the existing planets and **moons.** Moons are the smaller bodies that orbit the planets. Planets and moons are smaller and denser than the protoplanets. For example, the protoplanet that became the earth originally may have been 1,000 times bigger and 8 times less dense than the present-day earth.

The distance between a protoplanet and the developing sun influenced the composition of the planet that formed from the protoplanet.

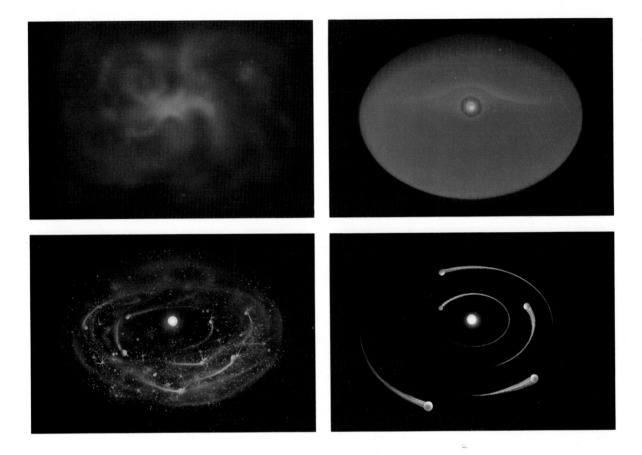

The four protoplanets closest to the sun became Mercury, Venus, Earth, and Mars. They contained large amounts of the heavier elements, such as iron.

The next four protoplanets became Jupiter, Saturn, Uranus, and Neptune. These four protoplanets formed in the cold regions of the solar nebula. The icy material of the outer protoplanets consisted of water and frozen gases, such as helium, hydrogen, methane, and ammonia. Far from the heat of the sun, the outer protoplanets developed into huge planets. Thick layers of ice surrounded small cores of heavy elements.

The inner planets may have lost their outer layers of gases because of an intense solar wind. Scientists think that the energy and matter from the solar wind may have stripped away the atmosphere of lighter elements from the inner planets. Because of their greater distance from the sun and their tremendous gravitational forces, the outer planets were able to retain much of their original gases. Today these planets are referred to as the *gas giants*. These planets are called gas giants because they are composed mostly of gases, and because they are such huge planets.

Pluto, the planet that is farthest from the sun and the smallest of the known planets, has a unique history. Astronomers theorize that Pluto may have formed as a moon of Neptune but later escaped to follow its own orbit around the sun.

Figure 28–7. The matter that forms into planets begins as planetesimals in a nebula. The planetesimals combine and form protoplanets, which condense into planets.

Formation of the Earth

When the earth first formed, it was very hot. Three sources of heat contributed to the high temperature on the new planet. First, the earth retained much of the heat produced when it collided with planetesimals. Second, the increasing weight of the outer layers compressed its inner layers. This compression generated more heat. Third, *radioactive* materials, which emit high-energy particles, were very abundant when the earth first formed. When these particles were absorbed by surrounding rocks, some of the energy of motion of the particles was converted to heat energy.

Career Focus: *Space*

"Having the chance to work on questions I've wondered about my whole life is what I find extremely rewarding."

Astronomer

I was one of those kids consumed with science. Rocks, biology, fossils, space—you name it. If it was science, I was interested," said Sandra Faber, professor of astronomy at the University of California, Santa Barbara, and staff member at the Lick Observatory in San Jose, California.

Faber, pictured left, eventually chose astronomy as her field of study for a very specific reason.

"I was interested in the big picture," she said. "Somewhere along the line, it occurred to me that the questions about the universe posed in astronomy were the most all-encompassing."

Faber has a bachelor's degree in physics from Swarthmore and a doctorate in astronomy from Harvard. She is now attempting to answer those questions about the nature of the stars, planets, galaxies, and the universe itself. To gather the necessary data, Faber uses various kinds of telescopes, including telescopes mounted on satellites.

"One of my most exciting experiences came in 1985 at the end of a long project with six other astronomers to study the expansion of the universe to find out how uniform and regular it is," she said. "After analyzing the data, we were shocked to discover enormous motions over a huge volume of space. This work

The Solid Earth

The temperature on the young earth was high enough to melt iron, the most common of the existing heavy materials. Gravity pulled the molten iron toward the center. Thus, as the earth developed, denser materials flowed to its center, and less-dense materials were forced to the outer layers.

The earth eventually separated into the three distinct layers. At the center is a dense *core*, composed mostly of iron and nickel. Around the core is the very thick layer of rock called the *mantle*. The outermost layer is a thin *crust* of less-dense solid materials.

could have a major impact on theories for the origin of structure in the universe."

Planetarium Technician

A planetarium technician, pictured lower left, installs, operates, and maintains sound and projection equipment for planetariums. Many technicians help to develop special effects and audiovisual displays. Technicians may also construct and install public education displays in the exhibit areas of planetariums.

Most planetarium technicians have vocational training in electronics, electromechanics, and construction.

Solar Energy Systems Designer

Solar energy systems designers develop solar heating systems for new and existing structures. Solar energy systems designers must have a knowledge of energy requirements, local weather conditions, thermodynamics, and solar technology. In developing a particular system, a designer must determine the location, type, and size of the solar components to be used. Solar components include solar panels for energy collection, pumps to circulate water, and water storage tanks. The choice of solar components is based on engineering principles, energy needs, and the angle at which the sun's rays strike the structure. The designer often inspects the system during construction to make sure it is built correctly.

The minimum requirement for entry into the field of solar energy systems design is a bachelor's degree in engineering.

For Further Information

For more information on careers in astronomy, write the American Institute of Physics, 335 East Forty-fifth Street, New York, NY 10017.

The Atmosphere

The protoplanet that became the earth had too little mass and too weak a gravitational force to hold gases. As collisions added more material to the protoplanet, the force of gravity increased. Eventually the protoplanet could capture some of the hydrogen and helium abundant in the solar nebula. By the time the protoplanet had evolved into the earth, its atmosphere consisted primarily of hydrogen and helium. Today these two elements, so common in the earth's first atmosphere, are rarely found in dry air.

The earth's first atmosphere was lost probably as a result of a solar explosion or the solar wind. The earth's second atmosphere resulted from explosions within the earth over 3 billion years ago. During the earth's early history, the heat in its interior caused volcanoes to form. The volcanic eruptions released large amounts of gases, such as water vapor, ammonia, carbon dioxide, and methane. These gases formed a new atmosphere. What would have happened to these gases if the earth had been much smaller?

The action of sunlight probably caused the ammonia and some of the water vapor in the atmosphere to change into nitrogen, hydrogen, and a little oxygen. Most of the hydrogen slowly escaped into space because it was too light to be permanently held by the earth's force of gravity. As green plants appeared, the amount of oxygen increased. Green plants use carbon dioxide and release oxygen during *photosynthesis,* the process by which plants use sunlight to make food. Slowly the amount of oxygen in the atmosphere increased to current levels. Some of the oxygen formed ozone, which collected in a layer around the earth. This ozone layer shielded the earth and its inhabitants from the harmful ultraviolet radiation of the sun. The series of events that resulted in the formation of the earth's atmosphere are pictured in Figure 28–8.

The Oceans

As the atmosphere was developing, the earth was cooling enough for liquid water to form. Between 3 billion and 3.5 billion years ago, water vapor began to condense. It fell to the earth as rain and

Figure 28–8. During Earth's early history, volcanic eruptions formed an atmosphere of water vapor, ammonia, carbon dioxide, and methane. (left) The action of sunlight on this primitive atmosphere is thought to have triggered chemical changes, which began to form Earth's high oxygen atmosphere and ozone layer. (right)

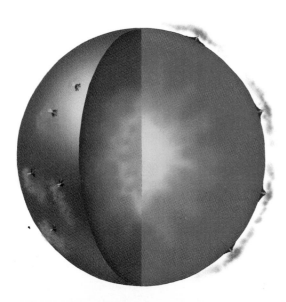

Science Notebook

The Solar Constant

Decades ago astronomers assumed that the sun's energy output remained steady. They called the amount of energy that hits each square centimeter of the earth per second the *solar constant*. This definition is based upon the average distance of the sun from the earth and upon the assumption that the earth has no atmosphere.

However, data from instruments sent outside the earth's atmosphere indicate that the solar "constant" is not constant. In 1980 the Solar Maximum Mission satellite, shown being approached by astronaut George D. Nelson, revealed that the solar constant may vary up to 0.2 percent from year to year.

Astronomers have two hypotheses to explain these changes in the sun's energy output. One states that the internal nuclear reactions of the sun vary.

A second hypothesis states that the variation in the solar constant is caused by changes in the sun's photosphere. Sunspots are known to be cooler regions of the photosphere. Therefore, it follows that the solar constant is lowest during peak sunspot activity, which occurs about every 11 years.

Fluctuations of the solar constant over thousands of years may have severe effects on the earth. Astronomers are trying to detect slight long-term changes that could be responsible for major climatic changes.

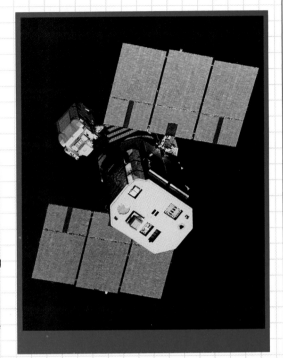

If the average temperature on the earth stayed a few degrees cooler than the current temperature for 1,000 years, what climate pattern might emerge?

---

formed oceans in the cratered surface. The ocean water absorbed much of the carbon dioxide from the atmosphere. Water naturally absorbs carbon dioxide whenever its concentration in the atmosphere is high enough. By 1.5 billion years ago, the chemical composition of the oceans was similar to what it is today.

Section 28.3 Review

1. What two forces caused the solar nebula to develop into the sun?
2. How were planetesimals different from protoplanets?
3. List three reasons why Earth was hot when it first formed.
4. What gas in the atmosphere was absorbed by the oceans as they formed?
5. Explain the following statements: Explosions cause the loss of an atmosphere. Explosions aid in the creation of an atmosphere.

Reviewing Chapter 28

Key Concepts

- The enormous pressure and heat of the sun converts matter into energy. See page 489.

- Energy from the sun's core moves through the radiative zone and the convective zone before it enters the sun's atmosphere. See page 492.

- Sunspots are caused by powerful magnetic fields on the sun. See page 494.

- Prominences and solar flares are two types of solar activity caused by disturbances in the solar atmosphere. See page 495.

- The solar wind, composed of electrically charged particles, can cause auroras when the particles enter the earth's atmosphere. See page 497.

- According to the nebular theory, the sun and the planets formed at the same time. See page 498.

- Planets formed from small bodies of matter in the outer regions of the solar nebula. See page 498.

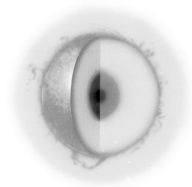

- **The sun's atmosphere is composed of the photosphere, the chromosphere, and the corona. See page 493.**

- As the earth cooled down after its formation, the current atmosphere and oceans developed. See page 502.

Key Terms

aurora (497)	moon (498)	prominence (495)	solar system (498)
chromosphere (493)	nebular theory (498)	protoplanet (498)	solar wind (493)
convective zone (492)	photosphere (492)	radiative zone (492)	sunspot (494)
corona (493)	planet (498)	solar flare (496)	sunspot cycle (495)
granulation (492)	planetesimal (498)	solar nebula (498)	

Review

On your paper, write the letter of the term that best completes each of the following statements.

1. According to Einstein's theory of relativity, in the formula $E = mc^2$, the c stands for (a) corona (b) core (c) the speed of light (d) the length of time.

2. A nuclear reaction in which two atomic nuclei combine is called (a) fission (b) fusion (c) magnetism (d) granulation.

3. A reaction between carbon, nitrogen, and oxygen nuclei in the sun's core produces (a) planetesimals and water (b) nebulas and hydrogen (c) mass and oxygen (d) energy and helium.

4. The portion of the sun in which energy moves from atom to atom in the form of waves is called the (a) radiative zone (b) convective zone (c) solar wind (d) chromosphere.

5. The portion of the sun normally visible from the earth is the (a) core (b) photosphere (c) corona (d) solar nebula.

6. The sunspot cycle lasts (a) 2 years (b) 5 years (c) 11 years (d) 19 years.

7. Sudden outward eruptions of electrically

charged atomic particles from the sun are called (a) planetesimals (b) coronas (c) sunspots (d) solar flares.
8. Gusts of solar wind can cause (a) protoplanets (b) magnetic storms (c) nuclear fission (d) nuclear fusion.
9. Northern lights and southern lights are other names for (a) prominences (b) auroras (c) granulations (d) protoplanets.
10. The hypothesis that the sun and the planets developed out of the same cloud of gas and dust is called the (a) nebular theory (b) theory of relativity (c) nuclear theory (d) theory of convection.
11. Of the original matter in the solar nebula, the amount that became part of the sun is about (a) 50 percent (b) 66 percent (c) 75 percent (d) 99 percent.

12. The small bodies of matter that filled the solar nebula are called (a) protoplanets (b) planetesimals (c) auroras (d) proton nuclei.
13. Compared with the size of the present-day planets, the protoplanets were (a) much smaller (b) slightly smaller (c) similar (d) larger.
14. The first atmosphere of the earth had a high percentage of (a) helium (b) oxygen (c) nitrogen (d) water vapor.
15. In the process of photosynthesis, green plants give off (a) oxygen (b) carbon dioxide (c) hydrogen (d) helium.
16. Water vapor began to condense into oceans about (a) 1 million to 1.5 million years ago (b) 1 billion to 1.5 billion years ago (c) 3 billion to 3.5 billion years ago (d) 15 billion to 15.5 billion years ago.

Application

On your paper, write answers to the following questions.
1. Explain the following statement: Solar hydrogen is responsible for life existing on the earth.
2. Explain how the transfer of energy in a pan of hot water is similar to the transfer of energy in the sun's convective zone.
3. If all the holes in the corona closed up, what would happen to the solar wind?
4. Predict what would happen to the number of sunspots if the sun's magnetic field suddenly increased in strength.
5. If a high number of solar flares occurred, what would happen to the number of auroras on the earth?

6. If the earth's magnetosphere shifted, what would happen to the area where auroras were most often visible?
7. Imagine that astronomers discovered that the protoplanets needed a nearby source of light and heat in order to develop into planets. How would such a discovery modify the nebular theory?
8. List three similarities between the formation of the sun and the formation of the planets.
9. How would the layers of the earth be different if the planet had never been hotter than it is today?
10. How would the atmosphere of the earth be different if the earth had formed from a much larger protoplanet?

Extension

1. Write an essay describing an imaginary trip to the center of the sun. Describe each layer and zone through which you would pass.
2. Draw a diagram showing the activities that take place in the atmosphere of the sun.
3. In the early 1900's, two scientists proposed an alternative to Laplace's theory about the origin of the solar system. These scientists, Thomas Chamberlin and Forest Moulton

proposed a dualistic theory. Find out what caused them to question Laplace's theory. Describe the theory they proposed.
4. The earth's atmosphere had very little oxygen before green plants existed. Do research to find out how important green plants, particularly in the rain forests, are for the earth's oxygen supply today.

Chapter 29

The Solar System

*V*isible *in the night sky are objects that shine with a steady light and wander across the sky. Ancient astronomers called these moving objects* planetae, *the Greek word for "wanderers." They are now called* planets. *In this chapter you will learn about the planets and other bodies in the solar system.*

Chapter Outline

Jupiter is the largest planet in the solar system.

29.1 Models of the Solar System

Section Objectives

- **Compare the models of the universe developed by Ptolemy and Copernicus.**
- **Summarize Kepler's three laws of planetary motion.**

Two thousand years ago many philosophers developed ideas about the universe based on what they saw. One of these was Aristotle, a Greek philosopher who lived from 384 to 322 B.C. Through observations and reasoning, Aristotle developed an earth-centered, or **geocentric,** model of the solar system. This inaccurate model stated that the sun, the stars, and the planets revolved around the earth.

Aristotle's model, however, did not explain why some planets appear to reverse direction occasionally, moving from east to west instead of the usual west to east. This backward motion is called **retrograde motion.** About 500 years after Aristotle developed his model, another Greek astronomer, Claudius Ptolemy, modified the model to include retrograde motion.

Ptolemy proposed a model of the universe in which each planet had two motions. One motion was the revolution of the planet around the earth. As the planet revolved, it also moved in a series of small circles, like the links of a chain. Ptolemy called these small circles **epicycles.** According to Ptolemy, the motion of the planet in its epicycle would make it appear to move backward at times.

Copernicus's Model

In the 1500's Nicolaus Copernicus, a Polish astronomer, challenged the ideas of Aristotle and Ptolemy. Copernicus proposed a **heliocentric,** or sun-centered, model of the solar system. According to this model, the earth and the other planets revolve around the sun. Copernicus also proposed that all planets orbit in the same direction but that each moves at a different speed and distance from the sun. The faster planets, therefore, pass the slower planets from time to time.

Confirmation of Copernicus's model finally came in the early 1600's. At that time the Italian scientist Galileo Galilei was able to observe the motions of the planets with the newly invented telescope. He collected evidence that proved the heliocentric model was correct one.

Kepler's Laws

While Galileo was making his observations, other scientists were collecting data that also supported a heliocentric solar system. Tycho Brahe, a Danish astronomer, devoted his life to making detailed observations of the positions of the stars and planets. Near the end of his life, Brahe hired a German astronomer, Johannes Kepler (1571–1630), as his assistant. Kepler was able to explain Brahe's precise observations in mathematical terms. Kepler developed three laws that explained most aspects of planetary motion. Kepler's laws were the first mathematically proven ideas in science.

Law of Ellipses

Kepler's first law states that each planet orbits the sun in a path called an **ellipse.** An ellipse is an oval whose shape is determined by two points within the figure, as shown in Figure 29–1. Each of these points is called a **focus** (pl: foci). The sun is one focus of the orbit of a planet. If you draw lines from each of the two foci to any point on the ellipse, the total length of the lines will be the same. Some ellipses look almost like circles. In fact, a circle is a special kind of ellipse in which the two foci are the same point. Other ellipses are more elongated ovals.

Because the orbits of the planets are ellipses, a planet is not always the same distance from the sun. The point where an orbit is closest to the sun is the *perihelion*; the point where it is farthest from the sun is the *aphelion*. Why would a circular orbit have neither a perihelion nor an aphelion?

The distance of a planet from the sun is usually defined as the average of the distances from the sun at its perihelion and its aphelion. For example, the aphelion of the earth's orbit is about 152 million km from the sun; the perihelion is about 147 million km from the sun. The average of 147 million and 152 million is 149.5 million. This average distance between the earth and the sun is known as one **astronomical unit,** or AU. The distance between the sun and other planets is usually measured in astronomical units.

Law of Equal Areas

Kepler's second law describes the speed at which planets travel at different points in their orbits. By studying Brahe's data, Kepler found that the orbit of the earth was a nearly perfect circle, with the sun off-center. He found that the earth moves fastest when it is closest to the sun. He calculated that a line from the center of the sun to the center of the planet sweeps through equal areas in equal periods of time.

Imagine a line that connects the center of the sun to the center of a planet. When the planet is near the sun, the imaginary line is relatively short. The planet is moving rapidly and in ten days, for example, the imaginary line sweeps through a short, wide triangular

Figure 29–1. You can draw an ellipse using two pins as the foci and a string (left). Kepler's second law states that the areas a planet sweeps out in a given period of time—the shaded areas between points *1* and *2* and points *3* and *4*—are equal.

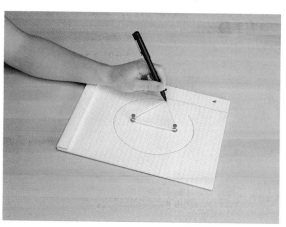

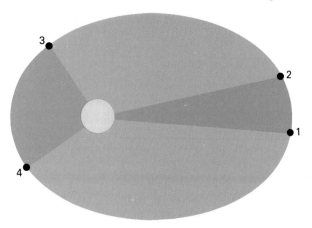

sector. When the planet is farther from the sun, the imaginary line is longer. However, the planet is moving more slowly and the imaginary line sweeps through a long, thin triangular sector in ten days. Kepler's second law states that the area of the long, thin sector is the same as the area of the short, wide sector.

Law of Periods

Kepler's third law describes the relationship between the average distance of a planet from the sun and the **period** of the orbit of the planet. The period of an orbit is the time required for the planet to make one revolution around the sun. According to Kepler's third law, the cube of the average distance of a planet from the sun (r) is always proportional to the square of the period (p). The mathematical formula that describes this relationship is $K \times r^3 = p^2$, where K is a mathematical constant. When distance is measured in AU's and the period is in earth-years, then $K = 1$ and $r^3 = p^2$.

For example, the radius of the earth's orbit, or its distance from the sun, is 1 AU, and its period is 1 year. Putting these numbers into the formula yields $1 \times 1^3 = 1^2$. This simplifies to $1 = 1$. Jupiter is 5.2 AU's from the sun, and its period is 11.9 years. The cube of 5.2 is 140.6. The square of 11.9 is 141.6. The two results, 140.6 and 141.6, are approximately equal. Apparent errors in the law of periods are caused by rounding off the distance or periods. When the distance and period are calculated with enough accuracy, the period squared always equals the distance cubed.

Newton's Application of Kepler's Laws

Kepler's laws explained how the planets orbit the sun. Isaac Newton asked why the planets move in this way. The explanation that Newton eventually gave described both the motion of objects on the earth and the motion of the planets in space. He hypothesized that a moving body will change its motion only if an outside force causes it to do so. For example, a ball rolling on a smooth surface will continue to move in a straight line unless something causes it to change direction. The tendency of a moving body to move in a straight line at a constant speed until an outside force acts on it is called **inertia.** Inertia also refers to the tendency of an object to remain at rest until an outside force acts on it.

Newton compared a planet to a rolling ball. Because a planet does not follow a straight path, an outside force must cause it to curve. Newton identified this force as *gravity,* the attractive force that exists between all objects in the universe. The gravitational pull of the sun keeps the planets in orbit around the sun.

Section 29.1 Review

1. What is the basic difference between Ptolemy's and Copernicus's models of the universe?
2. According to Kepler, what is the shape of planetary orbits?
3. Upon what observation did Kepler base his second law?
4. What is the period of a planet located 4 AU's from the sun?

29.2 The Inner Planets

The four planets closest to the sun are called inner planets. These planets are Mercury, Venus, the earth, and Mars. The inner planets are sometimes called the *rocky planets,* or **terrestrial planets,** because they are similar to the earth. Because these planets formed close to the heat of the sun, materials with low boiling points were driven off. The planets consist mostly of solid rock, with a metal core. Inner planets do not have rings. Each has a maximum of two moons and bowl-shaped depressions called *impact craters* on its surface. Impact craters result from collisions of the planets with objects made primarily of rock. These collisions took place during the later stages of the formation of the solar system.

Mercury

Mercury is the planet that is closest to the sun. Consequently Mercury has a shorter period than does any other planet—88 days. In fact, the ancient Romans named the planet *Mercurius* after the swift messenger of the gods. Mercury rotates very slowly on its axis—only once every 59 days. Mercury is so close to the sun that light from the sun usually obscures the planet. Even the best photographs taken from the earth show Mercury as only a fuzzy ball. Mercury has no moons.

In 1974 and 1979 *Mariner 10* visited Mercury. The spacecraft transmitted photographs to the earth that revealed a surface that was heavily cratered, like that shown in Figure 29–2. The large number of craters suggests that Mercury has changed little since the formation of the solar system. Some craters appear to be filled with hardened lava. If this is the case, Mercury was once volcanic. The photographs also showed a line of cliffs hundreds of kilometers long. These cliffs may be wrinkles in the crust, which developed when the once-molten core cooled and shrank.

Figure 29–2. The surface of Mercury, shown in this photograph taken by *Mariner 10,* probably looks very much like it did shortly after the solar system was formed.

Mercury has a thin atmosphere for two reasons: its closeness to the sun and its size. Because Mercury is so close to the sun, solar heat causes the gas molecules near the surface of the planet to move rapidly. Because the planet is so small, its gravitational pull is too weak to hold enough rapidly moving gas molecules to form a dense atmosphere. The absence of a dense atmosphere contributes to the huge daily temperature range on Mercury. During the day the temperature may reach as high as 500°C. At night the temperature may plunge to −200°C.

Before the *Mariner 10* expedition, astronomers thought Mercury was too small and rotated too slowly to have a magnetic field. However, the instruments on *Mariner 10* did detect a weak magnetic field, which suggests that Mercury has a core of iron.

Venus

The second planet from the sun is Venus. The period of Venus is 225 days. Like Mercury, Venus rotates very slowly on its axis— once every 243 days. The direction of rotation is opposite that of the other planets. For this reason, the sun, viewed from Venus, rises in the west and sets in the east. Like Mercury, Venus has no moons.

In some ways Venus is the earth's twin. The two planets are of almost the same size, mass, and density. However, Venus is much hotter than the earth because its atmosphere provides an insulating effect. The average surface temperature on Venus is approximately 480°C, and its atmospheric pressure is almost 100 times the atmospheric pressure on the earth.

The high temperature and dense atmosphere of Venus are closely related. Astronomers think that when Venus was formed, its temperature was lower and its atmosphere was less dense than they are today. Astronomers also think that extensive oceans may have flowed across the surface of Venus at that time, and there may have been much volcanic activity.

Figure 29–3. Venus has approximately the same size, mass, and density as the earth. Its surface is composed of basalt and granite rocks.

Astronomers theorize that as the sun became hotter, the oceans on Venus evaporated. Because there was no water to combine with the carbon dioxide released by erupting volcanoes, the level of carbon dioxide in the atmosphere steadily increased. As a result, the atmospheric pressure increased. Carbon dioxide still accounts for about 97 percent of the atmosphere of Venus. This blanket of carbon dioxide allows solar energy to penetrate the atmosphere of the planet but blocks the escape of heat. Consequently the planet stays too hot to support life. High above the surface of the planet, the temperature decreases dramatically. There is very little water vapor in the atmosphere of Venus. Therefore, instead of containing water vapor, the clouds of Venus are composed of droplets of sulfuric acid, which is an extremely strong acid.

In the 1970's the Soviet Union sent six probes, the *Venera* probes, to explore the surface of Venus. The spacecraft were refrigerated and heavily insulated before they descended onto the surface of the planet. The probes survived the heat and pressure of the atmosphere long enough to transmit several photographs of the rocky landscape of Venus. The photographs showed a relatively smooth plain, with some features that resemble the earth's mountains and valleys. Other instruments carried by the probes indicated that the surface of Venus is composed of basalt and granite. These are two types of rock commonly found in volcanic regions on the earth. What do these rocks suggest about the history of Venus?

Earth

The third planet from the sun is Earth. The period of the earth is 365.26 days, and the planet completes one rotation in 23 hours and 56 minutes. The earth is the fifth largest planet. It has one moon.

The earth has had an extremely active geologic history. Geologic records indicate that in just over 200 million years, the earth's continents separated from a single landmass and drifted to their present positions. Weathering and erosion have changed and continue to change the surface of the earth.

Life on the earth is possible because of the distance of the planet from the sun. The temperature is warm enough for water to exist as a liquid. Mercury and Venus are so close to the sun that they are too hot to retain liquid water. The outer planets are so far from the sun that water cannot exist in liquid form. Most of the water on these planets is in the form of ice. The earth is the only planet in the solar system that has oceans of water.

Geologists theorize that as oceans formed on the earth, water combined with carbon dioxide in the atmosphere. Since carbon dioxide did not build up in the atmosphere, solar heat was able to escape. Consequently, the earth maintained the moderate temperatures needed to support life, an average of 14°C. Plants contributed additional oxygen to the atmosphere. The earth is the only known planet with the proper combination of water, temperature, and oxygen to support plants and animals.

Figure 29–4. The earth is covered with oceans of water and an atmosphere that can support life.

Figure 29–5. Evidence of the active volcanism on Mars is found on the surface of the planet.

Mars

Mars is the fourth planet from the sun. Its period is 687 days; and its rate of rotation, 24 hours and 37 minutes. Therefore, the length of a day on Mars and on the earth is almost the same. Mars also has seasons much like the earth's because the tilt of its axis is nearly the same as that of the earth's axis. Mars has two moons.

Mars is geologically active, with large surface areas covered with lava. The volcanoes on Mars may be the largest in the solar system. The largest of these, *Olympus Mons,* is three times higher than Mount Everest, and its base is about the size of Nebraska.

A system of deep canyons also covers part of the surface. The largest, the *Valles Marineris,* is as long as the United States— 4,500 km. Great cracks in the crust of the planet may have caused mars-quakes that formed these canyons.

The atmospheric pressure and temperature on Mars are too low for water to exist as a liquid. However, two U.S. spacecraft, *Viking 1* and *Viking 2,* found evidence of erosion by water. Astronomers, therefore, can infer that Mars once had a warmer and wetter climate. The temperature near the equator of Mars now approaches 20°C during the summer. The temperature near the poles drops as low as −130°C during the winter. The little water that remains on Mars is trapped in polar ice caps or is possibly frozen beneath the surface.

One of the tasks of the *Viking* probes was to search for signs of life on Mars. The probes found no convincing evidence that life has ever existed on the planet.

Section 29.2 Review

1. What evidence suggests that Mercury has changed little since it was formed?
2. What aspect of the earth makes its temperature favorable for life?
3. Describe two ways in which Mars is similar to the earth.
4. What is unusual about the rotation of Venus?

29.3 The Outer Planets

The five planets farthest from the sun are called the *outer planets*. They are Jupiter, Saturn, Uranus, Neptune, and Pluto. The first four are called the *giant planets* because they are the largest planets in the solar system. They are also called the **Jovian planets** because they are similar to Jupiter. The smallest and usually the most distant planet in the solar system is Pluto. Pluto is different from all the other planets and may not have formed in the same way.

Although the Jovian planets are much larger and heavier than the inner planets, they are far less dense. Each of these planets has a thick atmosphere made up mostly of hydrogen and helium gases. Each also probably has a core of rock, metals, and water.

Jupiter

The first of the outer planets, Jupiter, is the fifth planet from the sun. Jupiter, shown in Figure 29–6, is the largest planet in the solar system. Its mass is twice that of the other eight planets combined. Its period is almost 12 years. Jupiter rotates faster than any other planet—once every ten hours. Jupiter has at least 16 moons and one ring made up of millions of particles.

Jupiter seems to have a liquid metallic core surrounded by lighter elements. The large mass of Jupiter causes the temperature and pressure in the interior of the planet to be much greater than

Figure 29–6. The Great Red Spot on Jupiter is similar to hot spots on the earth. The colored bands of Jupiter, however, are unlike any earthly phenomenon.

those within the earth. Temperatures in the interior of Jupiter rise as high as 30,000°C. The intense pressure has changed most of the interior of the planet into a sea of liquid hydrogen. Electrical currents arising in this hot liquid may be the source of the enormous magnetic field around Jupiter.

Jupiter is made up mostly of gases; 95 percent is hydrogen and helium. Due to high temperature and great pressure, the surface of the planet is not solid but rather is a mixture of hot gases and liquids. This composition makes Jupiter similar to a small star. Astronomers theorize that when Jupiter was formed about 4.6 billion years ago, it did not have enough mass to enable nuclear fusion to begin. As a result, Jupiter never became a star.

The surface of Jupiter is covered with a thin layer of colored bands. The orange, gray, blue, and white bands spread out parallel to the equator. The colors suggest the presence of ammonia, methane, and water vapor. Scientists think that the rapid rotation of Jupiter causes these gases to swirl around the planet, forming the bands. The average temperature of the atmospheric layers is −160°C. The lower layers are about 20°C, closer to the temperature of the earth's atmosphere. Jupiter also has lightning and thunderstorms. The combination of lightning, ammonia, methane, and water vapor is thought to be necessary for the development of life. However, no evidence of life has been found on Jupiter.

To an observer on the earth, the most distinctive feature of Jupiter is the Great Red Spot. Astronomers think that heated material rising to the surface from the interior of the planet causes this feature. Similar hot spots exist on the earth. Hawaii, for example, rests on the best known of 30 or more such hot spots. The earth's hot spots produce eruptions of magma, whereas the hot spot of Jupiter produces eruptions of gas. How does the different composition of the two planets account for these different types of hot spots?

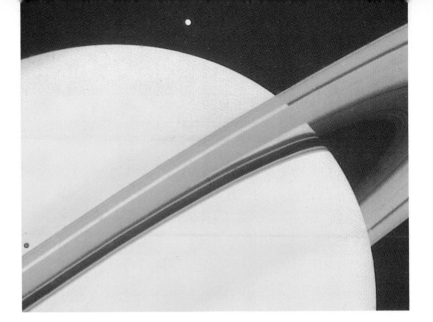

Figure 29–8. Saturn has at least 21 moons and a complex system of rings.

Saturn

The sixth planet from the sun, Saturn, is over half a billion kilometers farther from the sun than Jupiter is. Saturn is the second largest planet in the solar system. Its average temperature is −176°C. It has at least 21 moons and several rings.

Saturn spins very rapidly, rotating on its axis every 10 hours and 40 minutes. The rapid rotation of these two planets causes each to bulge out at its equator and to flatten at its poles. Like Jupiter, Saturn has bands of colored clouds that run parallel to the equator. Both planets also have small, rocky cores, interiors of liquid hydrogen, and dense atmospheres of hydrogen and helium gas.

Saturn differs from Jupiter in three ways. One difference is that Saturn is much less dense than Jupiter. In fact, Saturn is the least

Table 29–1: Planetary Data

	Average distance from sun (10^6 km)	Diameter (km)	Period of revolution (earth time)
Mercury	57.9	4,878	88 d
Venus	108.2	12,104	225 d
Earth	149.6	12,756	365.25 d
Mars	227.9	6,796	687 d
Jupiter	778.3	142,796	11.9 yr
Saturn	1,427	120,660	29.5 yr
Uranus	2,870	50,800	84 yr
Neptune	4,496	48,600	164.8 yr
Pluto	5,900	about 2,200	247 yr

* at least

Figure 29–9. The color variations in Saturn's rings, produced by using a computer-generated false-color photograph, indicate different compositions.

dense planet in the solar system, with a density less than that of water. Another difference is that the period of Saturn, which is 29.5 years, is nearly 20 years longer than that of Jupiter. Saturn also differs from Jupiter and from all the other planets in the solar system because it has a much more complex system of rings. These rings are discussed in Chapter 30.

Uranus

Uranus is the seventh planet from the sun and the third largest planet in the solar system. Uranus was discovered in 1781. It was the first planet to have been discovered since ancient times. Because Uranus is nearly 3 billion km from the sun, astronomers have great difficulty

Rate of rotation (earth time)	Number of satellites	Surface temperature (°C)	Dominant atmospheric gases
59 d	0	−200 to 500	He
243 d	0	480	CO_2
23 hr 56 min	1	14	N_2, O_2
24 hr 37 min	2	−130 to 20	CO_2
9 hr 50 min	16*	−160 to 400	H_2, He
10 hr 40 min	21*	−176	H_2, He
15 hr 30 min	15*	−215	H_2, He, CH_4
17 hr	2	−218	H_2, He, CH_4
6 d	1	−208 to −223	CH_4, NH_3

in trying to study the planet. Uranus has at least 15 moons and at least 11 small rings. Its period is 84 years.

The most distinctive feature of Uranus is its unusual rotation. Most planets, including the earth, rotate like tops as they revolve around the sun. Uranus, however, appears to rotate like a rolling ball. The axis of Uranus tilts so much that it almost lies in the plane of the orbit of the planet. Because of this odd tilt, the exact rotation rate of Uranus was not discovered until 1986. In that year, when *Voyager 2* passed by Uranus, astronomers were able to determine that Uranus rotated about once every 16 hours.

The greenish color of Uranus indicates that its atmosphere contains methane. Like the atmospheres of the other outer planets, the atmosphere of Uranus also contains hydrogen and helium. The average cloud-top temperature of Uranus is –215°C; however, astronomers believe that the temperature of the planet increases greatly below the clouds. Below the atmosphere of gas and liquid hydrogen, a region of water and ammonia may exist. Scientists also think that the center of Uranus may be a core of rock and metals, with a temperature of about 7,000°C.

Horizons

Technology: Uranus Revealed

I n January 1986 the 810-kg U.S. spacecraft *Voyager 2* streaked past Uranus, shown left. This spacecraft had been launched in 1977 to explore Jupiter and Saturn. However, a rare alignment of the outer planets allowed the *Voyager* team to expand the mission of the spacecraft. They used the gravitational force of Saturn as a slingshot to hurl *Voyager 2* toward Uranus and Neptune.

Voyager 2, shown far right, flew within 81,000 km of the methane clouds of Uranus. For six hours the spacecraft focused its many sophisticated instruments on Uranus and its moons and transmitted a wealth of information back to

the earth. Because a visit to Uranus was not part of the original *Voyager* mission, the project team had to anticipate and solve many problems. This had to be done at a distance of more than a billion and a half kilometers. The team did this by frequently reprogramming the two computers on *Voyager 2*.

One of the most critical problems facing the team was that of getting sharply focused pictures of the planet. Because

Neptune

The eighth planet from the sun, Neptune, is similar in size and mass to Uranus. Its period is 165 years, and it rotates about every 17 hours. Neptune has two moons and possibly one ring.

The existence of Neptune was predicted before it was discovered. After Uranus was discovered, astronomers noted variations from its expected orbit. They suspected that the gravity of an unknown planet beyond Uranus might be responsible. In the mid-1800's John Couch Adams, an English mathematician, and Urbain Leverrier, a French astromomer, independently calculated the position of such a planet. Three years later a German astronomer, Johann Galle, discovered a bluish-green disk where Leverrier had predicted it would be—4.5 billion km from the sun. Because of the color of Uranus, the astronomers named the planet *Neptune* after the Roman god of the sea. Data from the *Voyager* spacecraft indicate that Neptune is made up largely of ice.

The upper atmosphere of Neptune is composed of white clouds of frozen methane. These clouds appear as continually changing bands between the equator and the poles of Neptune. Distinctly

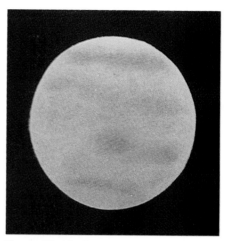

Figure 29–10. Clouds of frozen methane appear as continually changing bands between the equator and the poles of Neptune, as shown in this drawing of the planet.

Uranus is so far from the sun, there is very little light. The camera shutters would have to be left open four times longer than they were when photographing Saturn. As a result, the pictures would blur when the spacecraft jiggled, which it did when certain instruments were operating.

To solve this problem, the team programmed the attitude-control thrusters to fire briefly to counter the jiggling caused by the instruments.

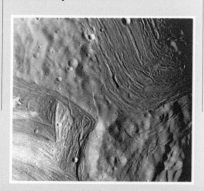

The thrusters kept the spacecraft steady, and the result was spectacular photos of Uranus and its many moons and small rings, shown left.

Miranda, one of the moons, shown near left, was photographed from 29,000 km as Voyager 2 sped by at 72,000 km/hr. As the photo was being taken, the thrusters had to be fired precisely so that the motions of the spacecraft and Miranda were counterbalanced.

What tendency of Voyager 2 did the gravitational force of Saturn have to overcome to hurl it toward Uranus and Neptune?

brighter clouds at the poles suggest to astronomers that Neptune may have a much more active weather system than Uranus has. These dynamic weather patterns probably mean that heat is circulating in the interior of the planet. The average temperature of Neptune is probably about –218°C.

Pluto

Pluto, the ninth planet from the sun, was discovered in 1930 after a long search. Clyde Tombaugh, an astronomer at Flagstaff Observatory in Arizona, had been looking for a Planet X. It was thought that Planet X could explain the wobbling in the orbits of Uranus and Neptune. What Tombaugh found was Pluto. But Pluto is far too small to cause the wobbling, so the search of space continues—for a tenth planet.

Investigation

The Solar System

You can compare the relative sizes of the planets in the solar system and the distances among them with a simple model.

Science Process Skills Focus: constructing models, comparing, naming and labeling

Materials
Planetary Data Table on pages 516–517; drawing compass; notebook paper; scissors; sheet of paper, about 1.5 m².

Procedure
1. Draw a table like the one shown. List the sun and 9 planets in the first column.
2. Study Table 29–1 on pages 516–517. Using the scale 1 cm equals 10,000 km, calculate the scale diameter and distance from the sun for each planet. Complete your table.
3. Using the information in your table and a drawing compass, draw a to-scale circle on a sheet of paper to represent each planet. Cut out each circle and label it with the name of the planet it represents.
4. Out of the large sheet of paper, cut a circle that approximates the scale diameter of the sun.
5. Compare the relative sizes of the sun and the

Table 29–I	Diameter (cm)	Distance from sun (cm)
Sun		
Mercury		
Venus		

planets you have constructed.
6. Study the scale distance from the sun to each planet. Write a planetary-model plan that uses the scale distances you have calculated.

Analysis and Conclusions
1. Where is the largest concentration of mass in the solar system?
2. Would it be practical to complete your model using the scale 1 cm equals 100,000 km for both distance from the sun and diameter?
3. Compare the distances among the inner planets with the distances among the outer planets.
4. Why are models of the solar system that are often displayed in classrooms inaccurate?

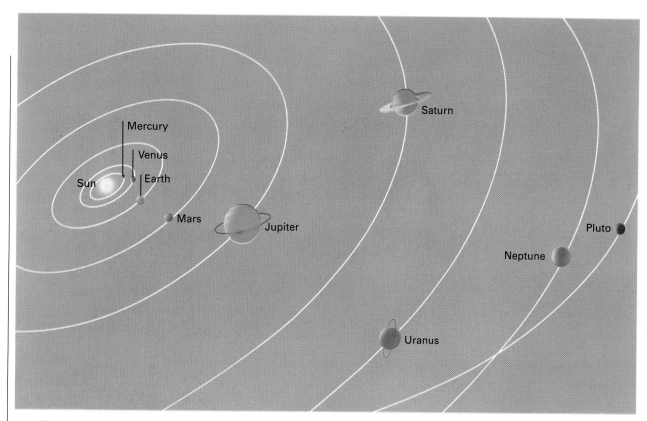

Figure 29–11. This diagram shows the relative positions of the bodies in the solar system. Note that the sun and planets are not drawn to scale.

Pluto is about 2,200 km in diameter, making it the smallest planet in the solar system. Pluto rotates on its axis every six days, and its period is 250 years. It orbits the sun in an unusually elongated ellipse, its distance from the sun varying from 4.4 billion km to 7.4 billion km and averaging 5.9 billion km. Pluto is sometimes inside the orbit of Neptune but is usually far beyond it. It is now inside Neptune's orbit and will remain there until about 1999.

Pluto appears to be made up mostly of frozen methane and ice, with an average temperature of between –208°C and –223°C. Infrared images of Pluto show that it has extensive methane ice caps and a very thin methane atmosphere. The planet has one moon and no rings. Charon, the moon of Pluto, was discovered in 1978. Charon is half as large as Pluto. The two bodies orbit so close together that they appear to be a double-planet system.

The small size, unusual orbit, and comparatively large moon of Pluto suggest to some astronomers that the planet was once a moon of Neptune. The gravity of a passing object may have pulled Pluto away from Neptune and into its own orbit.

Section 29.3 Review

1. List the names of the five outer planets.
2. What makes Jupiter similar to a star?
3. How dense is Saturn in comparison with other planets?
4. In what way is the axis of Uranus unusual?
5. Why is Pluto considered the ninth planet from the sun even though Neptune is sometimes farther from the sun?

29.4 Asteroids, Comets, and Meteoroids

In addition to the sun and the planets, the solar system includes millions of smaller bodies of matter. Some of these bodies are just bits of dust or ice floating in space. Others are as big as small moons. Many astronomers theorize that these small bodies were left over from the nebula from which the solar system formed.

Asteroids

The largest of the smaller bodies in the solar system are **asteroids,** or minor planets. Asteroids are fragments of rock that orbit the sun. Astronomers have plotted the orbits of more than 2,000 asteroids. As many as 500,000 asteroids may exist in the solar system. The orbits of most asteroids, like those of the planets, are ellipses. The largest known asteroid, Ceres, is about 1,000 km in diameter.

Most asteroids exist in a region between the orbits of Mars and Jupiter known as the **asteroid belt.** The asteroid belt begins about 100 million km beyond the orbit of Mars and stretches for about 150 million km toward the orbit of Jupiter.

Asteroids are usually classified into three types based on their composition. One type of asteroid is composed mostly of carbon materials. These materials give the asteroids a dark appearance. A second type is composed mostly of iron and nickel; these asteroids have a metallic appearance. The third and most common type of asteroid is made up mostly of silicate minerals. These asteroids look like ordinary earth rocks.

Figure 29–12. The illustration is an artist's impression of an asteroid shown in front of the sun. Asteroids are fragments of rock that orbit the sun.

Many astonomers think that asteroids in the asteroid belt are the remains of planetesimals that were not able to form a planet because of the strong gravitational force of Jupiter. The composition of the asteroids suggests that they formed from the same materials as the planets. For example, iron is common in the cores of the inner planets, and silicate minerals are common in their crusts. The high carbon content of some asteroids is similar to that of the outer planets. Scientists estimate that the mass of all asteroids, including Ceres, is not as great as that of the moon.

Some asteroids—the **Trojan asteroids** and the **earth-grazers**—orbit the sun, but they are not in the asteroid belt. The Trojan asteroids are concentrated in groups just ahead of and just behind the orbit of Jupiter. Earth-grazers have elongated orbits that sometimes bring them very close to the sun and the earth. Despite their name, earth-grazers infrequently collide with the earth.

Comets

Like earth-grazers, some **comets** orbit the sun in long ellipses. A comet is a body of rock, dust, methane, ammonia, and ice. The core, or *nucleus,* of a comet is made up of rock or metals and ice and is usually between 1 km and 100 km in diameter. A spherical cloud of gas and dust—the **coma**—surrounds the nucleus. The coma can extend up to 1 million km from the nucleus. The bright appearance of a comet may be due to sunlight reflected by the coma. The nucleus and the coma together form the head of the comet.

The most spectacular part of a comet is its tail. The tail of a comet is the gas and dust that streams out from the head. The tail forms as the solar wind—electrically charged particles released by

Figure 29–13. A comet, such as the Comet Halley shown here in 1986, consists of a nucleus, a coma, and a tail.

the sun—pushes gas and dust away from the head of the comet. Thus, regardless of the direction in which the comet is traveling, its tail always points away from the sun. Some of the larger comets have tails that are more than 80 million km long. Figure 29–13 shows the nucleus, coma, and tail of a typical comet.

Astronomers hypothesize that comets originate in the **Oort cloud,** which was named after the Dutch astronomer Jan H. Oort. The Oort cloud is a spherical cloud of dust and ice that may contain the nuclei of as many as 100 billion comets. The bodies within the cloud circle the sun at a speed of up to only about 140 m/sec, requiring a few million years to complete just one orbit. The cloud surrounds the solar system, beginning at a distance of one light-year from the sun and reaching halfway to the nearest star. The matter in the Oort cloud may have been left over from the formation of the solar system and captured later by the gravity of the sun. The gravity of a star passing near the solar system may cause a comet within the cloud to fall into a long elliptical orbit. The orbit stretches from the Oort cloud to the sun and back to the cloud. Because the Oort cloud is spherical, comets are not oriented in any particular direction and, when released, can orbit the sun at any angle.

Some comets, called **long-period comets,** have periods of several thousand or even several million years. Other comets, called **short-period comets,** have periods of up to 100 years. The gravity of one of the outer planets, such as Jupiter, can affect these comets as they orbit the sun. Halley's comet, which appears about every 76 years, is a short-period comet. It last appeared in 1986. Approximately when should Halley's comet appear next?

Meteoroids

In addition to asteroids and comets, smaller bits of rock or metal called **meteoroids** move throughout the solar system. Most meteoroids are less than 1 mm in diameter. Such small meteoroids are pieces of matter that become detached from passing comets. Larger meteoroids—more than 1 cm in diameter—are produced by collisions between asteroids.

A meteoroid that enters the earth's atmosphere is called a **meteor.** As the meteor passes through the atmosphere, friction heats the meteor and slows it down. Most meteors burn up in the atmosphere before reaching the earth's surface, producing a bright streak of light. You may have seen a meteor, often called a *shooting star,* streaking through the sky. Meteoroids sometimes vaporize very quickly in a brilliant flash of light called a **fireball.** Observers on the earth can sometimes hear a loud noise as a fireball disintegrates.

Large numbers of small meteoroids often enter the earth's atmosphere during a relatively short time. Once in the earth's atmosphere, the meteors burn up, creating a **meteor shower.** During the most spectacular of these showers, several meteors may be visible every minute. Meteor showers occur at the same time each year as the earth intersects the orbits of comets that have been broken up,

Figure 29–14. This meteorite was collected in 1979 near the United States McMurdo Base in Antarctica. Scientists estimate the 33.01 kg rock to be 1,300 million years old.

Figure 29–15. The straight vertical lines in this photograph are meteors. The lines appear thicker and brighter when meteors burn up in the earth's atmosphere. The curving lines are trails of stars created as the earth rotates during this long-exposure photograph.

leaving behind meteoroids. Astronomers estimate that about 1 million kg of matter from meteors falls to the earth each day.

Millions of meteors enter the earth's atmosphere each night. A few meteors do not burn up entirely in the atmosphere but fall to the earth. A meteor or any part of a meteor that is left after it hits the earth is called a **meteorite.** Most meteorites are small, with a mass of less than 1 kg. However, large meteorites occasionally strike the earth's surface with the force of an exploding bomb. These leave large craters. One of the best-known is Barringer Crater in Arizona, which was created by a meteor that struck about 20,000 years ago. The meteor, as it entered the atmosphere, was 50 meters in diameter and weighed 500,000 tons. It left a crater 1.5 km in diameter and 280 m deep and about 25 tons of fragments spread over the land.

Meteorites can be classified into three basic types: stony, iron, and stony-iron. **Stony meteorites** are similar in composition to the rocks found on the earth. Some stony meteorites contain carbon-bearing substances similar to the materials found in living organisms. Although most meteorites are stony, **iron meteorites** are easier to find. This is because stony meteorites look like ordinary earth rocks and often are not noticed. Iron meteorites have a distinctive metallic appearance. **Stony-iron meteorites** contain both iron and stone and are very rare.

Astronomers think that the oldest meteorites are about 100 million years older than the earth or its moon. Therefore, meteorites can provide information about the composition of the solar nebula that existed before the earth and its moon formed.

Section 29.4 Review

1. Between the orbits of what planets is the asteroid belt located?
2. In what direction does the tail of a comet point?
3. How is a meteor different from a meteorite?
4. In what way are the orbits of Trojan asteroids and earth-grazers unlike the orbits of other asteroids?

Chapter 29 Review

Key Concepts

- Geocentric models of the solar system, such as those developed by, Claudius Ptolemy and Nicolaus Copernicus, were replaced by the currently accepted heliocentric model. See page 507.

- Kepler's three laws describe the motion of the planets in their orbits around the sun. See page 507.

- Mercury and Venus are the two planets closest to the sun. See page 510.

- Both the earth and Mars have a history of geologic activity. See page 512.

- Jupiter and Saturn rotate rapidly. See page 514.

- Uranus is the third largest planet in the solar system. Because of its color, Neptune was named the Roman god of the sea. Pluto is the smallest planet in the solar system. See page 517.

- Most asteroids are found in the region known as the asteroid belt. Some comets orbit the sun in long periods; and others, in short periods. See page 522.

- Meteoroids move throughout the solar system. When they enter the earth's atmosphere, they are meteors; when they strike the earth's surface, they are meteorites. See page 524.

Key Terms

asteroid (522)	focus (508)	meteor shower (524)	stony meteorite (525)
asteroid belt (522)	geocentric (507)	meteorite (525)	stony-iron meteorite (525)
astronomical unit (508)	heliocentric (507)	meteoroid (524)	terrestrial planet (510)
coma (523)	inertia (509)	Oort cloud (524)	Trojan asteroid (523)
comet (523)	iron meteorite (525)	period (509)	
earth-grazer (523)	Jovian planet (514)	retrograde motion (507)	
ellipse (508)	long-period comet (524)	short-period comet (524)	
epicycle (507)	meteor (524)		
fireball (524)			

Review

On your paper write the letter of the term that best completes each of the following statements.

1. Ptolemy modified Aristotle's model of the universe to include (a) Oort clouds (b) retrograde motion (c) comets (d) shooting stars.

2. Copernicus's model of the solar system differed from Ptolemy's because it was (a) geocentric (b) lunocentric (c) based on observations made with a telescope (d) heliocentric.

3. Kepler's first law states that each planet orbits the sun in a path called (a) an ellipse (b) a circle (c) an epicycle (d) a period.

4. Kepler's law that describes how fast planets travel at different points in their orbits is called the law of (a) ellipses (b) equal speed (c) equal areas (d) periods.

5. The weak magnetic field around Mercury suggests (a) volcanic activity (b) a dense atmosphere (c) a core of iron (d) that it is located close to the sun.

6. The planet that rotates in a direction that is opposite the direction of the other planets is (a) Mercury (b) Venus (c) the earth (d) Mars.

7. The tilt of the axis of Mars is nearly the same

as that of (a) Mercury (b) Venus (c) the earth (d) Jupiter.
8. The planet that rotates faster than any other planet in the solar system is (a) the earth (b) Jupiter (c) Uranus (d) Pluto.
9. The most distinctive feature of Jupiter is its (a) Great Red Spot (b) rotation (c) rings (d) elongated orbit.
10. All of the outer planets in the solar system are large except (a) Saturn (b) Uranus (c) Neptune (d) Pluto.
11. Some astronomers think that Pluto was once a moon of Neptune because of its (a) age (b) period (c) unusual orbit (d) temperature.
12. The asteroid belt exists in a region between

the orbits of (a) Mercury and Venus (b) Venus and the earth (c) the earth and Mars (d) Mars and Jupiter.
13. The composition of asteroids suggests that they are (a) small moons (b) fragments of planetesimals (c) the nuclei of comets (d) environments that possibly can support life.
14. Most meteoroids are (a) small pieces of stars (b) pieces of matter left by passing comets (c) minor planets (d) fireballs.
15. Meteorites can provide information about (a) the composition of the solar nebula before the earth and its moon formed (b) the size of the earth (c) the end of the solar system (d) the size of the universe.

Application

On your paper write answers to the following questions.
1. Assume that an intelligent life form exists on Pluto—the planet with the longest period in the solar system. Would astronomers on Pluto be likely to propose a heliocentric model of the solar system. Explain your answer.
2. If you know the distance from the sun to a planet, what other information can you determine about the orbit of the planet? Explain your answer.
3. Suppose that a new planet is discovered. It has no rings or moons and has a surface pitted with impact craters. In what group of planets do you think this planet is located? Explain how you know.
4. What type of core do you predict the new planet mentioned in question 3 will have?

5. Some scientists think there is a tenth planet beyond Pluto. Do you predict this planet would be larger or smaller than Pluto? Explain your answer.
6. Imagine that when Jupiter formed, it was 100 times more massive than its actual mass. How might this fact have influenced the development of Jupiter?
7. The surfaces of some asteroids reflect only small amounts of light. Other asteroids reflect up to 40 percent of the light falling on them. Of what materials would each type of asteroid probably be composed?
8. Suppose you live in an unglaciated area and have found a chunk of rock that you suspect might be a stony meteorite. What data would help you verify your hypothesis?

Extension

1. Make a three-dimensional scale model of the solar system. Include all nine planets. Construct your model using coathangers, thread, and different sized polystyrene balls for the planets and sun.
2. Do research in your school or community library on the discovery and characteristics of Pluto. Collect evidence both supporting and opposing its classification as a planet. Present your findings to the class in the form of a

debate, first taking one side, then the other. Ask the class to vote on the issue.
3. Talk with the director of a local planetarium or natural history museum about meteorites that have been found in your area. If possible, go to look at any specimens that may be on display. Find out the age and composition of the meteorites and classify them according to the system discussed in the chapter. Report your findings to the class.

Chapter 30

Moons and Rings

On July 20, 1969, astronaut Neil Armstrong stepped onto the surface of the moon. This first visit to another body in the solar system marked a new era in history and provided an excellent opportunity for firsthand study. This chapter discusses the earth's moon and the moons and rings of the other planets.

Chapter Outline

Astronauts reached the moon for the first time on July 20, 1969.

30.1 The Earth's Moon

Section Objectives

- **List the five kinds of lunar surface features.**
- **Describe the interior of the moon.**
- **Summarize the four stages in the development of the moon.**

A body that orbits a larger body is called a *satellite*. In 1957 the Soviet Union launched *Sputnik 1*, the earth's first artificial satellite. In 1958 the United States launched its first satellite, *Explorer 1*. A natural satellite of any planet is called a *moon*. The earth's natural satellite and closest neighbor is referred to as *the moon*.

If you were to visit the moon, you would immediately notice that the force of gravity of the moon differs from that of the earth. The mass of an object determines its gravitational force. Since the moon has less mass than the earth does, its gravity is weaker. Its surface gravity is about one-sixth the surface gravity of the earth. As a result, a person who weighs 600 newtons (N) on the earth would weigh about 100 N on the moon. As you may recall from Chapter 2, on the earth's surface, one kilogram of mass weighs about 10 N.

Between 1969 and 1972, the United States sent six *Apollo* spacecraft to the moon, each with three astronauts. These astronauts found that the weak gravity affected the way they moved on the moon. Instead of walking, they bounced along in hops and leaps.

Gravity on the moon has never been strong enough to hold gases. Therefore the moon has no atmosphere and cannot support life. Because the moon has no atmosphere to act as insulation, the temperature variation is far greater than it is on the earth. The temperature ranges from 134°C during the day to −170°C at night. The dramatic temperature shifts are due in part to slow rotation. The moon completes a rotation on its axis once every 27.3 earth days. Thus each lunar day is equal to almost four earth weeks.

The Lunar Surface

The moon is close enough to the earth to be seen easily without a telescope. The next closest object to the earth is Venus. The distance between Venus and the earth is more than 100 times the distance between the moon and the earth.

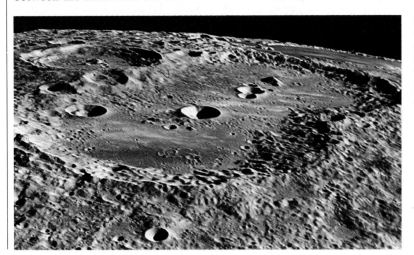

Figure 30–1. A clear but distant view of the moon is visible with unaided eyes from the earth. Modern space technology provides detailed views, such as this photograph taken from the Apollo 15 spacecraft in 1971.

Figure 30–2. The moon's rilles are long, deep channels that run through the maria (left). From a distance, the maria appear as dark areas (right). The light areas are rough lunar highlands.

An observer on the earth can see light and dark patches on the moon. The light areas are rough highlands, the sloping surfaces of which catch and reflect sunlight. The dark areas, called **maria** [(MAHR-ee-uh); sing., mare (mahr-AY)] are smooth and reflect little light. *Mare* is the Latin word for "sea." Early astronomers thought that the dark areas on the moon were bodies of water. Today astronomers know that the maria are dry plains of solidified lava. These lava plains are the remains of ancient volcanic eruptions on the surface of the moon.

Running through the maria are long, deep channels called **rilles.** A rille looks like a dry riverbed. Some rilles resemble narrow trenches and are as much as 240 km long. Yet there is no evidence that water has ever existed on the moon. The origin of these rilles remains a mystery.

Craters

The surface of the moon is covered with bowl-shaped depressions called **craters.** Astronomers think that most of these were formed about 4 billion years ago. At that time the moon, like other bodies in the solar system, was bombarded with matter left over from the formation of the solar system. As this matter struck the moon, it pushed surface material up and out, thus forming craters. The surface material splashed around the craters in streaks, or **rays.** Rays may be compared to the splashes you make in the water when you jump into a swimming pool.

Some craters have diameters of 100 km or more. The entire state of Rhode Island could easily fit into some of the craters. The largest craters are named for famous scholars and scientists, like *Tycho* and *Copernicus*. The lunar surface also has millions of smaller craters. Many of these craters overlap one another. Some smaller craters are located entirely within larger craters.

Geologists suggest that the earth once had many craters similar to the moon craters. The earth's craters have been almost completely eroded by the forces of wind and water. Because the moon lacks wind and water, its surface has changed little since it was formed.

Lunar Rocks

Meteorites of all sizes struck the moon. They crushed much of the rock on the surface into dust and small fragments. Today almost all of the lunar surface is covered by a layer of this dust and rock. The material in this layer is called *regolith*. The depth of the regolith layer varies from 1 m to 6 m. The number of meteorites that reached the surface of the moon was greater than the number that reached the earth's surface. Why was this?

Lunar rocks contain many of the same elements found in the earth's rocks but in very different amounts. Near the lunar surface, igneous rocks, like those in the earth's crust, are composed mainly of oxygen and silicon. Rocks from the lunar highlands are light-colored, coarse-grained rocks called *anorthosites*. These rocks are rich in calcium and aluminum. Rocks from the maria are dark-colored, fine-grained basalts, which contain large amounts of titanium, magnesium, and iron.

One type of rock found in both the maria and the highlands is *breccia*. Breccia contains fragments of other rocks that have melted together. Astronomers think that breccia formed when meteorites struck the moon. The force of the impact broke up rocks, and the heat from the impact melted the fragments together.

Lunar rocks do not contain some elements commonly found in the earth's rocks. For example, lunar surface rocks lack the elements with low melting points, such as sodium. These elements may have boiled off when the moon was still molten. Also, lunar rocks do not contain water.

The surface rocks on the moon are about as dense as those on the surface of the earth. Yet the overall density of the moon is only

Figure 30–3. Analysis indicates that the rock shown left is about 4.6 billion years old—the oldest rock found on the moon. The varied texture of the rock indicates that it had a complicated history. The rock shown right is the largest rock sample collected on the Apollo 15 mission. The rock, a basalt, measures 30 × 15 × 15 cm.

three fifths the density of the earth. The difference in overall density indicates that the interior of the moon is less dense than the interior of the earth.

The Interior of the Moon

Most of the information that astronomers have gathered about the interior of the moon comes from seismographs. These instruments were placed on the moon by the *Apollo* astronauts. Seismographs have recorded numerous weak quakes on the moon, which are similar to earthquakes. They have also revealed that, on the side facing the earth, the crust of the moon is 60 km thick. On the side away from the earth, it is up to 100 km thick. The surface on the far side is mountainous and has only a few small maria. The crust appears to consist of materials similar to that in the rocks in the highlands.

Beneath the crust of the moon is the mantle, which is made up of dense rock that is probably rich in silica, magnesium, and iron. The mantle reaches to a depth of about 1,000 km.

The lower portion of the mantle may be slightly molten. The moon may even have a small iron core. Scientists estimate that the

Horizons

Technology: Mining on the Moon

Ever since astronauts first landed on the moon in 1969, scientists have worked on plans for constructing and operating permanent lunar bases.

The biggest obstacle to building bases on the moon has been the expense. Shipping costs for construction materials and life-support systems needed to build and maintain lunar bases would literally be "out of this world." Therefore, scientists are developing ways of producing the materials they need on the moon itself. One way is mining the natural resources of the moon like the rock the scientist is examining at left.

One of the most promising moon-mining methods has been developed by Larry Haskin and David Lindstrom. Haskin and Lindstrom are researchers at the McDonnell Center for Space Sciences of Washington University in St. Louis, Missouri. Their method uses *electrolysis,* a process that changes the chemical composition of a substance by passing an electrical current through it. Using electrolysis, Haskin and Lindstrom have been able to extract iron and oxygen from rocks like those found on the moon.

The process they used to draw out the iron and oxygen included four simple steps. First, Haskin and Lindstrom

radius of the core is approximately 700 km. Scientists think that if any portion of the moon is liquid, it is probably the core. The moon is almost completely lacking in an overall magnetic field, although there are small areas of local magnetism.

Development of the Moon

The study of moon rocks and data gathered by astronauts have provided evidence to help astronomers understand the history of the moon. Astronomers generally support one of three hypotheses about the first stage of its origin. The first hypothesis suggests that the moon was once part of the earth. Early in the formation of the earth, a portion of the planet split off and became the moon. According to the second hypothesis, the earth and its moon were formed at the same time and from the same materials. The third view states that the moon was captured by the earth's gravity. This hypothesis suggests that the moon joined with the earth after having been formed elsewhere in the solar system, separate from the earth.

Scientists generally agree about the subsequent stages of the development of the moon. In the second stage of the development, the

combined chemicals to create silicate rock like the kind commonly found on the moon. The rock was then heated until it melted. Next, they put a positive and a negative electrode into the liquid rock. Finally, Haskin and Lindstrom sent an electrical current between the two electrodes, causing a change in the chemical composition of the liquid silicate. Iron separated out at one electrode, and oxygen separated out at the other electrode.

Haskin and Lindstrom think that solar energy could be used to melt the rock and to produce the electricity needed to produce iron. The iron could then be used to manufacture steel for building structures. The oxygen created through electrolysis could be used for human life support. An artist's conception of a lunar mining operation shown above depicts the production of liquid oxygen. Ilmenite, an oxygen-rich mineral is being mined from the lunar soil.

How would a decrease in the cost of space travel make mining on the moon more practical?

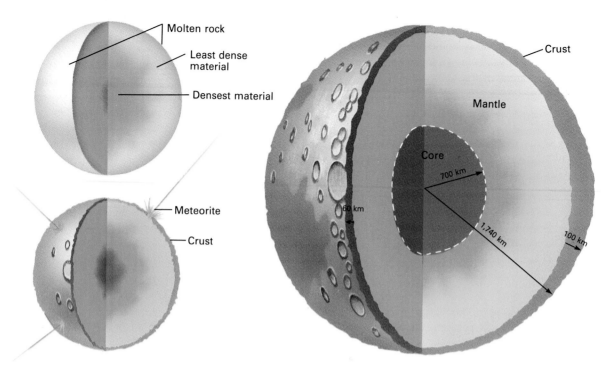

Figure 30–4. After the moon formed, it was covered by a layer of white-hot molten rock (top, left). As the moon cooled, a crust formed over the molten rock (bottom, left). By 3 billion years ago, the moon looked much as it does today (right).

surface of the moon was covered by an ocean of white-hot molten rock. Eventually the light and heavy minerals in the molten rock separated. The densest materials moved to the center and formed a small, partially molten core. The least dense materials formed an outer layer. Materials of intermediate density settled between the core and the outer layer, thus forming the mantle.

The third stage in the development of the moon began as the outer surface of the moon cooled. A thick, solid crust formed over the molten rock. Some of the meteorites that hit the moon broke through the crust. Molten rock from beneath the crust flowed onto the surface through these breaks. When the molten rock partially filled the craters on the surface, the smooth maria were formed.

Sometime before 3 billion years ago the number of loose particles in the solar system decreased. Therefore less material hit the lunar surface. This marked the beginning of the fourth stage of development of the moon. With fewer meteorites hitting the surface, virtually all geologic activity stopped. The moon today looks almost exactly as it did 3 billion years ago. Therefore the moon is a valuable source of information about the conditions that existed in the solar system long ago.

Section 30.1 Review

1. Describe the maria that cover most of the surface of the moon.
2. How thick is the crust of the moon on the side facing the earth?
3. During which stage of development did the maria form?
4. How would the surface of the moon differ today if meteorites had continued to hit it over the past 3 billion years?

30.2 Movements of the Moon

Section Objectives

- **Describe the orbit of the moon around the earth.**
- **Explain why eclipses occur.**

To observers on the earth, the moon appears to orbit the earth. To an observer in outer space, however, the earth and its moon appear to orbit each other. The earth and its moon together would appear to form a single system that orbits the sun. The center of this earth-moon system, not the center of the earth, follows a smooth orbit around the sun.

Half of the mass of the earth-moon system is on either side of the center, or balance point, as shown in Figure 30–5. The moon is not nearly as massive as the earth. Therefore the balance point of the earth-moon system is not halfway between the centers of the two bodies. Rather, at the balance point, part of the earth's mass is on the same side as the moon. This means that the balance point is located within the earth's interior, as shown in Figure 30–5.

Lunar Orbit and Rotation

The orbit of the moon around the earth forms an ellipse, not a circle. Therefore the distance between the earth and its moon varies. When the moon is farthest from the earth, it is at **apogee.** When the moon is closest to the earth, it is at **perigee.** The average distance of the moon from the earth is 384,000 km.

The rotation of the earth on its axis causes the moon to appear to rise and set. If you watch the moon rise on successive nights, you will note that it rises and sets approximately 50 minutes later each night. While the earth completes one rotation, the moon has moved 1/27 of the way around the earth. Therefore the difference in rising and setting time of the moon each day is equivalent to the time required to make 1/27 of a rotation. That time is about 50 minutes.

Figure 30–5. The balance point of the earth-moon system is in the interior of the earth. Only the balance point of the earth-moon system orbits the sun in a smooth ellipse.

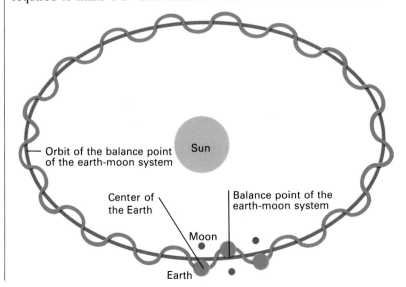

Orbit of the balance point of the earth-moon system

Sun

Center of the Earth

Balance point of the earth-moon system

Moon

Earth

Science Notebook

Effects of Tides

In Chapter 20 you read about how the moon and the sun cause tides on the earth. The gravitational attraction between the earth and the moon causes water levels to rise and fall along the shores of the ocean.

The tides also affect the moon. The earth's tides influence the speed at which the moon orbits the earth and the distance between the earth and the moon. The tides create a bulge of land and water on one side of the earth. As the earth rotates, this bulge moves eastward, slightly ahead of its expected position directly under the moon. As the bulge moves forward, the gravitational attraction between it and the moon pulls the moon along. The speed of the moon in its orbit around the earth increases. As the speed of the orbit of the moon increases, the moon moves slightly

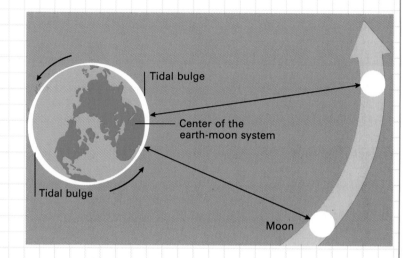

Tidal bulge

Center of the earth-moon system

Tidal bulge

Moon

farther from the earth.

The tides caused by the moon also affect the speed of the earth's rotation. As the water moves to form the tides, it creates friction against the ocean floor. This friction slows the earth's rotation. Consequently, each rotation of the earth takes slightly longer than the previous one. Astronomers predict that 1,000 years from now the length of each day will be 0.2 seconds longer than it is now.

At this rate, in about how many years will the day become one minute longer than it is at the present time?

In addition to revolving around the sun as part of the earth-moon system, the moon also spins on its axis. It spins very slowly, completing a rotation only once during each orbit around the earth. The moon makes one revolution around the earth in about 27.3 days. Because the rotation and the revolution of the moon take the same amount of time, observers on the earth always see the same side of the moon. The moon also rocks slightly. As a result, observers on the earth can see about 59 percent of the surface of the moon.

The rate at which the moon rotates is due to the effect of the earth's gravity. One side of the moon has more mass than the other. When the moon and the earth first formed, the gravitational pull of the earth attracted the more massive side of the moon. It is this more massive side of the moon that faces toward the earth as the moon revolves around the earth.

Eclipses

All the bodies moving around the sun, including the earth and its moon, cast long shadows into space. An **eclipse** occurs when one planetary body passes through the shadow of another. Shadows cast by the earth and the moon have two parts. In the inner, cone-shaped part of the shadow, the **umbra,** sunlight is completely blocked. In the outer part of the shadow, the **penumbra,** sunlight is only partially blocked.

Solar Eclipses

When the moon is between the earth and the sun, the shadow of the moon may fall upon the earth, causing a **solar eclipse.** People within the umbra see a *total solar eclipse*. That is, the sun's light is completely blocked by the moon. The umbra falls on the area of the earth that lies directly in line with the moon and the sun. Outside the umbra, but within the penumbra, people see a *partial solar eclipse*. The penumbra falls on the area immediately surrounding the umbra.

The umbra of the moon is too small to make a large shadow on the earth's surface. Consequently, a particular total eclipse of the sun covers only a small part of the earth and is seen by people in only that part of the earth. The earth's rotation causes the area under the shadow to move rapidly. Therefore a total solar eclipse never lasts more than seven minutes at any location.

At times the umbra of the moon is too short to reach the earth. If the moon is at or near apogee when it comes directly between the earth and the sun, its shadow does not reach the earth. If the umbra fails to reach the earth, a ring-shaped *annular eclipse* occurs. In an annular eclipse, the sun is not completely blocked out. Instead, a thin ring of sunlight is visible around the outer edge of the moon.

Figure 30–6. A total solar eclipse is a dramatic event (top). During a solar eclipse, the shadow of the moon falls on the earth (bottom).

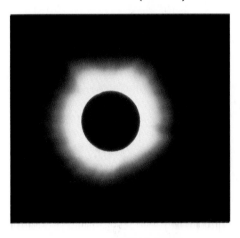

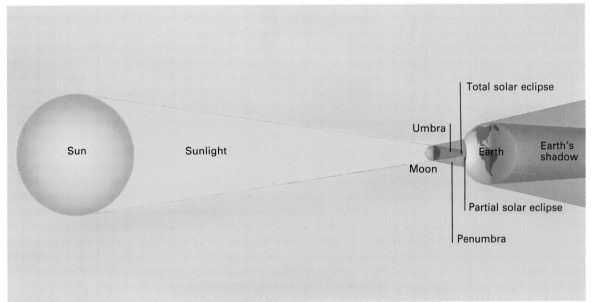

Total solar eclipse

Umbra

Sun

Sunlight

Earth

Earth's shadow

Moon

Partial solar eclipse

Penumbra

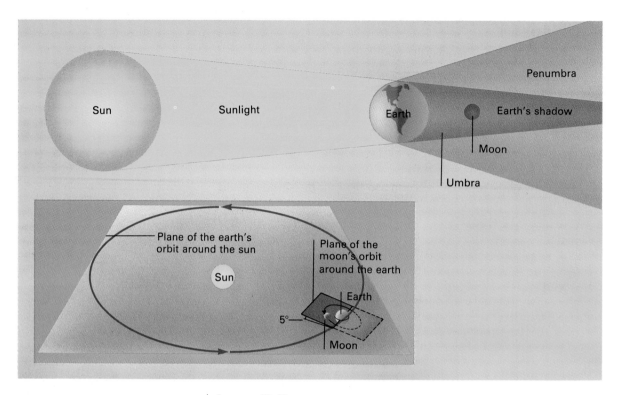

Lunar Eclipses

Solar eclipses occur when the moon is between the earth and the sun. As the moon orbits the earth, the earth's shadow can cause eclipses on the moon. A **lunar eclipse** occurs when the earth is positioned between the moon and the sun and the earth's shadow crosses the lighted half of the moon. To produce a *total lunar eclipse,* the moon must pass completely into the earth's umbra. A *partial lunar eclipse* occurs when only part of the moon passes through the earth's umbra. When the moon passes through the earth's penumbra only, a *penumbral eclipse* occurs. During a penumbral eclipse, the moon darkens slightly. Lunar eclipses may last several hours. Why do lunar eclipses last so much longer than solar eclipses?

Frequency of Eclipses

Eclipses of the moon and the sun occur almost equally often. However, lunar eclipses are seen by more people. An eclipse of the moon is visible everywhere on the dark side of the earth. A solar eclipse can be seen only by observers in the small path of the shadow of the moon as it moves across the earth's surface.

Solar and lunar eclipses do not occur in every lunar orbit. The orbit of the moon is not in the same plane as the orbit of the earth around the sun. The orbit of the moon is tilted at a 5° angle to the plane of the earth and the sun, as shown in Figure 30–7. Therefore the moon is usually above or below the plane of the earth's orbit. The moon crosses the plane of the earth's orbit only twice in each

Investigation

Eclipses

As the earth and moon revolve around the sun, each casts a shadow into space. An eclipse occurs when one planetary body passes through the shadow of another body. You can demonstrate how an eclipse occurs by using clay models of planetary bodies.

Science Process Skills Focus: constructing models, observing, predicting outcomes

Materials

modeling clay, sheet of notebook paper, penlight or small flashlight

Procedure

1. Make two clay balls, one about 4 cm in diameter and one about 1 cm in diameter.
2. Position the balls about 15 cm apart on the sheet of paper, as shown in the figure.
3. Turn off any nearby lights. Place the penlight approximately 15 cm in front of and almost on level with the large ball. Shine the light on the large ball. Sketch your model, noting the effect of the beam of light.
4. Reverse the positions of the two balls and repeat Step 3. Sketch your model, again noting the effect of the beam of light.

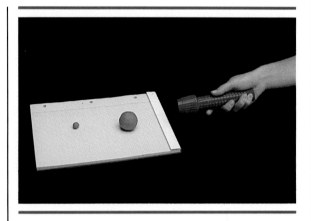

Analysis and Conclusions

1. What planetary body does the large clay ball represent? The small clay ball? The penlight?
2. As viewed from the earth, what event did you represent with your model in Step 3? As viewed from the moon?
3. As viewed from the earth, what event did you represent with your model in Step 4? As viewed from the moon?
4. What are the frequencies of the solar eclipse and the lunar eclipse in your model? How are the frequencies of these eclipses inaccurate? Explain your answer.

---

revolution around the earth. If this crossing occurs when the moon is between the earth and the sun, a solar eclipse will occur. If this crossing occurs when the earth is between the moon and the sun, a lunar eclipse may occur. Usually, however, no eclipse occurs when the moon crosses the plane of the earth's orbit.

Section 30.2 Review

1. Why do observers on the earth always see the same side of the moon?
2. Why does the moon rise and set about 50 minutes later each successive night?
3. What conditions must be present for a solar eclipse to occur?
4. Compare the alignment of the sun, the moon, and the earth during solar eclipses and lunar eclipses.

30.3 The Lunar Cycle

On some nights the moon shines brightly enough for you to read a book by its light. But moonlight is not produced by the moon. The moon reflects light from the sun. As the moon revolves around the earth, different parts of the lighted side of the moon face the earth. Thus, as shown in Figure 30–8, the shape of the visible portion of the moon varies. These varying shapes, lighted by reflected sunlight, are called **phases** of the moon.

Phases of the Moon

When the moon is between the sun and the earth, the side of the moon facing the earth is unlighted. At these times the moon is in the **new moon** phase. During the new moon phase there is no lighted area of the moon visible from the earth.

As the moon continues to move in its orbit around the earth, part of its lighted half becomes visible. When the size of the visible portion of the moon is increasing, the moon is **waxing.** When a sliver of the moon becomes visible from the earth, the moon enters the *waxing-crescent* phase.

When the moon has moved through one-quarter of its orbit after the new moon phase, the moon looks like a semicircle. Half of the lighted side of the moon is facing the earth. When a waxing moon becomes a semicircle, it enters the *first-quarter* phase. When the visible portion of the moon is larger than a semicircle and still increasing, the moon is in the *waxing-gibbous* phase.

The moon continues to wax until it appears as a full circle. At **full moon** the earth is between the sun and the moon, as shown in Figure 30–8. Consequently, the entire half of the moon reflecting the light of the sun is visible from the earth.

After the full moon phase, the portion of the moon visible from the earth decreases. The moon is then **waning.** After the full-moon phase, the moon enters the *waning-gibbous* phase. In this phase the visible portion of the moon is still larger than a semicircle, but this portion is decreasing in size. When the visible portion of the moon becomes a semicircle, the moon enters the *last-quarter* phase. When only a sliver of the moon is visible, the moon enters the *waning-crescent* phase.

After the waning-crescent phase, the moon again moves between the earth and the sun, as shown in Figure 30–8. The moon becomes a new moon, and the cycle of phases begins again.

In the crescent phases before and after a new moon, only a small part of the moon shines brightly. However, the rest of the moon is not completely dark. It shines dimly from sunlight that reflects off the earth's clouds and oceans and then reflects off the moon. Sunlight reflected off the earth is called **earthshine.**

Although the moon revolves around the earth in 27.3 days, a longer period of time is needed for it to go through a complete cycle

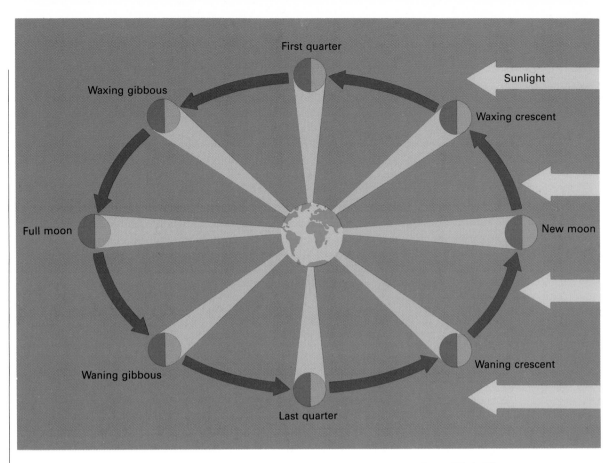

Figure 30–8. The moon goes through phases as the lighted portion of the moon visible from the earth changes during the lunar orbit.

of phases. The period from one new moon to the next is 29.5 days. The cause of this 2.2-day difference is the orbiting of the earth and the moon around the sun. In the 27.3 days in which the moon orbits the earth, the two bodies move slightly farther along their orbit around the sun. Therefore the moon must go a little farther to get directly between the earth and the sun. This is the position of the moon in the new moon phase. About 2.2 days are needed for the moon to travel this extra distance.

The Calendar

For a long time, people were able to measure the passage of time by keeping track of the changing phases of the moon. Eventually many societies developed systems of measuring the passage of time called **calendars.**

The three basic units of most calendars—day, month, and year—are determined by the movements of the earth and the moon. A **day** is the time required for the earth to make one rotation on its axis, about 23.95 hours. A **month** is the time required for the moon to go through one cycle of phases as it orbits the earth, about 29.5 days. A **solar year** is the time required for the earth to make one orbit around the sun, about 365.25 days.

The different lengths of time that are needed for the movements of the earth and the moon cause problems in creating any calendar. First, the earth does not make a whole number of rotations on its axis in the time that it makes one orbit around the sun. A calendar year of 365 days is too short, while a calendar year of 366 days is too long. Therefore the number of days in a calendar year must vary. The variation is to compensate for the extra one-quarter day required for the earth to orbit the sun.

The second problem is that the earth does not make a whole number of rotations on its axis in the time the moon orbits the earth. A month of 29 days is too short, but a month of 30 days is too long. Thus the number of days in a month must vary in order to account for the extra one-half day in a cycle of the phases of the moon.

Third, the moon does not make a whole number of orbits around the earth in the time that the earth orbits the sun. Therefore the number of months in every year cannot exactly correspond to actual movements of the earth and its moon. To remain accurate, a calendar must account for these differences.

The Julian Calendar

In 738 B.C. the leader Romulus supposedly introduced the first calendar used in Rome. It consisted of a year of 304 days divided into 10 months. Although later rulers made various modifications in the calendar, it was about 3 months behind the seasons by the time of Julius Caesar. For example, the winter solstice occurred in September. In 46 B.C. Caesar revised the calendar into what is now called the **Julian calendar.** Caesar divided the year into 12 months. Eleven of those months had 30 or 31 days. The other month, February, had 29 days. Later, one day was moved from February to August. This left February with only 28 days. Every fourth year was a **leap year,** a year with an extra day in it. During leap years February had 29 days. Having a leap year was necessary to make the average length of a year 365.25 days.

By the Julian calendar, which was used for over 1,500 years, the year was 365.25 days long. This, however, was about 11 minutes longer than the actual solar year. Each year the difference between the calendar and the solar year increased.

The Gregorian Calendar

By 1580 the calendar was about ten days ahead of the seasons. To correct this discrepancy, Pope Gregory XIII ordered that ten days be dropped from the month of October in 1582. This made up for the error that had built up over previous centuries.

To correct the calendar for the future, the pope ordered a revision of the Julian calendar. The revised system, called the **Gregorian calendar,** is the calendar currently used in most of the world. Compared with the Julian calendar, the Gregorian calendar has three fewer days every 400 years. Leap years normally come in years divisible by 4. However, years ending in *00* that are not divisible by 400 are not leap years. For example, 1700, 1800, and 1900 are not divisible by 400, so none of these years were leap years. The year 1600 ends in *00* and is divisible by 400. Therefore, 1600 was a leap year. Will the year 2000 be a leap year?

Figure 30–9. The Aztecs, who lived in what is now Mexico, used a stone calendar, such as the one shown here, to measure the passage of time.

| January | | | | | | |
S	M	T	W	T	F	S
1	2	3	4	5	6	7
8	9	10	11	12	13	14
15	16	17	18	19	20	21
22	23	24	25	26	27	28
29	30	31				

| February | | | | | | |
S	M	T	W	T	F	S
			1	2	3	4
5	6	7	8	9	10	11
12	13	14	15	16	17	18
19	20	21	22	23	24	25
26	27	28	29	30		

| March | | | | | | |
S	M	T	W	T	F	S
					1	2
3	4	5	6	7	8	9
10	11	12	13	14	15	16
17	18	19	20	21	22	23
24	25	26	27	28	29	30

| April | | | | | | |
S	M	T	W	T	F	S
1	2	3	4	5	6	7
8	9	10	11	12	13	14
15	16	17	18	19	20	21
22	23	24	25	26	27	28
29	30	31				

| May | | | | | | |
S	M	T	W	T	F	S
			1	2	3	4
5	6	7	8	9	10	11
12	13	14	15	16	17	18
19	20	21	22	23	24	25
26	27	28	29	30		

| June | | | | | | |
S	M	T	W	T	F	S
					1	2
3	4	5	6	7	8	9
10	11	12	13	14	15	16
17	18	19	20	21	22	23
24	25	26	27	28	29	30	W

| July | | | | | | |
S	M	T	W	T	F	S
1	2	3	4	5	6	7
8	9	10	11	12	13	14
15	16	17	18	19	20	21
22	23	24	25	26	27	28
29	30	31				

| August | | | | | | |
S	M	T	W	T	F	S
			1	2	3	4
5	6	7	8	9	10	11
12	13	14	15	16	17	18
19	20	21	22	23	24	25
26	27	28	29	30		

| September | | | | | | |
S	M	T	W	T	F	S
					1	2
3	4	5	6	7	8	9
10	11	12	13	14	15	16
17	18	19	20	21	22	23
24	25	26	27	28	29	30

| October | | | | | | |
S	M	T	W	T	F	S
1	2	3	4	5	6	7
8	9	10	11	12	13	14
15	16	17	18	19	20	21
22	23	24	25	26	27	28
29	30	31				

| November | | | | | | |
S	M	T	W	T	F	S
			1	2	3	4
5	6	7	8	9	10	11
12	13	14	15	16	17	18
19	20	21	22	23	24	25
26	27	28	29	30		

| December | | | | | | |
S	M	T	W	T	F	S
					1	2
3	4	5	6	7	8	9
10	11	12	13	14	15	16
17	18	19	20	21	22	23
24	25	26	27	28	29	30	W

Compared with previous calendars, the Gregorian calendar is so accurate that the calendar year is only 26 seconds longer than the solar year. As a result, the current Gregorian calendar accumulates an error of less than one day over a period of 3,000 years.

Figure 30–10. In the World calendar, a world day follows December 30. Every four years an additional world day follows June 30.

Current Calendar Reform

Many reforms have been proposed to simplify the current calendar. Some of these recommend that all months and years begin on the same day of the week. They also try to give all months nearly the same number of days.

One proposed calendar is the called the *Thirteen-Month calendar*. Each of the thirteen months would contain four weeks. A new month, called *Sol,* would come between June and July. One day, called a *year day,* belonging to no week or month, would fall at the end of each year. Every four years, a leap-year day would be added at the end of Sol.

Another proposal is the World calendar. This calendar would have 12 months of either 30 or 31 days. A world day would be placed at the end of each year. A leap-year day would be added at the end of June every four years.

Section 30.3 Review

1. Describe the relative locations of the sun, the earth, and the moon during a new moon.
2. To an observer on the earth, how does the appearance of the moon change when it is waxing?
3. List three problems in devising any calendar based on the movements of the earth and the moon.
4. How many fewer days are in the Gregorian calendar than in the Julian calendar?
5. If the earth were to orbit the sun in 365.20 instead of 365.25 days, how often would a leap year be necessary?

30.4 Satellites of Other Planets

Until the 1600's astronomers thought that the earth was the only planet with a moon. In 1610, however, Galileo discovered moons orbiting Jupiter. He also noticed what later were discovered to be the rings of Saturn. Since the time of Galileo, astronomers have found that all planets except Mercury and Venus have moons. In addition, Saturn, Jupiter, Uranus, and possibly Neptune have rings.

Moons of Mars

Mars has two moons, Phobos and Deimos. They revolve around Mars quite rapidly. Phobos completes its orbit in 7 hours and 40 minutes. Deimos completes its orbit in about 30 hours. The two moons of Mars may be the remains of a single moon that collided with a large meteoroid.

The moons of Mars differ in shape and in size from the earth's moon. Phobos and Deimos are irregularly shaped chunks of rock, whereas the earth's moon is spherical. Phobos is only 27 km across at its longest place and 19 km at its shortest. Deimos is about 15 km across at its longest place and 11 km at its shortest. Either moon could fit within a medium-size crater on the earth's moon.

The surfaces of Phobos and Deimos are dark like the maria on the earth's moon. Both of these moons have many craters. Phobos has one crater that is 8 km wide. This crater covers a large portion of the entire surface of Phobos. The large number of craters suggests that these moons have been hit by many meteorites and asteroids and are therefore fairly old.

Figure 30–11. Deimos, one of the two moons of Mars, is small and irregularly shaped.

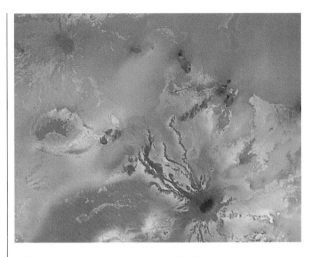

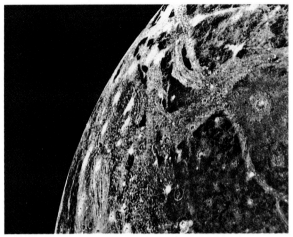

Ring and Moons of Jupiter

Jupiter has both a ring and many moons. The ring is about 6,000 km wide and less than 30 km thick. It is made up of very small particles of dark rock. Jupiter also has at least 16 moons. Most of the moons of Jupiter are very small. All except 4 moons are less than 200 km in diameter. The 4 largest satellites of Jupiter are called the **Galilean moons,** because they were first seen by Galileo in 1610.

Io, the Galilean moon closest to Jupiter, is about the size of the earth's moon. It is the only satellite in the solar system known to have active volcanoes. Io has at least ten active volcanoes. In addition, Io has lava flows, mountain ranges, and *calderas,* or large craters, similar to those on the earth.

The volcanoes on Io erupt with rocks and gases containing sulfur compounds. Thus astronomers infer that the atmosphere of Io is composed of sulfur dioxide. These compounds and the activity of the volcanoes make Io one of the most colorful moons in the solar system. Its colors include brilliant yellow, orange, and red. The surface of Io is probably covered with layers of sulfur and frozen sulfur dioxide. The interior of Io may still be molten rock.

Europa is the second closest Galilean moon to Jupiter. It is also about the size of the earth's moon, but it is less dense. Astronomers think that Europa has a rock core covered with a crust of ice about 100 km thick. An ocean of water may exist under this blanket of ice. If so, simple forms of life, similar to those in Antarctica, could exist there. Astronomers, however, have no evidence that life actually does exist on Europa.

Ganymede is the third Galilean moon from Jupiter. It is the largest moon in the solar system, larger than the planet Mercury. However, the density of Ganymede is very low. This moon is probably composed mostly of ice and has a rock core. Ganymede has dark, crater-filled areas, but it also has lighter areas. These light areas show marks that seem to be long ridges and valleys.

Callisto is the farthest of the four Galilean moons from Jupiter. Callisto is similar to Ganymede in size, density, and composition.

Figure 30–12. The four Galilean moons of Jupiter are Io, Europa, Ganymede, and Callisto. They are the largest moons of Jupiter. Io, shown left, is the only satellite in the solar system known to have active volcanoes. Ganymede, shown right, is the largest moon in the solar system.

However, it has a much rougher surface than Ganymede does. Callisto may be the most heavily cratered body in the solar system. What does this abundance of craters suggest about the age of the surface of Callisto?

Rings and Moons of Saturn

The rings of Saturn are larger and brighter than those of Jupiter or Uranus. Each of the several rings circling Saturn is divided into hundreds of smaller ringlets. The ringlets are composed of billions of pieces of ice. The pieces range in size from particles the size of dust to chunks the size of a house. Each piece follows its own individual orbit around Saturn. Altogether, the system of rings is about 67,000 km wide and 5 km thick.

Astronomers think that the rings of Saturn formed in one of two ways. One hypothesis is that the rings formed when a body orbiting Saturn was torn apart by the gravity of the planet. A second hypothesis is that the material in the ring was unable to condense into a moon while the other moons around Saturn were forming. These two hypotheses are similar to the hypotheses about the origins of the asteroid belt. In each case astronomers are unsure whether the bits of matter in the rings are the fragments of a destroyed body or simply loose material.

In addition to the billions of pieces of material in the rings, Saturn has at least 21 moons. Most of them are small, icy bodies with many craters. However, 5 of the moons of Saturn are fairly large. One, called *Titan,* has a diameter of more than 5,000 km. Unlike most of the moons of Saturn, Titan has a thick atmosphere that is composed mainly of nitrogen. The atmosphere is so thick that it conceals the surface of the moon.

Figure 30–13. In addition to its rings, Saturn has 11 moons. Titan is Saturn's largest moon and Saturn's only moon with substantial atmosphere. The moons are shown in a mosaic taken from several photographs.

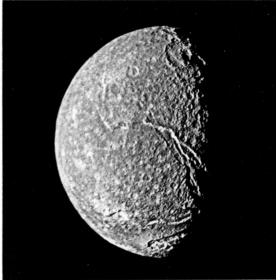

Satellites of Uranus, Neptune, and Pluto

Uranus has at least 15 small moons. The 4 largest, Oberon, Titania, Umbriel, and Ariel, were known by the mid-1800's. A fifth, Miranda, was found in 1948. The other 10 moons have been detected since 1985 by the U.S. space probe *Voyager 2*. Uranus also has at least 11 small rings. These rings are composed of small pieces of black rock. The inner rings are very narrow, varying from 0.5 km to 12 km wide. The outermost ring is the largest. It is between 10 km and 100 km wide.

Figure 30–14. Oberon (left) and Titania (right) are 2 of the 4 larger moons that orbit Uranus along with 11 smaller moons.

Neptune has two moons, Triton and Nereid. Triton, an icy moon, is almost as large as Titan, the moon of Saturn. Triton is unusual because it travels from east to west as it revolves around Neptune. Most planets and moons orbit from west to east. Some astronomers think that Triton, because of its unusual orbit, is not an original moon of Neptune. Triton may have been captured by the gravity of Neptune after forming elsewhere. The diameter of Triton, 4,800 km, is about 20 times larger than that of Nereid. Nereid orbits Neptune in the same direction that Neptune rotates.

Pluto, which is the smallest planet, has one moon. The moon, named *Charon*, was discovered in 1978. Charon is about half as large as Pluto and completes its orbit around Pluto in 6.4 days, the same length as a day on Pluto. Consequently, Charon is always above the same spot on Pluto.

Section 30.4 Review

1. Describe the size and shape of Phobos and Deimos.
2. List the four moons of Jupiter noticed by Galileo.
3. Which three planets have rings?
4. How does the number of craters on Phobos and Deimos suggest that these moons are very old?

Reviewing Chapter 30

Key Concepts

- The lunar surface features have changed little since they formed. See page 531.

- Astronomers have explored the structure of the interior of the moon using instruments that measure seismic waves. See page 532.

- After the moon formed, it separated into layers and cooled. Eventually it became geologically inactive. See page 534.

- The moon spins on its axis once during each orbit of the earth. See page 535.

- Eclipses occur when one planetary body passes through the shadow of another. See page 537.

- As the moon orbits the earth, the size of the lighted portion facing the earth increases and decreases. See page 540.

- Calendars are based on the rotation of the earth and the revolutions of the earth around the sun and of the moon around the earth. See page 541.

- Phobos and Deimos, the moons of Mars, differ in shape and size from the earth's moon. See page 544.

- The Galilean moons are the four largest satellites of Jupiter. Saturn, Uranus, and Jupiter all have rings. See page 545.

Key Terms

apogee (535)	Galilean moon (545)	month (541)	solar eclipse (537)
calendar (541)	Gregorian calendar	new moon (540)	solar year (541)
crater (530)	(542)	penumbra (537)	umbra (537)
day (541)	Julian calendar (542)	perigee (535)	waning (540)
earthshine (540)	leap year (542)	phase (540)	waxing (540)
eclipse (537)	lunar eclipse (538)	ray (530)	
full moon (540)	maria (530)	rille (530)	

Review

On your paper write the letter of the term that best completes each of the following statements.

1. Dark areas on the moon that are smooth and reflect little light are called (a) rilles (b) maria (c) rays (d) breccia.

2. Craters were probably formed (a) when fragments of matter struck the moon (b) by wind and rain (c) because the rilles went dry (d) because heat caused rock fragments to melt together.

3. Most of the information astronomers have gathered about the interior of the moon has come from (a) telescopes (b) satellites (c) spectrographs (d) seismographs.

4. Soon after the moon formed, it was covered with (a) water (b) anorthosites (c) frozen hydrogen (d) molten rock.

5. In the fourth stage in the development of the moon (a) the densest material sank to the core (b) the crust began to break (c) the earth's gravity captured the moon (d) the number of meteorites hitting the moon decreased.

6. The moon is closest to the earth at (a) new moon (b) full moon (c) perigee (d) apogee.

7. During each orbit around the earth, the moon spins on its axis (a) 1 time (b) about 27 times (c) about 29 times (d) 365 times.

8. In a lunar eclipse, the moon (a) casts a shadow on the earth (b) is in the earth's shadow (c) is between the earth and the sun (d) blocks part of the sun from view.

9. When the size of the visible portion of the moon is decreasing, the moon is (a) full (b) annular (c) waxing (d) waning.
10. In the crescent phases, the entire moon shines dimly because of (a) light produced by the earth (b) sunlight reflected off the earth (c) hydrogen fusion in the core of the moon (d) energy produced by the rotation of the moon.
11. The number of days in a year must vary because (a) the earth orbits the sun in 365.25 days (b) the moon orbits the earth in 27.3 days (c) the earth spins on its axis (d) the moon spins so slowly.
12. A year with an extra day in it is called a (a) solar year (b) lunar year (c) leap year (d) Julian year.
13. The two moons of Mars are (a) Io and Europa (b) Titan and Charon (c) Phobos and Deimos (d) Triton and Nereid.
14. Compared with the other moons of Jupiter, the four Galilean moons are (a) larger (b) farther from Jupiter (c) denser (d) younger.
15. The rings of Saturn are composed of (a) regolith (b) small pieces of black rock (c) several hundred small moons (d) billions of pieces of ice.

Application

On your paper, write answers to the following questions.
1. How would the craters on the moon be different today if the moon had developed an atmosphere that had wind and contained water?
2. Compare the structure of the interior of the moon with that of the earth. How are they similar?
3. If meteorites had stopped hitting the moon before the outer surface of the moon cooled, why would the maria not have developed?
4. Suppose that the moon spun twice on its axis during each orbit around the earth. How would study of the moon from the earth be easier?
5. Venus does not cause a solar eclipse even though it passes between the earth and the sun. Instead, it appears as a black dot moving across the sun. Explain why this happen.
6. Would a satellite orbiting the earth go through phases like those of the moon? Explain your answer.
7. Imagine a planet that orbits the sun in 100 days and has one moon. The moon goes through a complete set of phases in 20 days. On this planet, how many months would one year have?
8. What characteristic of Phobos and Deimos suggests that they have no highlands such as those found on the earth's moon?
9. If it is found that Ganymede is very dense, what might be concluded about its core?
10. Suppose that astronomers noticed that the moons of the earth, of Mars, and of Pluto were beginning to break into pieces. What would this suggest about the origin of the rings of Saturn, Jupiter, and Uranus?

Extension

1. Draw diagrams that illustrate each of the three hypotheses about how the moon formed.
2. Do research on how one of these ancient or traditional cultures explained solar eclipses: Chinese, Persian, Greek, Roman, Mayan, Sioux, aboriginal, and Tahitian. Present your findings to the class.
3. Imagine that astronomers have discovered a planet that spins on its axis in 20 hours and orbits the sun in 245 days and 4 hours. The imaginary planet has one moon that goes through a complete set of phases in 24 days and 10 hours. Design a calendar for this planet. Indicate how many months are in a year, how many days are in each month, and how often leap years occur.
4. Write a paper about one of the moons of Jupiter, Saturn, Uranus, or Neptune. In your paper describe the latest research about the features of that moon.

Reference

Contents

Metric System and SI Units

The Système Internationale, or SI, has been the international standard for measurement since 1960. SI is a decimal system. The ease of converting among units in the SI system is a great advantage over the English system. For example, to convert kilometers to meters, you can multiply by 1,000 or simply move the decimal point three places to the right: 2.8 km equals 2,800 m. In contrast, to convert miles to yards in the English system, you must multiply the number of miles by 1760: 2.8 miles equals 4,928 yards.

SI Units

Length   SI-English Equivalents

kilometer (km)	= 1,000 m	1 km = 0.621 mile (mi)	1 mi = 1.61 km
meter (m)	= 100 cm	1 m = 3.28 feet (ft)	1 ft = 0.305 m
centimeter (cm)	= 0.01 m	1 cm = 0.394 inch (in)	1 in = 2.54 cm
millimeter (mm)	= 0.001 m	1 mm = 0.039 inch	
micrometer (μm)	= 0.000001 m		
nanometer (nm)	= 0.000000001 m		

Area

square km (km^2)	= 100 hectare	1 km^2 = 0.3861 mi^2	1 mi^2 = 2.590 km^2
hectare (ha)	= 10,000 m^2	1 ha = 2.471 acre (a)	1 a = 0.4047 ha
square meter (m^2)	= 10,000 cm^2	1 m^2 = 1.1960 yd^2	1 yd^2 = 0.8361 m^2
square centimeter (cm^2)	= 100 mm^2	1 cm^2 = 0.155 in^2	1 in^2 = 6.4516cm^2

Mass

kilogram (kg)	= 1,000 g	1 kg = 2.205 pound (1b)	1 lb = 0.4536 kg
gram (g)	= 1,000 mg	1 g = 0.0353 ounce (oz)	1 oz = 28.35 g
milligram (mg)	= 0.001 g		
microgram (μg)	= 0.000001 g		

Volume of Solids

1 cubic meter (m^3)	= 1,000,000 cm^3	1 m^3 = 1.3080 yd^3	1 yd^3 = 0.7646 m^3
1 cubic centimeter (cm^3)	= 1,000 mm^3	= 35.315 ft^3	1 ft^3 = 0.0283 m^3
		1 cm^3 = 0.0610 in^3	1 in^3 = 16.387 cm^3

Volume of Liquids

kiloliter (kL)	= 1,000 L	1 kL = 264.17 gallon (gal)	1 gal = 3.785 L
liter (L)	= 1,000 mL	1 L = 1.06 quart (qt)	1 qt = 0.94 L
milliter (mL)	= 0.001 L	1 mL = 0.034 fluid ounce (fl oz)	1 pint (pt) = 0.47 L
microliter (μl)	= 0.000001 L		1 fl oz = 29.57 mL

Temperature

Celsius (C) = 5/9 (°F-32)
Fahrenheit (°F) = 9/5°C + 32

0°C = 32°F = Freezing point of water
100°C = 212°F = Boiling point of water

Guide to Common Minerals

Mineral	Chemical Formula	Color	Luster
Apatite	$Ca_5(PO_4)_3F$	Green	Glassy
Biotite	Complex substance containing Fe, Mg, Si, O, and other elements	Black, brown, dark green	Pearly, glassy
Chalcopyrite	$CuFeS_2$	Brass, yellow	Metallic
Copper	Cu	Copper red to black	Metallic
Corundum	Al_2O_3	Usually brown	Brilliant to glassy
Dolomite	$CaMg(CO_3)_2$	Pink, white, gray, green, brown, black	Glassy or pearly
Feldspar	$(K, Na, Ca)(AlSi_3O_8)$	Colorless, white, various colors	Glassy
Fluorite	CaF_2	Light green, yellow bluish green, other colors	Glassy
Galena	PbS	Lead gray	Metallic
Graphite	C	Black to gray	Metallic
Hematite	Fe_2O_3	Reddish brown to black	Metallic
Hornblende	Complex substance containing Fe, Mg, Si, O, and other elements	Dark green, brown, black	Glassy, silky
Magnetite	Fe_3O_4	Iron black	Metallic
Pyrite	FeS_2	Brass, yellow	Metallic
Quartz	SiO_2	Colorless, white, any color when not pure	Glassy, greasy
Serpentine	$Mg_3Si_2O_5(OH)_4$	Green	Glassy, waxy, silky
Sphalerite	ZnS	Brown to black	Resinous
Sulfur	S	Yellow	Resinous
Talc	$Mg_3(OH)_2Si_4O_{10}$	Gray, white, greenish	Pearly to greasy

Mineral	Streak	Hardness	Specific Gravity	Cleavage
Apatite	White	5	3.15–3.2	Hexagonal
Biotite	White to light brown	2.5–3	2.8–3.2	Thin sheets
Chalcopyrite	Greenish black	3.5–4	4.1–4.3	Uneven
Copper	Copper red	2.5–3	8.5–9	Fracture
Corundum	White	9	4.02	Hexagonal
Dolomite	Colorless	3.5–4	2.85	Conchoidal fracture
Feldspar	Colorless, white	6	2.55–2.75	In two planes at or near 90°
Fluorite	Colorless	4	3.18	Octahedral
Galena	Lead gray	2.5	7.4–7.6	Cubic
Graphite	Black	1–2	2.3	Basal cleavage (scales)
Hematite	Light to dark red	5.5–6.5	5.26	Uneven
Hornblende	Gray to white	5–6	3.2	Long prism
Magnetite	Black	5–6	5.18	Partly octahedral
Pyrite	Greenish, brownish, black	6–6.5	5.02	Uneven
Quartz	Colorless, white	7.5–8	2.65	Hexagonal
Serpentine	White	2.5	2.2–2.65	Parallel fibers
Sphalerite	White, yellow, brown	3.5–4	3.9–4.1	Dodecahedral
Sulfur	White	1.5–2.5	2.07	Conchoidal fracture
Talc	White	1	2.7–2.8	Uneven fracture

Star Maps

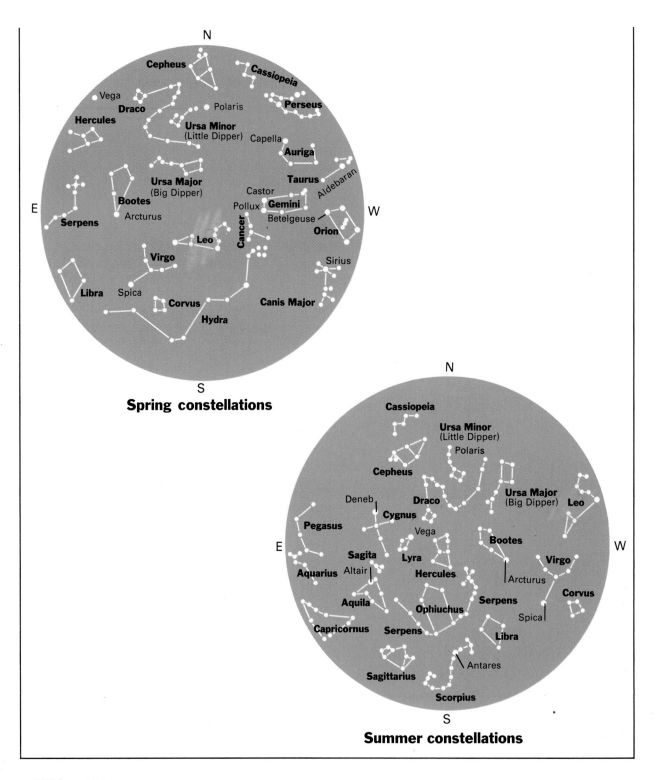

Spring constellations

Summer constellations

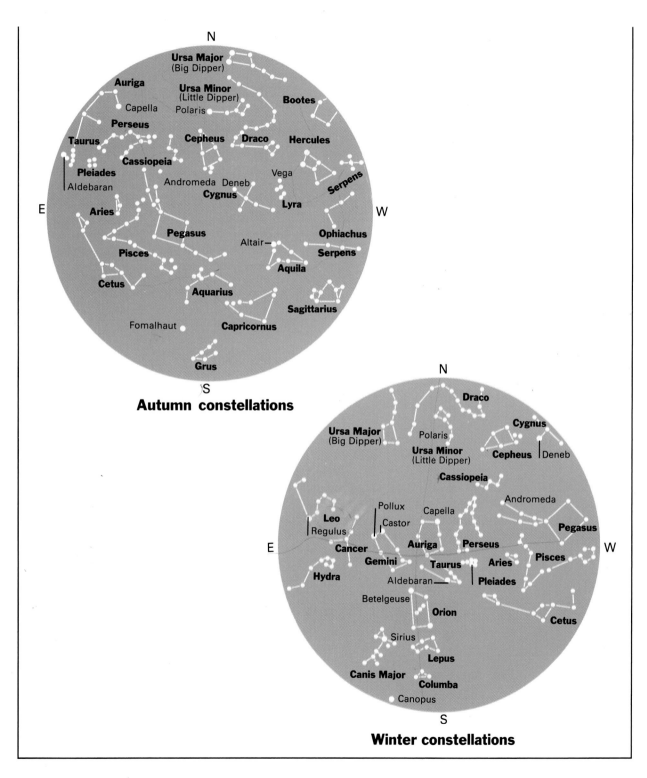

Autumn constellations

Winter constellations

Analyzing Science Terms

You can often unlock the meaning of an unfamiliar science term by analyzing its word parts. Prefixes and suffixes, for example, each carry a meaning that derives from a word root—usually Latin or Greek. The prefixes and suffixes listed below provide clues to the meanings of many science terms.

Word Part or Root	Meaning	Application
astr-, aster-	star	astronomy
bar-, baro-	weight, pressure	barometer
batho-, bathy-	depth	batholith, bathysphere
circum-	around	circum-Pacific, circumpolar
-cline	lean, slope	anticline, syncline
eco-	environment	ecology, ecosystem
epi-	on	epicenter
ex-, exo-	out, outside of	exosphere, exfoliation, extrusion
geo-	earth	geode, geology, geomagnetic
-graph	write, writing	seismograph
hydro-	water	hydrosphere
hypo-	under	hypothesis
iso-	equal	isoscope, isostasy, isotope
-lith, -lithic	stone	Neolithic, regolith
-logy	study of	ecology, geology, meteorology
magn-	great, large	magnitude
mar-	sea	marine
meso-	middle	mesosphere, Mesozoic
meta-	among, change	metamorphic, metamorphism
micro-	small	microquake
-morph, -morphic	form	metamorphic
nebula-	mist, cloud	nebula
neo-	new	Neolithic
paleo-	old	paleontology, Paleozoic
ped-, pedo-	ground, soil	pediment
peri-	around	perigee, perihelion
seism-, seismo-	shake, earthquake	seismic, seismograph
sol-	sun	solar, solstice
spectro-	look at, examine	spectroscope, spectrum
-sphere	ball, globe	geosphere, lithosphere
strati-, strato-	spread, layer	stratification, stratovolcano
terra-	earth, land	terracing, terrane
thermo-	heat	thermosphere
top-, topo-	place	topographic
trop-, tropo	turn, respond to	tropopause, troposphere

Glossary

The glossary contains all of the key terms in the text and their definitions. The number of the page where each term is introduced in the text is listed in parentheses after the definition.

A

aa jagged chunks of lava formed by rapid cooling on the surface of a lava flow (104)

abrasion mechanical weathering in which rocks collide and scrape against each other, thus wearing away the exposed surfaces (187)

absolute age specific age of an object (279)

absolute magnitude brightness of a star as it would appear if located 32.6 light-years from the earth (473)

abyssal plain extremely level area of the ocean floor (338)

abyssal zone ocean-life zone that extends from beneath the bathyal zone to a depth of 6,000 m (356)

acid rain rain formed by a combination of pollutants and water in the atmosphere (188)

adiabatic describing a change in temperature resulting from the expansion or compression of air (414)

advection fog condensation of water vapor that results from the cooling of warm, moist air as it moves across a cold surface (418)

advective cooling decrease in the temperature of a mass of air that results as it moves over a cold surface (416)

aftershock tremor that follows and is smaller than a major earthquake (85)

air mass large body of air with uniform temperature and moisture content (427)

air pollutant any substance in the atmosphere that is harmful to living organisms or to other materials (394)

albedo percent of solar radiation reflected by a surface (398)

alloy solid solution of two or more metals (129)

alluvial fan fan-shaped deposit of sediments at the base of a slope on land (216)

amber hardened tree sap in which fossils such as insects may be preserved (287)

anemometer instrument used to measure wind speed (437)

angular unconformity boundary between horizontal and tilted layers of rock (277)

anthracite hardest form of coal (171)

anticline upcurved fold in horizontal rock layers (73)

anticyclone storm that spirals outward from a high-pressure center (433)

aphelion point in the orbit of a planet at which it is farthest from the sun (27)

apogee point in the orbit of a satellite at which it is farthest from the earth (535)

apparent magnitude brightness of a star as it appears from the earth (472)

aquaculture the farming of the ocean (360)

aquifer body of rock that can store much water and from which water flows freely (223)

arête sharp, jagged ridge formed between cirques (242)

artesian formation sloping layer of permeable rock sandwiched between two layers of impermeable rock and exposed at the surface (229)

artesian spring natural flow of water to the earth's surface from an artesian formation (229)

artesian well hole dug through the cap rock of an artesian formation through which water flows freely with no pumping necessary (229)

asteroid chunk of rock that orbits the sun (522)

asteroid belt the region between the orbits of Mars and Jupiter in which most asteroids are found (522)

asthenosphere the layer of the mantle beneath the lithosphere that consists of slowly flowing solid rock (62)

astronomical unit the average distance between the earth and the sun, approximately 149.5 million km (508)

astronomy the study of the universe beyond the earth (6)

atmosphere the thick blanket of gases surrounding the earth (21)

atmospheric pressure weight of the air pressing on a given surface area (389)

atoll nearly circular coral reef surrounding a shallow lagoon (269)

atom smallest unit of an element (118)

atomic number number of protons in an atom (119)

aurora sheets of colored light produced by a magnetic storm in the earth's upper atmosphere (497)

autumnal equinox the beginning of the fall season (29)

axis imaginary straight line running through the earth from pole to pole (21)

B

background radiation low levels of energy evenly distributed throughout the universe (16)

barchan crescent-shaped dune with the open side facing away from the wind direction (258)

barometer instrument that measures atmospheric pressure (390)

barred spiral galaxy type of spiral galaxy with a bar of stars that runs through its center (483)

barrier island long, narrow ridge of sand that lies parallel to the shore (268)

barrier reef type of coral reef that surrounds the remnant of a partially submerged volcanic island (269)

batholith largest type of igneous intrusion, covering hundreds of square kilometers and reaching a depth of thousands of kilometers (154)

bathyal zone ocean-life zone that begins at the end of the continental shelf and extends to a depth of 4,000 m (355)

bathyscaph self-propelled, free-moving submersible used for deep-ocean research (333)

bathysphere spherical submersible that remains attached to a research ship for communications and support (332)

beach deposit of rock fragments along an ocean shore or a lakefront (262)

bed load coarse sediment carried along a stream bottom by sliding, rolling, and saltation (213)

bedding plane boundary between two sedimentary rock layers (276)

bedrock solid, unweathered rock beneath the regolith (193)

benthos organisms that live on the ocean floor (355)

berm raised midsection of a beach, the part usually used for recreation (263)

big bang theory the theory that the universe began with an explosion of a hot, dense sphere billions of years ago and is still expanding (7)

bimetal thermometer instrument used to measure temperature, consisting of a bar made of two strips of different metals that curves when heated and straightens when cooled (437)

binary star pair of stars that revolve around each other (484)

biodegradable able to be broken down into component parts by microorganisms (8)

biosphere ecosystem encompassing all the life on earth and the physical environment around it (7)

bituminous coal soft coal (171)

black dwarf star that results when a white dwarf star no longer gives off energy (479)

black hole hole in space with a gravity so great that not even light can escape, formed by the collapse of a very large supernova (480)

blue shift apparent shortening of the light waves emitted by a star moving towards the earth (469)

bora cold northern wind that blows down the mountains of Yugoslavia toward the Adriatic Sea (454)

breaker foamy mass of water that washes onto the shore (374)

breccia clastic sedimentary rock composed of angular fragments cemented together by minerals (156)

Bright Angel shale rock layer of the Grand Canyon deposited during the Cambrian Period (322)

butte elevated, narrow, flat-topped area (203)

C

calcareous ooze type of ooze that is mostly calcium carbonate (343)

caldera large basin-shaped depression formed when an explosion destroys the upper part of a volcanic cone (108)

calendar system used to measure the passage of time (541)

Canadian Shield the exposed portion of the craton around which North America has been built up (318)

cap rock top layer of impermeable rock in an artesian formation (229)

capillary fringe region of soil just above the water table that receives moisture from the zone of saturation by capillary action (225)

carbonation chemical weathering process in which minerals react with carbonic acid (188)

carbonization process in which plant materials are changed into carbon (170)

cast replica of the outer surface of an organism that has decayed (289)

cementation process in which dissolved minerals left by water passing through sediments binds the sediments together (156)

Cenozoic Era the most recent geologic era, beginning 65 million years ago; the Age of Mammals (296)

Cepheid variable star that brightens and fades in a regular cycle (472)

channel path that a stream follows (211)

chemical bond force that holds together the atoms that make up a compound (126)

chemical formula symbols indicating the elements a compound contains and the relative numbers of each element (128)

chemical property characteristic that describes how a substance interacts with other substances to produce different kinds of matter (117)

chemical sedimentary rock rock formed from minerals that have been dissolved in water (156)

chemical weathering process in which rock is broken down as a result of chemical reactions (185)

Chinook warm, dry wind that blows down the eastern slopes of the Rocky Mountains (454)

chromosphere the thin layer of the sun's atmosphere that lies above the photosphere and glows with a reddish light (493)

cinder cone steep-sloped deposit of solid fragments ejected from a volcano (106)

circum-Pacific belt mountain belt that surrounds the Pacific Ocean (76)

circumpolar describing any star that is always visible in the night sky and, from the Northern Hemisphere, can be seen circling Polaris (469)

cirque bowl-shaped depression produced by a valley glacier (242)

cirrus cloud feathery cloud composed of ice crystals that has the highest altitude of any cloud in the sky (416)

clastic sedimentary rock rock made up of fragments from pre-existing rocks (156)

cleavage splitting of a mineral along smooth, flat surfaces (140)

climate average weather conditions over a long period of time (387)

cloud seeding addition of freezing nuclei to supercooled clouds in an attempt to induce or increase precipitation (422)

coalescence combination of different-sized cloud droplets to form larger droplets (421)

Coconino sandstone rock layer of the Grand Canyon deposited during the Permian Period (323)

cold front boundary formed where a cold air mass overtakes and lifts a warm air mass (430)

coma spherical cloud of gas and dust that surrounds the nucleus of a comet (523)

comet body of rock, dust, methane, ammonia, and ice that revolves around the sun in a long, elliptical orbit (523)

compaction process in which air and water are squeezed out of sediments, resulting in the formation of sedimentary rock (156)

composite cone also called *stratovolcano*, steep-sloped volcanic deposit with alternating layers of hardened lava flows and tephra (107)

compound two or more atoms chemically combined (125)

compression stress that squeezes crustal rocks together (72)

concretion nodule of rock with a different composition than that of the main rock body (160)

condensation process by which water vapor changes to liquid water (208)

condensation nuclei solid particles in the atmosphere, such as ice and dust, that provide the surfaces on which water vapor condenses (414)

conduction type of energy transfer in which vibrating molecules pass heat along to other vibrating molecules by direct contact (400)

cone of depression lowered area of a water table produced by pumping water from a well (228)

conglomerate sedimentary rock composed of rounded gravel or pebbles cemented together by minerals (156)

conic projection map projection in which the meridians converge at the poles; the parallels appear as equally spaced, concentric curves (42)

constellation pattern of stars (481)

contact metamorphism change in the structure and mineral composition of rock surrounding an igneous intrusion (161)

continental crust material that makes up landmasses (62)

continental drift the hypothesis stating that the continents once formed a single landmass, broke up, and drifted to their present locations (57)

continental glacier massive moving ice sheet that covers large land areas (238)

continental margin the part of the ocean basin made of continental crust (335)

continental polar describing a cold, dry air mass that is formed over land in polar regions (428)

continental rise accumulation of sediments at the base of a continental slope (337)

continental shelf edge of a continent covered by shallow ocean water (335)

continental slope steep incline at the edge of a continental shelf (336)

continental tropical describing a warm, dry air mass that is formed over land in tropical regions (428)

contour interval difference in elevation between one contour line and the next (47)

contour line line on a map connecting points with the same elevation (46)

contour plowing system of plowing soil in bands that follow the shape of the land as a means of preventing erosion (198)

convection transfer of heat through the movement of a fluid material (64)

convection current movement in a fluid caused by uneven heating (64)

convective cooling decrease in the temperature of a mass of air that results as it rises and expands (414)

convective zone region around the sun's radiative zone in which moving gases transfer energy (492)

convergent boundary border formed by the direct collision of two lithospheric plates (63)

coprolite fossilized waste material from an animal (289)

coral reef ridgelike coastal feature made of millions of coral skeletons (269)

core center of a planetary body, such as the earth (22)

core sample cylindrical sample of sediments from the deep-ocean floor (341)

Coriolis effect the deflection of wind and ocean currents caused by the earth's rotation (365)

corona the outermost layer of the sun's atmosphere (493)

covalent bond bond based on the attraction between atoms that share electrons (128)

covalent compound compound formed from atoms that share electrons (128)

crater funnel-shaped pit at the top of a volcanic cone (108); bowl-shaped depression on the surface of a planetary body (530)

craton large area of Precambrian rocks found on all continents (314)

creep slow downhill movement of weathered rock material (202)

crest highest point of a wave (371)

crevasse large crack that forms between pressure ridges on the surface of a glacier (240)

crop rotation practice of planting different crops in alternating years (199)

crude oil unrefined petroleum (172)

crust outermost layer of the solid earth (22)

crystal natural solid substance that has a definite geometric shape (135)

cumulus cloud thick, billowy cloud that forms above stratus clouds and below cirrus clouds (416)

current steady movement in one direction, such as that of water in the ocean (365)

D

day time required for the earth to make one rotation on its axis, about 23.95 hours (541)

daylight saving time system in which clocks are set one hour ahead of standard time from April to October (32)

deep current streamlike movement of water beneath the surface of the ocean (365)

deep-focus earthquake earthquake that occurs at a depth of 300 km to 650 km (86)

deflation most common form of wind erosion, in which fine, dry soil particles are blown away (256)

deflation hollow shallow depression left after the wind has eroded a layer of exposed soil (256)

deformation bending, tilting, and breaking of the earth's crust (71)

delta fan-shaped deposit of sediments at the mouth of a stream (216)

density ratio of the mass of a substance to its volume, expressed as g/cm^3 (143)

desalination process of removing salt from ocean water (210)

desert pavement surface of closely packed small rocks left after the top layer of soil has been removed by deflation (256)

dew type of condensation formed when air that is in contact with a cool surface loses heat until it reaches saturation (413)

dew point temperature to which air must be cooled to become saturated (413)

diatomic consisting of two atoms (125)

dike igneous intrusion that cuts across rock layers (155)

discharge volume of water moved by a stream in a given time (213)

disconformity boundary between layers of rock that have not been deposited continuously (277)

dissolved load mineral material carried by a stream as ions or molecules (213)

distillation process in which ocean water is heated until it evaporates, leaving behind dissolved salts (357)

divergent boundary boundary formed by two lithospheric plates that are moving apart (63)

divide elevated region that separates two watersheds (211)

doldrum narrow zone of low air pressure at the equator characterized by weak and unpredictable winds (403)

dome mountain landform created when molten rock pushes up rock layers on the earth's surface and the layers then are worn away in places, leaving separate high peaks (81)

Doppler effect apparent shift in the wavelength of a sound wave or a light wave emitted by a source moving away from or toward an observer (15)

double refraction property exhibited by transparent minerals that produce a double image of any object viewed through them (144)

drift weak, slow-moving ocean current (367)

drumlin long, low, tear-shaped mound of till (245)

dune mound of wind-blown sand (257)

E

earth science the study of the earth and the universe around it (3)

earth-grazer asteroid that orbits the sun in an elongated ellipse that may pass close to the earth and sun (523)

earthquake vibration of the earth's crust (85)

earthshine sunlight reflected off the surface of the earth (540)

eclipse passing of one planetary body through the shadow of another (537)

ecology the study of the complex relationships between living things and their environment (7)

ecosystem community of organisms and the environment they inhabit (7)

elastic rebound theory the theory that rocks strained past a certain point will fracture and spring back to their original shape (85)

electrical thermometer instrument used to measure temperature based on the increased flow of electricity through certain materials when they are heated (437)

electromagnetic spectrum the complete range of wavelengths of radiation (395)

electron subatomic particle with a negative electrical charge (119)

electron cloud region of space around the nucleus of an atom in which electrons may be found (119)

element substance that cannot be broken down into a simpler form by ordinary chemical means (15)

elevation height above sea level (46)

ellipse oval whose shape is determined by two points within the figure (508)

elliptical galaxy type of galaxy with a very bright center that contains little dust and gas and is spherical to disklike in shape (483)

emergent coastline coast along which sea level falls or the land rises (267)

energy level arrangement of electrons within the electron cloud of an atom (125)

epicenter point on the earth's surface directly above the focus of an earthquake (85)

epicycle small circular motion of the planets within their orbits proposed as an explanation for retrograde motion (507)

epoch subdivision of a geologic period (296)

era largest unit of geologic time (296)

erosion process by which the products of weathering are transported (197)

erratic large boulder transported and deposited by a glacier (244)

esker long, winding ridge of gravel and coarse sand deposited by a glacier (246)

estuary wide, shallow bay formed where ocean water submerges the mouth of a river (266)

Eurasian-Melanesian belt mountain belt that runs through Asia, northern Africa, and southern Europe (76)

evaporation process by which liquid water changes to water vapor (207)

evaporite sedimentary rock formed from solids left after water evaporates (157)

evapotranspiration processes by which water enters the atmosphere; evaporation and transpiration combined (207)

exfoliation process in which sheets of rock peel or flake as a result of weathering (188)

exosphere the layer of the atmosphere above the ionosphere that merges with interplanetary space (393)

experimentation process by which a scientific procedure is carried out according to certain guidelines (12)

extrusion mass of igneous rock that has hardened on the earth's surface (155)

extrusive igneous rock melted rock that hardens on the earth's surface (152)

fault   break in rock along which rocks on either side of the break move (74)

fault plane   surface of a fault along which movement of rocks occurs (74)

fault zone   group of interconnected faults (87)

fault-block mountain   mountain formed where faulting breaks the earth's crust into large blocks and the blocks are uplifted and tilted (80)

felsic lava   silicon-rich lava (104)

fetch   distance that wind can blow across open water (373)

fiord   narrow, deep, steep-walled bay formed by flooding of a glacial valley due to a rise in sea level (267)

fireball   brilliant flash of light produced by a meteor that vaporizes quickly (524)

firn   grainy ice in a glacier that has been partially melted and refrozen (237)

fissure   crack in a rock surface through which lava flows (103)

floodplain   part of a river valley floor usually covered during a flood (217)

floodway   permanent overflow channel used to control flooding (219)

fluorescence   ability to glow under ultraviolet light (144)

focus   area along a fault at which slippage first occurs, initiating an earthquake (85); one of two points within an ellipse that determines the shape of the figure (508)

foehn   warm, dry wind that blows down the northern slopes of the Alps Mountains (454)

folded mountain   landform created when tectonic movements bend and uplift rock layers (78)

folding   permanent deformation or bending of a rock under stress (73)

foliated   describing a metamorphic rock with visible layers or bands (161)

footwall   rock below a fault plane (74)

fossil   trace or remains of a plant or an animal in sedimentary rock (160)

fossil fuel   fuel formed from the remains of living organisms, such as coal, petroleum, and natural gas (170)

fracture   break in rock (74)

freezing nuclei   condensation nuclei with a crystalline structure like that of ice (421)

fringing reef   type of coral reef that forms around a volcanic island (269)

front   boundary between air masses of different temperatures (430)

frost   ice crystals formed when the dew point is below 0°C and water vapor sublimates directly to the solid state (413)

full moon   phase of the moon during which the entire half of the moon facing the earth is visible (540)

galaxy   large-scale group of stars (482)

Galilean moon   any one of the four largest satellites of Jupiter, which were first seen by Galileo (545)

gas   physical form of matter that does not have a definite volume or shape (123)

gastrolith   fossilized stone found within the digestive system of a dinosaur or other reptile (289)

gemstone   nonmetallic mineral that is brilliant and colorful when cut (168)

geocentric   earth-centered (507)

geode   crystal-filled cavity in sedimentary rock (160)

geologic column   arrangement of rock layers based on the ages of the rocks (295)

geology   the study of the origin, history, and structure of the solid earth and the processes that shape it (4)

geomagnetic pole   point on the earth's surface above a pole of the earth's imaginary internal magnet (41)

geosphere   the solid earth, its hydrosphere, and its atmosphere (7)

geosynchronous orbit   orbit directly above the earth's equator and moving in the direction of the earth's rotation (33)

geosyncline   thick accumulation of sediment deposited on the ocean floor near the shore (72)

geothermal energy   energy contained in and available from water heated by magma or gases within the earth (177)

geyser   hot spring that erupts periodically (230)

glacial drift   sediments deposited by a glacier (245)

glacial period   time when a glacier advances during an ice age (249)

glacier   mass of moving ice (237)

glaze ice   thick layer of ice formed when rain freezes as it contacts a surface (419)

globular cluster spherically shaped group of hundreds of stars located around the core of the Milky Way Galaxy (484)

gnomonic projection map projection in which the parallels appear as unevenly spaced, concentric circles, the meridians appear as straight lines radiating from a central point, and all other great circles appear as straight lines (42)

graben long, narrow valley formed by faulting and downward slippage of a crustal block (80)

gradient change in elevation over a distance (213)

granulation grainy appearance of the photosphere caused by the rising and sinking of gases (492)

gravity the force of attraction between all objects in the universe (25)

great circle any imaginary line that circles the earth and divides it in half (40)

greenhouse effect the process by which the atmosphere traps infrared rays over the earth's surface (399)

Gregorian calendar revision of the Julian calendar by Pope Gregory XIII, which is currently used in most of the world (542)

ground moraine unsorted material left beneath a glacier when the ice melts (245)

groundwater water that soaks deep into soil and rock (207)

Gulf Stream the swift, warm Atlantic current that flows from the Gulf of Mexico, around Florida, and up the east coast of North America (366)

gullying erosion process in which furrows in soil are enlarged as the soil is washed away (197)

guyot flat-topped, submerged seamount (340)

H

H-R diagram Hertzsprung-Russell diagram, graph showing the relationship of the surface temperature and absolute magnitude of a star (473)

hachure line short, straight line drawn along the inside of a contour loop pointing toward its center that indicates a depression (51)

hadal zone bottom environment of the ocean deeper than 6,000 m (356)

hail type of precipitation in the form of lumps of ice (420)

hair hygrometer instrument used to measure relative humidity, based on how human hair stretches as humidity increases (412)

half-life time required for half the mass of a radioactive element to decay into its daughter element (282)

hanging valley small abandoned glacial valley suspended on a mountain above the main glacial valley (243)

hanging wall rock above a fault plane (74)

hard water water that contains relatively large amounts of dissolved minerals (231)

hardness measure of the ability of a mineral to resist scratches (141)

headland projection of a shore into the water (264)

headward erosion lengthening and branching of a stream (211)

headwater beginning of a stream (213)

heliocentric sun-centered (507)

Hermit shale rock layer of the Grand Canyon deposited during the Permian Period (323)

horizon layer of a soil profile (194)

horn sharp, curved peak formed where several aretes join (242)

horse latitude subtropical high; belt of high air pressure around 30° latitude (403)

hot spot area of volcanism within a lithospheric plate (103)

hot spring hot groundwater that rises to the surface before cooling (229)

humid continental climate middle-latitude climate occurring between 35° to 50° north latitude, with warm, humid summers and cold winters (460)

humid subtropical climate very wet middle-latitude climate that occurs in southeastern coastal areas around 30° north and south latitude, with warm, humid summers and generally mild winters (460)

humidity amount of water vapor in the atmosphere (410)

humus dark, organic material formed from the decayed remains of plants and animals (193)

hurricane severe storm that develops over tropical oceans, with strong winds that spiral in toward the storm center (433)

hydraulic gradient slope of a water table (226)

hydrocarbon compound made up of atoms of carbon and hydrogen (170)

hydroelectric power power produced by running water (178)

hydrolysis chemical reaction between water and another substance (187)

hydrosphere all the earth's water (21)

hypothesis possible explanation of a problem that is based on facts (11)

ice age long period of climatic cooling during which ice sheets cover large areas of the earth's surface (249)

ice wedging mechanical weathering caused by the freezing and thawing of water that seeps into cracks in rocks (185)

igneous rock rock formed from cooled and hardened magma (149)

impermeable rock or sediment through which water cannot flow (225)

index contour every fifth contour line on a topographic map that is printed dark for reference (49)

index fossil guide fossil; fossil found in the rock layers of only one geologic age that is used to establish the relative age of the rock layers (290)

inertia tendency of a moving body to remain in motion or a stationary body to remain at rest until an outside force acts on it (509)

infrared wavelength longer than that of visible light (395)

inorganic not produced by or made up of living organisms or the remains of living organisms (133)

intensity amount of damage caused by an earthquake (90)

interglacial period time of relatively warm temperatures between glacial periods (249)

intermediate-focus earthquake earthquake that occurs at a depth between 70 km and 300 km (86)

internal flow slow movement of a glacier in which ice crystals slip over each other (239)

international date line the imaginary line running from north to south through the Pacific Ocean where the date changes from one day to the next (32)

intertidal zone ocean-life zone that lies between the low-tide and high-tide lines (355)

intrusion mass of igneous rock that has hardened beneath the earth's surface (154)

intrusive igneous rock rock formed from the cooling of magma within the earth's crust (152)

invertebrate animal without a backbone (299)

ion atom or group of atoms that carries an electrical charge (127)

ionic bond bond in which electrons are transferred from one atom to another (126)

ionic compound compound formed through the transfer of electrons (126)

ionosphere the lower region of the thermosphere, at an altitude of 80 to 550 km (393)

iron meteorite type of meteorite made of iron, with a characteristic metallic appearance (525)

irregular galaxy type of galaxy with no identifiable shape and an uneven distribution of stars within it (483)

island arc chain of volcanic islands formed along an ocean trench (64)

isobar line drawn on a weather map connecting points of equal atmospheric pressure (441)

isostasy balancing of the forces pressing up and down on the earth's crust (71)

isostatic adjustment up-and-down movements of the earth's crust to reach isostasy (71)

isotope atom of an element that has the same atomic number but different atomic mass than another atom of that element (123)

J

jet streams bands of high-speed, high-altitude westerly winds (404)

joint sheeting process in which rock breaks into curved sheets that peel away from underlying rock (185)

joints long, curved cracks in a rock parallel to its surface (185)

Jovian planet any one of the first four outer planets—Jupiter, Saturn, Uranus, or Neptune—with properties similar to those of Jupiter (514)

Julian calendar calendar devised by Julius Caesar consisting of 12 months, 11 with 30 or 31 days, one with 28 days, and an extra day added every four years (542)

K

Kaibab limestone rock layer of the Grand Canyon deposited during the Permian Period (323)

karst topography region where the effects of chemical weathering due to groundwater, such as sinkholes and caverns, are clearly visible (233)

kettle depression on a glacial outwash plain (246)

L wave   surface or long wave, the slowest wave generated by an earthquake and the last to be recorded by a seismograph (89)

laccolith   a flat-bottomed intrusion that pushes overlying rock layers into an arc (155)

lagoon   narrow region of shallow water between a barrier island and the shore (268)

land breeze   warm, gentle wind that flows from the land to the water (405)

landform   physical feature of the earth's surface (202)

landslide   sudden movement of loose rock and soil down a slope (199)

lapilli   tephra particles between 2 mm and 64 mm in diameter (105)

latent heat   energy stored in molecules (410)

lateral moraine   unsorted material deposited along the sides of a valley glacier (245)

laterite   thick infertile soils produced in tropical climates (195)

latitude   angular distance north or south of the equator (39)

lava   magma that reaches the earth's surface (101)

lava flow   erratically shaped masses of extrusive rock (155)

lava plateau   raised, flat-topped area made of layers of hardened lava (155)

law of crosscutting relationships   the principle that a fault or intrusion is always younger than the rock layers it cuts through (278)

law of gravitation   the principle that the force of attraction between two objects depends on the masses and the distance between the objects (26)

law of superposition   the principle that a sedimentary rock layer is older than the layers above and younger than the layers below it (276)

layered drift   glacial deposit that has been sorted and layered by the action of streams or meltwater (245)

leaching   process in which water carries dissolved minerals to lower layers of rock (187)

leap year   year with an extra day in it, occurring every four years (542)

legend   list of map symbols and their meanings (44)

light-year   distance that light travels in one year, about 9.5 trillion km (470)

lignite   brown coal (171)

liquid   physical form of matter with a definite volume but no definite shape (123)

lithosphere   rigid outer shell of the earth consisting of the crust and part of the upper mantle (62)

Local Group   the cluster of galaxies containing the Milky Way Galaxy and approximately 17 other galaxies within a distance of 3 million light-years (483)

lode   deposit formed by thick mineral veins (167)

loess   accumulation of wind-blown dust (259)

longitude   angular distance east or west of the prime meridian (40)

long-period comet   comet with a period of several thousand or million years (524)

lunar eclipse   passing of the earth between the moon and the sun during which the earth's shadow falls on the lighted half of the moon (538)

luster   light reflected from the surface of a mineral (140)

mafic lava   lava rich in magnesium and iron (104)

magma   molten rock produced deep inside the earth (104)

magnetic declination   angle between the direction of the earth's geographic pole and the direction a compass needle points (40)

magnetosphere   area around the earth that is affected by the earth's magnetic field (24)

main-sequence star   star with characteristics that places it within a band running through the middle of the H-R diagram (474)

mantle   layer of rock below the crust of the earth (22)

map projection   flat map that represents a three-dimensional curved surface (42)

maria   (sing: mare) dark areas of dry, solidified lava on the moon that are smooth and reflect little light (530)

marine west-coast climate   wet middle-latitude climate that occurs in western coastal areas located between 40° and 60° latitude, with relatively cool summers and mild winters (459)

maritime polar   describing a cold, moist air mass that is formed over the ocean in polar areas (428)

maritime tropical   describing a warm, moist air mass that is formed over the ocean in tropical areas (428)

mass   amount of matter in an object (22)

mass movement   movement of rock fragments down a slope (199)

mass number   sum of the numbers of protons and neutrons in an atom (119)

matter substance that takes up space and has mass (117)

mean sea level point midway between the highest and lowest tide levels of the ocean (46)

meander wide curve in a stream channel (215)

measurement comparison of a property of an object or phenomenon with a standard unit (9)

mechanical weathering process that changes the physical form of rocks (185)

medial moraine ridge of unsorted glacial material along the center of a valley glacier (245)

Mediterranean climate middle-latitude climate that occurs in coastal areas located between 30° and 40° latitude, with dry summers and wet winters (459)

meltwater melted ice flowing from a glacier (245)

Mercalli scale scale that expresses the intensity of an earthquake by a Roman numeral and a description (90)

Mercator projection map projection in which the meridians appear as straight, parallel, evenly spaced lines and form a grid with the parallels, which appear as straight, parallel, and unevenly spaced lines (42)

meridian imaginary semicircle on the earth that runs north and south from pole to pole (39)

mesa elevated, flat-topped area smaller than a plateau (203)

mesopause the upper boundary of the mesosphere, marked by an increase in temperature (393)

mesosphere the layer of the atmosphere that extends upward from the stratosphere to an altitude of about 80 km (393)

Mesozoic Era the geologic era that lasted from 225 million to 65 million years ago, the Age of Reptiles (296)

metamorphic rock rock formed from other rocks as a result of intense heat, pressure, or chemical processes (150)

metamorphism changing of one type of rock to another by heat, pressure, and chemical processes (161)

meteor meteoroid that enters the earth's atmosphere (524)

meteor shower phenomenon caused by the burning up of large numbers of meteors as they enter the earth's atmosphere (524)

meteorite meteor or part of a meteor left after it hits the earth's surface (523)

meteoroid small bit of rock or metal moving through the solar system left by a comet or produced by a collision between asteroids (524)

meteorology the study of the earth's atmosphere (5)

microquake earthquake with a magnitude less than 2.5 on the Richter scale (90)

Mid-Atlantic Ridge undersea mountain range with a steep, narrow valley along its center (58)

middle-latitude climate climate with a maximum average temperature of 18°C in the coldest month and a minimum average temperature of 10°C in the warmest month (455)

middle-latitude-desert climate middle-latitude climate that is very dry, with both a cold winter and a warm to very hot summer (459)

middle-latitude-steppes climate middle-latitude climate with slightly more precipitation than a middle-latitude-desert climate and a high yearly temperature range (459)

mid-ocean ridges system of undersea mountain ranges that wind around the earth (58)

Milankovitch theory the theory that regular changes in the earth's orbit and in the tilt of the earth's axis caused the ice ages (250)

Milky Way Galaxy the galaxy in which the earth and solar system are located (482)

mineral natural inorganic solid found in the earth's crust (133)

mineralogist scientist who specializes in the study of minerals (139)

minute 1/60 of a degree of latitude, 1.85 km (39)

mistral cold, northern wind that blows down the Alps Mountains toward the Mediterranean Sea (454)

mixture material that contains two or more substances that are not chemically combined (129)

Moho the Mohorovicic discontinuity, boundary between the earth's crust and mantle (22)

Mohs' scale standard against which the hardness of a mineral is tested (141)

mold cavity in sedimentary rock, left by a decayed or dissolved organism, that retains the shape and surface markings of the organism (288)

molecule smallest complete unit of a compound (125)

monadnock knob of rock that protrudes above a peneplain (202)

monocline gently dipping bend in horizontal rock layers (73)

monsoon seasonal wind that blows toward the land in summer, bringing heavy rains, and away from the land in the winter, bringing dry weather (453)

month time required for the moon to go through one set of phases as it orbits the earth, about 29.5 days (541)

moon body that is smaller than a planet and orbits the planet (498)

mountain belt large group of mountain systems (76)

mountain breeze gentle wind that flows down mountain slopes at night (405)

mountain range group of adjacent mountains with the same general shape and structure (76)

mountain system group of adjacent mountain ranges (76)

Muav limestone rock layer of the Grand Canyon deposited during the Cambrian Period (322)

mud fine particles of rock combined with water (343)

mud crack feature formed when muddy deposits dry and shrink (160)

mud pot weathered rock around a hot spring that mixes with the hot water to form liquid clay that bubbles at the surface (230)

mudflow rapidly moving large mass of mud (199)

multiple-star system group of stars in which two stars may revolve around a common center with other stars revolving around them (484)

mummification preservation of a dead organism by drying (287)

N

natural bridge arch of rock formed by groundwater erosion (233)

natural levee raised river bank that results when a river deposits its load at the river's edges (218)

neap tide tide with minimum daily tidal range that occurs during the first and third quarters of the moon (378)

nebula cloud of gas and dust in space; first stage in the development of a star (475)

nebular theory the theory that the big bang spread matter throughout the expanding universe and that the sun and the planets condensed out of a spinning cloud of gas and dust (498)

nekton forms of ocean life that swim, such as fish, dolphins, and squid (355)

neritic zone ocean-life zone above the sublittoral zone filled with marine life (355)

neutron subatomic particle with no electrical charge (119)

neutron star collapsed core of a supernova consisting of a small, extremely dense ball of neutrons (479)

new moon phase of the moon during which the side of the moon facing the earth is unlighted (540)

nitrogen cycle process in which nitrogen moves from the air to the soil to animals and back to the air (389)

nodule lump of minerals on the ocean floor (343)

nonconformity unconformity in which stratified rock rests upon unstratified rock (277)

nonrenewable resource substance of limited supply that cannot be replaced (167)

nonsilicate mineral mineral that does not contain silicon (134)

normal fault fault in which the fault plane is almost vertical (74)

nova white dwarf star that explodes as it cools, temporarily becoming thousands of times brighter (479)

nuclear fission splitting of the nucleus of a large atom into smaller nuclei (174)

nuclear fusion combination of the nuclei of small atoms to form a larger nucleus (175)

nucleus region in the center of an atom that contains the protons and neutrons (119)

O

observation act of using the senses to gather information (9)

occluded front boundary formed where a fast-moving cold air mass overtakes and lifts a warm air mass, completely cutting it off from the ground (431)

ocean basin continental crust and oceanic crust that lie beneath the ocean (332)

oceanic crust material that makes up the ocean floor (62)

oceanic zone ocean life zone that extends seaward beyond the continental shelf (356)

oceanography the study of the earth's oceans (5)

Oort cloud spherical cloud of dust and ice that surrounds the solar system and is thought to be the place of origin of comets (524)

ooze soft organic sediment on the ocean floor (343)

open cluster loosely-shaped group of hundreds of stars (484)

ordinary spring natural flow of groundwater to the earth's surface (228)

ordinary well hole dug below the water table that fills with groundwater (228)

ore deposit of minerals from which metals and nonmetals can be removed profitably (167)

organic sedimentary rock rock formed from the remains of decaying organisms (156)

outcrop area of exposed rock (322)

outwash plain fan-shaped deposit of layered drift in front of a glacier (246)

oxbow lake water remaining in an isolated meander in a floodplain (215)

oxidation chemical combination of metallic elements with oxygen (188)

ozone form of atmospheric oxygen that has three atoms per molecule (387)

P

P wave primary wave, the fastest wave generated by an earthquake and the first to be recorded by a seismograph (89)

pack ice floating layer of ice that completely covers an area of the ocean surface (350)

pahoehoe solidified mafic lava with a wrinkled surface (104)

paleontologist scientist who studies fossils (286)

paleontology the study of fossils (286)

Paleozoic Era the geologic era that followed Precambrian time, lasting from 600 million to 225 million years ago (296)

Pangaea the single landmass thought to have been the origin of all continents (57)

Panthalassa giant ocean surrounding Pangaea (57)

parallax method of determining the distance from the earth to a star based on the shift in the apparent position of the star when viewed from different angles (470)

parallel any imaginary circle that runs east and west around the earth parallel to the equator (39)

peat brownish-black material produced by partial decomposition of plant remains (170)

pedalfer soil found in temperate regions receiving more than 65 cm of rain a year (196)

pedocal soil found in temperate regions receiving less than 65 cm of rain a year (196)

peneplain low, almost level surface of a mountain in its old stage (202)

penumbra outer part of the shadow cast by the earth or the moon in which sunlight is only partially blocked (537)

perched water table secondary water table formed by a layer of impermeable rock above the main water table (226)

perigee point in the orbit of a satellite in which it is closest to the earth (535)

perihelion point in the orbit of a planet in which it is closest to the sun (27)

period subdivision of a geologic era (296); time required for a planet to make one revolution around the sun (509)

permeability the ease with which water flows through the open spaces in a rock or sediment (224)

petrification process in which organic materials are replaced by minerals (287)

petrochemical chemical derived from petroleum (172)

phase varying shape of the visible portion of the moon (540)

phosphorescence ability to glow during and after exposure to ultraviolet light (144)

photosphere the innermost layer of the solar atmosphere (492)

physical property characteristic that is observable in a substance without changing the chemical composition of the substance (117)

phytoplankton microscopic ocean plants (354)

pillow lava lava that flows out of fissures on the ocean floor and cools rapidly (104)

placer deposit fragments of native metals that are concentrated in layers at the bottom of a stream bed (168)

planet any one of the nine major bodies that orbit the sun (498)

planetary nebula expanding shell of gases shed by a dying star (478)

planetesimal small body of matter that formed in the outer regions of the solar nebula while the sun was forming in its center (498)

plankton free-floating, microscopic ocean plants and animals (354)

plate tectonics the theory that the lithosphere is made up of plates that float on the asthenosphere and that the plates possibly are moved by convection currents (62)

plateau large area of flat-topped rocks high above sea level (79)

polar climate climate with a maximum average monthly temperature of 10°C (455)

polar desert area with a very dry tundra climate located between 60° latitude and the poles (458)

polar easterlies weak global winds located north of 65° north latitude and south of 65° south latitude that blows away from the poles (404)

polar front boundary formed where cold polar air meets the warmer air of the middle latitudes (431)

polar orbit orbit that passes over the earth's North and South poles (33)

pollution contamination of the environment with waste products or impurities (8)

polyconic projection map made by fitting together a series of conic projections of adjoining areas (44)

porosity percentage of open spaces in a rock or sediment (223)

porphyry igneous rock composed of large and small grains (152)

Precambrian time the earliest and longest geologic era, lasting from 4.6 billion to 600 million years ago (296)

precession wobbling of the earth on its axis in which the axis traces a circle in space every 26 thousand years (30)

precipitation process by which water falls from clouds to the earth as rain, snow, sleet, and hail (208)

pressure ridge arch of ice caused by buckling of the surface of a glacier (240)

prevailing westerlies global winds located between 40° and 60° latitude that blow from the southwest in the Northern Hemisphere and blow from the northwest in the Southern Hemisphere (403)

prime meridian the meridian that passes through Greenwich, England, designated as 0° (40)

prominence cloud of glowing gases that arches high above the sun's surface (495)

proton subatomic particle with a positive electrical charge (119)

protoplanet large body of matter that formed from the coalescence of planetesimals in the solar nebula (498)

protostar center of a shrinking, spinning nebula; the second stage in the development of a star (475)

psychrometer instrument used to measure relative humidity (410)

pulsar neutron star that emits two beams of radiation that sweep across space (480)

pyroclastics also called *tephra,* all the rock fragments ejected from a volcano (104)

Q

quasar starlike object that gives off radio waves and X rays (485)

R

radar device that can detect objects and weather conditions in the upper atmosphere by sending and receiving radio beams (439)

radiation all forms of energy that travel through space as waves (395)

radiation fog condensation of water vapor that results from the cooling of air that is in contact with the ground (418)

radiative zone region surrounding the core of the sun, in which energy is transferred in the form of waves (492)

rain gauge instrument used to measure the amount of rainfall (423)

rain-forest climate wet tropical climate that occurs within 5° to 10° on either side of the equator (455)

ray streak of rock material radiating from a crater (530)

red clay type of mud that is 40 percent clay particles, mixed with silt, sand, and organic material (343)

red giant very large, cool, bright star (474)

red shift apparent lengthening of the light waves emitted by a star moving away from the earth (470)

Redwall limestone rock layer of the Grand Canyon deposited during the Mississippian Period (322)

refraction bending of a light ray as it passes from one substance to another (144); bending of a wave as it reaches shallow water (375)

regional metamorphism metamorphism that affects rocks over large areas during periods of tectonic activity (161)

regolith layer of weathered rock fragments covering much of the earth's surface (193)

rejuvenated describing a river with a gradient that has been made steeper by a movement of the earth's crust (215)

relative age age of an object compared with the ages of other objects (276)

relative humidity ratio of the amount of water vapor in the air to the amount of water vapor the air can hold when saturated (410)

relief difference in elevation between the highest and lowest points of an area (47)

renewable resource substance that can be replaced (167)

retrograde motion apparent periodic reversal in the motion of some planets as viewed from the earth (507)

reverse fault fault in which the hanging wall moves up relative to the footwall (74)

revolution movement of a planet around the sun (27)

Richter scale scale that expresses the magnitude of an earthquake (90)

rift valley steep, narrow valley along the center of a mid-ocean ridge (63)

rille long, deep channel that runs through the maria on the moon (530)

Ring of Fire major earthquake zone that forms a ring around the Pacific Ocean (86)

rip current swift movement of water caused by the return of water to the ocean through breaks in underwater sandbars (375)

ripple mark wavy mark formed by the action of wind or water on sand (159)

roche moutonnée rounded knob of rock produced by glacial erosion (243)

rock cycle series of processes in which rock changes from one type to another (150)

rockfall fall of rock from a steep cliff (199)

rock-forming mineral any common mineral that forms the rocks of the earth's crust (133)

rotation spinning of a planet on its axis (27)

runoff water that flows over the land into streams and rivers (207)

S

S wave secondary wave, a wave generated by an earthquake and the second to be recorded by a seismograph (89)

salinity number of grams of dissolved salt in 1 kg of ocean water (348)

saltation movement of sand by short jumps, caused by wind or water (213)

sand bar long ridge of sand deposited offshore (263)

satellite object in orbit around a body with a larger mass (33)

saturated describing air that contains all the water vapor it can hold at a specific temperature (410)

savanna climate tropical climate in which a rainforest and a tropical-desert climate alternate, producing very wet summers and very dry winters (455)

scale relationship between distance shown on a map and actual distance (45)

scientific law rule that correctly describes a natural phenomenon (14)

scientific methods organized, logical approaches to scientific research (9)

sea arch bridge of rock formed as waves cut completely through a sea cave in a rock projection (261)

sea breeze cool, gentle wind that flows from the water to the land (405)

sea cave large hole at the base of a sea cliff (261)

sea cliff steep structure produced by the action of waves striking directly against the rock of a shoreline (260)

sea stack isolated column of rock left after the center of a sea arch has been eroded (261)

seafloor spreading movement of the ocean floor away from either side of a mid-ocean ridge (60)

seamount isolated volcanic mountain on the ocean floor (339)

second 1/60 of a minute of latitude, 0.03 km (39)

sediment fragments that result from the breaking of rocks, minerals, and organic matter (149)

sedimentary rock rock formed from hardened deposits of sediment (149)

seismic gap zone of rock where a fault is locked and unable to move and in which no major earthquake has occurred for at least 30 years (96)

seismic wave vibration that travels through the earth (21)

seismograph instrument used to detect and record seismic waves (89)

shadow zone location on the earth's surface where no seismic waves or where only P waves can be detected (23)

shallow-focus earthquake earthquake that occurs within 70 km of the earth's surface (86)

shearing stress that pushes rocks in opposite horizontal directions (72)

sheet erosion process in which parallel layers of topsoil are stripped away, exposing the surface of the underlying subsoil or partially weathered bedrock (198)

shield cone volcanic deposit of hardened lava with a broad base and gentle slopes (106)

shoreline place where the ocean and the land meet (260)

short-period comet comet with a period of up to 100 years (524)

silicate mineral mineral that contains atoms of silicon and oxygen (134)

siliceous ooze type of ooze that is mostly silicon dioxide (343)

silicon-oxygen tetrahedron four oxygen atoms arranged in a pyramid with one silicon atom in the center (137)

sill sheet of hardened magma that forms between and parallel to layers of rock (155)

sinkhole circular depression caused when the roof of a cavern collapses (232)

sleet ice pellets that form when rain falls through a layer of freezing air (419)

sliding movement of a glacier caused by melting of the ice in contact with the ground (239)

slump downhill movement of a large block of soil under the influence of gravity (200)

smog air pollution formed from a mixture of dust and chemicals (129)

snowfield almost motionless mass of permanent snow and ice (237)

snowline elevation above which ice and snow remain throughout the year (237)

soft water water that contains few dissolved minerals (231)

soil complex mixture of minerals, water, gases, and the remains of dead organisms (193)

soil profile cross section of soil layers and bedrock (194)

solar collector device for capturing solar energy (176)

solar eclipse passing of the moon between the earth and the sun during which the shadow of the moon falls on the earth (537)

solar flare sudden, violent eruption of electrically charged atomic particles from the sun's surface (496)

solar nebula the cloud of gas and dust that developed into the solar system (498)

solar system the sun and the bodies that revolve around it (498)

solar wind electrically charged atomic particles that stream out into space through holes in the sun's corona (493)

solar year time required for the earth to make one orbit around the sun, about 365.25 days (541)

solid physical form of matter with a definite shape and volume (123)

solifluction slow downslope flow of wet, muddy topsoil over frozen or clay-rich subsoil (201)

solution mixture in which one substance is uniformly dispersed in another substance (129)

sonar acronym for **so**und **na**vigation and **r**anging, method of mapping the ocean basin by reflected sound waves (333)

sorting uniformity in the size of the particles of a rock or sediment (223)

specific heat amount of heat needed to raise the temperature of 1 g of a substance 1°C (452)

specific humidity actual amount of moisture in the air (413)

spectroscope instrument that splits white light into a band of colors (15)

spectrum band of the various colors of light (14)

spiral galaxy type of galaxy with a nucleus of bright stars and flattened arms that swirl around the nucleus (483)

spit long, narrow deposit of sand connected at one end to the shore (264)

spring tide tide with maximum daily tidal range that occurs during the new and full moons (378)

squall line long line of heavy thunderstorms that may advance just ahead of a fast-moving cold front (430)

stalactite cone-shaped calcite deposit suspended from the ceiling of a cavern (232)

stalagmite cone-shaped calcite deposit built up from the floor of a cavern (233)

standard atmospheric pressure the atmospheric pressure measured at sea level, 760 mm of mercury (391)

standard time zone one of 24 regions of the earth in which noon is set as the time when the sun is highest over the center of the region (31)

star body of gases that gives off a tremendous amount of radiant energy in the form of light and heat (467)

station model cluster of weather symbols plotted on a map indicating the weather conditions at a particular reporting station (440)

stationary front boundary formed where two air masses meet and neither is displaced (431)

steam fog condensation of water vapor that results when cool air moves over warm water (418)

stock igneous intrusion with an area less than 100 km^2 (154)

stony meteorite most common type of meteorite, similar in composition to rocks found on the surface of the earth (525)

stony-iron meteorite rare type of meteorite that contains both iron and stone (525)

strain change in shape and volume of rocks that occurs due to stress (72)

stratification layering of sedimentary rock (159)

stratopause the high-temperature zone that marks the upper boundary of the stratosphere (392)

stratosphere   the layer of the atmosphere that extends upward from the troposphere to an altitude of 50 km (392)

stratovolcano   also called *composite cone,* steep-sloped volcanic deposit with alternating layers of hardened lava flows and tephra (107)

stratus cloud   cloud with a sheetlike form that is the lowest cloud in the sky (416)

streak   color of a mineral in powder form (140)

stream load   sediments carried by a stream (213)

stream piracy   capture of a stream in one watershed by a stream in another watershed (211)

stress   force that applies pressure to rocks in the earth's crust (72)

strike-slip fault   fault in which the rock on either side of a fault plane moves horizontally (75)

stripcropping   process in which different types of crops are planted in alternating bands (199)

subarctic climate   type of polar climate that occurs in areas between 55° and 65° north latitude, with little precipitation and a large yearly temperature range (458)

subduction zone   region where one lithospheric plate is forced under another (63)

sublimation   process in which a solid changes directly into a vapor or a vapor changes directly into a solid (409)

sublittoral zone   shallow ocean-life zone that is continuously submerged and that contains the largest number of bottom-zone-dwelling organisms (355)

submarine canyon   deep valley in the continental slope (337)

submergent coastline   coast along which sea level rises or the land sinks (266)

submersible   underwater research vessel (332)

subpolar low   belt of low air pressure around 65° north and 65° south latitude (404)

summer solstice   the beginning of summer (29)

sunspot   cool, dark area of gas within the photosphere caused by powerful magnetic fields (494)

sunspot cycle   variation in the number of sunspots that occurs periodically approximately every 11 years (495)

Supai formation   rock layer of the Grand Canyon deposited during the Permian Period, consisting of red sandstone and shale (323)

supercooling   process in which water droplets are induced to remain liquid at temperatures below 0°C (421)

supergiant   extremely large red giant star (474)

supernova   star that blows apart with a tremendous explosion (479)

surface current   streamlike movement of water on or near the surface of the ocean (365)

suspended load   particles of sand, silt, and mud carried by a stream that do not sink to the stream bed (213)

swell   one of a group of long, rolling waves that are all the same size (372)

syncline   down-curved fold in horizontal rock layers (73)

T

talus   pile of rock fragments that accumulates at the base of a slope (199)

Tapeats sandstone   rock layer of the Grand Canyon deposited during the Cambrian Period (322)

temperature inversion   atmospheric condition in which warm air traps polluted, cooler air near the earth's surface (394)

temperature range   difference between the highest and lowest temperatures of a particular time period (449)

tension   stress that pulls rocks apart (72)

tephra   also called *pyroclastics,* all the rock fragments ejected from a volcano (105)

terminal moraine   till deposited at the leading edge of a melting glacier (246)

terracing   construction of steplike ridges that follow the contours of a sloped field (199)

terrane   piece of land with a geologic history distinct from that of the surrounding land (66)

terrestrial planet   any one of the four planets closest to the sun—Mercury, Venus, Earth, or Mars—with properties similar to those of the earth (510)

theory   hypothesis or set of hypotheses supported by the results of experimentation and observation (14)

theory of evolution   the theory that organisms change over time and that new organisms are derived from ancestral types (297)

theory of suspect terranes   the theory that continents are a patchwork of pieces of land that have individual geologic histories (66)

thermocline   zone of rapid temperature change that begins just below the surface of the ocean (350)

thermograph   instrument that measures temperature changes by recording the movement of the bar of a bimetal thermometer (437)

thermosphere   the atmospheric layer above the mesosphere (393)

thrust fault   type of reverse fault in which the fault plane is nearly horizontal rather than vertical (74)

thunderstorm   storm accompanied by thunder, lightning, and strong winds (434)

tidal bore   surge of water that rushes upstream in a river as the tide rises (381)

tidal current   movement of water toward and away from the coast due to the rise and fall of the tides (381)

tidal flat   muddy or sandy part of a lagoon that is visible at low tide (268)

tidal oscillation   slow, rocking motion of ocean water that occurs as tidal bulges move around the earth (380)

tidal range   difference between the levels of the high and low tides at a specific location (378)

tide   daily change in the level of the ocean surface (377)

till   unsorted rock material deposited by a glacier (245)

tombolo   chain of spits that connects islands to each other or to the shore (264)

topographic map   map that shows the surface features of the earth (46)

topography   surface features of the earth (46)

tornado   whirling, funnel-shaped cyclone (435)

trade wind   global wind moving toward the equator between 30° and 0° latitude (367)

transform fault boundary   boundary formed where two lithospheric plates grind past each other (64)

transpiration   process by which plants give off water vapor (207)

travertine   form of calcite that is deposited in terraces around the mouths of hot springs (230)

trench   deep valley in the ocean floor (337)

tributary   feeder stream that flows into a main stream (211)

Trojan asteroid   asteroid that orbits the sun just ahead of or behind the orbit of Jupiter (523)

tropical climate   climate with a minimum average monthly temperature of 18°C (455)

tropical-desert climate   dry tropical climate that occurs in regions between the Tropic of Cancer and the Tropic of Capricorn (455)

tropopause   the upper boundary of the troposphere in which the temperature remains almost constant (392)

troposphere   the atmospheric layer closest to the earth's surface (392)

trough   lowest point between two wave crests (371)

true north   direction of the geographic North Pole (40)

tsunami   giant ocean wave that often occurs after a major earthquake with an epicenter on the ocean floor (94)

tundra climate   polar climate that occurs in areas near the ocean at the latitude of the Arctic Circle, with a small yearly temperature range and very little precipitation (458)

turbidity current   dense current that carries large amounts of sediment down the continental slopes (337)

typhoon   hurricane that forms over the Pacific Ocean (434)

U

umbra   inner, cone-shaped part of the shadow cast by the earth or the moon in which sunlight is completely blocked (537)

unconformity   break in the geologic record created when rock layers are removed by erosion (277)

undertow   irregular current that pulls water from a beach back to the ocean (375)

unfoliated   describing a metamorphic rock without visible layers or bands (161)

uniformitarianism   the theory that geologic processes that are at work in the present were also at work in the past (275)

upslope fog   condensation of water vapor that results from the lifting and adiabatic cooling of air rising up a slope of land (418)

upwelling   process in which surface water moves farther out into the ocean and deep water moves upward to replace the surface water (354)

V

valley breeze   gentle wind that flows up mountain slopes during the day (405)

valley glacier   long, narrow, wedge-shaped mass of ice that usually moves through a mountain valley (238)

variable   factor in an experiment that can be changed (12)

varve annual layer of sedimentary deposit on a lake bed (280)

vein narrow band of mineral deposits in rock (167)

vent opening through which molten rock flows onto the earth's surface (101)

ventifact any stone smoothed by wind abrasion (256)

vernal equinox the beginning of spring (30)

vertebrate animal with a backbone (302)

Vishnu schist rock layer of the Grand Canyon deposited during Precambrian time (322)

volcanic ash tephra particles between 0.25 mm and 2 mm in diameter (105)

volcanic block the largest tephra, formed from solid rock blasted from a fissure (106)

volcanic bomb large, spindle-shaped clot of lava thrown out of a volcano (106)

volcanic dust tephra particles less than 0.25 mm in diameter (105)

volcanic mountain mountain formed when molten rock erupts onto the earth's surface (80)

volcanic neck solidified central shaft of a volcano (155)

volcanism any activity that includes the movement of magma toward or onto the earth's surface (101)

volcano lava and tephra built up on the earth's surface around a vent (101)

W

waning describing the phase of the moon during which the size of its visible portion is decreasing (540)

warm front boundary formed where a warm air mass overtakes and rises above a cold air mass (430)

water budget gains and losses of water from a region (208)

water cycle continuous movement of water from the air to the earth and back again (207)

water gap deep notch left where a stream erodes through a mountain as it is uplifted (213)

water table upper surface of the zone of saturation (225)

watershed land from which water runs off into a stream (211)

waterspout tornado that occurs over the ocean (436)

wave periodic up-and-down movement of water (371)

wave cyclone large storm that develops along cold or stationary fronts, with winds that spiral in toward a central region of low air pressure (431)

wave height vertical distance between the crest and the trough of a wave (371)

wave period time required for a complete wavelength to pass a given point (371)

wave-built terrace extension of a wave-cut terrace that results from deposition of eroded material offshore (261)

wave-cut terrace nearly level platform of rock left beneath the water after the erosion of a sea cliff (261)

wavelength distance between wave crests (14)

waxing describing the phase of the moon during which the size of its visible portion is increasing (540)

weather general condition of the atmosphere at a particular time and place (387)

weathering change in the physical form or chemical composition of rock materials exposed at the earth's surface (185)

weight measure of the pull of gravity on an object (26)

white dwarf small, hot, dim star (474)

whitecap crest of a wave that is blown off by high winds (373)

wind cell looping pattern of flowing air (402)

wind gap water-eroded notch in a mountain through which water no longer flows (213)

wind vane instrument used to determine the direction of the wind (437)

winter solstice the beginning of winter (29)

Z

zone of aeration upper region of groundwater between the water table and the earth's surface (225)

zone of saturation lower region of groundwater where all the pores in a rock or sediment are filled with water (225)

zooplankton microscopic ocean animals (354)

Index

Italicized page numbers denote definitions; boldface page numbers denote illustrations.

Green Mountains, 76
greenhouse effect, **398,** *398-399,*
456
Greenland, continental glaciers
of, 238, 241
Gregorian calendar, *542-543*
Gregory XIII, Pope, 542
ground fog. *See* radiation fog.
ground moraines, *245,* **245**
groundwater, 160, *207, 208, 223-*
227, 237, 287; movement of,
226; weathering by, 231-233;
in wells and springs, 228-
230. *See also* water.
guide fossils. *See* index fossils.
Gulf Stream, *366-367,* 369-370,
443, 452
gullies, *211*
gullying, *197, 198*
guyots, *340*
gypsum, 134, 135 [table], 142
[table], 157, 169 [table]
gyres, 368

H

hachure lines, *51*
hadal zone, *356*
hail, 419, *420*
hair hygrometers, *412*
half-life, *282-283,* 285
halides, 134 [table], 135
halite, 134, 134 [table], 140, 142
[table], 157, 169 [table]
Halley's comet, **524**
hanging valleys, *243*
hanging wall, *74-75*
hard water, *231*
hardness, *141*
Haskin, Larry, 532-533
Hawaiian Islands, 80, 103, 106,
262, 283, 457
headlands, *264,* **264,** 266-268
headward erosion, *211-212*
headwaters, *213,* 217
heat, and changes in physical
forms of matter, 124; latent,
410; and metamorphic rock
formation, 161; and stress, 73;
and water evaporation, 409-
410. *See also* temperature.
Heffernan, Mary, 444-445
heliocentric model, *507*
helium, 21, 175; bright-line
spectrum of, 15, **15;** in outer

planet atmospheres, 515, 516,
518; in protoplanets, 499; in
solar nebula, 502; in stars,
468, 477-479; in sun, 489-
491
hematite, 135 [table], 142
[table], 167, 169 [table]
herbivores, 305-308
Hermit shale, *323*
Hertzsprung, Ejnar, 473-474
Hess, Harry, 59, 60, 61
Hicks, Doug, 358
highland rocks, 163
hills, ordinary springs in, 228-
229
Himalaya Mountains, 71, 76;
formation of, 63, 77-80, 306,
317
Hipisodus, **306**
Holocene Epoch, 306, 308-309
homeostasis, *11*
homeostatic ability, *11*
Homo sapiens, 308-309
Hood, Mount, 107
horizons, **194,** *194-195,* 198
hornblende, 134, 153
horns, *242,* **242**
horse latitudes, *403*
Horsehead Nebula, 482, **482**
horses, development of, 307, 308,
308
Hot Dry Rock, 178-179
hot spots, 80, *103,* 515
hot springs, *229-230*
H-R diagram, *473-474,* **474**
humans, origins of, 308-309
humid continental climates,
460, **460**
humid subtropical climates, **460,**
460-461
humidity, **410,** *410-413,* 411
[table]; of air masses, 427-
429; measuring of, 439-442
humus, *193,* 195, 196, 198
Huntley, D. J., 282
hurricanes, **433,** *433-434,* 444
Hutton, James, 275
hydrated minerals, **208**
hydraulic gradient, *226*
hydrocarbons, *170,* 171, 387, 394
hydroelectric power, *178*
hydrogen, 187, 188, 196, 353, 499;
in atmosphere, 502; atomic
number of, 119; bright-line
spectrum of, 15, **15;** energy
level of, 125, 126; in

hydrocarbons, 170; isotopes of,
123; in molecules, 125, 128; in
outer planet atmospheres, 515,
516, 518; in solar nebula, 498,
502; in stars, 468, 477-479;
in sun, 175, 489-491
hydrologic cycle. *See* water cycle.
hydrologists, 201
hydrolysis, *187*
hydrosphere, 7, *21*
hypocenter. *See* focus of an
earthquake.
hypotheses, *11-14*
Hyracotherium, 307, **308**

I

ice, 409; and cloud formation,
414; in frozen dew, 413; as
precipitation, 419-422; and
sedimentary rock formation,
149, 156. *See also* crystals.
ice ages, **249,** *249-251,* 298,
306-308, 321
ice caps. *See* continental
glaciers.
ice shelves, 241
ice wedging, **185,** *185-186,* 191,
192
icebergs, *241,* 342
Iceland, 60, 80, 103
ichthyosaurs, *305*
Ichthyostega, 303
igneous rock, *149,* **150,** 152-155,
278; and fossils, 286; on moon,
531; and rock cycle, 150-151,
151, 161; stratification of,
276, 277; weathering of, 190-
191. *See also* rocks.
impact craters, *510*
Imperial Valley, 80
impermeable rock, *225,* 226, 227,
229
imprints, fossil, 288, **288**
index contours, *49*
index fossils, *290,* **290,** 295, 299,
302, 305
Indian landmass, formation of, 77-
78, 315, 316
Indian Ocean, 261, 331, 332, 366,
368
inertia, *509*
information, gathering of, 9-10
infrared rays, 350, *395,* 397, 398-399
inorganic substances, 133

runoff, *207*, 208, 211, **211**, 214
Russell, Henry, 473-474
rust. *See* iron oxide.

Credits

Art

Ligature, Inc. 21, 22, 23 (b), 32, 33, 39, 40, 40, 41, 41, 42, 43, 43, 44, 45, 45, 46 (c), 46 (r), 46 (b), 48 (t), 51 (b), 86 (l), 87, 88, 89, 90, 90 (t), 97, 101, 126, 173, 173, 215, 215, 215, 223, 223, 223, 225 (t), 225 (c), 241, 247, 247, 247, 248, 249, 251, 255, 265, 281, 282, 282, 285, 296, 300, 303, 313, 319, 319, 323, 331, 333, 348, 350, 356, 366, 367, 377, 378, 387, 392, 393, 395, 399, 401, 404, 410, 428, 431, 431, 431, 432, 432, 432, 437, 441, 441, 442, 473, 474, 481, 481, 508, 535, 536, 543, **Nadine Sokol** 318, 320, 354, **Precision Graphics** 4, 12, 16, 23 (t), 26, 27, 28, 29, 30, 46 (t), 48 (b), 51 (t), 57, 57, 62, 71 (r), 72 (c), 72 (t), 72 (b), 73, 74, 74, 74, 74, 74, 76, 79, 81, 85, 106, 106, 106, 117, 119, 122, 123, 123, 124, 127, 128, 137, 138, 138, 138, 142, 142, 151, 155, 168, 168, 172, 172, 175, 178, 185, 191, 193, 195, 195, 196, 209, 209, 209, 225 (b), 226, 228, 229, 230, 276, 277, 277, 277, 278, 295, 304, 305, 308, 315, 315, 316, 316, 317, 349, 359, 369, 370, 371, 372, 374, 388, 389, 394, 394, 397, 398, 405, 405, 409, 414, 417, 421, 430, 433, 450, 450, 452, 454, 455, 460, 470, 480, 484, 490, 493, 499, 499, 499, 499, 502, 502, 502, 521, 534, 534, 537, 538, 538, 541, **Sarah Woodward** 61, 63, 63, 64, 64, 67, 67, 71 (l), 86 (r), 102 (l), 102, 103, 140, 140, 207, 239, 242, 242, 245, 257, 257, 257, 263, 264, 266, 266, 269, 269, 269, 335, 337

Photographs

0: Jeffrey Rotman, NASA; **1:** NASA, Hansen Planetarium (c); **2:** J. Brandenburg, Click/Chicago; **3:** Ginzy Schaefer, Click/Chicago; **5:** S. Ferguson, William E. Ferguson (l), Woods Hole Oceanographic Institute (r); **6:** Milton Mann, Marilyn Gartman (r); Alan Carey, The Image Works (l); **7:** B. Miller, Terraphotographics/BPS; **8:** Doug Wedhsler, Earth Scenes (t); Adele Hodde, Gamma-Liaison (b); **9:** David Falconer, West Stock; **10:** Anthony Howarth, Photo Researchers (br); Bruce Bohor, U.S.G.S. (tl); **11:** G.R. Roberts Frank Whitney, The Image Bank; **13:** Francis Gohier, Photo Researchers; **14:** Ray Reiss, Click/Chicago; **15:** Sargent Welch Co.; **17:** Ralph Brunke; **20:** NASA; **24:** Energy Technology Visuals (b); Ken Pierce, David Swift (t); **25:** Grant Heilman; **31:** William Means, Click/Chicago; **34:** Grant Heilman; **35:** Ralph Brunke; **38:** R. Perron, Nawrocki Stock Photo; **41:** Doug Wilson, The Stock Solution (r); **48:** Tom Tracy, After Image (br); NASA (tl); **50:** Ralph Brunke; **56:** John Ford, Marilyn Gartman; **58:** Charles Palek, Earth Scenes; **59:** W.R. Normark, U.S.G.S. (r); Kenneth Garrett, Woodfin Camp & Associates (l); **60:** Pete Turner, The Image Bank; **65:** Ralph Brunke; **66:** Charles McNulty, Click/Chicago; **70:** Barbara Von Hoffmann, Tom Stack and Associates; **73:** Breck Kent, Earth Scenes; **75:** Ralph Brunke; **77:** Galen Rowell, Mountain Light Photography; **78:** University of Texas at Austin (r); NASA (l); **79:** Wally Hampton, Marilyn Gartman; **80:** E.R. Degginger, Earth Scenes (l); Jerome Wyckoff, Earth Scenes (t); Mary Kay Desotell, Marilyn Gartman (r); **84:** Nancy Simmerman, Click/Chicago; **89:** John Yates, After Image; **91:** Ralph Brunke; **93:** John Barr, Gamma-Liaison; **94:** Dr. Wayne Thatcher, Gamma-Liaison (t); B.F. Bohor, U.S.G.S. (b); **95:** G.R. Roberts; **96:** George Hall, Woodfin Camp & Associates; **100:** David Weintraub, Photo Researchers; **102:** David Falconer, West Stock (tr); **104:** Joseph Holmes; **105:** Phil Degginger, Click/Chicago (l); Kal Muller, Woodfin Camp & Associates (r); **106:** Willard Clay, Click/Chicago; **107:** Ralph Brunke; **108:** Willard Clay, Click/Chicago; **109:** Wolfgang Kaehler Photography; **110:** NASA (l); NASA (r); **111:** NASA; **114:** Warren Faubel, Bruce Coleman, Larry Lee, West Light (in); **116:** Spencer Swanger, Tom Stack and Associates; **118:** HRW photo by Russell Dian, Ralph Brunke; **122:** Fermilab; **123:** Fermilab; **129:** Grant Heilman, T.J. Florian, Nawrocki Stock Photo (l); **132:** E.R. Degginger, Earth Scenes; **133:** E.R. Degginger (l); E.R. Degginger (r); Mary Root, Root Resources (c); **134:** Breck Kent (ml); E.R. Degginger (mr); E.R. Degginger (bl); Breck Kent, Earth Scenes (br); E.R. Degginger, Earth Scenes (tr); E.R. Degginger, Earth Scenes (tl); **135:** Breck Kent (bl); E.R. Degginger (tr); E.R. Degginger (tl); E.R. Degginger (c); Loren and Mary Root, Root Resources (br); **136:** Colin Mulvany (l); J. Steere, Nawrocki Stock Photo (r); **137:** Ted Coringley, Nawrocki Stock Photo; **139:** Runk/Schoenberger, Grant Heilman (r); Mary Root, Root Resources (l); **140:** E.R. Degginger (b); E.R. Degginger (t); **141:** Breck Kent (b); **144:** E.R. Degginger (l); E.R. Degginger (b); E.R. Degginger (r); **145:** Roger Beck, Nawrocki Stock Photo; **148:** David Muench, The Image Bank; **149:** E.R. Degginger; **150:** Breck Kent (c); E.R. Degginger (r); E.R. Degginger (l); **152:** E.R. Degginger; **153:** Breck Kent (t); Breck Kent (br); Mike Chuang, FPG (l); **154:** Ralph Brunke; **155:** James Rowan, Hillstrom Stock Photography (b); **156:** Breck Kent, Earth Scenes (r); Ruth Dixon, Earth Scenes (l); **157:** S.L. Eckert, Click/Chicago (b); J.N.A. Lott, Terraphotographics/BPS (t); **158:** © CNES, Spot Image Corporation; **159:** Joe Viesti, FPG (tr); © CNES, Spot Image Corporation (b); **160:** Breck Kent (bl); Jim Brandenburg, Click/Chicago (tr); Steve Traudt, Tom Stack and Associates (tl); **162:** Breck Kent (tl); Breck Kent (b); E.R. Degginger, Earth Scenes (tr); **163:** Martin Rogers, Click/Chicago (r); NASA (l); **166:** Gary Ladd; **167:** William E. Ferguson; **170:** G.R. Roberts; **171:** G.R. Roberts; **174:** Pierre Kopp, West Light; **176:** William E. Ferguson; **177:** Ralph Brunke; **179:** Department of Energy; Ray Nelson, Phototake; **182:** Jeff Gnass Photography; Porterfield-Chickering, Photo Researchers (in); **184:** Joseph Holmes; **185:** Nancy Simmerman, Click/Chicago; **186:** Ligature, Inc.; **187:** Jeff Gnass Photography; **188:** Willard Clay, Click/Chicago (b); Runk/Shoenberger, Grant Heilman (t); **189:** J.N.A. Lott, Terraphotographics/BPS; **190:** Doris DeWitt, Click/Chicago; **192:** John Jenkins III; John Jenkins III (inb); New York Historical Society (int); **194:** William E. Ferguson; **197:** John Gerlach, Earth Scenes; **198:** Larry LeFever, Grant Heilman (l); B. Crader, Hillstrom Stock Photography (r); **199:** Breck P. Kent, Earth Scenes; **200:** Cameron Davidson, After Image (r); Soil Conservation Service, (l); **201:** John Running, After Image; **202:** Gerry Ellis, Manhattan Views (l); Ruth Dixon (r); **203:** Jim Brandenburg, Click/Chicago (l); O'Neill, Gary Ladd (r); **206:** Lee Foster, FPG; **208:** Breck Kent (r); William Boehm, West Stock (l); **210:** Gary Milburn, Tom Stack and Associates; **211:** G.R. Roberts; **212:** Ralph Brunke; **213:** G.R. Roberts; **214:** G.R. Roberts (r); Thomas Wanstall, The Image Works (l); **216:** G.R. Roberts (b); G.R. Roberts (in); Brett Froomer, The Image Bank (t); **217:** D.W. Hamilton, The Image Bank; **219:** E.R. Degginger (l); Suzi Barnes, Tom Stack and Associates (r); **222:** Royce Blair, Royce Blair & Associates; **224:** Ralph Brunke; **227:** William E. Ferguson; **231:** John Lemker, Earth Scenes; **232:** Phil Degginger, E.R. Degginger; **233:** Gary Ladd; **236:** Nancy Simmerman, Alaska Photo/Aperture; **237:** Wilson Goodrich, Marilyn Gartman; **238:** Nancy Simmerman, Click/Chicago (l); Phil Degginger, Click/Chicago; **240:** Myron Wright, Alaska Photo/Aperture (l); John T. Turner, Aperture; **243:** Jack Wilburn, Earth Scenes; **244:** Ralph Brunke ; **245:** Grant Heilman; **246:** George Gersler, Photo Researchers; **250:** Bob Waterman, West Light; **254:** Joseph Holmes; **256:** E.R. Degginger; Ralph Brunke; **259:** E.R. Degginger; **260:** Thomas Wanstall, The Image Works; **261:** D.C. Lowe, Aperture; **262:** Phil Degginger, Click/Chicago; **267:** F. Roiter, The Image Bank; **268:** Robert Perron, Nawrocki Stock Photo; **272:** Doug Wechsler, Earth Scenes (in); Gary Ladd; **274:** Gary Ladd; **275:** Gary Ladd; **279:** Mary Root, Root Resources; **280:** E.R. Degginger, Earth Scenes; **282:** Los Alamos National Laboratory; **283:** Gerald Corsi, Tom Stack and Associates; **284:** Ralph Brunke; **285:** Edward Ross; **286:** Richard Kolar, Earth Scenes; **287:** Field Museum of Natural History, Chicago; **288:** Gary Ladd (t); Gary Ladd (b); **289:** Grant Heilman; **290:** Louise Brown, Root Resources; **291:** Eastcot/Momtiuk, The Image Works; **294:** Breck Kent; **297:** Kevin Schafer, Tom Stack and Associates; **298:** Breck Kent; **299:** Grant Heilman; **302:** Kitt Peak National Observatory; **305:** E.R. Degginger, Earth Scenes; **306:** E.R. Degginger; **307:** Alan G. Nelson, Hillstrom Stock Photography; **312:** Chuck Place Photography; **314:** Ralph Brunke; **321:** Tim Thompson Photography; **322:** James Tallon; **324:** Field Museum of Natural History, Chicago (tl); S. McNutt, Phototake (br); **325:** Frank Staub, The Picture Cube; **328:** Jeffrey Rotman; John Telford, The Stock Solution; **330:** Fred Bavendaum, Peter Arnold; **332:** Jason Ion, Woods Hole Oceanographic Institute; **333:** Rod Catavah, Woods Hole Oceanographic Institute; **334:** Ralph Brunke; **336:** C.C. Lockwood, The Image Works; **338:** Chris Newbert, Marine Life Photography; Werner Braun, FPG (r); **339:** F. Stuart Westmorland, Aperture; **340:** Marie Tharp/Oceanographic Cartographer; **341:** David Ross, Woods Hole Oceanographic Institute; **342:** Manfred Kage, Peter Arnold (r); M. Abbey, Photo Researchers (l); **343:** Steinhart, Photo Researchers; **346:** Carl Roessler; **351:** Ralph Brunke; **352:** Al Grotell; **353:** Peter David, Photo Researchers; **355:** E.R. Degginger (t); E.R. Degginger (b); **358:** UOP Fluid Systems; **360:** University of Hawaii; **361:** Tom Pantages; **364:** Christopher Arend, Aperture; **368:** E.R. Degginger, Earth Scenes; **373:** Ralph Brunke; **374:** Steve Lissau, Marilyn Gartman; **375:** John Shelton; **378:** Nova Scotia Power Corp.; **379:** Nova Scotia ower Corp.; **380:** Breck Kent, Earth Scenes (t); Breck Kent, Earth Scenes (b); **381:** Gregg Stockey; **384:** D. Dietrich, FPG; Gary Ladd; **386:** Pierre Kopp, West Light; **390:** Ralph Brunke; **391:** Breck Kent; **396:** John Livzey, DOT; **402:** Robert Frerck, Click/Chicago (r); Mark Antman, Phototake (l); **403:** Ruth Dixon; **408:** Brett Froomer, The Image Bank; **411:** Breck Kent; **412:** Ralph Brunke; **413:** Craig Blouin, Manhattan Views (t); Randy Duchane, Manhattan Views (b); **415:** Philip Amdal, West Stock; **416:** Frank Mitchell, The Stock Solution (t); I.P.A., The Image Works (c); Ric Noyle, Visual Impact (b); **418:** Thomas Wan, The Image Works; **419:** Robert Essey, Manhattan Views; **420:** Nuridsany & Perennou, Photo Researchers; **421:** Royce L. Blair, The Stock Solution; **422:** Gary Ladd, National Optical Astronomy Observatories (in); **423:** E.R. Degginger; **426:** Gary Milburn, Tom Stack and Associates; **427:** NASA; **436:** E.R. Degginger; **438:** Ralph Brunke; **439:** Lawrence Migdale; **443:** Peter Breck, Uniphoto; **444:** Lawrence Migdale (l); Waverly Person, Natural Earthquake Information Center (r); **445:** James Kay, The Stock Solution; **448:** James Kay Photography; **451:** Brian Munoz, Scripps Institution of Oceanography; **456:** Chip Isinhart, Tom Stack and Associates; **457:** Michael Fogden, Earth Scenes; **458:** David Fritts, Earth Scenes; **459:** Chuck Place Photography; **464:** NASA; **465:** NASA; **466:** Hale Observatory; **468:** Anglo-Australian Telescope Board; **469:** National Optical Astronomy Observatories (in); Nicholas Foster, The Image Bank; **471:** Ralph Brunke; **472:** Yerkes Observatory; **475:** Palomar Observatory; **476:** Ad Lib; **477:** Jet Propulsion Laboratory; **478:** Hansen Planetarium; **479:** National Optical Astronomy Observatories; **482:** National Optical Astronomy Observatories; **483:** Hale Observatory (l); Kitt Peak National Observatory (r); National Optical Astronomy Observatories (c); **485:** National Optical Astronomy Observatories; **488:** NASA; **491:** Ralph Brunke; **492:** Dan McCoy, Rainbow; **494:** National Optical Astronomy Observatories; **495:** NASA; **496:** NASA; **497:** Jeff Poonan, Nawrocki Stock Photo; **500:** Two-Twenty-Two Productions (r); Sandra Faber, Dan Coyro (l); **501:** E.I. Robinson, Photo Researchers; **503:** NASA; **506:** Jet Propulsion Laboratory; **508:** Ligature, Inc.; **510:** Jet Propulsion Laboratory; **511:** Jet Propulsion Laboratory; **512:** Jet Propulsion Laboratory; **513:** Jet Propulsion Laboratory; **514:** Jet Propulsion Laboratory; **515:** Jet Propulsion Laboratory; **516:** Jet Propulsion Laboratory; **517:** Jet Propulsion Laboratory; **518:** Jet Propulsion Laboratory (r); Jet Propulsion Laboratory (l); **519:** Charles Capen, Hansen Planetarium (t); Jet Propulsion Laboratory (bl); Jet Propulsion Laboratory (c); **522:** Julian Baum/Sci. Photo Library, Photo Researchers; **523:** Hansen Planetarium; **524:** NASA; **525:** Pearson/Milon/Sci. Photo Library, Photo Researchers; **528:** NASA; **529:** NASA; **530:** NASA (r); NASA (l); **531:** NASA (r); NASA (l); **532:** NASA; **533:** NASA; **537:** Kitt Peak National Observatory; **539:** Ralph Brunke; **542:** Aldona Sabalis, Photo Researchers; **544:** Jet Propulsion Laboratory; **545:** Jet Propulsion Laboratory; Jet Propulsion Laboratory; **546:** Jet Propulsion Laboratory; **547:** Jet Propulsion Laboratory (l); Jet Propulsion Laboratory (r)